AF776681

Impeller Pumps

Wiley-ASME Press Series

Thermal Spreading and Contact Resistance: Fundamentals and Applications
Yuri S. Muzychka and M. Michael Yovanovich

Analysis of ASME Boiler, Pressure Vessel, and Nuclear Components in the Creep Range
Maan H. Jawad and Robert I. Jetter

Voltage-Enhanced Processing of Biomass and Biochar
Gerardo Diaz

Pressure Oscillation in Biomedical Diagnostics and Therapy
Ahmed Al-Jumaily and Lulu Wang

Robust Control: Youla Parameterization Method
Farhad Assadian and Kevin R. Mellon

Metrology and Instrumentation: Practical Applications for Engineering and Manufacturing
Samir Mekid

Fabrication of Process Equipment
Owen Greulich and Maan H. Jawad

Engineering Practice with Oilfield and Drilling Applications
Donald W. Dareing

Flow-Induced Vibration Handbook for Nuclear and Process Equipment
Michel J. Pettigrew, Colette E. Taylor, and Nigel J. Fisher

Vibrations of Linear Piezostructures
Andrew J. Kurdila and Pablo A. Tarazaga

Bearing Dynamic Coefficients in Rotordynamics: Computation Methods and Practical Applications
Lukasz Brenkacz

Advanced Multifunctional Lightweight Aerostructures: Design, Development, and Implementation
Kamran Behdinan and Rasool Moradi-Dastjerdi

Vibration Assisted Machining: Theory, Modelling and Applications
Li-Rong Zheng, Dr. Wanqun Chen, and Dehong Huo

Two-Phase Heat Transfer
Mirza Mohammed Shah

Computer Vision for Structural Dynamics and Health Monitoring
Dongming Feng and Maria Q. Feng

Theory of Solid-Propellant Nonsteady Combustion
Vasily B. Novozhilov and Boris V. Novozhilov

Introduction to Plastics Engineering
Vijay K. Stokes

Fundamentals of Heat Engines: Reciprocating and Gas Turbine Internal Combustion Engines
Jamil Ghojel

Offshore Compliant Platforms: Analysis, Design, and Experimental Studies
Srinivasan Chandrasekara and R. Nagavinothini

Computer Aided Design and Manufacturing
Zhuming Bi and Xiaoqin Wang

Pumps and Compressors
Marc Borremans

Corrosion and Materials in Hydrocarbon Production: A Compendium of Operational and Engineering Aspects
Bijan Kermani and Don Harrop

Design and Analysis of Centrifugal Compressors
Rene Van den Braembussche

Case Studies in Fluid Mechanics with Sensitivities to Governing Variables
M. Kemal Atesmen

The Monte Carlo Ray-Trace Method in Radiation Heat Transfer and Applied Optics
J. Robert Mahan

Dynamics of Particles and Rigid Bodies: A Self-Learning Approach
Mohammed F. Daqaq

Primer on Engineering Standards, Expanded Textbook Edition
Maan H. Jawad and Owen R. Greulich

Engineering Optimization: Applications, Methods and Analysis
R. Russell Rhinehart

Compact Heat Exchangers: Analysis, Design and Optimization using FEM and CFD Approach
C. Ranganayakulu and Kankanhalli N. Seetharamu

Robust Adaptive Control for Fractional-Order Systems with Disturbance and Saturation
Mou Chen, Shuyi Shao, and Peng Shi

Robot Manipulator Redundancy Resolution
Yunong Zhang and Long Jin

Stress in ASME Pressure Vessels, Boilers, and Nuclear Components
Maan H. Jawad

Combined Cooling, Heating, and Power Systems: Modeling, Optimization, and Operation
Yang Shi, Mingxi Liu, and Fang

Applications of Mathematical Heat Transfer and Fluid Flow Models in Engineering and Medicine
Abram S. Dorfman

Bioprocessing Piping and Equipment Design: A Companion Guide for the ASME BPE Standard
William M. (Bill) Huitt

Nonlinear Regression Modeling for Engineering Applications: Modeling, Model Validation, and Enabling Design of Experiments
R. Russell Rhinehart

Geothermal Heat Pump and Heat Engine Systems: Theory and Practice
Andrew D. Chiasson

Fundamentals of Mechanical Vibrations
Liang-Wu Cai

Introduction to Dynamics and Control in Mechanical Engineering Systems
Cho W.S. To

Thermal Spreading and Contact Resistance: Fundamentals and Applications
Yuri S. Muzychka and M. Michael Yovanovich

Mixed-flow Pumps: Modeling, Simulation, and Measurements
Wei Li, Leilei Ji, Ramesh Agarwal, Weidong Shi, and Ling Zhou

Thermal Explosion: Theory and Application
Vasily B. Novozhilov

Rotating Disc Cavity Flow and Heat Transfer
John W. Chew

Impeller Pumps: Design and Performance Evaluation using Computational Fluid Dynamics
Wei Li, Leilei Ji, and Ramesh K. Agarwal

This Work is a co-publication between John Wiley & Sons, Inc. and ASME Press

Impeller Pumps

Design and Performance Evaluation using Computational Fluid Dynamics

Wei Li
Jiangsu University, Zhenjiang, China

Leilei Ji
Jiangsu University, Zhenjiang, China

Ramesh K. Agarwal
Washington University in St. Louis, St. Louis, USA

ASME PRESS

WILEY

Copyright © 2026 by John Wiley & Sons, Inc. All rights reserved.

Published by John Wiley & Sons, Inc., Hoboken, New Jersey.

This Work is a co-publication between John Wiley & Sons, Inc. and ASME Press.

No part of this publication may be reproduced, stored in a retrieval system, or transmitted in any form or by any means, electronic, mechanical, photocopying, recording, scanning, or otherwise, except as permitted under Section 107 or 108 of the 1976 United States Copyright Act, without either the prior written permission of the Publisher, or authorization through payment of the appropriate per-copy fee to the Copyright Clearance Center, Inc., 222 Rosewood Drive, Danvers, MA 01923, (978) 750-8400, fax (978) 750-4470, or on the web at www.copyright.com. Requests to the Publisher for permission should be addressed to the Permissions Department, John Wiley & Sons, Inc., 111 River Street, Hoboken, NJ 07030, (201) 748-6011, fax (201) 748-6008, or online at http://www.wiley.com/go/permission.

The manufacturer's authorized representative according to the EU General Product Safety Regulation is Wiley-VCH GmbH, Boschstr. 12, 69469 Weinheim, Germany, e-mail: Product_Safety@wiley.com.

Trademarks: Wiley and the Wiley logo are trademarks or registered trademarks of John Wiley & Sons, Inc. and/or its affiliates in the United States and other countries and may not be used without written permission. All other trademarks are the property of their respective owners. John Wiley & Sons, Inc. is not associated with any product or vendor mentioned in this book.

Limit of Liability/Disclaimer of Warranty: While the publisher and the authors have used their best efforts in preparing this work, including a review of the content of the work, neither the publisher nor the authors make any representations or warranties with respect to the accuracy or completeness of the contents of this work and specifically disclaim all warranties, including without limitation any implied warranties of merchantability or fitness for a particular purpose. No warranty may be created or extended by sales representatives, written sales materials or promotional statements for this work. The fact that an organization, website, or product is referred to in this work as a citation and/or potential source of further information does not mean that the publisher and authors endorse the information or services the organization, website, or product may provide or recommendations it may make. This work is sold with the understanding that the publisher is not engaged in rendering professional services. The advice and strategies contained herein may not be suitable for your situation. You should consult with a specialist where appropriate. Further, readers should be aware that websites listed in this work may have changed or disappeared between when this work was written and when it is read. Neither the publisher nor authors shall be liable for any loss of profit or any other commercial damages, including but not limited to special, incidental, consequential, or other damages.

For general information on our other products and services or for technical support, please contact our Customer Care Department within the United States at (800) 762-2974, outside the United States at (317) 572-3993 or fax (317) 572-4002.

Wiley also publishes its books in a variety of electronic formats. Some content that appears in print may not be available in electronic formats. For more information about Wiley products, visit our web site at www.wiley.com.

Library of Congress Cataloging-in-Publication Data applied for:

Hardback ISBN: 9781394324163

Cover Design: Wiley
Cover Image: Surasak_Photo/Shutterstock

Set in 9.5/12.5pt STIXTwoText by Straive, Chennai, India

Contents

Preface

Pumps are among the most essential and energy-intensive components in modern industrial systems, serving as the core devices for energy conversion and fluid transportation in sectors such as energy, water resources, manufacturing, and environmental engineering. Among the diverse types of pumps, impeller pumps, including centrifugal pump, mixed-flow pump, axial-flow pump, vortex pump, and multistage pump, represent the most widely applied class due to their mechanical simplicity, high reliability, and broad operational adaptability. The performance of impeller pumps directly influences the efficiency and stability of large-scale engineering systems, making their design and optimization a subject of enduring academic and industrial interest.

The book provides with the continuous improvement in performance requirements and the advent of digital engineering technologies, traditional empirical and semi-empirical approaches to pump design have gradually become insufficient for addressing complex flow phenomena and multi-objective optimization tasks. In particular, the internal flow within impeller pumps is inherently three-dimensional, unsteady, and highly turbulent, featuring secondary flows, tip leakage vortices, rotor–stator interactions, and cavitation instabilities. These complex flow structures are difficult to capture through experimental means alone due to limitations in measurement accessibility, cost, and temporal resolution. Consequently, computational fluid dynamics (CFD) has emerged as a powerful and indispensable tool for analyzing, predicting, and optimizing the internal flow behavior and performance characteristics of impeller pumps.

This book may provide a comprehensive and systematic exposition of the fundamental theories, hydraulic design methodologies, and numerical simulation techniques applicable to impeller-type pumps. The book integrates classical hydraulic design principles with modern CFD analysis to build a coherent framework that spans theoretical foundations, modeling procedures, simulation implementation, and practical engineering applications. It is intended to serve as both a reference for researchers and a guidebook for graduate students and practicing engineers engaged in the study, design, and optimization of hydraulic machinery.

This book is organized into 11 chapters. Chapter 1 introduces the basic concepts, classifications, and industrial applications of impeller pumps, emphasizing their engineering importance. Chapter 2 presents the fundamental hydraulic principles, including velocity triangle analysis, energy conversion theory, and similarity laws, which form the theoretical backbone of pump design. Chapter 3 describes in detail the hydraulic design methodologies for different types of impeller pumps. Chapter 4 lays the theoretical foundation of CFD as applied to impeller pumps, including governing equations, numerical discretization, turbulence modeling, wall functions, and solution algorithms. Chapter 5 introduces three-dimensional geometric modeling techniques for various pump components, providing a bridge between design theory and computational practice. Chapters 6 through 10 systematically describe the numerical simulation and performance analysis of impeller pumps. Each chapter presents detailed workflows encompassing geometry modeling, mesh generation, steady and unsteady computations, and postprocessing techniques such as streamline visualization, contour analysis, and performance characterization. Finally, Chapter 11 offers concluding remarks and an outlook on future developments in pump technology. It discusses the current challenges and opportunities in the field, such as the integration of artificial intelligence, data-driven design, digital twin modeling, and sustainable energy solutions, which are expected to guide the evolution of impeller pump technology in the coming decades.

It is acknowledged that this book cannot encompass all aspects of the rapidly advancing field of pump research. Both the depth and scope of data collection are constrained by the available literature and computational resources. Inevitably, certain omissions and errors may exist, for which the authors sincerely invite constructive criticism and suggestions from readers and fellow researchers. It is our hope that this work will serve as a valuable reference for those engaged in the theoretical, experimental, and computational study of hydraulic machinery and will contribute to the continued advancement of impeller pump science and engineering.

25 March 2026

Wei Li
Jiangsu University, Zhenjiang, China

Leilei Ji
Jiangsu University, Zhenjiang, China

Ramesh K. Agarwal
Washington University in St. Louis, St. Louis, USA

Acknowledgments

This work was sponsored by the National Natural Science Foundation of China (No. 52179085), the National Natural Science Foundation of China (No. 52309112), the China Postdoctoral Science Foundation (No. 2025M780566), the Key International Cooperative Research Program of the National Natural Science Foundation of China (No. 52120105010), the Sixth "333 High Level Talented Person Cultivating Project" of Jiangsu Province, and projects of the "Blue Project" in Jiangsu Colleges and Universities, and National Key Laboratory on Ship Vibration and Noise (6142204240505). The author Leilei Ji would like to thank Chinese Scholarship Council (CSC) for the financial support at Imperial College London as a sponsored researcher (No. 202308320258), and Shandong Xinchuan Mining Equipment Co., Ltd. for support of the joint postdoctoral research.

The publication of this book was made possible by the help of several colleagues and students at the National Engineering Research Center for Pumps and Systems of Jiangsu University and the CFD Laboratory of Washington University in St. Louis, USA, to whom the authors would like to express their sincere gratitude.

In particular, the authors extend special thanks to Ms. Qiaoyue Yang, whose dedicated effort and meticulous coordination were instrumental in integrating the chapters into a coherent whole. Sincere appreciation is also expressed to Yunfei Chen, Jin Ye, Yunfei Wang, and Yongkang Li for their substantial work in manuscript preparation, technical review, and verification. The chapter preparation efforts were distributed as follows: Jin Ye and Yunfei Chen were responsible for Chapters 4 and 11; Wei Pu contributed to Chapter 6; Wanyun Yang and Jiabao Li handled Chapter 7; Qiaoyue Yang covered Chapters 1, 2, and 8; Yongkang Li prepared Chapter 9; and Yunfei Wang worked on Chapter 10. It is also noted that Chapter 3 was collaboratively developed by all the individuals named above.

The authors would like to dedicate this book to their respective families for their unwavering support, perseverance, and encouragement during the preparation of this book.

1

Introduction

1.1 Definition and Importance of Pumps

1.1.1 Definition of Pumps

A pump is a machine that converts the mechanical energy of a prime mover into the energy of a liquid. Pumps are used to increase the potential energy, pressure energy, and kinetic energy (in high-speed fluid flow) of a liquid. The prime mover drives the pump impeller through the pump shaft, performing work on the liquid and increasing its energy. This allows the required amount of liquid to be transported from the suction pool through the flow components of the pump to a higher elevation or a location requiring a specific pressure.

Figure 1.1 shows a simple schematic diagram of a centrifugal pump system. The prime mover drives the impeller to rotate, drawing water into the pump from point A and discharging it at point B. The impeller plays a dominant role in the pump's operation. The blades within the impeller force the liquid to rotate, causing it to be thrown outward due to centrifugal force. This process is similar to water droplets being flung outward from a spinning umbrella.

Once the liquid inside the pump is expelled, new liquid enters the pump under atmospheric pressure. This continuous process allows liquid (water) to be drawn from point A and delivered to point B.

Before starting the pump, it must be primed with water. If the pump is not filled with water, the impeller will only rotate the air. Since the mass per unit volume of air is very small, the centrifugal force generated is minimal, making it impossible to expel air from the pump and the discharge pipeline. Consequently, a sufficient vacuum cannot be created inside the pump, and water will not be drawn in. The foot valve is used for priming the water, while the control valve on the pump outlet side is used to regulate the flow rate [1].

1.1.2 The Importance of Pumps

Pumps, as crucial engineering devices, play an indispensable role in various fields. First, in many industrial applications, pumps are widely used for transporting liquids and gases. Whether handling water, oil, chemicals, or gases, pumps ensure the transfer of these fluids from one location to another, maintaining the normal operation of production lines. In industrial manufacturing, pumps are essential for transportation of raw material, circulation systems within the production processes, and final product distribution. Their high efficiency in fluid movement not only enhances the production efficiency but also ensures the stability of product quality [2].

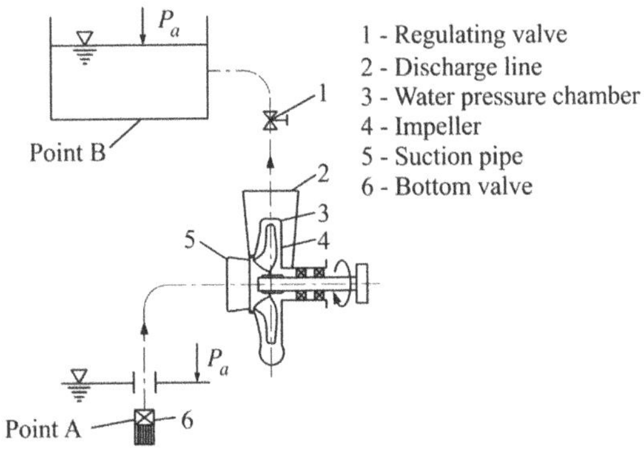

Figure 1.1 Schematic diagram of a centrifugal pump system.

In the municipal and environmental sectors, the role of pumps is also irreplaceable. Municipal infrastructure relies on pumps to maintain the proper functioning of water supply and wastewater treatment systems. Water treatment plants use pumps to extract water resources from sources and distribute them across cities through pipeline networks. Similarly, wastewater treatment plants use pumps to transport sewage to treatment facilities, where it is purified before being discharged into the environment. Additionally, pumps play a critical role in flood control, drainage, and irrigation systems. The stability and efficiency of pump operations directly impact the reliability of urban infrastructure and environmental health [3].

In energy and resource extraction, pumps serve as a key component. In the oil and gas industry, pumps are extensively used to extract crude oil and natural gas from underground reservoirs. Various types of pumps, such as centrifugal pumps and reciprocating pumps, support drilling, pipeline transportation, and enhanced oil recovery processes. In power generation, pumps are essential for the normal operation of power plants. They are used for circulating water, steam, or other fluids to facilitate heat transfer, generate electricity, and maintain optimal operating conditions for equipment. Pumps play an irreplaceable role in energy extraction and power generation, providing stable support for energy production.

In conclusion, as a vital component of modern engineering, pumps play a significant role in industrial production, municipal infrastructure, energy extraction, and environmental protection. Their efficiency, reliability, and versatility make them indispensable tools across various industries. In the future, as technology continues to evolve and societal demands shift, the role of pumps will further expand, driving industrial innovation, improving resource utilization efficiency, and maintaining environmental balance.

1.2 Classification and Structural Forms of Pumps

1.2.1 Pump Classification

According to the way in which the fluid interacts with the pump, pumps can be divided into positive displacement pumps and centrifugal pumps. In positive displacement pumps, the working medium is in one or more closed working chambers, and the volume of the working chamber changes. The main interaction between the pump and the fluid is due to static pressure. In centrifugal pumps, energy conversion occurs between the rotor with blades and the fluid medium with continuous flow, and the interaction force between the blades and the fluid is inertial. See Table 1.1 for specific classifications.

Table 1.1 Pump classification.

Impeller pump	Centrifugal pump	Single stage (single suction, double suction, self-priming, non-self-priming)
		Multistage (segmental, volute)
	Mixed-flow pump	Volute pump, guide vane (fixed blade, adjustable blade)
	Axial-flow pump	Fixed blade, adjustable blade
	Vortex pump	Single stage, multistage, self-priming, non-self-priming
Positive displacement pump	Reciprocating pump	(Piston type, plunger type) steam double-acting (single cylinder, double cylinder)
		Electric reciprocating—single acting, double acting (single cylinder, double cylinder)
	Rotary pump	Screw type (single screw, double screw, triple screw, quintuple screw, etc.), roots type, gear type (internal gear, external gear), vane type, plunger type (axial, radial)
Other types of pumps	—	Jet pump, air lift pump, hydraulic ram pump, electromagnetic pump, water turbine pump, etc.

1.2.2 Impeller Pump Overflow Components and Structural Forms

1.2.2.1 Overflow Components of Impeller Pump

The main overflow components of impeller pumps include the suction chamber, impeller, and pressure chamber (including guide vanes).

The pump suction chamber is located in front of the impeller, and its function is to direct the liquid toward the impeller. It comes in three forms: straight conical, curved pipe, and spiral (see Figure 1.2).

The pressure chamber is located at the periphery of the impeller, and its function is to collect the liquid flowing out of the impeller and deliver it to the discharge pipe. The pressure chamber mainly includes the volute chamber (spiral pressure chamber), radial guide vanes, spatial guide vanes, and axial pump guide vanes (see Figure 1.3).

The impeller is the most important working component of the pump and the core of the overflow components. The impeller is composed of a cover plate and intermediate blades. Depending on the

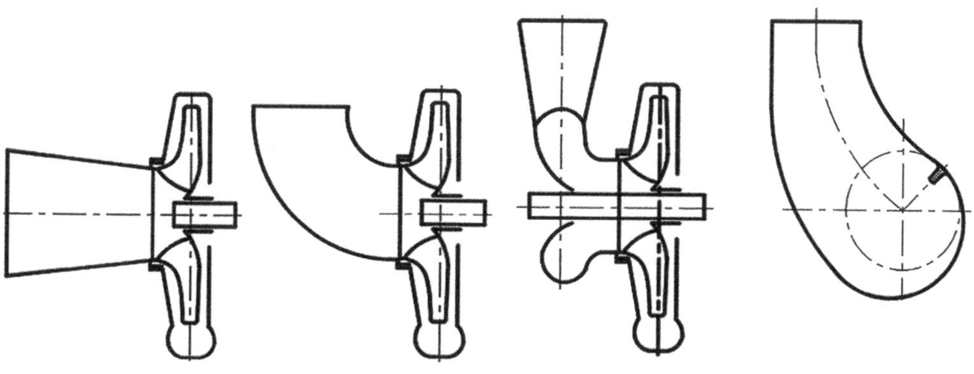

(a) (b) (c)

Figure 1.2 Classification of suction chambers: (a) Straight conical, (b) curved shape, (c) spiral shape.

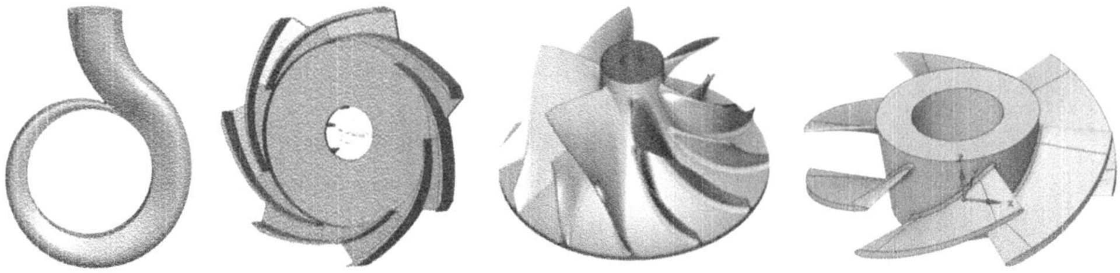

(a) (b) (c) (d)

Figure 1.3 Velocity circulation distribution of impeller inlet under the initial stall condition: (a) Volute, (b) radial guide-leaf, (c) space guide vane, (d) axial pump guide vane.

(a) (b) (c)

Figure 1.4 Velocity circulation distribution of impeller inlet under the initial stall condition: (a) Centrifugal, (b) mixed flow, (c) axial flow.

direction in which the liquid flows out of the impeller, the impeller can be divided into three types: centrifugal (radial), mixed flow (oblique), and axial flow (see Figure 1.4).

In the centrifugal (radial) impeller, the direction in which the liquid flows out of the impeller is perpendicular to the axis, i.e., it flows out along the radial direction.

In the mixed-flow (oblique) impeller, the direction in which the liquid flows out of the impeller is inclined to the axis.

In the axial-flow impeller, the direction in which the liquid flows out of the impeller is parallel to the axis, flowing out along the axis direction.

1.2.2.2 Structural Forms of Blade Pumps

Blade pumps are classified according to their structural forms as follows:

1) According to the main shaft direction:
   1) Horizontal: the main shaft is horizontally positioned;
   2) Vertical: the main shaft is vertically positioned;
   3) Inclined: the main shaft is inclined.
2) According to the type of impeller:
   1) Centrifugal: equipped with a centrifugal impeller;
   2) Mixed flow: equipped with a mixed-flow impeller;
   3) Axial flow: equipped with an axial-flow impeller.
3) According to the suction method:
   1) Single suction: equipped with a single suction impeller;
   2) Double suction: equipped with a double-suction impeller.

4) According to the number of stages:
   1) Single-stage: equipped with one impeller;
   2) Multistage: two or more impellers are mounted on the same shaft.
5) According to the blade installation method:
   1) Adjustable blades: the angle of the impeller blades can be adjusted;
   2) Fixed blades: the angle of the impeller blades is fixed.
6) According to the shell partition method:
   1) Sectional: the shell is divided by a plane perpendicular to the main shaft;
   2) Segmental: in the sectional structure, each stage of the shell is separate;
   3) Semi-open: the shell is split along the plane passing through the axis centerline;
   4) Horizontal semi-open: in the semi-open structure, the splitting plane is horizontal;
   5) Vertical semi-open: in the semi-open structure, the splitting plane is perpendicular to the horizontal plane;
   6) Inclined semi-open: in the semi-open structure, the splitting plane is inclined.
7) According to the pump body form:
   1) Volute casing: the shell on the impeller discharge side has a volute chamber;
   2) Double volute casing: the shell on the impeller discharge side has double volute chambers;
   3) Turbine type: a centrifugal pump with guide vanes;
   4) Barrel type: a pressure-resistant shell outside the inner shell;
   5) Double-shell type: a pump with a double shell outside the simple barrel.
8) According to the pump body support method:
   1) Suspended: the pump body has pump feet below, fixed on the base, with the bearing body suspended at one end;
   2) Bracket type: the lower part of the bearing body is fixed on the base, with the pump body supported by the bearing body at one end;
   3) Center-supported: the pump body is fixed on the base on both sides at a plane passing through the axis center.
9) Special structural blade pumps:
   1) Submersible electric pump: a pump in which the motor driving the pump is used in water along with the pump;
   2) Through-flow pump: a pump with a motor and other driving devices installed inside the pump body;
   3) Shielded pump: a pump directly connected to the motor (sharing a shaft), with the motor fixed inside with a shield to prevent liquid from entering the stator;
   4) Magnetic pump: a pump in which the motor drives an outer magnetic steel to rotate, causing the inner magnetic steel connected to the pump impeller to rotate through magnetic induction, with isolation sleeves between the inner and outer magnetic steels, completely preventing liquid leakage;
   5) Self-priming pump: a pump that does not require water filling when restarted;
   6) Pipeline pump: a pump that is a part of the pipeline and can be installed without special changes to the pipeline;
   7) Nonclogging pump: a pump that does not cause blockages in the pump when pumping liquid contains solid substances.

For the names, shapes, and characteristics of some pump overflow components, see Table 1.2, and for the combination forms of pump bodies, see Table 1.3.

Table 1.2 Selected pump overflow components.

Name			Shape of overcurrent part	Feature
Forms of overcurrent components	Centrifugal pumps	Impeller volute	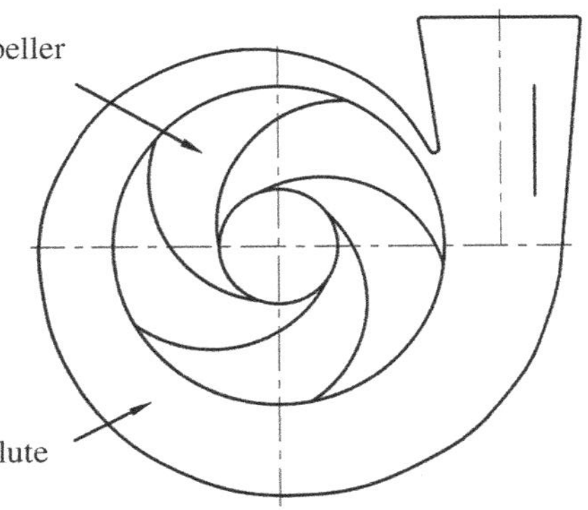	Wider range of applications
		Impeller guide vanes	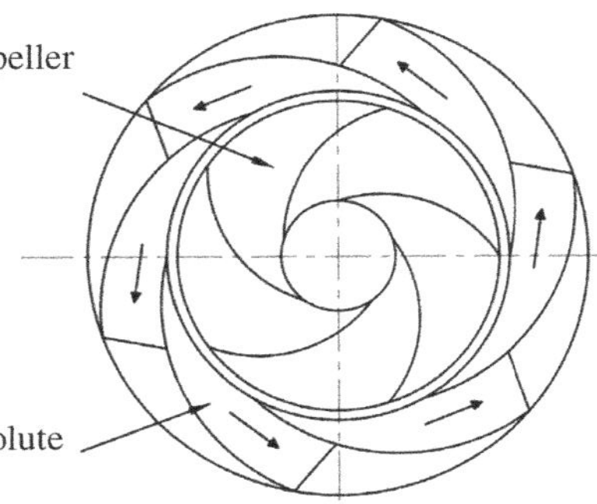	Sectional multistage pumps

	Impeller guide vane volute	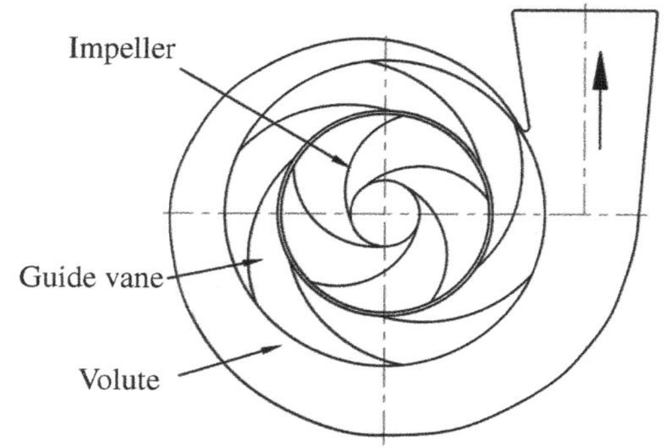	Larger volute pumps with low radial forces
Mixed-flow pump	Impeller volute	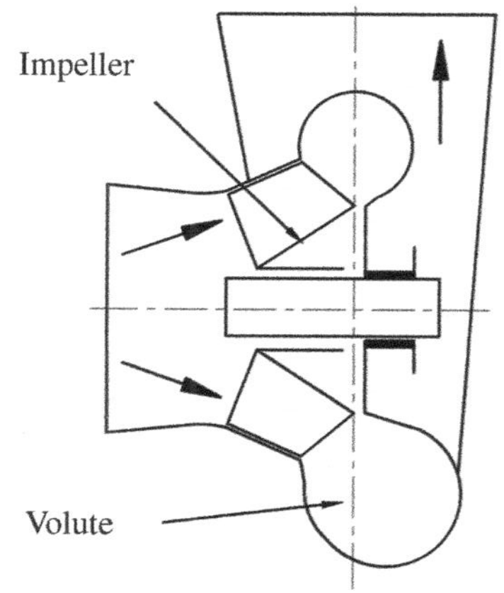	Medium-low head, large flow rate, simple structure

(Continued)

Table 1.2 (Continued)

Name		Shape of overcurrent part	Feature
	Impeller guide vane	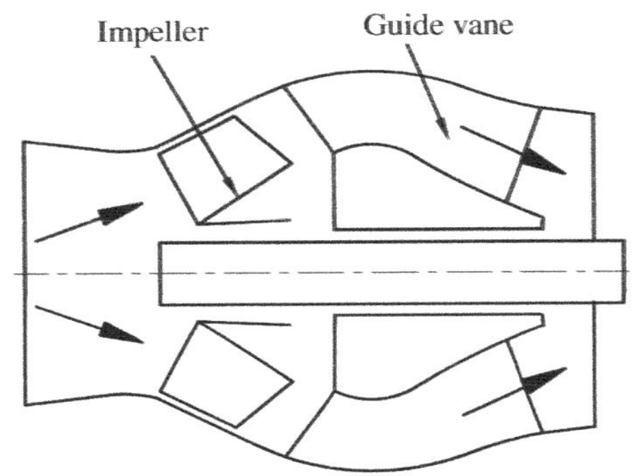	Head mostly between 10 and 20 m, small radial size
Axial pump	Impeller guide vane	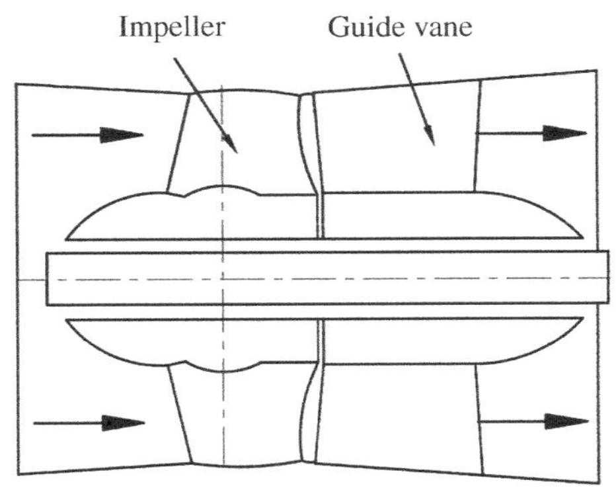	Low head, high flow

Table 1.3 Combined form of pump body.

Name	Form of construction	Features
Single-stage, single-suction centrifugal pumps	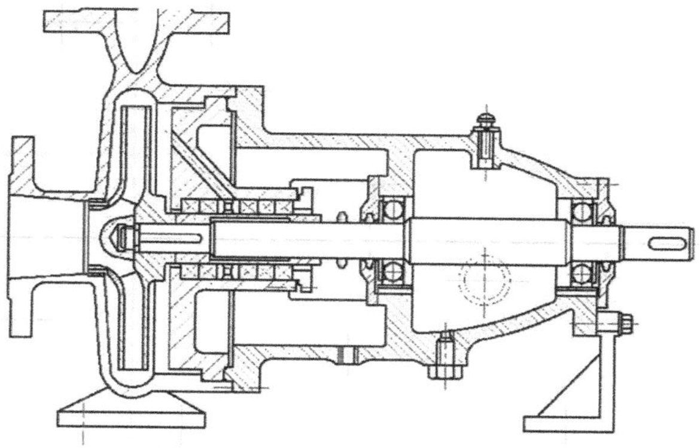	The most versatile pump
Double-suction pumps	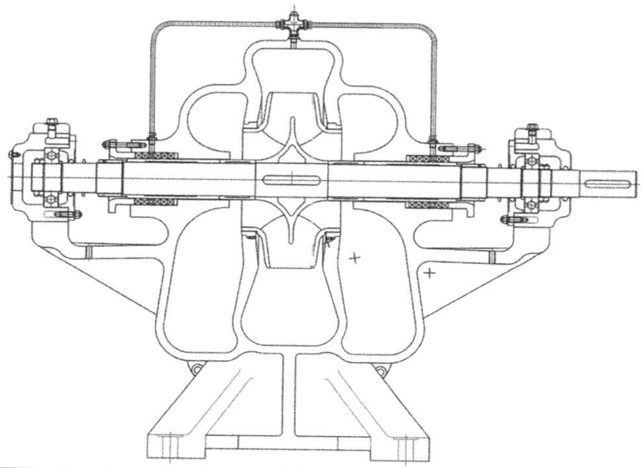	High flow *Horizontal split* Ease of inspection

(Continued)

Table 1.3 (Continued)

Name	Form of construction	Features
Guide vane multistage pumps	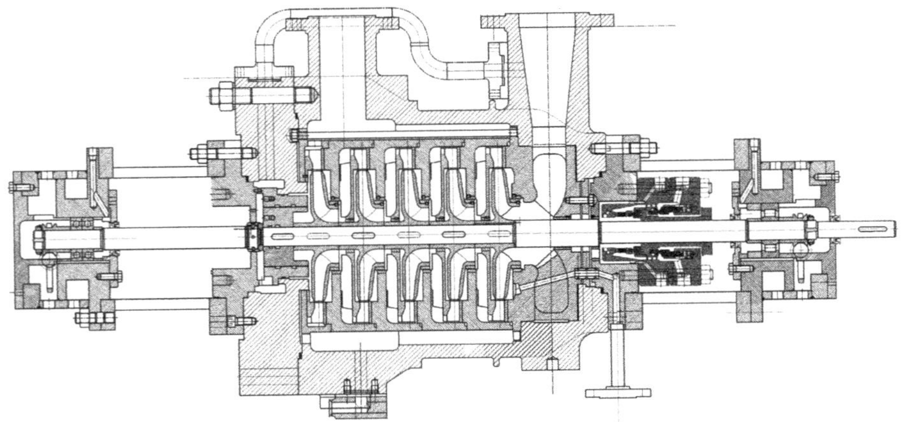	High head Sectional Variable number of stages
Volute multistage pumps	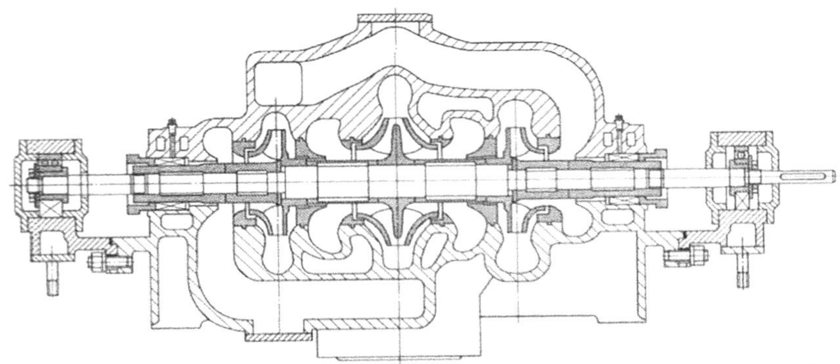	High head Horizontal cutaway Easy maintenance

1.3 Applications of Pumps

Pumps are universal machines with a wide variety of types and a broad range of applications. Wherever there is liquid flow in various sectors of the national economy, pumps are at work. Their primary applications include: agricultural irrigation, petrochemical industry, power industry, urban water supply and drainage, mining, and shipbuilding industries. Additionally, pumps are also used in rocket fuel supply and ship propulsion [4].

As an example, consider the application of pumps in thermal power plants in the power industry as shown in Figure 1.5. A power plant operates on a steam-water cycle, where the boiler heats water to steam which drives a turbine that powers a generator to produce electricity. In this process, a boiler feed water pump is used to supply water from the heater to the boiler. The exhaust gas from the turbine is condensed into water in the condenser, and the condensor pump is used to pump the condensed water back to the heater for recirculation. The cooling circulating water for the condenser is supplied by the circulating water pump. The pump used for the boiler ash discharge is the ash slurry pump, and to remove the sulfur dioxide (SO_2) in the flue gas of the boiler, a desulfurization pump is used to transport lime slurry for desulfurization.

The types of pumps used for various purposes are listed in Table 1.4.

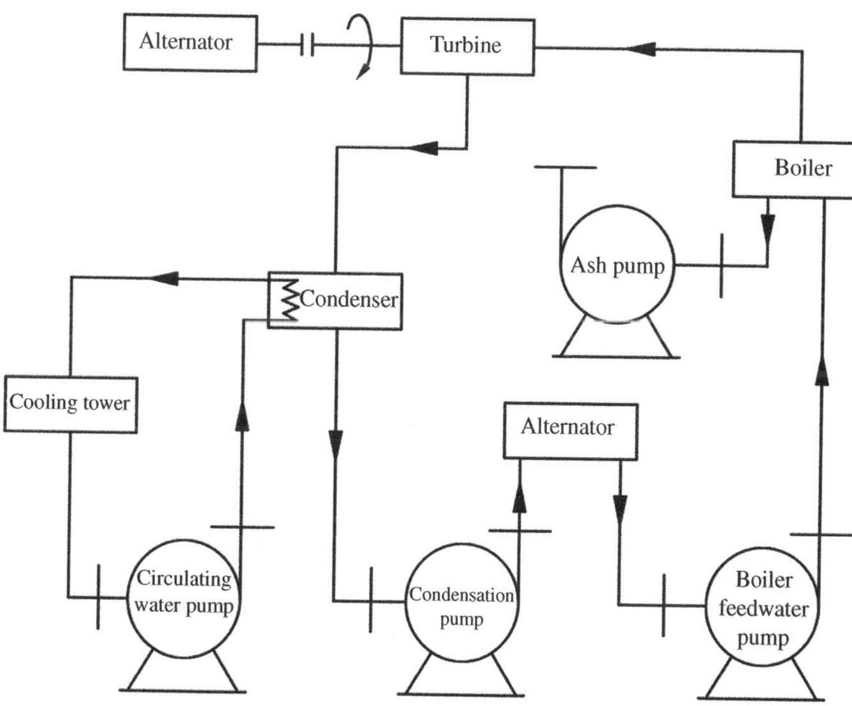

Figure 1.5 Thermal power plant using various pumps.

Table 1.4 Types of pumps for various applications.

Application	Main use	Summary	Types of pumps used	
Agricultural pumps	Irrigation	Irrigation of paddy fields or dry fields	Low-lift centrifugal pump	Deep well submersible pump
			Mixed-flow pump	Small submersible pump
	Drainage	Soil improvement or flood drainage	Axial-flow pump	Submersible axial-flow pump
			Self-priming pump	Submersible mixed-flow pump
Water supply engineering	Diversion	Diversion of raw water to the water purification site	Low-head centrifugal pumps, vertical swash pumps	
	Delivery	Delivery of water from the water purification site to the distribution tanks	Vertical swash pumps, double-suction pumps	
	Distribution of water	Distribution of water from the distribution tanks to the demand end	Vertical swash pumps, double-suction pumps	
	Boosting	Boosting the pressure of the water in the distribution line	Vertical centrifugal pumps, double-suction pumps	
Sewerage	Drainage	Discharge of sewage, rainwater	Sewage pumps, mixed-flow pumps, axial-flow pumps	
	Relay	Lift sewage to a higher level where it naturally goes down	Mixed-flow pumps, axial-flow pumps	
	Sewage treatment	Conveying sewage dirt	Sewage submersible pumps	
Mines, steel mills	Drainage of pits	Removal of gushing water from pits	Multistage centrifugal pumps, submersible pumps	
	Hydraulic coal mining	High-pressure water injection from nozzles to pulverize coal seams	Horizontal multistage, high-pressure centrifugal pumps	
	Hydro transportation	Hydro transportation of pelletized or coarser-grained coal	Mud pumps	
	Descaling of iron and steel	Descaling the surface of rolled steel with high-pressure water	Horizontal, high-head, multistage centrifugal pumps	
Thermal power plants	Boiler feed water	Feed water to the boiler	High-temperature and high-pressure multistage centrifugal pumps	
	Condensate	Pumping condensate in the condenser	High cavitation resistance centrifugal pumps	
	Circulating water	Cooling circulating water to the condenser	Vertical swash flow pumps, axial-flow pumps	

	Ash removal	Hydraulic conveyance of boiler slag	Slurry pumps, ash pumps
	Desulfurization	Removal of sulfur dioxide from flue gas ash	Desulfurization pumps (single-stage, low-head centrifugal pumps)
Oil industry	Drilling	Drill cuttings cooling bits	Piston mud pumps, centrifugal mud pumps
	Oil pumping	Pumping oil from wells	Piston pumps, submersible pumps
	Water injection	Injection of water around the wells to increase the pressure in the formation	High-pressure multistage centrifugal pumps, high-pressure piston pumps
	Transmission	Collective oil transport and long-distance pipeline transport	Centrifugal oil pumps
	Oil refining	Transportation of all kinds of oils in refineries in various unit processes	Multistage open-end pumps, double-suction oil pumps
Chemical industry	Chemical process	Conveying liquids in chemical installations	Single-stage corrosion-resistant centrifugal pumps
	High-temperature liquids	Conveying heat-mediated salts above 400°C, etc.	Subliquid pumps
	Molten metals	High-temperature media	High-temperature pumps
	Low-temperature liquids	Conveying liquid nitrogen, etc., below −100°C	Low-temperature pumps
Marine industry	Boiler feed water	Feed water to boilers	Turbine-driven medium-open multistage pumps
	Pumping condensate	Pumping condensate from condensers	Vertical multistage (first-stage double-suction) pumps
	Conveying circulating water	Conveying circulating water for condensers	Vertical single-suction (two-stage) centrifugal pumps
	Cargo oil clearing	Loading, unloading, and clearing of ships	Vertical reciprocating pumps, vertical self-priming pumps
	Balancing propulsion	Balancing of hulls and water-jet propulsion	Axial-flow pumps and guided vane mixing flow pumps
Light industry	Pulp	Transporting pulp of various concentrations	Clog-free centrifugal pump
	Chemical liquid	Transporting black liquor	Corrosion-resistant centrifugal pump
	Ceramic slurry	Transporting ceramic slurry	Diaphragm plunger pump

(Continued)

Table 1.4 (Continued)

Application	Main use	Summary	Types of pumps used
Food	Beverages	Conveying beverages, dairy products	Stainless steel centrifugal pumps
	Yeast	Conveying yeast	Rotor pumps, gear pumps
	Homogenization	Refining fruit tea, dairy products, soy products, etc.	High-pressure plunger homogenizing pumps
Building construction	Drainage	Drainage of surges in construction	Submersible and self-priming pumps for construction
	Dredging	Dredging	Sewage pumps
	Concrete	Compacting concrete, supplying water to mixers	Diaphragm piston pumps, micro-centrifugal pumps
Construction equipment	Water supply to boilers	Water supply to boilers	Small vertical (horizontal) multistage centrifugal pumps
	Water supply to buildings	Water supply to high-rise buildings	Vortex pumps, pipeline pumps, vertical multistage pumps
	Discharge of sewage water	Drainage of sewage water filth in hotels, restaurants, etc.	Submersible pumps for sewage water
	Recirculation of warm water	Warm water for hot water supply or for warming rooms	Pipeline pumps
Construction equipment	Cold room (cooling, circulation, misting)	Warm water from the freezer condenser is pumped to a cooling tower for circulation to circulate the water between the freezer evaporator and the air conditioner coil	Shielded pumps
Nuclear power plants	Main circulation pumps	Circulate reactor fluid to steam generators	Vertical swash flow pumps
	Charging pump	Supply top charge and main pump seal cooling water to the primary circuit	Horizontal multistage pumps
	Safety spray pumps	Supply cooling water to the reactor containment sprayers	Horizontal multistage pumps
	Waste heat removal pumps	Remove the waste heat from the reactor during a shutdown	Horizontal single-stage centrifugal pumps
	Main feed pumps	Supply circulating water to the steam generators	Multistage centrifugal pumps or double-suction pumps

References

1 Guan, X. (2011). *Modern Pump Theory and Design.* Beijing: China Aerospace Press.

2 Wong, Y. and Sumathy, K. (1999). Solar thermal water pumping systems: a review. *Renewable and Sustainable Energy Reviews* 3 (2–3): 185–217.

3 Ram, M., Bogdanov, D., Aghahosseini, A., Solomon, O., Gulagi, A., Child, M., Fell, H., and Breyer, C. (2017). *Global Energy System Based on 100% Renewable Energy-Power Sector.* Lappeenranta: Lappeenranta University of Technology and Energy Watch Group.

4 Rahman, M.M., Khan, I., Field, D.L., Techato, K., and Alameh, K. (2022). Powering agriculture: present status, future potential, and challenges of renewable energy applications. *Renewable Energy* 188: 731–749.

2

Pump Fundamentals and Physical Concepts

2.1 Calculation of Velocity Triangles

2.1.1 Motion Analysis

A pump is a machine where liquid and mechanical forces interact. The performance characteristics of a pump are the external manifestations of the liquid's motion within the pump, and the liquid's motion is determined by the geometric shape of the flow passage. Therefore, analyzing the liquid's motion within the pump is fundamental to determining the pump's performance characteristics and designing its geometry. The impeller is the core component of the pump's operation, and therefore a detailed analysis of the liquid's motion within the impeller is particularly important.

The motion of the impeller itself is simple; it simply rotates with the shaft. However, due to the action of the impeller blades, the motion of the liquid within the impeller is relatively complex and represents a composite motion. On one hand, the liquid undergoes transport motion by rotating with the impeller, and on the other hand, it continuously flows out of the rotating impeller, i.e., it moves relative to the impeller. When observed from a coordinate system fixed to the ground, the motion of the liquid in the impeller is termed the absolute motion, which is a combination of the transport motion and the relative motion, expressed as $v = u + w$.

To further analyze the liquid's motion within the impeller, we adopt a method that decomposes the overall flow into localized components, i.e., dividing the flow within the impeller into layers. Assuming that the liquid between the layers does not mix when the number of layers is sufficiently large, we obtain micro-element flow layers. These flow layers exhibit the properties of stream surfaces—the velocity at any point on the surface is tangential to it. Clearly, the inner rotating surfaces of the front and rear shrouds of the impeller represent the two boundary stream surfaces, and any number of similar stream surfaces (typically 1–3) can be defined between them. Thus, the study of flow within the impeller is simplified to the study of the flow on several stream surfaces. While the flow on these stream surfaces may not be identical, the analytical methods remain the same. Therefore, once the flow on one stream surface is thoroughly understood, the flow on other stream surfaces can be similarly resolved.

Figure 2.1a illustrates the rear shroud flow surface, on which the intersection lines of the blades with the flow surface (blade profiles) are depicted. These intersection lines represent the streamlines of the liquid's relative motion within the impeller. Between any two adjacent blades, an arbitrary number of such streamlines can be drawn. Assuming the blades are infinitely numerous and infinitely thin, the shapes of these streamlines are identical, and thus only one streamline needs to be analyzed. By arranging the relative streamlines from several stream surfaces in a systematic manner and adding thickness, the blade surfaces within the impeller are formed. Therefore, the

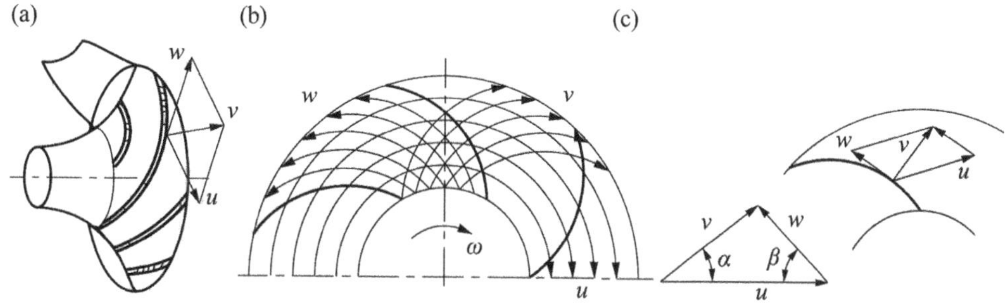

Figure 2.1 Analysis of liquid movement in the impeller: (a) Flow surface, (b) Flow streamlines, (c) Velocity triangle.

blades can be considered as composed of several streamlines in relative motion. Consequently, the study of flow within the impeller is simplified to the analysis of the flow of streamlines in relative motion. The streamlines of transport motion, relative motion, and absolute motion of the impeller are shown in Figure 2.1b.

2.1.2 Velocity Triangle

The velocity triangle is an essential tool for studying the flow along streamlines during the composite motion of fluid through the impeller. The transport motion is the rotational motion of the impeller, whose trajectory is circular. The relative motion streamline aligns with the shape of the blade. The absolute motion can be determined using the parallelogram law. In practical applications, only half of the parallelogram is drawn, forming the velocity triangle (Figure 2.1). Since the velocities u (transport velocity) and w (relative velocity) are both tangent to the stream surface, the plane of the velocity triangle is also tangent to the stream surface at that point. While the velocity triangle is typically drawn on a plane (Figure 2.1c), its spatial orientation is determined by the shape of the stream surface.

To solve for the velocity triangle, the components of velocities involved must be analyzed in detail.

① Circumferential Velocity u

The circumferential velocity at any point in the impeller is tangent to the circle at that point and is calculated as:

$$u = \frac{D\pi n}{60}$$

where n is the impeller rotational speed (rpm), and D is the diameter at the point of interest.

② Relative Velocity w

Assuming that the blades are infinitely numerous, the direction of the relative velocity at any point aligns with the tangent to the blade surface at that location. While the direction is known, the magnitude of w cannot be determined immediately.

③ Absolute Velocity v

The absolute velocity is the resultant velocity, and neither its direction nor its magnitude is immediately apparent. To analyze it, we decompose v into two mutually perpendicular components: $v = v_{\mathrm{u}} + v_{\mathrm{m}}$. One is the velocity v_{u} along the circumference; since its direction is the same as u, it is called the circumferential component or circumferential component of the absolute velocity. Since only one plane can be constructed perpendicular to a given line at a point, therefore, there is only

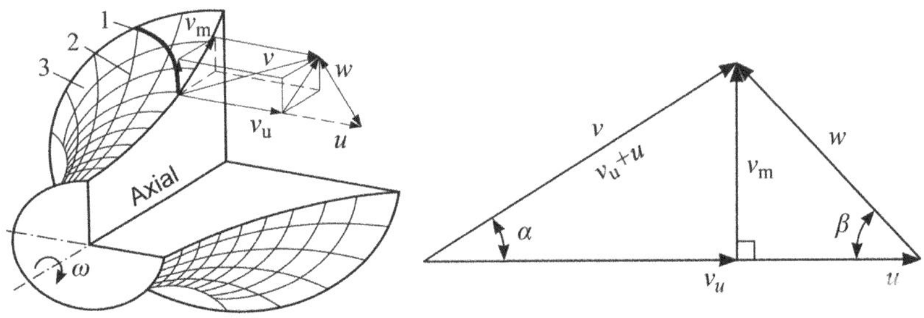

Figure 2.2 Velocity triangle: 1. Blade, 2. Axial flow line, 3. Flow surface.

one plane perpendicular to the velocity vector u at that point. Furthermore, since the velocity u is perpendicular to the radius at that point, this plane must coincide with the meridional plane containing the radius. Since v_m lies in this plane, it is referred to as the meridional component of the absolute velocity as shown in Figure 2.2.

The intersection of the meridional plane and the stream surface is called the meridional streamline, which represents the projection of any streamline on the stream surface onto the meridional plane. Since the meridional velocity v_m lies both on the stream surface and the meridional plane, its direction must be tangent to the meridional streamline. That is, $v = v_u + v_m$.

Now, consider the magnitudes of v_u and v_m. The circumferential component v_u is related to the rotational motion, and its magnitude is associated with the head (energy transfer) but not directly with the flow rate through the impeller.

The meridional component v_m represents the flow of liquid moving along the meridional plane toward the impeller outlet. Its magnitude depends on the flow rate Q_t through the impeller. Once Q_t is given, v_m can be calculated. To calculate v_m, we assume a meridional liquid flow and determine its area of cross-section.

As shown in Figure 2.3a, a curve MN is drawn perpendicular to all meridional streamlines. This curve is called the forming line of the meridional flow cross-section. Rotating this line around the axis generates the cross-sectional surface of the meridional flow. Its area is calculated as:

$$F_0 = 2\pi R_c b$$

where F_0 is the cross-sectional area, b is the length of the forming line and R_c is the radius of the centroid of the forming line (calculation methods are discussed in the impeller design section).

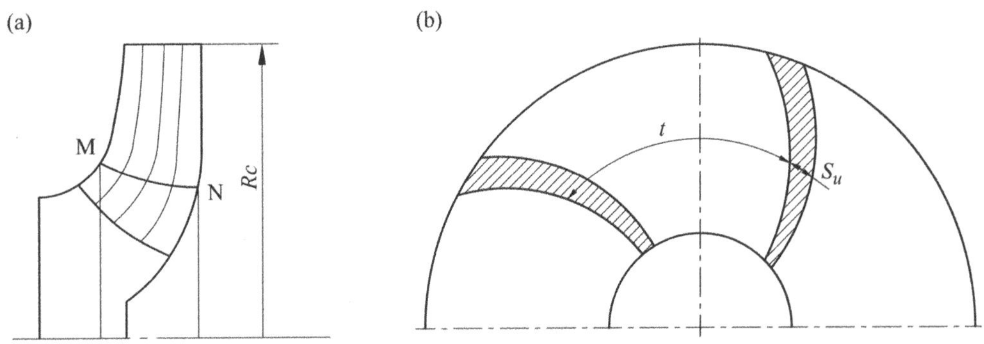

Figure 2.3 Axial surface liquid flow through the water section and exclusion coefficient: (a) Axial liquid flow through the water section, (b) Blade blockage coefficient.

The outlet flow area of the impeller is typically a cylindrical surface concentric with the axis, and its value is given by:

$$F_0 = 2\pi R_2 b_2$$

where R_2 is the impeller outer circle radius and b_2 is the blade outlet width.

Due to the obstruction caused by the blades, the effective cross-sectional area F is smaller than F_0. As shown in Figure 2.3b, the blade blockage coefficient can be calculated as:

$$\psi = \frac{t - s_u}{D} = 1 - \frac{Zs_u}{D\pi}$$

where D is the diameter at the calculation point, Z is the number of blades and s_u is the circumferential thickness of the blade at the calculation point.

The effective cross-sectional area is then given by:

$$F = F_0 \psi$$

The meridional velocity v_m can be expressed as:

$$v_m = \frac{Q_1}{F} = \frac{Q}{\eta_v 2\pi R_c b \psi}$$

Based on the above analysis, the velocity triangle can be constructed for given conditions. This completes the description of the method for solving the velocity triangle at any arbitrary point.

2.1.3 Velocity Triangle at Blade Inlet and Outlet

When analyzing the pump flow problems, one primarily uses the velocity triangles at the inlet and outlet of the blades. For clarity, we define the following notation: Parameters at the inlet of the blade are denoted with a subscript 1. Parameters at the outlet of the blade are denoted with a subscript 2. Parameters just before the inlet are denoted with a subscript 0. Parameters just after the outlet are denoted with a subscript 3.

2.1.3.1 Blade Inlet Velocity Triangle

① Circumferential Velocity u_1

$$u_1 = \frac{D_1 \pi n}{60}$$

where D_1 is the diameter of the calculation point at the blade inlet.

② Axial surface Velocity v_{m1}

$$v_{m1} = \frac{Q_1}{F_{01}\psi_1} = \frac{Q}{\eta_v 2\pi R_{e1} b_1 \psi_1}$$

where F_{01} is the area of the water crossing section at the inlet of the blade perpendicular to the flow line of the axial plane of the study point and ψ_1 is the blade inlet blockage coefficient given by:

$$\psi_1 = 1 - \frac{Zs_{u1}}{D_1 \pi}$$

③ Circumferential component velocity v_{u1}

v_{u1} is the circumferential component of the absolute velocity of the liquid at the inlet of the blade. Its magnitude is related to the shape and size of the suction chamber in front of the impeller. Since the liquid has just entered the blade, the blade cannot yet fully control the direction of the relative

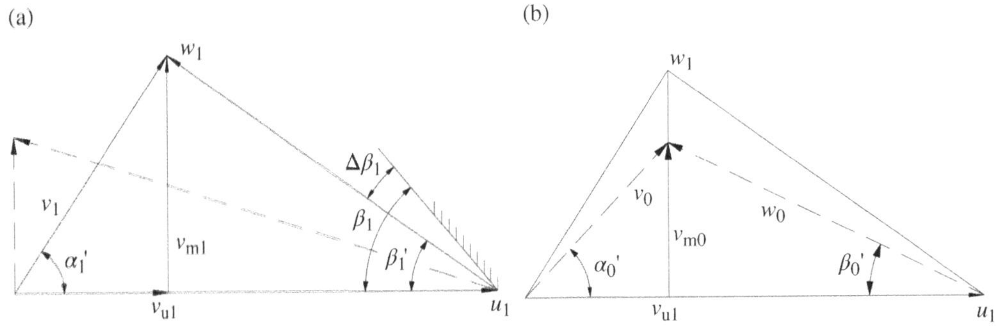

Figure 2.4 Velocity triangle at blade inlet: (a) Blade inlet velocity triangle, (b) Velocity triangle located slightly behind the blade inlet.

velocity w_1. The direction of the absolute velocity v_1 depends on the shape of the flow path prior to the inlet. For a straight conical suction chamber, the liquid flows without any circumferential component, therefore $v_{u1} = 0$ and $v_{u1} = v_{m1}$, that is v_1 is perpendicular to the circumferential velocity u_1. For other shapes of suction chambers, the method for calculating v_{u1} will be discussed in the design section.

Based on the magnitude and direction of v_1 and v_{m1}, the velocity triangle at the blade inlet can be constructed as shown in Figure 2.4a. The dotted line in this figure represents the velocity triangle when $v_{u1} = 0$. The angle between the absolute velocity and the circumferential velocity is denoted as α' and is called the absolute flow angle. The angle between the relative velocity and the circumferential velocity is denoted as β' and is called the relative flow angle. Typically, the difference $\Delta\beta = \beta_1 - \beta_1'$ is called the incidence angle.

Slightly before the blade inlet, due to the absence of blade blockage, the axial velocity is smaller than v_{m1}. This velocity is denoted as v_{m0}. The circumferential velocity u_1 can be considered equal to v_{m0}. The velocity triangle slightly before the blade inlet is shown by the dotted line in Figure 2.4b. In this figure, the direction of the absolute velocity is primarily determined by the structure of the suction chamber and remains constant. The magnitudes and directions of u_1 and u_0 differ only due to the effect of blade blockage and can be approximately considered the same. When analyzing problems, it can be assumed that for a given suction chamber, the direction of v_1 remains unchanged and does not vary with the flow rate. Based on this, the inlet velocity triangle for different flow rates can be constructed, which is crucial for the problem analysis [1].

2.1.3.2 The Velocity Triangle at the Blade Outlet

① Circumferential Velocity u_2

$$u_2 = \frac{D_2 \pi n}{60}$$

where D_2 is the Diameter at the calculation point of the blade outlet.

② Axial Velocity v_{m2}

$$v_{m2} = \frac{Q_t}{F_{02}\psi_2} = \frac{Q}{\eta_v 2\pi R_{c2} b_2 \psi_2} \qquad \psi_2 = 1 - \frac{Z s_{u2}}{D_2 \pi}$$

③ Relative Velocity w_2

At the blade outlet, the relative velocity of the liquid is constrained by the blade, and its direction depends on the blade's orientation. Assuming an infinite number of blades, the direction of w_2 aligns with the tangential direction of the blade's outlet surface and is fixed.

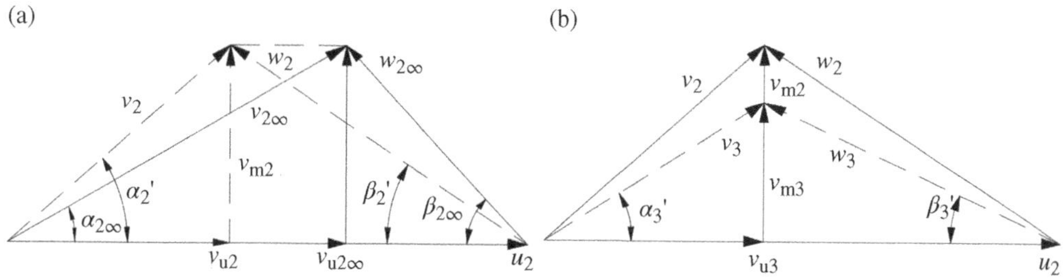

Figure 2.5 Velocity triangle at blade outlet: (a) Blade outlet velocity triangle, (b) Velocity triangle located slightly behind the blade outlet.

Based on the magnitude and direction of u_2 and v_{m2}, the velocity triangle at the blade outlet can be constructed as shown in Figure 2.5. Here, it should be noted that the parameters for the case of an infinite number of blades are denoted with the subscript ∞ to distinguish them from those for a finite number of blades.

Assuming an infinite number of blades, $\beta'_{2\infty} = \beta_2$, meaning that the relative flow angle at the outlet coincides with the blade outlet angle. In reality, the number of blades is finite, and in this case $\beta'_2 < \beta_2$. The corresponding velocity triangle is shown by the dotted line in Figure 2.5a. When analyzing problems, it can be assumed that the direction of the relative velocity at the outlet remains unchanged. This allows for the construction of the velocity triangle for different flow rates.

Slightly after the blade outlet, as the liquid is no longer constrained by the blade blockage, the axial velocity v_{m3} is smaller than the outlet axial velocity v_{m2}. The corresponding velocity triangle is shown by the dotted line in Figure 2.5b.

It is worth noting that the velocity triangle can be constructed for different geometry and flow conditions. When analyzing a problem, the geometric parameters of the impeller can be used to create the velocity triangle; in design problems, the velocity triangle is constructed based on the given motion parameters, and finally a geometric shape that meets these motion conditions is manufactured.

2.2 The Theory of Energy Conversion in Pumps

During the process of converting mechanical energy into liquid energy, a pump experiences various losses, which are represented by corresponding efficiency values. To improve the efficiency of the pump, it is essential to analyze the distribution of pump power, understand its origins and destinations, and identify ways for reducing losses and enhancing the pump efficiency.

Below, the energy input and output within the pump are introduced sequentially according to the energy transfer process inside the pump [2, 3].

2.2.1 Mechanical Loss and Mechanical Efficiency

The power transmitted from the prime mover to the pump shaft (shaft power) is first partially consumed to overcome the friction losses in the bearings and sealing devices. The remaining shaft power is used to drive the rotation of the impeller. However, not all of the mechanical energy from the impeller rotation is transferred to the liquid passing through the impeller. A portion is consumed in overcoming the friction between the front and rear cover plates of the impeller and the liquid in the pump casing (pump cavity). This portion of lost power is called the disc friction loss. The principle of energy loss in the pump is illustrated in Figure 2.6.

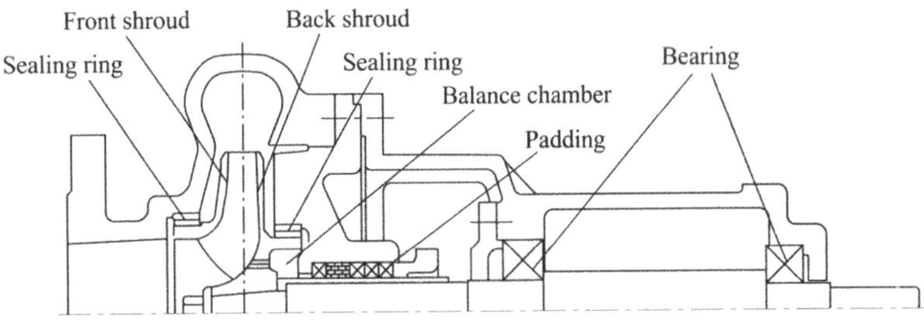

Figure 2.6 Sources of energy loss in the pump.

The sum of the aforementioned bearing loss power P_{m1}, sealing loss power P_{m2}, and disc friction loss power P_{m3} is referred to as the mechanical loss power P_m, whose magnitude is represented by the mechanical efficiency η_m.

The power remaining after subtracting the mechanical loss power from the shaft power is used to perform work on the liquid passing through the impeller and is called the input hydraulic power, denoted as P'. The mechanical efficiency is the ratio of the input hydraulic power to the shaft power, i.e.,

$$\eta_m = \frac{P - P_m}{P} = \frac{P'}{P}$$

2.2.2 Volume Loss and Volume Efficiency

The input hydraulic power is used to perform work on the liquid passing through the impeller. As a result, the pressure of the liquid at the impeller outlet is higher than that at the inlet. The pressure difference between the outlet and inlet causes a portion of the liquid to flow back from the pump cavity through the impeller sealing ring (wear ring) clearance to the impeller inlet. This results in the flow rate through the impeller Q_t (also known as the theoretical flow rate of the pump) not being entirely delivered to the pump outlet. The leakage flow q consumes the energy obtained from the impeller during the leakage process, converting it from high-pressure (outlet pressure) liquid to low-pressure (inlet pressure) liquid. Therefore, the essence of volumetric loss is also an energy loss, and its magnitude is calculated using the volumetric efficiency. The volumetric efficiency is the ratio of the power of the liquid passing through the impeller after subtracting the leakage (i.e. the actual flow rate Q) to the power of the liquid passing through the impeller (theoretical flow rate Q_t) i.e. the input hydraulic power, and is given by

$$\eta_v = \frac{\rho g Q_t H_t - \rho g q H_t}{\rho g Q_t H_t} = \frac{\rho g Q H_t}{\rho g Q_t H_t} = \frac{Q}{Q_t}$$

where Q_t is the theoretical flow rate of the pump (flow rate through the impeller), that is $Q_t = Q + q$, H_t is the theoretical head of the pump (energy imparted to the unit mass of liquid by the impeller), and q is the leakage flow rate.

For single-stage pumps, leakage mainly occurs at the sealing ring. For multistage pumps, in addition to this leakage, inter-stage leakage also occurs. Furthermore, the leakage from axial force balancing devices, sealing devices, etc., should also be included in the volumetric loss of the pump.

2.2.3 Hydraulic Loss and Hydraulic Efficiency

The effective liquid passing through the impeller (after accounting for the leakage) receives energy H_t from the impeller, but not all of this energy is delivered to the outlet. This is because the liquid experiences hydraulic friction losses (due to flow resistance along the path) and hydraulic losses

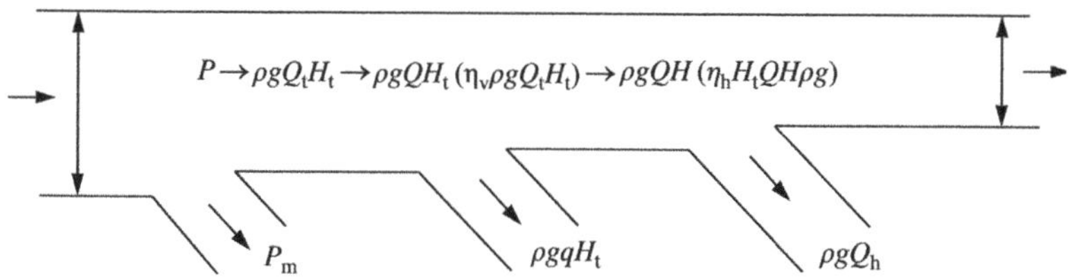

Figure 2.7 Power and energy balance in the pump.

caused by impacts, flow separation, changes in velocity direction and magnitude, and other local resistances as it flows through the pump's internal passages from the inlet to the outlet. These losses consume a portion of the energy. The energy lost by the unit mass of liquid during its flow through the pump's passages is called the hydraulic loss, denoted as h. Due to hydraulic losses, the energy increase H of the unit mass of liquid after passing through the pump is less than the energy H_t imparted by the impeller, i.e., $H = H_t - h$.

The magnitude of hydraulic losses is measured by the hydraulic efficiency. Hydraulic efficiency is the ratio of the power of the liquid after removing the hydraulic losses to the power of the liquid before removing the hydraulic losses, i.e.,

$$\eta_h = \frac{\rho g Q H}{\rho g Q H_t} = \frac{H}{H_t}$$

The sum of all losses within the pump is represented by the total efficiency (referred to as the pump efficiency). The total efficiency is the ratio of the effective output power P_e to the input power (shaft power) P, i.e.,

$$\eta = \frac{P_e}{P}$$

Through a transformation, one obtains

$$\eta = \frac{\rho g Q H}{P} = \frac{\rho g Q H}{P}\frac{Q_t H_t}{Q_t H_t} = \frac{\rho g Q_t H_t}{P}\frac{Q}{Q_t}\frac{H}{H_t}$$

Thus,

$$\eta = \eta_m \eta_v \eta_h$$

The efficiency of the pump is equal to the product of its mechanical efficiency, volumetric efficiency, and hydraulic efficiency.

The power balance in the pump can be illustrated as shown in Figure 2.7.

Before designing a pump, the efficiency of the pump to be designed can only be roughly determined based on the statistical data (empirical formulas or curves) or similar actual products. After the pump design is completed, the efficiency of the designed pump can be approximately estimated. Only after the pump is manufactured can its efficiency be precisely determined through testing.

2.3 Similarity Laws

2.3.1 Basic Concepts of Similarity Laws

The theory of similarity is widely applied in the design and testing of pumps. The common practice of scaling designs based on models and conducting model tests is guided by this theory. Using the similarity theory, the results of model tests can be converted to full-scale pumps, and the parameters of full-scale pumps can be scaled down to design and test models. Conducting tests on small models is much more economical than testing full-scale machines. Additionally, due to practical limitations, it is often difficult to test full-scale machines when their size is too large, their speed is too high, or when handling special liquids such as high-temperature fluids. In such cases, model tests are used as a substitute.

The similarity theory states that the two fluid flows are mechanically similar only if the following three conditions are met.

2.3.1.1 Geometric Similarity

The corresponding angles of the model and the full-scale pump must be equal, and the ratios of corresponding linear dimensions must be the same. Strictly speaking, surface roughness should also be similar, but this is difficult to achieve in practice and is usually corrected based on empirical data. Geometric similarity is a prerequisite for mechanical and kinematic similarity; without it, discussions of dynamic and kinematic similarity are meaningless. Using the subscript M for the model and no subscript for the full-scale pump (prototype), the condition for geometric similarity can be expressed as:

$$\frac{D}{D_M} = \frac{b}{b_M} = \cdots = \frac{L}{L_M} \tag{2.1}$$

where L, L_M are the corresponding linear dimensions of the prototype and the model.

2.3.1.2 Kinematic Similarity

The ratios of corresponding velocities of the liquid at corresponding points in the flow passages of the model and the prototype must be the same, i.e.,

$$\frac{v}{v_M} = \frac{w}{w_M} = \frac{u}{u_M} = \frac{Dn}{D_M n_M} \tag{2.2}$$

Kinematic similarity is a natural consequence of geometric and dynamic similarity.

2.3.1.3 Dynamic Similarity

The ratios of corresponding forces acting on the liquid at corresponding points in the flow passages of the model and the prototype must be the same. This means that the ratio of the external force F acting on the flow to the inertial force F_i caused by the fluid itself under external forces must be the same. This ratio is called the Newton number, denoted as N_e, i.e.,

$$\frac{F}{ma} = N_e \tag{2.3}$$

The value of N_e represents the general dynamic similarity condition of the flow. If N_e values are equal, the flows are dynamically similar. The external forces acting on the liquid include viscous forces (friction), pressure, gravity, surface tension, and elastic forces. It is impossible to satisfy the similarity conditions for all these forces simultaneously. In practical problems, only the dominant forces are considered for similarity, while the minor forces are neglected. Below are several commonly used dynamic similarity criteria in pumps:

2.3.1.4 Similarity Criterion for Pipe Flow—Reynolds Criterion

When studying flow resistance in pipes, viscous friction is the dominant force, while gravity and pressure have little effect and can be neglected. Since the inertial force can be expressed as: $ma = m\frac{v}{t} = m\frac{v^2}{L} = \frac{\rho L^3 v^2}{L} = \rho L^2 v^2 \left(t = \frac{L}{v}\right)$ and the viscous force can be expressed as: $F = zA = \mu\frac{v}{L}L^2 = \mu L\left(A = L^2, \tau = \mu\frac{dv}{dL}\right)$, the Newton number becomes:

$$N_e = \frac{\mu \omega L}{\rho L^2 v^2} = \frac{v}{vL} = \frac{1}{Re} \tag{2.4}$$

From this, it can be concluded that the condition for dynamic similarity of two flows is that their Reynolds numbers are equal, i.e., $Re = Re_M$.

2.3.1.5 Similarity Criterion for Free Surface Flow—Froude Criterion

Flow in open channels and suction pools falls into this category. Since the flow has a free surface, the hydrostatic pressure and gravity balance each other, making gravity the dominant force. Viscous forces and surface tension can be neglected. Since gravity can be expressed as: $F = \rho L^3 g$, the Newton number becomes:

$$N_e = \frac{\mu\omega L}{\rho L^2 v^2} = \frac{v}{vL} = \frac{1}{Re} \text{ or } N_e = \frac{\sqrt{Lg}}{v} = \frac{1}{Fr} \tag{2.5}$$

From this, it can be concluded that the condition for dynamic similarity of two flows is that their Froude numbers are equal.

2.3.1.6 Similarity Criterion for Flow in Pump Impellers—Euler Criterion

The dominant force in the flow within an impeller is the pressure generated by the blades acting on the liquid, while viscous forces and gravity are secondary and can be neglected. When viscous forces are neglected, $H_t = H, p = H, \rho g = \rho g \rho_i H$. The pressure per unit volume of the liquid in the impeller can be expressed as:

$$\frac{P}{v} = \frac{\rho g H \rho_i D_2 \pi b_2}{D_2 \pi b_2 \Delta R} = \frac{\rho g}{\Delta R} \quad \rho_i H = \frac{\rho g}{\Delta R} \frac{u_2 v_{u2}}{g} \left(1 - \frac{1}{2} \frac{v_{u2}}{u_2}\right)$$

where ΔR is the radial distance the liquid flows through in the impeller, and ρ_i is the reaction coefficient.

Next, consider the inertial force. Since the blade's action on the liquid is along the circumferential direction, only the tangential component of the flow's inertia is considered. If the tangential component of the liquid flow is Δy^2 over time Δt, and because the flow is kinematically similar, Δv_{u2} is proportional to v_{u2}. Thus, v_{u2} can be used as the characteristic velocity representing the inertial force. The tangential acceleration of the flow is:

$$a_u = \frac{\Delta v'_{u2}}{\Delta t} \propto \frac{v_{u2}}{\Delta t} = \frac{v_{u2} v_{m2}}{\Delta R} \quad \left(v_{m2} = \frac{\Delta R}{\Delta t}\right)$$

The inertial force per unit volume of liquid is:

$$\frac{ma_*}{v} = \rho a_* = \rho \frac{v_{*2} v_{*2}}{\Delta R}$$

The ratio of pressure to inertial force is called the Euler number:

$$Eu = \frac{\frac{P}{v}}{\frac{ma_u}{v}} = \frac{\rho \frac{v_{u2} u_2}{\Delta R}}{\rho \frac{v_{u2} v_{m2}}{\Delta R}} \left(1 - \frac{1}{2} \frac{v_{u2}}{u_2}\right) \tag{2.6}$$

Thus, $Eu = \frac{u_2}{v_{m2}} \left(1 - \frac{1}{2} \frac{v_{u2}}{u_2}\right)$

This equation shows that the Euler number depends only on the ratio of velocities. Therefore, impellers that are geometrically and kinematically similar naturally satisfy dynamic similarity. Conversely, if two impellers satisfy the condition of dynamic similarity, i.e., their Euler numbers are equal ($Eu = Eu_M$), then the flows within the two impellers are kinematically similar.

2.3.2 Similarity Law of Pumps

Strictly speaking, the mechanical similarity of fluid flow in full-scale and model pumps must satisfy geometric, kinematic, and dynamic similarity conditions.

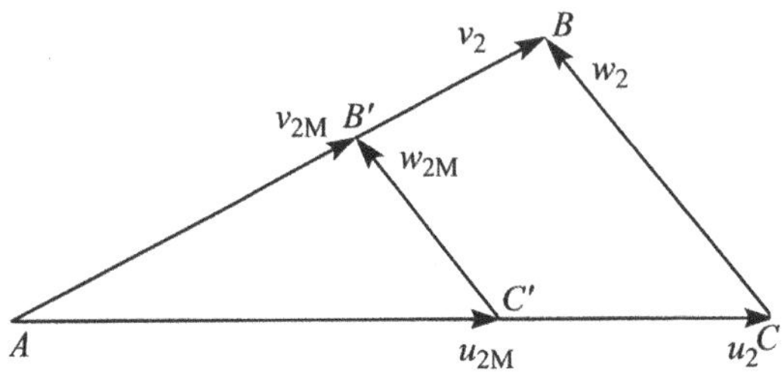

Figure 2.8 Velocity triangle similarity.

However, as long as geometric and kinematic similarity are satisfied, the flow within the impeller can naturally be assumed to satisfy pressure similarity. Additionally, in the flow channels of the pump's volute casing and other components, although viscous forces dominate, the flow velocity is typically very high, and the Reynolds number of the flow is large, placing it in the resistance square zone. Within this range, the frictional resistance (viscous force) of the flow is independent of the Reynolds number and varies only with surface roughness. Therefore, under geometric similarity (including roughness similarity), viscous force similarity is naturally and approximately satisfied. For this reason, dynamic similarity is usually not considered in pump analysis, and the similarity laws are derived based solely on geometric and kinematic similarity (i.e., the velocity triangle similarity, as shown in Figure 2.8). Kinematic similarity is also referred to as the operating condition similarity. Only when geometric similarity exists, can the kinematic similarity be achieved making the geometric similarity the prerequisite.

2.3.2.1 First Similarity Law—Flow Rate Similarity

The flow rate in a pump can be expressed as:

$$Q = D_2 \pi b_2 \psi_2 v_{m2} \eta_v$$

Thus, the ratio of the flow rates for the model and the prototype can be written as:

$$\frac{Q_M}{Q} = \frac{(D_2 \pi b_2 \psi_2 v_{m2} \eta_v)_M}{D_2 \pi b_2 \psi_2 v_{m2} \eta_v}$$

Since: $\frac{b_{2M}}{b_2} = \frac{D_{2M}}{D_2} \qquad \frac{v_{m2M}}{v_{m2}} = \frac{u_{2M}}{u_2} = \frac{D_{2M} n_M}{D_2 n}$

The displacement coefficient is:

$$\psi = \frac{t - s_u}{t} = 1 - \frac{s_u}{t}$$

For the model and prototype, due to geometric similarity, $\frac{s_{uM}}{t_M} = \frac{s_u}{t}$, thus the displacement coefficients are equal, $\psi_M = \psi$. Therefore:

$$\frac{Q_M}{Q} = \left(\frac{D_{2M}}{D_2}\right)^3 \frac{n_M}{n} \frac{\eta_{zM}}{\eta_v} \tag{2.7}$$

This equation shows that for geometrically similar pumps operating under similar conditions, the ratio of their flow rates is proportional to the cube of the impeller outer diameter and is directly proportional to the speed and volumetric efficiency.

2.3.2.2 Second Similarity Law—Head Similarity

The head of a pump can be expressed as:

$$H = \frac{(u_2 v_{u2} - u_1 v_{u1})}{g} \eta_h$$

where η_h is the hydraulic efficiency of the pump. Therefore:

$$\frac{H_M}{H} = \frac{u_{2M}v_{u2M} - u_{1M}v_{u1M}}{u_2v_{u2} - u_1v_{u1}}\frac{\eta_{hM}}{\eta_h}$$

Due to kinematic similarity, the velocity ratios of the model and prototype are the same. Based on the proportionality formula:

$$\frac{u_{2M}v_{u2M} - u_{1M}v_{u1M}}{u_2v_{u2} - u_1v_{u1}} = \frac{u_{2M}v_{u2M}}{u_2v_{u2}} = \frac{u_{1M}v_{u1M}}{u_1v_{u1}} = \left(\frac{D_{2M}n_M}{D_2n}\right)^2$$

Thus:

$$\frac{H_M}{H} = \left(\frac{D_{2M}}{D_2}\right)^2\left(\frac{n_M}{n}\right)^2\frac{\eta_{hM}}{\eta_h} \tag{2.8}$$

This equation indicates that for geometrically similar pumps operating under similar conditions, the ratio of their heads is proportional to the square of the impeller outer diameter and the square of the speed, as well as the hydraulic efficiency.

2.3.2.3 Third Similarity Law—Shaft Power Similarity

The shaft power of a pump can be expressed as:

$$P = \rho gQH\frac{1}{\eta} \qquad \eta = \eta_m\eta_h\eta_v$$

Thus: $\frac{P_M}{P} = \frac{\rho_M gQ_M H_M}{\rho gQH}\frac{\eta_m\eta_h\eta_v}{\eta_{mM}\eta_{hM}\eta_{uM}}$

Substituting Eqs. (2.7) and (2.8) into the above equation yields:

$$\frac{P_M}{P} = \frac{\rho_M}{\rho}\left(\frac{D_{2M}}{D_2}\right)^5\left(\frac{n_M}{n}\right)^3\frac{\eta_m}{\eta_{mM}} \tag{2.9}$$

This equation shows that for geometrically similar pumps operating under similar conditions, the ratio of their shaft powers is proportional to the fluid density, the fifth power of the impeller outer diameter, the cube of the speed, and inversely proportional to the mechanical efficiency.

In the similarity laws, the dimensions D_2 and speed n are easily determined for both the model and the prototype pumps. However, the precise determination of the various efficiencies of the model and the prototype pumps is challenging. In practical applications, if the size and speed differences between the prototype and model pumps are not too large, it can be assumed that the efficiencies are equal under similar operating conditions. In this case, Eqs. (2.7)–(2.9) simplify to:

$$\frac{Q_M}{Q} = \left(\frac{n_M}{n}\right)\left(\frac{D_{2M}}{D_2}\right)^3 \tag{2.10}$$

$$\frac{H_M}{H} = \left(\frac{n_M}{n}\right)^2\left(\frac{D_{2M}}{D_2}\right)^2 \tag{2.11}$$

$$\frac{P_M}{P} = \left(\frac{n_M}{n}\right)^3\left(\frac{D_{2M}}{D_2}\right)^5\left(\frac{\rho_M}{\rho}\right) \tag{2.12}$$

2.3.3 Specific Speed

2.3.3.1 Derivation of the Specific Speed Formula

The similarity laws establish the relationship between the performance parameters of geometrically similar pumps under similar operating conditions. In other words, if the performance parameters

Table 2.1 Definition and conversion of unit specific speed values in different countries.

Country	United States	China (USSR)	United Kingdom	Japan	Germany
Formula	$\frac{3.65n\sqrt{\text{U. S. gal/min}}}{(\text{ft})^{3/4}}$	$\frac{3.65n\sqrt{\text{m}^3/\text{s}}}{\text{m}^{3/4}}$	$\frac{n\sqrt{\text{Imp.gal/min}}}{\text{m}^{3/4}}$	$\frac{3.65n\sqrt{\text{m}^3/\text{min}}}{\text{m}^{3/4}}$	$\frac{n\sqrt{\text{m}^3/\text{s}}}{\text{m}^{3/4}}$
Conversion factor	1	0.0706	0.91	0.15	0.26
—	1.1	0.0776	1	0.165	0.28
—	6.68	0.4709	6.079	1	1.72
—	3.88	0.274	3.53	0.58	1

of pumps satisfy the above relationships, the pumps are geometrically and kinematically similar. However, using the similarity laws to determine whether pumps are geometrically and kinematically similar is neither convenient nor intuitive.

Based on the similarity laws, a comprehensive parameter can be derived for a series of geometrically similar pumps. If this parameter is equal for all pumps, they are geometrically and kinematically similar, and the similarity laws can be used to convert performance relationships. This comprehensive parameter is called the specific speed, denoted as n_s, and is calculated as:

$$n_s = \frac{3.65n\sqrt{Q}}{H^{3/4}} \tag{2.13}$$

The units for the quantities in Eq. (2.13) are generally defined as:

Q—m^3/s (for double-suction pumps, use $Q/2$)
H—m (for multistage pumps, use the head of a single stage)
n—r/min (Table 2.1 gives the specific speed formula and units used in different countries).

2.3.3.2 Explanation of Specific Speed

1) A single pump operating at different conditions will have different values of n_s. The specific speed n_s used as a similarity criterion refers to the value at the best efficiency point (BEP).
2) Specific speed is derived from similarity theory and can be used as a similarity criterion. It implies that geometrically similar pumps operating under similar conditions will have equal n_s values. Conversely, pumps with equal n_s values are generally geometrically and kinematically similar. However, equal n_s does not always guarantee geometric similarity, as pumps with different geometric parameters (e.g., axial-flow vs. mixed-flow designs, or different numbers of blades) can have the same specific speed. For pumps of the same type, however, equal n_s values typically indicate geometric similarity, especially when the performance is optimized.
3) Specific speed is a dimensional quantity, but this does not affect its significance as a similarity criterion. For geometrically similar pumps operating under similar conditions, the calculated n_s values using consistent units will be equal.

Recently, a dimensionless specific speed, known as the type number, has been introduced. Since the dimensions of specific speed are related to the dimensions of gravitational acceleration g, the dimensionless specific speed is obtained by dividing the conventional specific speed by g. This eliminates the dimensional dependency, making it a pure number.

References

1 Yuan, S. (2014). *Pump Theory and Technology*. Beijing: Mechanical Engineering Press.
2 Guan, X. (1995). *Modern Pump Technology Manual*. Beijing: Aerospace Press.
3 Bai, J.J. and Xu, M.M. (2014). Research status and development trend of vane pump. *Mechatronics Engineering, Computing and Information Technology* 556: 1143–1146.

3

Traditional Design Methodologies for Impeller Pumps

3.1 Hydraulic Design of the Centrifugal and Mixed-Flow Pump

3.1.1 Determination of Design Parameters and Hydraulic Structural Solutions

3.1.1.1 Parameters and Requirements for the Design

In pump design, the first step is to determine the design point of the performance parameters and design requirements. These design parameters and requirements include: design point flow, design point head, design point cavitation margin, design point efficiency, pump speed, the nature of the transport medium, and the shape of the desired pump characteristic curve [1, 2].

3.1.1.1.1 The Nature of the Medium The nature of the conveying medium directly affects the performance of the pump, materials, and structure, and is one of the main factors to be considered in the design of the pump. The nature of the medium usually includes the viscosity of the medium, temperature, density, impurities, corrosiveness, etc.

3.1.1.1.2 Flow and Head Pump design point flow and head values are generally determined by the pump selector during the pump design planning process. The conveying room water properties, for example, temperature $\leq 40\,°C$, viscosity $\leq 1.75 \times 10^{-6}\ m^2/s$, density $\leq 1050\ kg/m^3$, nonabsorbent free solids content less than $2.5\ kg/m^3$, and dissolved in water solids content $\leq 50\ kg/m^3$ are usually specified for the hydraulic design based on the pump flow rate and head.

In the design of pumps conveying high viscosity special liquids, it is usually necessary to convert the design parameters Q_v (the flow rate of pumps conveying viscous liquids) and H_v (the head of pumps conveying viscous liquids) into the flow rate of conveying room-temperature water Q and head H. Since there is no universally applicable law that can accurately reflect the change in performance of pumps conveying special liquids of all kinds and conveying room-temperature water, how to convert Q_v and H_v into Q and H is a problem still under discussion in the industry. Here, we recommend using a similar viscous conversion chart as shown, for example for oil in Figure 3.1 [3].

The correction factors given in Figure 3.1 are generally used to determine the performance of a centrifugal pump when conveying a highly viscous liquid that is still a Newtonian fluid, based on the pump's performance in conveying clear water at room temperature. It is used for the conversion from high viscosity performance to clear water performance; the results are obviously approximate. In order to obtain a smaller error conversion result, the two-way correction method can be used.

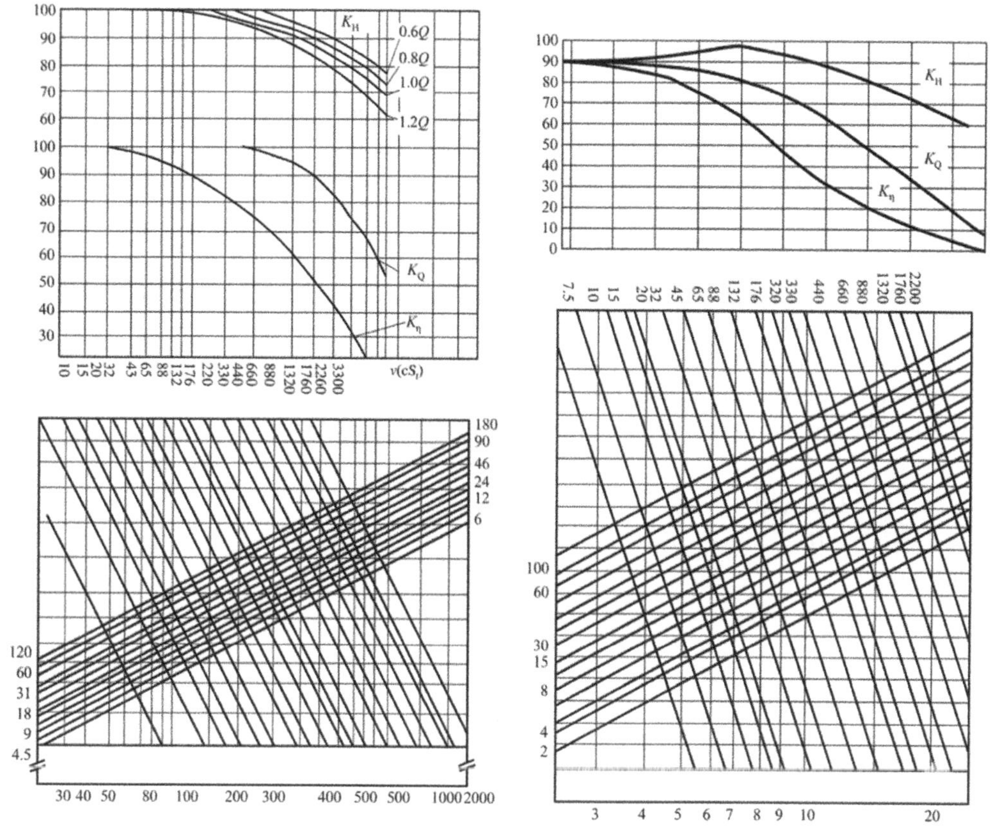

Figure 3.1 Conversion charts for pump design when conveying oil.

3.1.1.1.3 Determination of Cavitation Margin The necessary cavitation margin required for a pump is determined by the net positive suction head $(\text{NPSH})_r$ needed to achieve the desired performance parameters of cavitation. The report issued by the national standard GB/T13006 "centrifugal pumps, mixed flow pumps and axial pumps corrosion allowance," provides the provisions for the critical corrosion allowance (NPSH) [4]. This value is given or calculated by the pump designer to obtain the pump flow and suction device conditions

$$(\text{NPSH})_r = \frac{P_c}{\rho g} \mp h_g - h_c - \frac{P_v}{\rho g} - k \tag{3.1}$$

or

$$(\text{NPSH})_r = \left(\frac{P_c}{\rho g} \mp h_g - h_c - \frac{P_v}{\rho g}\right) / (1.1 - 1.3) \tag{3.2}$$

In the above equations, $\frac{P_c}{\rho g}$ is the pressure head of the liquid surface in the suction tank (m).

h_g—The geometric installation height of the pump (m). For suction installation height, use $-h_g$; for flooded suction installation, use $+h_g$.

h_c—The total hydraulic loss head of the pump suction pipeline is estimated based on the design flow rate (m).

$\frac{P_v}{\rho g}$—The vapor pressure head at the temperature of the conveyed liquid.

k—Safety margin. If no specific requirement for the k value is provided for the pump, $k = 0.3$ m may be used.

When using Eq. (3.2) to determine $(NPSH)_r$, for critical installations or pumps frequently operating at high flow rates, a larger value should be taken for the denominator.

3.1.1.1.4 Determination of Pump Speed, Specific Speed, and Hydraulic Structure The pump speed is sometimes specified directly by the design task assigner; otherwise, it must be determined by the designer. The following factors should be considered when determining the pump speed:

1) Higher pump speeds result in a smaller pump size and lighter weight. Therefore, the highest feasible speed should be selected.
2) Increasing the speed accelerates wear of the flow-passing components and amplifies vibration and noise of the unit.
3) Speed is related to specific speed, which in turn influences the pump's hydraulic structure and efficiency. Hence, the speed should be determined in conjunction with the hydraulic structure type and the specific speed of the pump. From the specific speed formula:

$$n_s = \frac{3.65n\sqrt{Q}}{H^{3/4}}$$

It follows that:

$$n = \frac{n_s H^{3/4}}{3.65\sqrt{Q}}$$

Based on the calculated value from the above formula and the speed range of the prime mover or variable-speed transmission device, a preliminary pump speed can be selected.

The following factors should also be considered when determining the specific speed from the above formula:

① In the range of n_s = 120–210, pump efficiency is the highest. When $n_s < 60$, pump efficiency decreases significantly. Figure 3.2 shows the relationship between the highest efficiency currently achievable for vane pumps and n_s [5, 6].

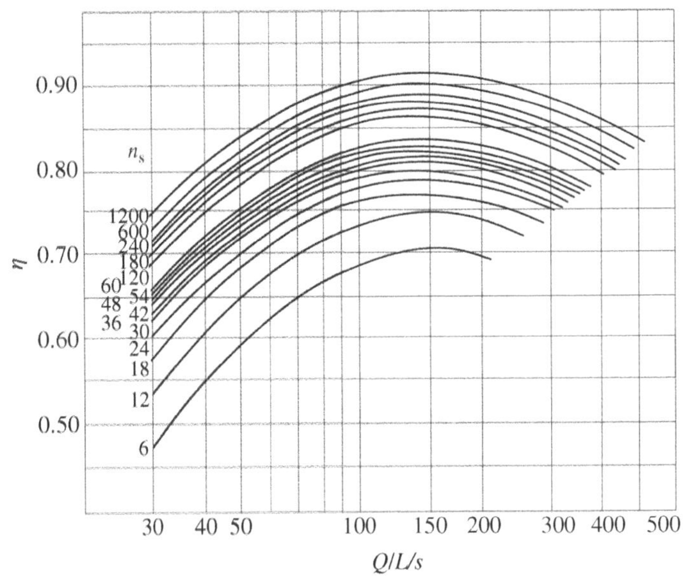

Figure 3.2 Relationship between efficiency, specific speed, and flow rate.

② The pump characteristic curve is also related to the value of n_s.
③ If the specific speed n_s is too high when using a single-suction impeller, switching to a double-suction impeller can be considered. Conversely, if n_s is too low when using a double-suction impeller, switching to a single-suction impeller may be appropriate.
④ Specific speed is related to the number of pump stages. The more is the number of stages, higher is the n_s. Horizontal pumps generally do not exceed 10 stages, while vertical deep-well pumps and submersible pumps can have dozens to hundreds of stages.

4) Increasing the pump speed is limited by the cavitation conditions. From the cavitation specific speed formula:

$$C = \frac{5.62n\sqrt{Q}}{(\mathrm{NPSH})_r^{3/4}}$$

It can be seen that for a given flow rate Q and required $(\mathrm{NPSH})_r$, if the hydraulic structure of the pump is fixed, increasing the speed n will increase the value of C. If the value of C already exceeds the range achievable by the designed pump, it will be difficult to meet the required $(\mathrm{NPSH})_r$. Therefore, the selected pump speed must satisfy the following requirement:

For single-suction pumps:

$$n \leq \frac{C(\mathrm{NPSH})_r^{3/4}}{5.62n\sqrt{Q}}$$

For double-suction pumps:

$$n \leq \frac{C(\mathrm{NPSH})_r^{3/4}}{3.974n\sqrt{Q}} \tag{3.3}$$

The value of C in the formula should be taken as the achievable value, which can be determined using Figure 3.3. When the rotational speed n is already given as a design parameter, the hydraulic

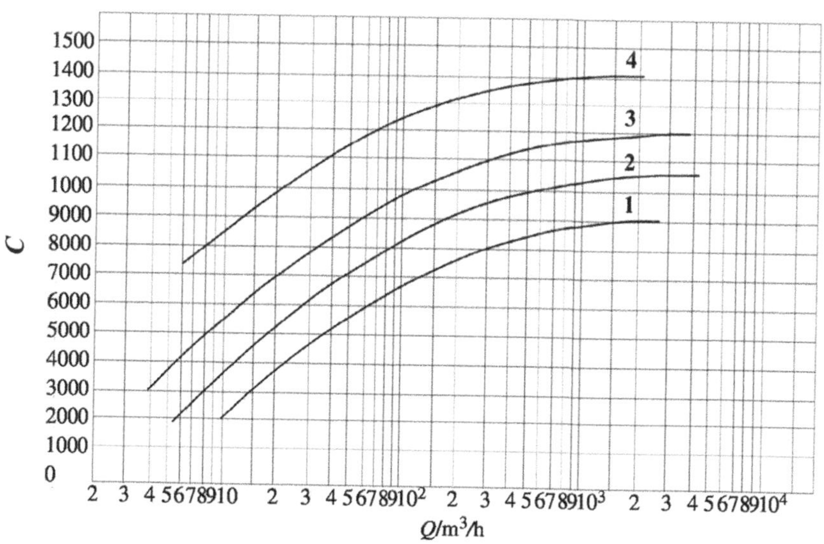

Figure 3.3 Relationship between C, Q, and n.

structure of the pump should be determined based on the relevant factors considered above, and then the specific speed n_s of the pump should be calculated. As shown below Fig. 3.3,

1) $K_0 = 4\text{–}4.5$—Generally achievable C values when $n \leq 1480$ r/min
2) $K_0 = 4\text{–}4.5$—Generally achievable C values when $n \geq 2900$ r/min; higher C values when $n \leq 1480$ r/min
3) $K_0 = 4.1\text{–}4.5$—Higher C values when $n \geq 2900$ r/min
4) $K_0 = 4.8\text{–}5.5$—Higher C values when $n \geq 2900$ r/min

3.1.1.1.5 Pump Efficiency

Overall Efficiency The minimum efficiency η to be achieved at the design operating point of the pump may be specified directly by the design task assigner. If the efficiency value is not provided in the design parameters, the value specified in relevant national standards, such as GB/T 13007-2001 "Efficiency of Centrifugal Pumps" may be adopted, or it may be determined based on Figure 3.2. Efforts should be made to achieve this efficiency in the design.

Estimation of Hydraulic Efficiency η_h

$$\eta_h = 1 + 0.0835 \lg \sqrt[3]{\frac{Q}{n}} \tag{3.4}$$

where

Q—Flow rate (m^3/h); for double-suction pumps, use $Q/2$
n—Rotational speed (r/min)

Estimation of Volumetric Efficiency η_v

$$\eta_v = \frac{1}{1 + 0.68\, n_s^{-2/3}} \tag{3.5}$$

This volumetric efficiency is applicable only when considering the leakage at the front wear ring of the impeller. For cases with balancing holes, interstage leakage, or leakage through balancing discs, the volumetric efficiency should be correspondingly reduced.

Figure 3.4 shows the relationship between the volumetric efficiency and the specific speed n_s for single-stage, multistage, and double-suction pumps, which can be used as a reference.

Estimation of Mechanical Efficiency η_m

$$\eta_m = 1 - \frac{P_m}{P} = \frac{P - \Delta P_f - \Delta P_d}{P} = 1 - \frac{\Delta P_f}{P} - \frac{\Delta P_d}{P}$$

The losses at bearings and sealing devices are generally small and can be estimated using the following formula:

$$\Delta P_f = (0.01 - 0.03)P$$

Smaller values apply to high-power pumps, while larger values apply to low-power pumps.

Figure 3.5 shows the relationship between the ratio of the disc friction power loss to shaft power and the specific speed.

It is also possible to estimate the mechanical efficiency considering only the disc friction losses using the Eq. (3.6).

$$\eta'_m = 1 - 0.07\frac{1}{\left(\frac{n_s}{100}\right)^{7/6}} \tag{3.6}$$

$$\eta_m = \eta'_m - (0.01 - 0.03)$$

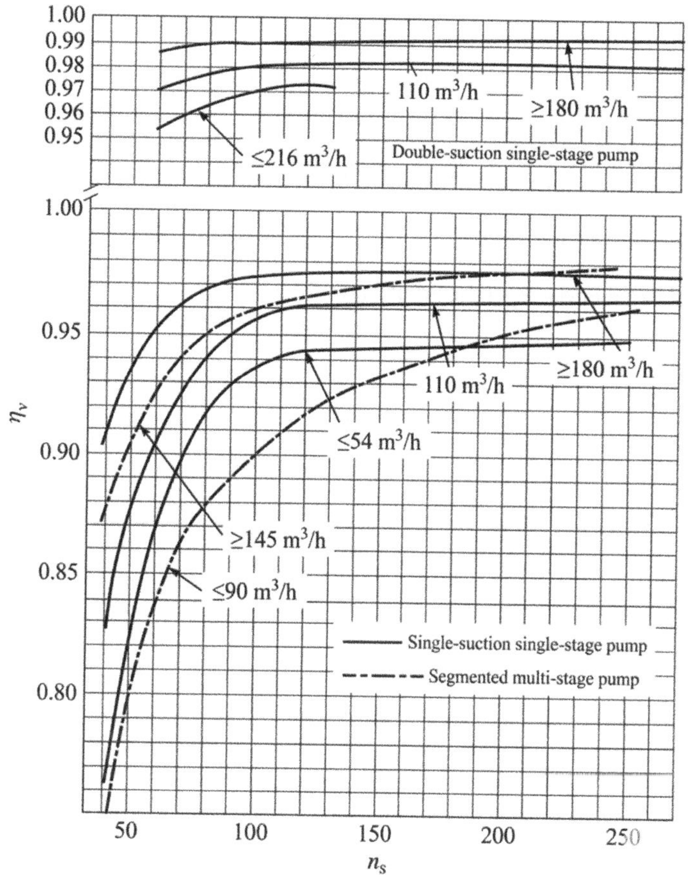

Figure 3.4 Volumetric efficiency vs. specific speed of centrifugal pumps.

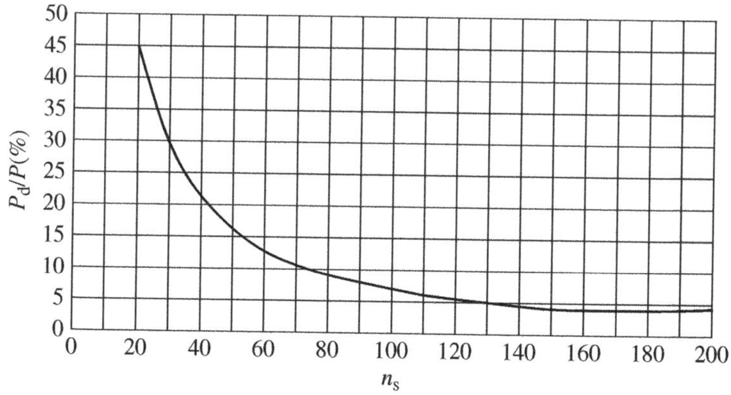

Figure 3.5 Relationship between the ratio of disc friction loss and shaft power and the specific speed.

By utilizing the relationships described above, it is straightforward to estimate the mechanical efficiency of the pump.

Following the methods described above, and considering the specific conditions of the pump being designed, the three efficiencies—η_h, η_v, and η_m—can be estimated. The overall pump efficiency is then obtained as $\eta = \eta_h \eta_v \eta_m$. Based on this estimated η value, and in combination with the η value provided in the design parameters or specified by the designer, the final pump efficiency can be determined.

3.1.1.2 Pump Shaft Power *P* and Rated Power P'_g of the Prime Mover

$$P = \frac{\rho g Q H}{1000\,\eta}$$

$$P'_g = \frac{KP}{\eta_t}$$

In the above formulas:

K—Margin coefficient;
η_t—Transmission efficiency.

3.1.1.3 Determination of Pump Inlet and Outlet Diameters

3.1.1.3.1 Pump Inlet Diameter D_s The pump inlet diameter is determined based on an appropriate inlet flow velocity. The typical inlet flow velocity for pumps is approximately 3 m/s. From the perspective of manufacturing, larger pumps tend to use higher flow velocities to reduce overall size and to enhance the flow capacity. However, to improve cavitation resistance, it is necessary to reduce the flow velocity by selecting a larger inlet diameter.

The relationship between commonly used pump suction diameters, flow rates, and flow velocities is provided in Table 3.1. For pumps requiring high cavitation resistance:

- When the suction diameter is less than 250 mm, the inlet flow velocity v_s may be set within the range 1.0–1.8 m/s.
- When the suction diameter is greater than 250 mm, v_s may be set within the range 1.4–2.2 m/s.

After selecting the inlet flow velocity, D_s can be determined using the following formula:

$$D_s = \sqrt{\frac{4Q}{\pi v_s}}$$

Table 3.1 Relationship between pump inlet diameter, flow velocity, and flow rate.

D		40	50	65	80	100	150	200	250	300	400
Single-stage pump	v_s/m/s	1.375	1.77	2.1	2.76	3.53	2.83	2.65	2.83	—	—
	Q/m³/h	6.25	12.5	25	50	100	180	300	500	—	—
Multistage pump	v_s/m/s	1.375	1.77	2.1	2.54	3	2.44	2.48	2.54	2.84	3.42
	Q/m³/h	6.25	12.5	25	46	85	155	280	450	720	1500

3.1.1.3.2 Pump Outlet Diameter D_d For most pumps, the outlet diameter is typically determined by the equation:

$$D_{\mathrm{d}} = (1 - 0.7)D_{\mathrm{s}}$$

The values of D_{d} and D_{s} calculated from the above formulas are rounded to standard pipe diameters, which then define the inlet and outlet diameters of the designed pump. Commonly used pump inlet diameters are listed in Table 3.1.

3.1.2 Hydraulic Design of the Impeller

The hydraulic design of a pump involves the design of its flow-passing components, including the impeller, discharge casing, and suction casing. Generally, the impeller is designed first, followed by the matching discharge and suction casings. The impeller is the core component of a pump, and its hydraulic design critically influences the pump's flow rate, head, efficiency, cavitation performance, and the shape of its characteristic curves.

The main geometric parameters of the impeller include:

- Impeller inlet diameter D_{j}
- Blade inlet diameter D_1
- Impeller hub diameter d_{h}
- Blade inlet width b_1
- Blade inlet angle β_1
- Impeller outlet diameter D_2
- Blade outlet width b_2
- Blade outlet angle β_2
- Number of blades z
- Blade wrap angle φ

The geometric parameters at the impeller inlet significantly affect cavitation, while those at the outlet primarily influence pump performance (head H and flow rate Q). Both sets of parameters impact the pump's overall efficiency. Figure 3.4 illustrates the geometry and key dimensional parameters of a through-shaft impeller and an overhung impeller.

3.1.2.1 Pump Shaft Diameter and Impeller Hub Diameter

The diameter of the pump shaft is determined based on the external loads (tension, compression, bending, and torsion), stiffness, and critical speed conditions. Since torque is the primary load on the pump shaft, the minimum shaft diameter (typically at the coupling location) may initially be calculated based on the torque. Considering the specific conditions of the pump being designed, factors affecting stiffness and critical speed should be taken into account. The preliminary calculated shaft diameter may be appropriately adjusted and rounded to a standard diameter. After completing the design of the pump rotor, a detailed verification of the shaft's strength, stiffness, and critical speed should be performed.

The equation for calculating the pump shaft diameter based on the torque is as follows:

$$d = \sqrt[3]{\frac{M_{\mathrm{n}}}{0.2[\tau]}} \tag{3.7}$$

$$M_{\mathrm{n}} = 9550\frac{P_{\mathrm{c}}}{n} \tag{3.8}$$

Table 3.2 Allowable shear stress of commonly used materials for pump shafts.

Material	Heat treatment requirement	(τ/MPa)	Application
35	Normalizing	34.3–44.1	General single-stage pumps
45	Quenching and tempering, HB = 241–286	44.1–53.9	General single-stage pumps
40Cr	Quenching and tempering, HB = 241–302	63.7–73.5	High-power high-pressure pumps
3Cr13	Quenching and tempering, HB = 264–302	53.9–68.7	Corrosion-resistant pumps
35CrMo	Quenching and tempering, HB = 241–285	68.7–78.5	High-temperature pumps (T = 200–400 °C)

In the above equations:

M_n—Torque, in N·m;
p_c—Calculated power, where $p_c = 1.2$ P may be used;
$[\tau]$—Allowable shear stress of the material, in Pa, obtained from Table 3.2.

The value of $[\tau]$ determines the thickness of the shaft. A thinner shaft saves material and improves the hydraulic and cavitation performance of the impeller, while a thicker shaft enhances the stiffness of the pump and increases operational reliability.

After determining the minimum diameter of the pump shaft, a preliminary structural sketch of the shaft is drawn with reference to the shafts of pumps with similar configurations. Based on the structural and process requirements of each shaft section, the shaft diameter at the impeller mounting location d_B and the hub diameter d_h are determined. The impeller hub diameter must ensure sufficient wall thickness after machining the keyway to maintain adequate strength. Typically, $d_h = (1.2 - 1.4)d_B$. While meeting the structural strength requirements of the hub, minimizing d_h is beneficial for improving the flow conditions.

When drawing the preliminary structural sketch of the pump shaft (Figure 3.6), the following points should be noted:

1) Shaft diameters for all sections should preferably adopt standard dimensions.
2) Threads on the shaft generally use standard fine-pitch threads; their minor diameter should be larger than the diameter of the shaft section preceding the threads.
3) The height of the shaft positioning shoulders is typically 2–3 mm.

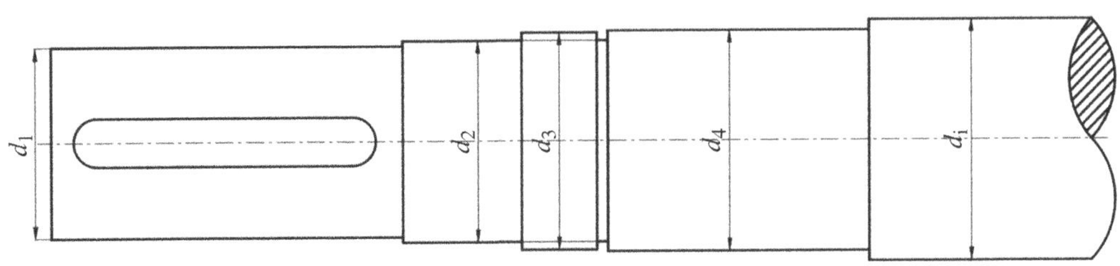

Figure 3.6 Structural diagram of the pump shaft.

3.1.2.2 Similarity Conversion Method

This method is simple and reliable. It involves selecting a pump similar to the one to be designed and scaling all flow-passing dimensions proportionately. Only the impeller part is discussed here. The steps are as follows:

1) Calculate the specific speed n_s based on the design parameters of the pump.
2) Select a model pump. The requirements for the model pump are:
   1) The specific speed of the model pump n_{sM} should be equal or close to that of the design pump n_s.
   2) The model pump should exhibit high efficiency (with a wide high-efficiency range), good cavitation resistance, and a characteristic curve shape that meets the requirements.
   3) The technical documentation of the model pump should be comprehensive and reliable.
   4) To maintain similarity, the ratio of Reynolds numbers between the design pump and the model pump $\mathrm{Re}/\mathrm{Re}_M$ should ideally fall within the range 1.0–1.5.

$$\mathrm{Re} = \frac{u_2 D_2}{v}$$

In the above formula:

D_2—Impeller outer diameter (m);
u_2—Impeller outlet circumferential velocity (m/s);
v—Kinematic viscosity of the conveyed liquid (m^2/s).

3) Determination of the Scaling Factor
   Let the corresponding operating point of the model pump be denoted as (Q_M, H_M), and the design operating point of the target pump as (Q, H). Assuming the volumetric efficiency and hydraulic efficiency of the model pump and the target pump are equal, the following relationships can be derived from the similarity laws:

$$\lambda = \frac{D}{D_M} = \sqrt{\frac{Q}{Q_M}\frac{n_M}{n}}$$
$$\lambda = \frac{D}{D_M} = \frac{n_M}{n}\sqrt{\frac{H}{H_M}} \tag{3.9}$$

   The values of λ obtained from the two formulas above may differ slightly. In such cases, the larger value or the average of the two values may be used.
4) Calculation of the Dimensions of the Design Pump
   Multiply all flow-passing dimensions of the model pump by the scaling factor λ to obtain the corresponding dimensions of the design pump.
   The blade angles of the design pump should be equal to the corresponding angles of the model pump, i.e.:

$$D = D_M \lambda$$
$$\beta = \beta_M$$

   For parameters such as blade thickness, similarity scaling may not always be applicable; they should be determined based on specific requirements.

5) Performance Curve Conversion for the Design Pump
Select multiple points on the performance curve of the model pump and convert them to the corresponding parameters of the design pump using the similarity conditions, then plot the performance curve accordingly as follows.

$$\frac{Q}{Q_M} = \frac{n}{n_M}\left(\frac{D}{D_M}\right)^3 \qquad Q = \lambda^3 \frac{n}{n_M} Q_M$$

$$\frac{H}{H_M} = \left(\frac{n}{n_M}\right)^2\left(\frac{D}{D_M}\right)^2 \qquad H = \lambda^2\left(\frac{n}{n_M}\right)^2 H_M$$

$$\frac{P}{P_M} = \left(\frac{n}{n_M}\right)^3\left(\frac{D}{D_M}\right)^5 \qquad P = \lambda^5\left(\frac{n}{n_M}\right)^3 P_M$$

$$\eta = \frac{\rho g Q H}{1000\,P}$$

6) Drawing the Design Pump Blueprint
Based on the converted dimensions and angles, make appropriate modifications to the parameters such as blade thickness and sealing gaps according to the specific requirements. Finally, the design pump blueprint should be drafted. The blade profile drawing can also be recreated under the condition of ensuring the key converted dimensions (for details, refer to the blade profiling section).

7) Considerations in the Similarity Design Method
① Efficiency Correction
In similarity-based design, it is typically assumed that the efficiency of the model pump and the target pump are equal. However, in practice, due to differences in the relative manufacturing precision, clearances, blade thickness, and other geometric characteristics between large and small pumps, the hydraulic efficiency and volumetric efficiency of larger pumps are generally higher than those of smaller pumps, and the mechanical efficiency is also often slightly higher. Therefore, when the size difference between similar pumps is significant, the correction for scale effects should be considered.
The modified scaling factor λ', accounting for scale effects, can be calculated as follows:

$$\lambda' = \lambda\sqrt{\frac{\eta_{hM}}{\eta_h}} = \lambda\sqrt{\frac{f_{NL}}{f}}$$

where
λ—Scaling factor without considering scale effects;
f—Scale correction factor for the target pump;
f_{NL}—Scale correction factor for the model pump.

$$f = \frac{1}{1 + 3.15/D_j^{1.6}}, \quad f_M = \frac{1}{1 + 3.15/D_{jM}^{1.6}}$$

where D_{jM} and D_j are the impeller inlet diameters (in cm) of the model pump and the target pump, respectively.
Alternatively, λ' can also be calculated using the following formula:

$$\lambda' = \lambda\sqrt{\frac{1 + 0.08351\,g\sqrt[3]{Q_M/n_M}}{1 + 0.08351\,g\sqrt[3]{Q/n}}} \tag{3.10}$$

where

Q—Flow rate of the pump (m^3/s);

N—Rotational speed of the pump (r/min).

② Model Modification

If an existing model pump performs well but its specific speed n_s differs from that of the design pump, and the difference is small, the model pump can be modified to adjust its performance parameters so that its modified specific speed n_{sM} matches the target specific speed n_s of the design pump. Similarity conversion can then be performed based on the modified model dimensions and performance. The following two scenarios describe how the model can be modified:

1) Keeping D_2 constant while changing the flow passage width (Figure 3.7)

If the impeller outer diameter D_2 and the blade shape remain unchanged, and the position of the front shroud is uniformly adjusted, the flow passage width can be modified. Assuming that the meridional velocity remains constant after the width change, when the width increases, the inlet diameter also increases. The inlet velocity triangle at the front shroud streamline is shown in Figure 3.8. The incidence angle and the relative velocity increase, i.e., $\Delta\beta_1' > \Delta\beta_1, w_1' > w_1$. Since w_1 increases, the cavitation resistance may slightly decrease. Assuming a uniform distribution of v_m, the change in flow rate is given by:

$$\frac{Q'}{Q} = \frac{D_2 \pi v_{m2} b_2'}{D_2 \pi v_{m2} b_2} = \frac{b_2'}{b_2}$$

Since D_2, β_2, and v_{m2} remain unchanged, and the head can be expressed as $H_t = \frac{v_2}{g}\left(\sigma u_2 - \frac{v_{m2}}{\tan\beta_2}\right)$, the head remains constant. The change in power is given by

$$P' = \frac{\rho g Q' H'}{1000\,\eta} = \frac{\rho g Q H}{1000\,\eta}\left(\frac{b_2'}{b_2}\right) = P\left(\frac{b_2'}{b_2}\right)$$

The change in specific speed is given by:

$$n_s' = \frac{3.65\, n\sqrt{Q'}}{H'^{3/4}} = \frac{3.65\, n\sqrt{Q}(b_2'/b_2)^{1/2}}{H^{3/4}} = n_s\left(\frac{b_2'}{b_2}\right)^{1/2}, b_2' = \left(\frac{n_s'}{n_s}\right)^2 b_2$$

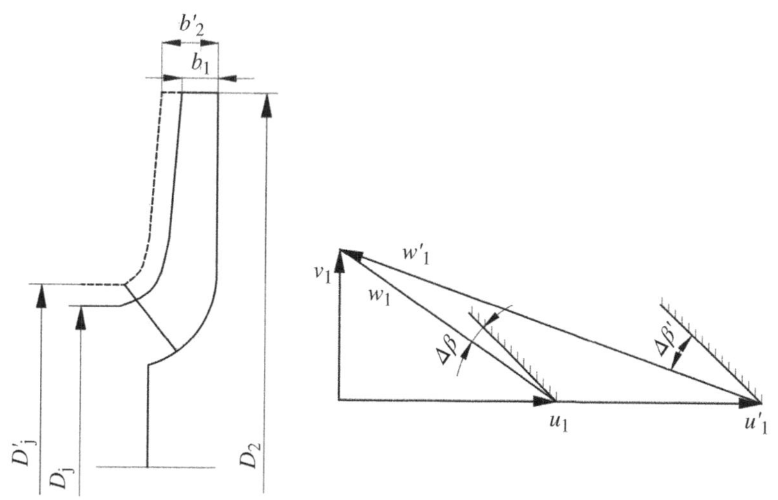

Figure 3.7 Modifying the flow passage width keeping the outer diameter constant.

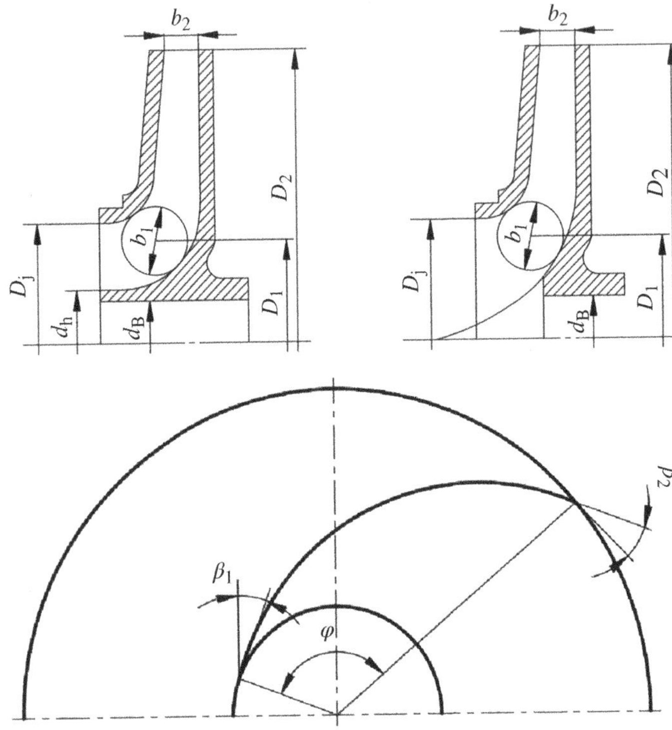

Figure 3.8 Key dimensions of the impeller.

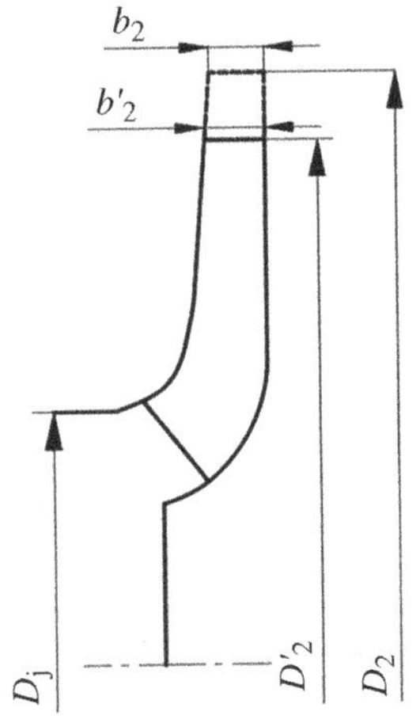

Figure 3.9 Modifying the outlet diameter.

2) Changing the outlet diameter D_2 (Figure 3.9)
After modifying D_2, it is assumed that $b_2' = b_2$ and $v_{m2}' = v_{m2}$.

$$\frac{H'}{H} = \frac{u_2' v_{u2}'}{u_2 v_{u2}} = \left(\frac{D_2'}{D_2}\right)^2 \frac{Q'}{Q} = \frac{D_2' \pi b_2' v_{m2}'}{D_2 \pi b_2 v_{m2}} = \frac{D_2'}{D_2}$$

$$n_s' = \frac{3.65\, n\sqrt{Q'}}{H'^{3/4}} = \frac{\frac{3.65\, n\sqrt{Q}}{H^{\frac{3}{4}}}\left(\frac{D'2}{D_2}\right)^{\frac{1}{2}}}{\left(\frac{D'2}{D_2}\right)^{\frac{6}{4}}} = n_s\left(\frac{D_2}{D'2}\right) \quad D'2 = D_2\frac{n_s}{n_s'}$$

8) Similarity in Cavitation Performance
For two similar pumps, the theoretical values of C or σ should be equal. Based on this consideration, the required $(NPSH)_r$ for the design pump can be derived. However, achieving complete geometric similarity at the pump inlet is challenging in practice. Moreover, cavitation performance is highly sensitive to the geometric parameters of the pump inlet. Therefore, when the size and speed differences between pumps are significant, the error in converting their cavitation performance increases.

Specifically:

- As the impeller outlet diameter D_2 and rotational speed n increase, the actual C value of the pump tends to be higher than the converted value.
- When scaling from a smaller pump or a pump with lower speed to a larger pump or a pump with higher speed, the actual cavitation resistance of the larger/higher-speed pump is generally better than the converted value.
- Conversely, when scaling from a larger pump or a pump with a higher speed to a smaller pump or a pump with a lower speed, the actual cavitation resistance of the smaller/lower-speed pump is worse than the converted value, making such conversions unreliable.

Similarity conversion is based on similarity theory and is inherently an approximate method. From the perspective of ensuring performance, parameters such as D_2, b_2, β_1, β_2, and z play a dominant role. Therefore, there are two approaches to similarity-based design:

① The entire flow-passing component, including the blade profile, is drawn strictly according to the similarity conversion results.
② Ensure similarity in the main parameters, while appropriately modifying the secondary parameters during the drafting process.

3.1.2.3 Calculation of Key Impeller Dimensions Using the Velocity Coefficient Method

The velocity coefficient method is essentially a different form of similarity-based design, sharing the same fundamental principles as the model conversion method. The key difference lies in the fact that the model conversion method relies on a single specific similar pump as the basis for design, whereas the velocity coefficient method is built upon a series of similar pumps [5–7]. In other words, the velocity coefficient method applies the principle of similarity and utilizes statistical coefficients to calculate the dimensions of various flow-passing components.

3.1.2.3.1 General Form of the Velocity Coefficient Formula Based on the principle of similarity, the general expression for determining the dimensions using velocity coefficients can be derived. For a series of geometrically and kinematically similar pumps, the following relationship holds:

For $\frac{Q}{nD^3}$ = constant, obtain

$$D = k_1\sqrt[3]{\frac{Q}{n}} \tag{3.11}$$

Since $v \propto nD$, obtain

$$v = k_2\sqrt[3]{Qn^2} \tag{3.12}$$

For $\frac{H}{n^2D^2}$ = constant, obtain

$$D = k_3\frac{\sqrt{H}}{n} \text{ or } D = k_4\frac{\sqrt{2gH}}{n} \tag{3.13}$$

Since $v \propto nD$, obtain

$$v = k_5\sqrt{2gH} \tag{3.14}$$

The coefficient k is referred to as the velocity (or dimensional) coefficient. For similar pumps, these coefficients are equal, and thus each can be expressed as a function of the specific speed n_s, i.e., $k = f(n)$.

Using relationships with velocity coefficients (such as formulas, curves, or data), the coefficient k can be determined. Based on the above formula, various dimensions can then be calculated.

Although the forms of the formulas used in different countries may vary, their essence remains the same [5–7]. Since these coefficients are statistical values, the values provided in different sources may not be entirely identical. Therefore, specific analysis should be applied when using them. The following sections introduce some commonly used formulas and curves for determining velocity coefficients.

3.1.2.3.2 *Formulas for Calculating Key Impeller Dimensions Using the Velocity Coefficient Method*

Impeller Inlet Diameter D_j Since some impellers have a hub (through-shaft impellers) while others do not (overhung impellers), the concept of the equivalent inlet diameter D_0 is introduced to generalize the problem by eliminating the influence of the hub. The circular area with diameter D_0 is equal to the effective inlet area of the impeller after excluding the hub area, i.e.,

$$\frac{D_0^2\pi}{4} = \frac{\left(D_j^2 - d_h^2\right)\pi}{4}$$

D_0 is determined using Eq. (3.11)

$$D_0 = k_0\sqrt[3]{\frac{Q}{n}}D_j = \sqrt{D_0^2 + d_h^2} \tag{3.15}$$

where k_0 is the velocity coefficient, selected based on statistical data:

For prioritizing efficiency: $k_0 = 3.5–4.0$
For balancing efficiency and cavitation: $k_0 = 4.0–4.55$
For prioritizing cavitation resistance: $k_0 = 4.5–5.5$

Alternative method to determine D_j is using Eq. (3.12). The impeller inlet diameter D_j can also be calculated by first determining the inlet flow velocity v_0 as follows:

$$v_0 = k_{v_0}\sqrt[3]{Qn^2} \tag{3.16}$$

where k_{v_0} is the velocity coefficient, selected as follows:

For prioritizing efficiency: $k_{v_0} = 0.08–0.10$
For balancing efficiency and cavitation: $k_{v_0} = 0.06–0.08$
For prioritizing cavitation resistance: $k_{v_0} = 0.04–0.06$

Once v_0 is obtained, D_j can be calculated by Eq. (3.17):

$$D_j = \sqrt{\frac{4Q}{v_0\pi\eta_v} + d_h^2} \tag{3.17}$$

Alternatively, D_j can be determined using Eq. (3.14): First, obtain k_{v_0} from the velocity coefficient curve shown in Figure 3.7, then calculate v_0 by Eq. (3.18):

$$v_0 = k_{v0}\sqrt{2gH} \tag{3.18}$$

Finally, substitute v_0 into Eq. (3.17) to compute D_j.

Impeller Outlet Diameter D_2 Using Eq. (3.11):

$$D_2 = k_{D2}\sqrt[3]{\frac{Q}{n}} \qquad k_{D2} = (9.35 - 9.6)\left(\frac{n_s}{100}\right)^{-1/2} \tag{3.19}$$

Using Eq. (3.13):

$$D_2 = k_{D2}\frac{\sqrt{2gH}}{n} \qquad k_{D2} = 19.2\left(\frac{n_s}{100}\right)^{1/6} \tag{3.20}$$

Using Eq. (3.14):

$$u_2 = k_{u2}\sqrt{2gH} \qquad D_2 = \frac{60u_2}{n\pi} \tag{3.21}$$

Here, k_{u2} is obtained from Figure 3.10.

For mixed-flow pumps, the calculated D_2 represents the average effective diameter at the impeller outlet:

$$D_2 = \sqrt{\frac{D_{2a}^2 + D_{2e}^2}{2}} \tag{3.22}$$

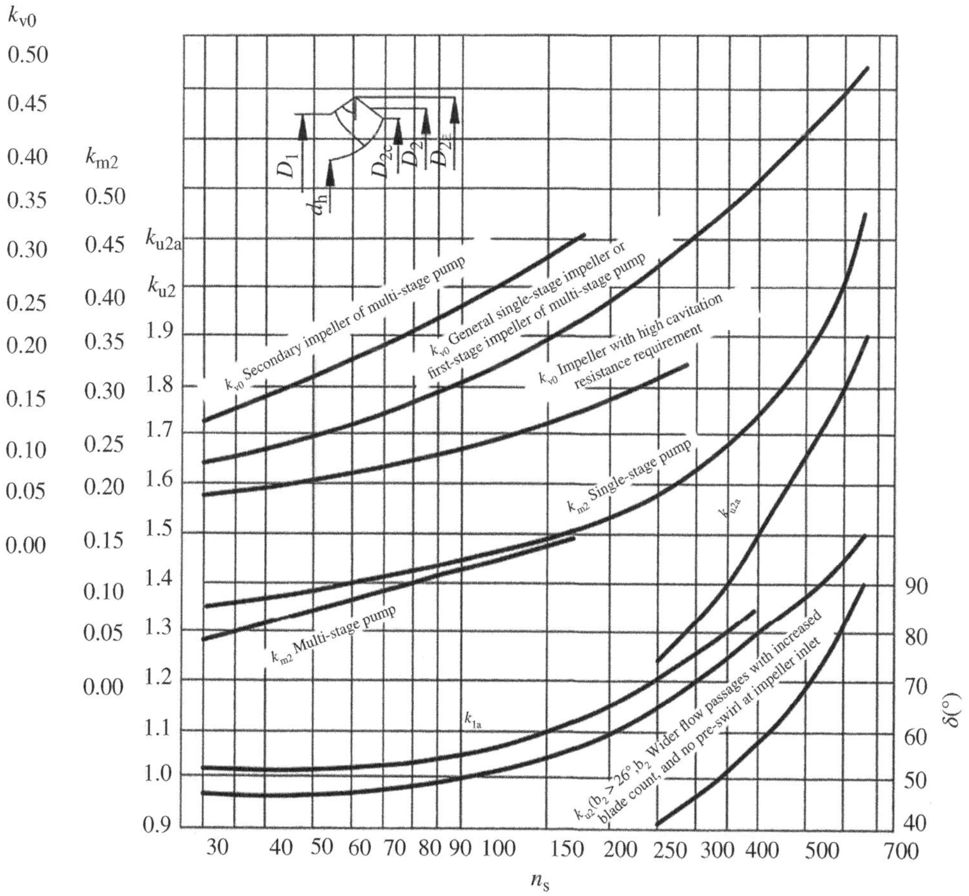

Figure 3.10 Coefficients used in determination of outlet diameter for various types of pumps.

The outer diameter at the front shroud streamline D_{2a} is determined using Eq. (3.23):

$$D_{2a} = \frac{60u_{2a}}{n\pi} \qquad u_{2a} = k_{u2a}\sqrt{2gH} \tag{3.23}$$

Here, k_{u2a} is obtained from Figure 3.7. After calculating D_2 and D_{2a}, the outer diameter at the rear shroud streamline D_{2e} can be derived from Eq. (3.22):

$$D_{2e} = \sqrt{2D_2^2 - D_{2a}^2}$$

To prevent excessive inclination of the blade outlet edge (which may cause significant secondary flow) and an overly large outlet angle at the rear shroud streamline, the calculated u_{2e} should generally satisfy:

$$\frac{u_{2e}}{\sqrt{2gH}} \geqslant 1 - 1.1 \tag{3.24}$$

Blade Outlet Width b_2 Using Eq. (3.11):

$$b_2 = k_{b2}\sqrt[3]{\frac{Q}{n}} \qquad k_{b2} = (0.64 - 0.7)\left(\frac{n_s}{100}\right)^{5/6} \tag{3.25}$$

Using Eq. (3.12):

$$b_2 = k_{b2}\frac{\sqrt{2gH}}{n} \qquad k_{b2} = 1.30\left(\frac{n_s}{100}\right)^{3/2} \tag{3.26}$$

Using Eq. (3.13):

$$b_2 = \frac{Q}{\eta_v \pi D_2 \psi_2 v_{m2}} \qquad v_{m2} = k_{m2}\sqrt{2gH} \tag{3.27}$$

Here:

k_{m2}—Velocity coefficient (obtained from Figure (3.7));
ψ_2—Blade outlet blockage coefficient (typically 0.85–0.95; smaller values for smaller pumps, larger values for larger pumps).

For impellers with low specific speed (n_s), b_2 may be widened based on manufacturing feasibility. When $n_s > 200$, k_{b2} should be multiplied by a correction factor $k < 1$. Refer to Table 3.3 for specific values.

For mixed-flow pump impellers, the term D_2 in Eq. (3.22) should represent the diameter $\overline{D_2}$ of the inscribed circle formed by the front and rear shroud streamlines (or their extensions) at the blade outlet edge, with a circle diameter equal to b_2, as illustrated in Figure 3.11.

$$\overline{D_2} \approx \frac{D_{2a} + D_{2e}}{2} \tag{3.28}$$

Since the front and rear shrouds are not parallel, D_2 may sometimes be slightly smaller than $(D_{2a} + D_{2e})/2$, and b_2 is smaller than the meridional projection length b_2' of the blade outlet edge. For mixed-flow pump impellers, the blade outlet blockage coefficient ψ_2 typically ranges from 0.9 to 0.95.

Table 3.3 Relationship between n_s and k.

n_s	200	250	300	350	400	450	500
k	0.95	0.9	0.85	0.8	0.75	0.7	0.65

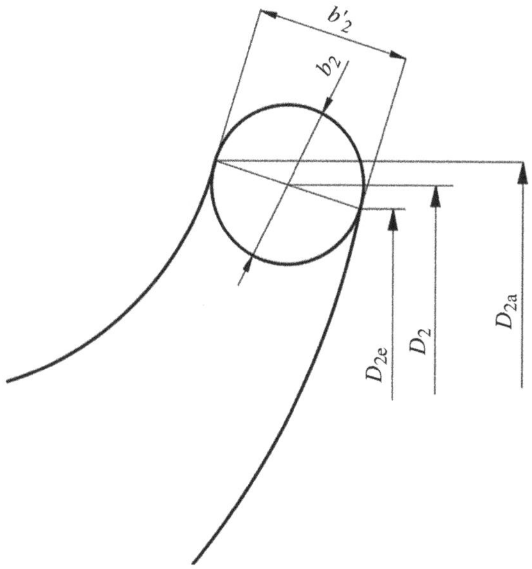

Figure 3.11 Meridional dimensions at the outlet of a mixed-flow pump impeller.

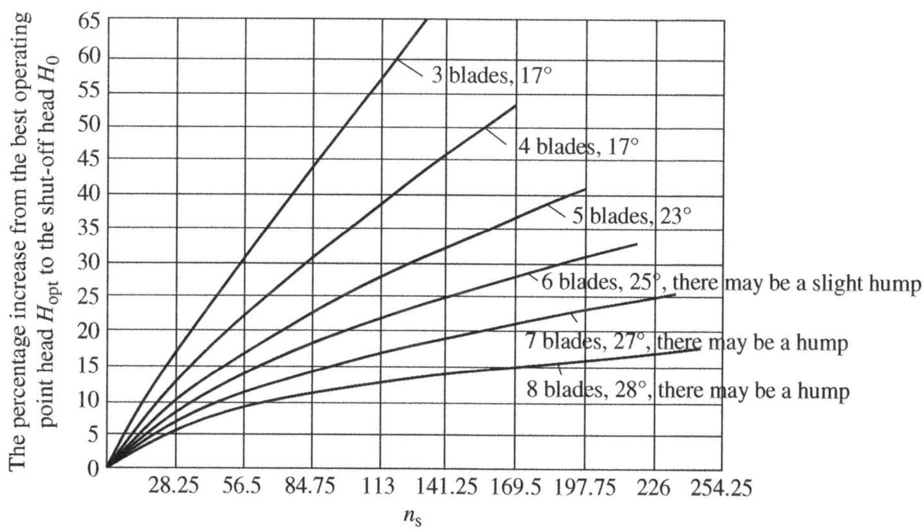

Figure 3.12 Determination of blade number and outlet angle based on head variation characteristics.

Meridional Dimensions of the Impeller for Specific Characteristic Curve Requirements The blade outlet angle β_2 and the number of blades Z are the most critical geometric parameters influencing the shape of the H–Q and P–Q curves. Figure 3.12 provides a statistical representation of the relationship between the shape of the H–Q curve (expressed as the percentage increase in head from the best efficiency point head H_0/H_{opt} to the shut-off head H_0) and the values of n_s, Z, and β_2. If the design requires specific characteristics in the H–Q curve, the following steps can be used to determine the relevant geometric dimensions of the impeller:

1) Select Z and β_2:
   Based on the requirements for the shape of the H–Q or P–Q curve (e.g., a flat P–Q curve indicating minimal power variation with flow, or a steeply dropping H–Q curve with a high H_0/H_{opt} ratio), determine Z and β_2 from Figure 3.14.
   Note: When selecting larger values of Z and β_2 that result in a smaller H_0/H_{opt} ratio, the H–Q curve may exhibit a saddle (hump). In such cases, the values of other relevant geometric parameters should be determined based on a comprehensive understanding of the factors influencing the H–Q curve shape.
2) Calculate the impeller outlet diameter D_2:
   Using Eq. (3.13):

$$D_2 = k_{D2}\frac{\sqrt{H}}{n} \tag{3.29}$$

$$k_{D2} = 84.65k_u$$

   where k_u is obtained from Figure 3.13 based on the specific speed n_s and the selected number of blades Z.
3) Determine the blade outlet width b_2: Calculate b_2 using Eq. (3.27). The velocity coefficient k_{m2} in the formula is obtained from Figure 3.14 based on the specific speed n_s and the selected number of blades Z.
4) Determine the impeller inlet diameter D_j:
   Obtain the ratio D_j/D_2 from Figure 3.15 according to n_s, then calculate D_j. Select D_j/D_2 ratio within the acceptable region shown in the Figure 3.15:

- For prioritizing efficiency, choose a smaller ratio.
- For prioritizing cavitation resistance, choose a larger ratio.
- For balancing efficiency and cavitation resistance, choose an intermediate value.

Note: In all the formulas provided above for the velocity coefficient method, the units of the parameters are as follows: Q-m^3/s; H-m; n-r/min; D_j-m; D_2-m; b_2-m.

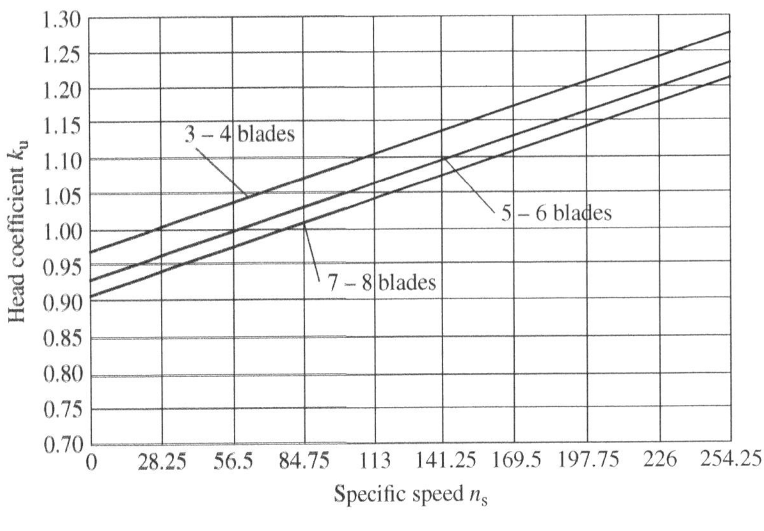

Figure 3.13 Relationship between k_u, n_s, and Z.

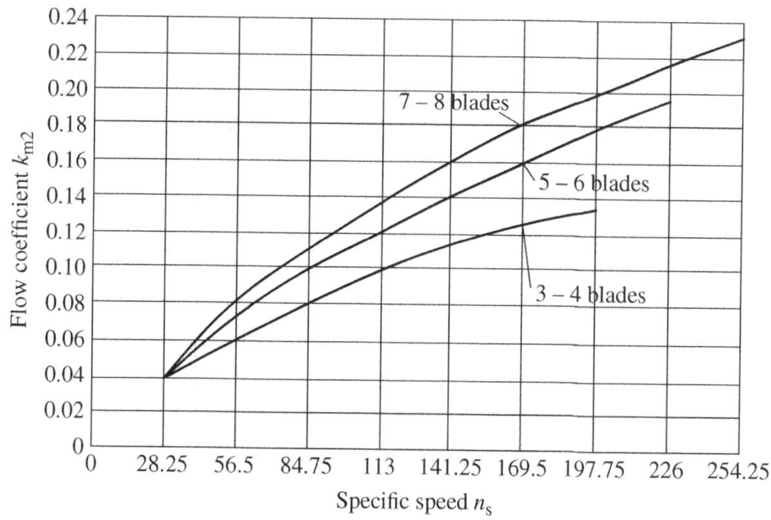

Figure 3.14 Relationship between k_{m2}, n_s, and Z.

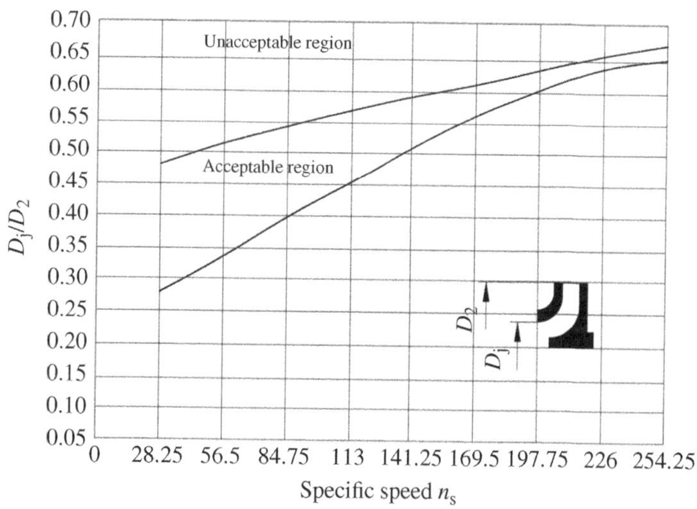

Figure 3.15 Relationship between D_j/D_2 and n_s.

3.1.2.3.3 Drafting the Meridional Projection and Meridional Streamlines of the Impeller For designing centrifugal pump impellers with $v_{u1} = 0$, this task may be performed after precisely calculating D_2.

Drafting the Meridional Projection Based on the calculated meridional dimensions of the impeller, the meridional projection can be drawn, and the blade inlet edge can be defined. It is advisable to reference impeller drawings with similar specific speeds, n_s, and good performance, adapting them to the specific conditions of the designed pump. The shape of the meridional projection is critical, and the following considerations should be taken into account during drafting:

① The outlet sections of the front and rear shrouds should maintain a parallel or symmetrically varying profile.

② The flow passage should not turn too sharply; a larger radius of curvature is preferred, subject to axial structural constraints.
③ On the meridional projection, the radial dimension of point M, where the blade inlet edge intersects the front shroud, may be equal to or slightly larger than the impeller inlet diameter D_j.

For low-specific-speed pumps, the inlet edge can be parallel to the axis or inclined to extend appropriately toward the impeller suction inlet, which helps avoid a saddle (hump) in the H–Q curve. For medium- and high-specific-speed pumps, the angle between the line connecting point M and point N (where the inlet edge intersects the rear shroud) and the axis should not exceed 30°–45°. For mixed-flow pumps, this angle may range from 40° to 65°. Excessively large angles may result in an overly large inlet angle at the rear shroud streamline. The meridional projection of the blade inlet edge is defined by a straight line or a smooth curve convex toward the impeller inlet, passing through points M and N.

When drafting the meridional projection, the following practices may also be referenced, with parameters selected based on Figure 3.16:

$$Z_E = (D_{2a} - d_h)\left(\frac{n_s}{74}\right)^{1.07} \qquad R_{DS} = (0.6 - 0.8)b_1$$

$$b_1 = \frac{D_j - d_h}{2}$$

The front shroud outlet angle ε_{DS} of the impeller influences the velocity distribution at the impeller outlet and is typically determined based on the specific speed n_s.

- For $n_s < 20$, ε_{DS} should be small, sometimes even zero.
- For high-specific-speed centrifugal impellers, ε_{DS} may range from 15° to 20°.

Similarly, the rear shroud outlet angle ε_{TS} can be determined based on the specific speed n_s:

- For $n_s < 30$, ε_{TS} should be small, sometimes zero or even negative.
- For high-specific-speed centrifugal impellers, ε_{TS} is generally positive and satisfies $\varepsilon_{TS} < \varepsilon_{DS}$.

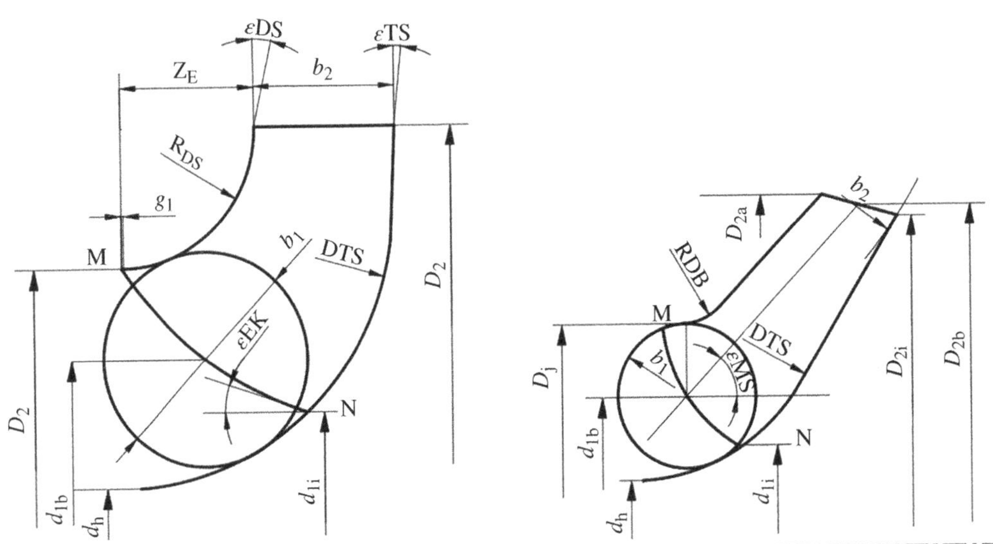

Figure 3.16 Design dimensions for impeller meridional projection.

Moving the inlet edge toward the impeller inlet can improve flow performance by reducing blade loading, lowering peak pressure fluctuations, and mitigating cavitation. The inlet edge should not be placed in the high-curvature region of the front shroud but should instead be positioned at a distance $g_1 = (0.2\text{–}0.3)\ b_1$ from the starting point of the front shroud curve. For smaller pumps or designs requiring minimal axial extension, the values of Z_E and R_{DS} should be smaller than those calculated by the above formulas.

Checking the Variation of Meridional Flow Passage Cross-Sections After completing the meridional projection, the variation in cross-sectional areas along the flow passage should be checked, as illustrated in Figure 3.17.

The curve AEB in the above diagram is perpendicular to all meridional streamlines and serves as the formation line of the flow cross-section. Its construction method is as follows:

Within the meridional projection, draw the incircle of the inner and outer crown streamlines, with points of tangency at A and B. Connect points A and B to the center O to form triangle AOB. Divide the altitude OD of triangle AOB into three equal segments: OE, EC, and CD. Draw a smooth curve AEB through points A, E, and B, ensuring the curve is tangent to line OA at point A and tangent to line OB at point B. This curve, AEB, is the formation line of the flow cross-section. Its length b can be measured with a flexible ruler or approximately calculated using the following formula:

$$b = \frac{2}{3}(s + \rho)$$

In the above formula:

s—the length of chord AB of the incircle;
ρ—the radius of the incircle.

The centroid of the flow cross-section formation line is approximately considered to coincide with point C, the centroid of triangle AOB, with a centroid radius of R_c.

The flow cross-section of the meridional flow is a surface of revolution formed by rotating the formation line of the flow cross-section around the axis by one full revolution. Its area is calculated using Eq. (3.30).

$$S = 2\pi R_c b \tag{3.30}$$

Figure 3.17 Cross-sectional area of meridional flow passage.

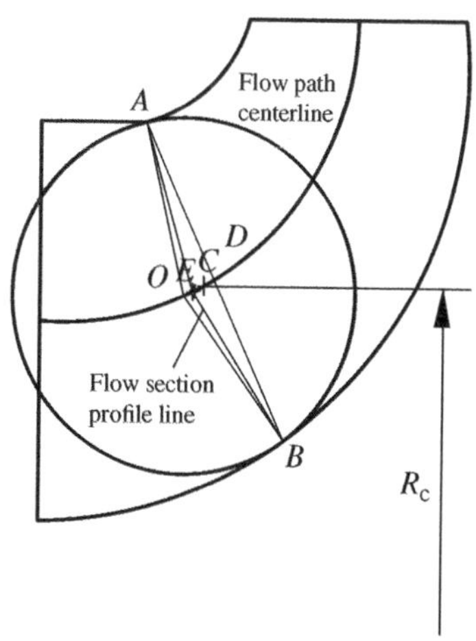

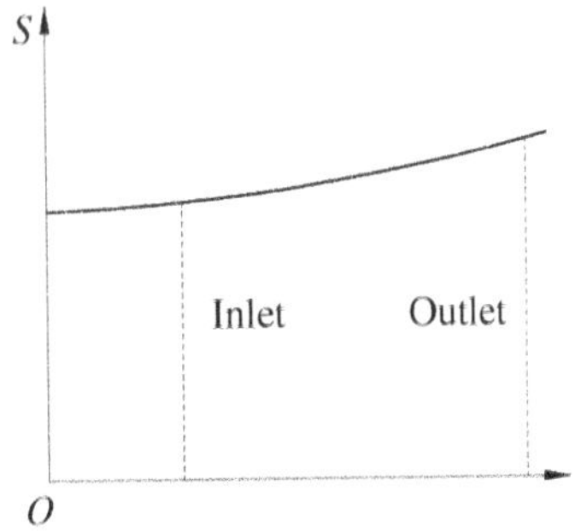

Figure 3.18 Meridional flow cross-section.

After determining a series of flow cross-sectional areas along the flow path, the variation curve of the flow cross-sectional area along the central line of the flow path (the line connecting the centers of the circles) can be plotted, as shown in Figure 3.18. This curve should be flat or smooth. Considering cavitation performance, the curve is generally convex in the inlet section. If the shape of the curve is unsatisfactory, the meridional projection shape should be modified until the requirements are met.

Drawing Meridional Streamlines Meridional streamlines are the intersection lines between flow surfaces and the meridional plane. The surface formed by rotating a streamline around the axis by one full revolution is a flow surface. To divide the flow passage, it is only necessary to split the entire impeller flow path into several smaller flow passages carrying equal flow rates. According to the one-dimensional theory, where velocity is uniformly distributed along the same flow cross-section, the total flow cross-section can simply be divided into several smaller equal-area cross-sections.

The area of each small flow cross-section is given by $\Delta S_i = 2\pi R_{ci} b_i$. Along the same flow cross-section, the condition $R_{ci} b_i$ = "constant" must be satisfied.

When practically constructing the streamlines graphically, points on the intermediate streamline between the front and rear shroud can be located on the formation line of the flow cross-section, such that $R_{c1} b_1 = R_{c2} b_2$, thereby defining the intermediate streamline. For pumps with lower specific speed, it is sufficient to draw only one intermediate streamline between the front and rear shroud streamlines. If the specific speed is less than 40, intermediate streamlines may be omitted, and the front and rear shrouds can be directly treated as the meridional streamlines. For centrifugal pumps and mixed-flow pumps with higher specific speeds, further subdivision of streamlines between adjacent existing streamlines is required, following the same method as described above, as illustrated in Figure 3.19.

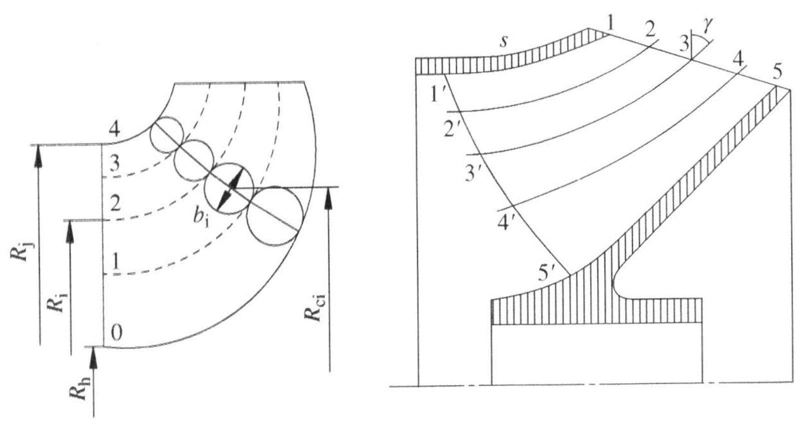

Figure 3.19 Meridional streamlines.

3.1.2.3.4 Selection of the Number of Blades The number of blades has a certain influence on the head, efficiency, and cavitation performance of a pump. It is generally considered that the number of blades Z in a centrifugal pump (or mixed-flow pump) is closely related to the specific speed n_s. The lower the n_s, the larger the Z value should be to avoid excessive load per unit area on the blades, as shown in Table 3.4.

However, Z interacts with other impeller geometric parameters such as the outlet diameter D_2, the blade outlet angle β_2, the outlet width b_2, and the blade wrap angle φ, collectively influencing pump performance. Therefore, to meet the requirements of the H–Q curve shape, Z must be determined in correlation with β_2, D_2, b_2, and D_j according to the patterns illustrated in Figures 3.12–3.15. Additionally, to achieve an optimal diffusion rate in the inter-blade flow passage, the product $Z\varphi$ (number of blades multiplied by the wrap angle) should be determined with reference to Table 3.5.

As can be seen from Figure 3.12, for low-specific-speed (low n_s) pumps with non-drooping H–Q curves, both domestic and international designs, typically adopt a blade count Z of less than 6. In order to increase the hydraulic radius of the inter-blade flow passages, reduce inlet blockage at the impeller eye, thereby minimizing hydraulic losses within the impeller, and avoid the formation of a droop (hump) in the H–Q curve; current domestic practice for low-specific-speed centrifugal pumps ($n_s < 80$) generally also uses a blade count in the range of $Z = 4$ to 6.

Within this range, the specific value of Z can be selected by referring to the Z values of existing, well-performing pumps with similar specific speeds. Furthermore, after the blade inlet and outlet angles (β_1 and β_2) are determined, Z can also be calculated using Eq. (3.31).

$$Z = 6.5\frac{R_2 + R_1}{R_2 - R_1}\sin\frac{\beta_1 + \beta_2}{2} \tag{3.31}$$

Based on the estimated value of Z, necessary corrections should be made to the initially selected Z value. In Eq. (3.31), R_2 represents the outlet radius of the impeller, while R_1 denotes the radius at the intersection point between the meridional flow channel centerline and the blade leading edge. As shown in Figure 3.20, for pumps with lower specific speeds, the addition of short blades between long blades can effectively wash away wake flows, preventing the generation and development of wakes. This approach increases the finite blade coefficient, thereby enhancing the pump's head, reducing the impeller outer diameter, improving the flow velocity distribution, and ultimately boosting pump performance.

3.1.2.3.5 Blade Thickness The actual blade thickness δ (i.e., the thickness measured perpendicular to the blade surface) can be approximately calculated using Eq. (3.32) based on strength requirements.

$$\delta = AD_2\sqrt{\frac{H}{Z}} \tag{3.32}$$

Table 3.4 Relationship between Z and n_s.

n_s	30–45	45–60	60–120	120–300	>300
Z	8–10	7–8	6–7	4–6	3–5

Table 3.5 Relationship between Z_φ and n_s.

n_s	30–50	55–70	80–120	130–220	230–280
$Z_\varphi/360°$	2.3–2.1	2.1–1.9	1.9–1.7	1.8–1.5	1.65–1.4

Figure 3.20 Schematic diagram of long and short blades.

where

H—Single-stage head (m);
Z—Number of blades;
D_2—Impeller outlet diameter (m);
A—Coefficient, whose value is selected from Table 3.6 based on specific speed and material.

The value of δ obtained from Eq. (3.32) represents the maximum blade thickness δ_{max}. If this calculated value is less than the minimum thickness achievable with the chosen manufacturing process, the process-allowed minimum value shall be adopted as δ_{max}.

For the inlet portion of the blade, the thickness is generally tapered uniformly from δ_{max} (located at a point 1/3 to 1/2 of the total blade length from the leading edge) down to the inlet thickness δ_1 at the leading edge. For cast impellers, δ_1 should be set to the minimum value allowed by the casting process whenever possible. The outlet thickness δ_2 at the trailing edge may be set to a value $\leq \delta_{max}$.

In impeller design, blade thicknesses measured along different directions are required. The calculation formulas for various thickness measurements are provided in Table 3.7.

Within the local region of the blade thickness calculation point, by approximating curved surfaces with planes and curves with straight lines, the various thicknesses and angles given in Table 3.7 are approximately defined as follows:

Flow surface thickness (s): The perpendicular distance from the pressure side to the suction side of the blade on the interface between the blade and the flow surface (which has a shape of a surface of revolution) (AC in Figure 3.21).

Table 3.6 Values of coefficient for common materials.

n_s materials	40	60	70	80	90	130	190	280
Cast iron, copper	3.2	3.5	3.8	4	4.5	6	7	10
Steel	3	3.2	3.3	3.4	3.5	5	6	8

Table 3.7 Various blade thicknesses.

Thickness type	Calculation formula
Actual thickness	δ
Flow surface thickness	$s = \delta\sqrt{1 + \cot^2 \lambda \cos^2 \beta}$
Circumferential thickness	$s_u = \delta\sqrt{1 + \cot^2 \beta / \sin^2 \lambda}$
Meridional thickness	$s_m = \delta\sqrt{1 + \cot^2 \lambda + \tan^2 \beta}$
Meridional normal thickness	$s'_m = \delta\sqrt{1 + \tan^2 \beta \sin^2 \lambda}$

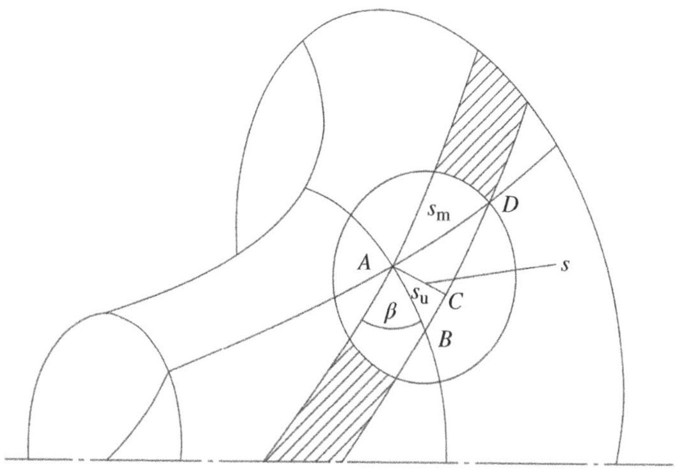

Figure 3.21 Blade thickness on the flow surface.

Circumferential thickness s_u: On the interface between the blade and the flow surface, the arc length along the circumferential direction from the calculation point on the pressure side to the suction side (arc *AB* in Figure 3.21).

Meridional thickness s_m: In the meridional projection, the thickness is measured along the tangent direction to the meridional streamline (the intersection line between the meridional plane and the flow surface) (*AD* in Figure 3.21, which can be approximated by the meridional streamline segment *AD*′).

Meridional normal thickness s'_m: In the meridional projection, the thickness is measured along the direction perpendicular to the meridional intersection line (the intersection line between the blade surface and the meridional plane), i.e., along the normal direction *EF* (*AP* in Figure 3.22).

β—Blade angle (the angle between the intersection line of the blade and the flow surface and the circumferential direction).

λ—Angle between the meridional streamline and the meridional intersection line.

3.1.2.3.6 Determination of the Blade Outlet Angle in Centrifugal Pumps Theoretical analysis indicates that the blade outlet angle β_2 of a centrifugal pump impeller is typically less than 90°. Within this range, for a given number of blades Z, there exists an optimum outlet angle β_2 that maximizes the theoretical head H_1. For centrifugal pump impellers, which usually have a blade count $Z = 4$–8, the corresponding optimum β_2 value generally falls within the range of approximately 18°–40°.

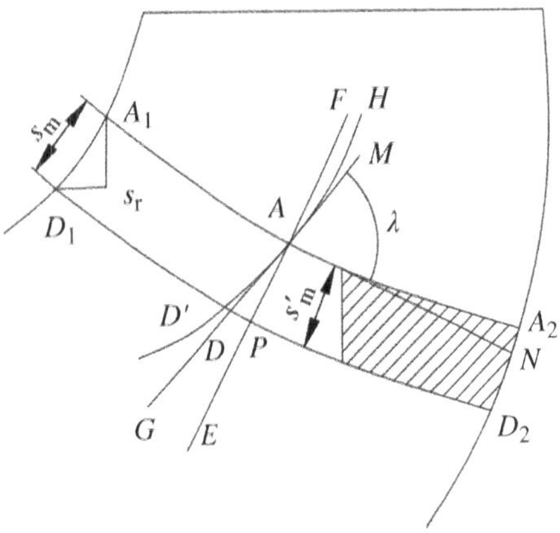

Figure 3.22 Blade thickness in the meridional view. A_1AA_2, Pressure side meridional intersection line; D_1DD_2, suction side intersection line; GAH, meridional streamline.

When determining the design value of β_2 within this range, the following factors should be considered:

1) For low-specific-speed pumps, a larger β_2 angle is often selected to increase the head and reduce the impeller outlet diameter D_2. This helps decrease disk friction losses and improve pump efficiency.
2) Increasing β_2 raises the outlet velocity v_2 at the same flow rate, leading to higher hydraulic losses in the volute or diffuser. It also increases shock losses at low flow rates, which may cause a droop (hump) in the performance curve. Therefore, to achieve a stable, droop-free characteristic curve, an excessively large β_2 angle should be avoided.
3) A larger β_2 can exacerbate flow diffusion within the blade passages, potentially leading to flow separation and reduced hydraulic efficiency.

$$w_1 \approx \frac{v_{m1}}{\sin\beta_1} \quad w_2 \approx \frac{v_{m2}}{\sin\beta_2} \quad \frac{w_2}{w_1} \approx \frac{v_{m2}\sin\beta_1}{\sin\beta_2 v_{m1}}$$

In pumps, $w_2 < w_1$, indicating that the relative flow undergoes diffusion. For a given flow rate, a larger blade outlet angle β_2 results in a smaller relative velocity w_2, thereby exacerbating flow diffusion losses.

3.1.2.3.7 *Calculation of the Blade Inlet Angle* The blade inlet angle β_1 should be greater than the relative flow angle β_1', and the difference $\Delta\beta = \beta_1 - \beta_1'$ is referred to as the incidence angle. The incidence angle typically falls within the range of $\Delta\beta = 3°–15°$. Using a positive incidence angle can improve cavitation resistance with minimal impact on efficiency.

The relative flow angle at the inlet β_1' is calculated using Eq. (3.33).

$$\tan\beta_1' = \frac{v_{m1}}{u_1 - v_{u1}} \tag{3.33}$$

where

u_1—Circumferential velocity of the liquid at the calculation point (i.e., the intersection points of each meridional streamline and the blade leading edge);
v_{m1}—Meridional velocity of the liquid at the calculation point;
v_{u1}—Circumferential component of the liquid velocity at the calculation point.
v_{u1} is determined by the structure of the suction chamber (or the outlet angle of the return guide vanes in multistage pumps).

For conical suction chambers, $v_{u1} = 0$; For semi-spiral suction chambers, it can be determined using the following empirical formula:

$$v_{u1} = \frac{m\sqrt[3]{Q^2 n}}{R_1} \tag{3.34}$$

where

R_1—Radius of the calculation point (m);
m—Empirical coefficient, $m = 0.055$–0.08. Smaller values are selected for lower specific speeds (n_s).

When return guide vanes are installed upstream of the impeller, $v_{u1} = v_{u6}R_6/R_1$, where R_6 is the radial position at the outlet of the return guide vanes, v_{u6} is the circumferential velocity component at the outlet of the return guide vanes, as illustrated in Figure 3.23.

$$v_{\alpha_6} = v_6 \cos \alpha_6$$

where

α_6—Outlet setting angle of the return guide vanes, typically $\alpha_6 = 60°$–$90°$;
v_6—Absolute velocity at the outlet of the return guide vanes, generally $v_6 = (0.85$–$1.0)\ v_j$;
v_j—Inlet velocity of the impeller.

The meridional velocity at the blade inlet is determined using the following formula:

$$v_{m1} = \frac{Q}{\eta_v S_1 \psi_1}$$

where

S_1—Flow cross-sectional area at the calculation point;
ψ_1—Blade blockage coefficient at the calculation point.

$$\psi_1 = 1 - \frac{Zs_{u1}}{D_1\pi} = 1 - \frac{Zs_1}{D_1\pi \sin\beta_1} = 1 - \frac{\delta_1 Z}{D_1\pi}\sqrt{1 + \left(\frac{\cot\beta_1}{\sin\lambda_1}\right)^2} \tag{3.35}$$

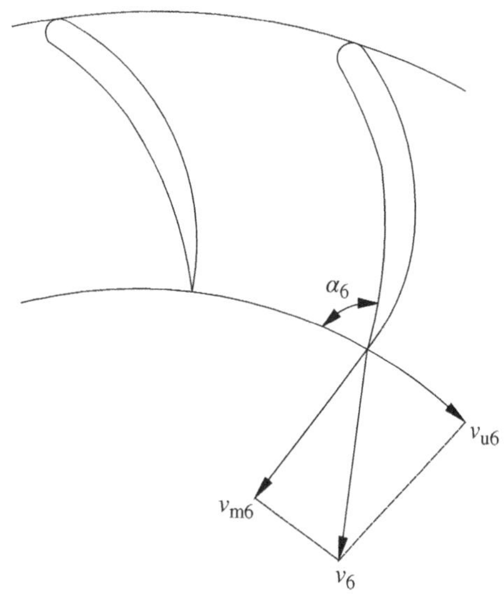

Figure 3.23 Outlet angle of return guide vanes.

where

z—Number of blades;
D_1—Diameter at the calculation point;
s_{u1}—Circumferential thickness of the blade at the calculation point;
s_1—Flow surface thickness of the blade at the calculation point;
δ_1—Actual thickness of the blade at the calculation point;
β_1—Inlet blade angle at the calculation point;
λ_1—Angle between the meridional intersection line and the meridional streamline at the calculation point, typically $\lambda_1 = 60°–90°$.

The blade inlet angle β_1 is obtained by adding the incidence angle to the calculated flow angle for each streamline. However, determining the blade angle initially requires calculating the flow angle, which itself depends on the blade angle for computing the blockage coefficient. Therefore, it is necessary to first assume an initial value for β_1 or the blockage coefficient. The final determined blade angle or blockage coefficient should be equal or close to the initially assumed value. Otherwise, the assumption must be revised, and an iterative calculation should be performed. Given the relatively wide acceptable range of incidence angles at the blade inlet, iterative calculations are generally unnecessary in practice.

Recent design practices indicate that applying a relatively larger incidence angle is preferable. Regarding the method of applying the incidence angle, the following approaches can be adopted:

1) Apply the same incidence angle to all streamlines.
2) Gradually decrease or increase the incidence angle from the front shroud streamline to the rear shroud streamline.
3) Select the incidence angle for the intermediate streamline, determine its β_1 value, and then determine the β_1 values for other streamlines based on the condition that $R_1 \tan \beta_1$ remains constant across the streamlines.

3.1.2.3.8 Theoretical Calculation of Impeller Outlet Diameter D_2 or Blade Outlet Angle β_2 The geometric parameters of centrifugal pump impellers previously discussed, such as D_j, d_h, D_2, b_2, Z, β_1, β_2, and δ, are typically calculated or selected individually based on statistical data. In reality, however, these parameters interact collectively to influence pump performance. To ensure that the final combination of geometric parameters better meets design requirements, a theoretical verification calculation should be performed for the most critical parameters, D_2 or β_2, based on the fundamental pump equation, which reflects the relationship between the impeller's geometric parameters and pump performance. This calculation is carried out under the assumption that all other geometric parameters except D_2 or β_2 are predetermined. The objective is to theoretically guarantee that the final combination of D_2 or β_2 with other geometric parameters satisfies the design specifications.

Meanwhile, the velocity coefficient method relies on numerous empirical coefficients to determine impeller dimensions. Since these coefficients are derived from practical experience, they are generally reliable under normal conditions. However, most velocity coefficients are obtained based on general assumptions (e.g., $\beta_2 = 22.5°$). During pump design, parameter values can vary significantly—for instance, if $\beta_2 \neq 22.5°$, which increases the approximation inherent in the velocity coefficients. Given that D_2 is the most critical dimension, it is advisable to use the D_2 value obtained via the velocity coefficient method as a basis for subsequent theoretical calculations.

Theoretical calculations are grounded in the fundamental pump equation and are, in principle, more rigorous. However, parameters such as hydraulic efficiency and the finite blade number correction coefficient used in these calculations can only be estimated using empirical formulas. Thus, the theoretical method also involves a degree of approximation. Practical experience has shown that theoretical calculation results are generally reliable. The following section describes a precise method for calculating D_2 and β_2.

From the fundamental equation:

$$H_t = \frac{u_2 v_{u2} - u_1 v_{u1}}{g}$$

Substituting:

$$v_{u2} = \sigma u_2 - \frac{v_{m2}}{\tan \beta_2}$$

and rearranging, obtain:

$$\sigma u_2^2 - \frac{v_{m2}}{\tan \beta_2} u_2 - u_1 v_{u1} - gH_t = 0$$

Solving the above quadratic equation for u_2

$$u_2 = \frac{v_{m2}}{2\sigma \tan \beta_2} + \sqrt{\left(\frac{v_{m2}}{2\sigma \tan \beta_2}\right)^2 + \frac{1}{\sigma}(gH_t + u_1 v_{u1})} \tag{3.36}$$

$$D_2 = \frac{u_2 60}{\pi n}$$

When using the above formula to calculate D_2, it is necessary to know v_{m2}. However, calculating v_{m2} requires D_2 itself, since:

$$v_{m2} = \frac{Q}{\eta_v D_2 \pi \psi_2 b_2}$$

Therefore, an iterative successive approximation method must be used. The value of D_2 initially used for drawing the meridional projection is taken as the starting value for the first iteration. If the calculated D_2 matches the initial D_2, then the initial value is correct. If the calculated D_2 differs from the initial value, the calculated D_2 is used as the new starting value for the next iteration. This process is repeated until the calculated D_2 is identical or sufficiently close to the starting value used in that iteration.

The relevant parameters in Eq. (3.36) are calculated as follows.

$$\psi_2 = 1 - \frac{Z s_{u2}}{D_2 \pi}$$

where

ψ_2—Blade outlet blockage coefficient, typically ranging from $\psi_2 = 0.8$ to 0.9.

s_{u2}—Circumferential thickness at the blade outlet, calculated as:

$$s_{u2} = \delta_2 \sqrt{1 + \cot^2 \beta_2 / \sin^2 \lambda_2}$$

where

λ_2—Angle between the meridional intersection line and the streamline at the impeller outlet, typically taken as $\lambda_2 = 70°–90°$;

δ_2—The predetermined actual blade thickness at the outlet.

$$H_1 = \frac{H}{\eta_h}$$

$$\sigma = 1 - \frac{\pi}{Z} \sin \beta_2$$

The value of v_{u1} is related to the structural type of the suction chamber:

For a conical suction chamber, $v_{u1} = 0$.

For a semi-spiral suction chamber, v_{u1} is calculated using Eq. (3.34) along the intermediate streamline.

When return guide vanes are installed upstream of the impeller, v_{u1} is determined using the same method as described earlier for calculating β_1.

$$u_1 = \frac{\pi R_1 n}{30}$$

When designing an impeller with $v_{u1} = 0$, the precise calculation of D_2 can be performed before drawing the meridional projection.

For centrifugal pumps, the blade outlet angle β_2 is typically selected first, followed by a precise calculation of D_2. The outlet edge position in the already-drawn meridional projection is then adjusted based on the calculated D_2.

In mixed-flow pumps, the blade outlet edge is inclined, and the outer diameter varies along different streamlines. To achieve the same head, streamlines with a smaller D_2 should be assigned a larger outlet angle β_2. In such cases, based on the outlet edge of the already-prepared meridional projection, the outlet blade angle β_2 for each streamline can be calculated using Eq. (3.37) through an iterative successive approximation method.

$$\tan \beta_2 = \frac{u_2 v_{m2}}{\sigma u_2^2 - H_1 g - u_1 v_{u1}} \tag{3.37}$$

3.1.2.4 Blade Drawing

After completing the design of the shape and dimensions of the impeller meridional projection, the subsequent task is to determine the circumferential development shape of the blade with thickness. The finalized blade is then represented in a drawing for impeller manufacturing. This process is referred to as blade profiling.

3.1.2.4.1 Conformal Mapping Method Using a Grid of Squares

Profiling Principle Figure 3.24a shows a flow surface with a single streamline. This surface is intersected by a set of meridional planes I, II, … with an angular spacing of $\Delta\theta$, and a set of planes

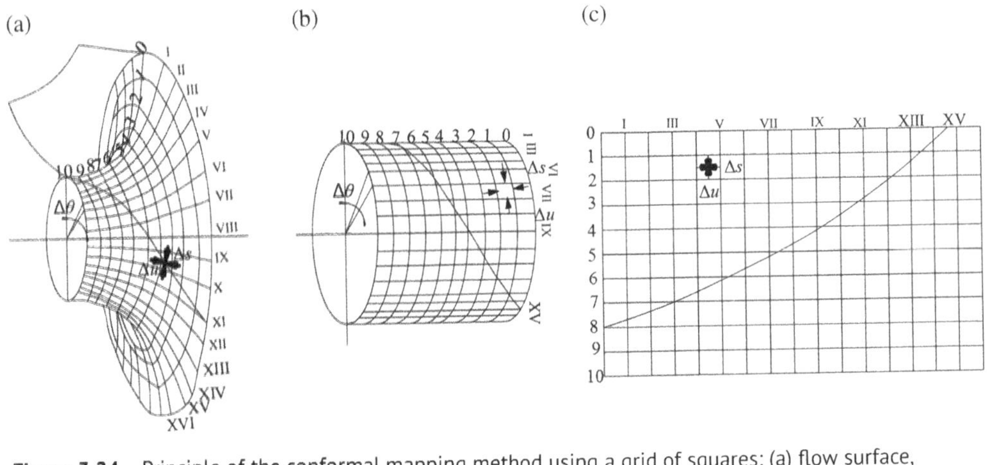

Figure 3.24 Principle of the conformal mapping method using a grid of squares: (a) flow surface, (b) distorting the flow surface into a cylindrical surface, and (c) developing the cylindrical surface into a plane.

1, 2, … perpendicular to the axis. This creates a grid of small curvilinear sectors on the flow surface, where the meridional streamline length Δs is made equal to the circumferential length Δu (i.e., $\Delta s = \Delta u$). When these sectors are sufficiently small, the curved sectors on the flow surface can be approximated as small squares on a plane.

Although the size of the sectors increases from the inlet to the outlet on the flow surface, conformal mapping, as the name implies, ensures that the angle between the spatial streamline and the circumferential direction is preserved. On the developed plane, the expanded streamline is only required to have the same angle with the circumferential direction as the spatial streamline; its length and shape may differ from the actual streamline. Since the goal is similarity rather than exact equality, one can imagine distorting the flow surface into a cylindrical surface (Figure 3.24b), then cutting the cylindrical surface along a generatrix and developing it into a plane (Figure 3.24c).

As shown in Figure 3.24, the spatial streamline intersects the small sectors on the flow surface, dividing each sector's sides into two segments. On the planar grid of squares, the expanded streamline similarly divides the sides of each square into proportional segments. Due to this similarity relationship, the corresponding angles remain equal—i.e., the angles are preserved.

Blade design follows the reverse process: streamlines drawn on the developed planar diagram are mapped back onto the flow surface (represented in the planar and meridional projections) using characteristic lines while preserving the angles. Since all methods for designing twisted blades are also applicable to cylindrical blades, the following description uses a twisted blade as an example.

Profiling Procedure

1) Dividing Points Along the Meridional Streamline

The essence of point division is to draw characteristic lines on the flow surface, forming a sector grid. Since the flow surface is generated by rotating a meridional streamline about the axis, it is axisymmetric. All meridional streamlines on the same flow surface are identical. Therefore, dividing points along one corresponding meridional streamline is equivalent to constructing the entire grid of squares on the flow surface.

Besides the meridional projection, draw two rays with an included angle of $\Delta\theta$ (generally taken as $\Delta\theta = 5°–10°$), as shown in Figure 3.25. These two rays represent two meridional planes separated by angle $\Delta\theta$. Starting from the outlet, tentatively select a segment length Δs along the meridional

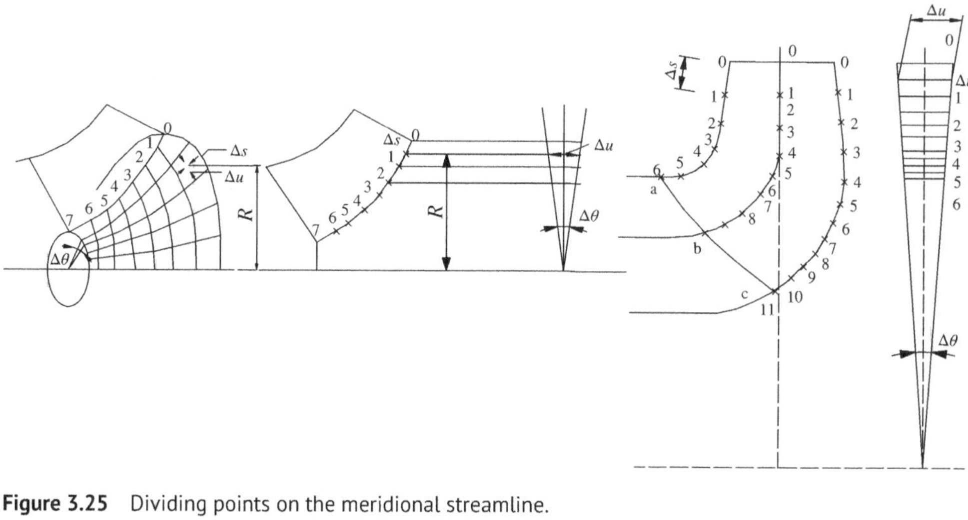

Figure 3.25 Dividing points on the meridional streamline.

streamline. If the arc length Δu between the two rays at the midpoint radius of Δs equals the tentatively selected Δs, then the point division is correct. If not, select a new Δs until $\Delta s = \Delta u$ is achieved.

After the first point is determined, use the same method to divide the second, third, and subsequent points. When the streamline is parallel to the axis, Δu remains constant, and equal segments can be directly obtained using the corresponding Δs. It is important to note that Δs is the length along the meridional streamline, not the perpendicular distance between the rays (these two are equal only when the meridional streamline is perpendicular to the axis).

As can be seen from Figure 3.25, dividing points along one meridional streamline is equivalent to covering the entire flow surface with a grid of squares.

2) Develop the Flow Surface (Planar Grid) and Draw the Streamlines

As shown in Figure 3.26, because the conformal mapping method using a grid of squares is based on local similarity rather than local equality [8, 9], a single planar grid can be used to draw the streamlines for several flow surfaces. The size of the squares in the planar grid can be chosen arbitrarily. The horizontal lines of the grid represent the corresponding division points on the meridional streamline, and the vertical lines represent the meridional planes with an angular spacing of $\Delta\theta$ corresponding to the division points used.

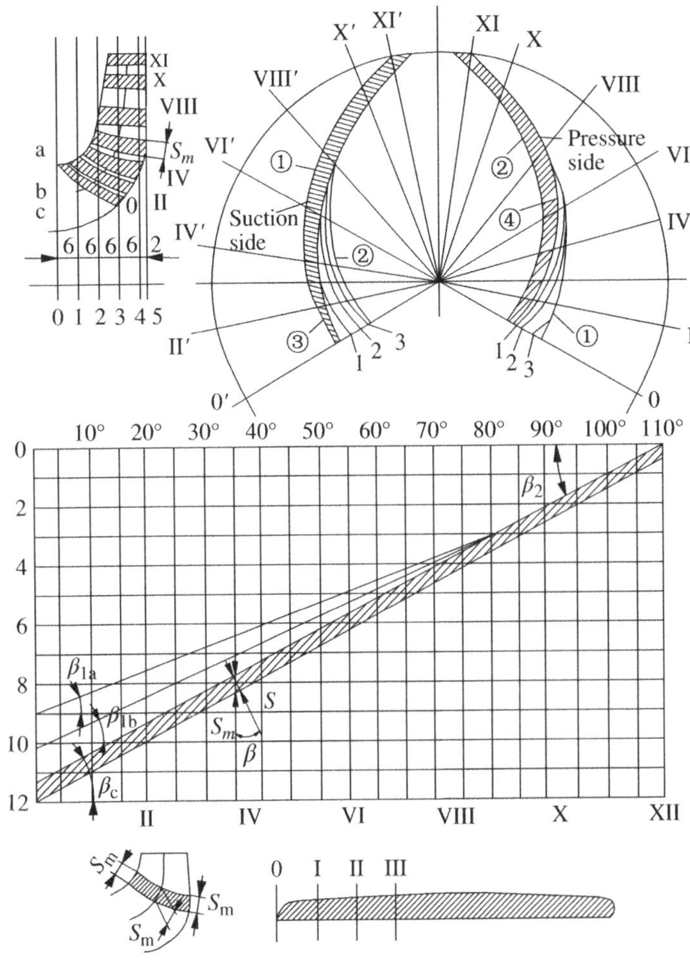

Figure 3.26 Blade profiling.

After drawing the grid (or using graph paper), the streamlines can be drawn on it. Usually, the intermediate streamline is drawn first. The inlet and outlet positions of the streamline on the grid should correspond to the division points of the meridional streamline. The wrap angle can be selected with reference to Table 3.5. Draw straight lines through the inlet and outlet points with angles equal to the calculated blade inlet and outlet setting angles, then create a smooth curve (using a French curve, steel ruler, or triangle template for circular arcs) tangent to the inlet and outlet lines. In some cases, the profile can be drawn freely while ensuring that the inlet and outlet setting angles are maintained. Other streamlines can be drawn using a similar method. The profile should be smooth and fair; it can be curved in one direction only (concave or convex), but should not exhibit an S-shape.

The shape of the profile is extremely important. If unsatisfactory, it should be resolutely modified. If necessary, adjustments can be made, such as appropriately changing the blade wrap angle, not aligning the blade leading and trailing edges on the same meridional plane, adjusting the position of the blade leading edge, or modifying the inlet setting angle, and then redrawing the profile.

3) Draw the Meridional Intersection Lines on the Meridional Projection

The three relative streamlines drawn on the grid represent the three profile lines of the blade surface. Intersecting these three streamlines with meridional planes (represented by the vertical lines in the grid) is equivalent to cutting the blade with meridional planes. The line connecting the three resulting points is the meridional intersection line of the blade.

The intersections of the vertical lines (representing meridional planes at regular angular intervals) in the grid with the three streamlines transferred to the corresponding division points (labeled 1, 2, 3, …) on the three streamlines in the meridional projection using interpolation. These points are then connected with smooth curves to form the meridional intersection lines of the blade pressure surface.

The meridional intersection lines should be smooth and vary regularly. The angle "λ" between the meridional intersection line and the streamline should ideally be close to 90° and generally not less than 60°. If the "λ" angle is too small, the actual angle between the shroud and the blade will be too small, leading to significant blockage, difficulties in casting, and poor flow passage shape (increased wetted perimeter). The relationship between the actual angle γ between the shroud and the blade and the angle "λ" is given by $\cot\gamma = \cot\lambda\cos\beta$.

4) Blade Thickening

The meridional intersection lines represent the blade without thickness. Since actual blades have thickness, thickening must be applied. Thickening can be performed on the developed flow surface, the meridional plane, or the planar view. A commonly used method is to apply thickening in the meridional projection using the meridional thickness. The value of the meridional thickness s_m can be calculated using the following formula:

$$s_m = s/\cos\beta = \delta\sqrt{1 + \tan^2\beta + \cot^2\lambda}$$

where s is the flow surface thickness, β is measured at the corresponding point on the developed grid, λ is measured at the corresponding point on the meridional intersection line in the meridional projection, and δ is the actual thickness. The value of δ at each thickening calculation point (i.e., the intersection of the meridional streamline and the meridional intersection line) can be determined based on the previously defined thickness distribution along the streamline (represented by the positions of meridional planes I, II, …, as shown in Figure 3.26).

Table 3.8 Thickness calculation.

Streamline	Parameter	Meridional plane 0	Meridional plane I	Meridional plane II	......
a	δ				
	β				
	λ				
	s_m				
b	δ				
	β				
	λ				
	s_m				
c	δ				
	β				
	λ				
	s_m				

For small pumps, the same thickness may be used for all streamlines. For large pumps, draft angle considerations should be taken into account, with the thickness gradually decreasing from the rear shroud to the front shroud.

Thickness calculations should be performed systematically using a table (see Table 3.8).

An alternative method for determining the meridional thickness is to directly apply thickening along the streamline on the developed grid based on a given flow surface thickness s. Here, s is the flow surface thickness measured perpendicular to the streamline. The meridional thickness s_m is obtained by measuring s vertically on the grid and can be directly applied to the corresponding point in the meridional projection (along the direction of the meridional streamline), as shown in Figure 3.26.

It should be noted that the streamlines on the grid are constructed based on the principle of local similarity, so their lengths are not equal to the actual streamline lengths. The thickness variation along these streamlines also differs from the actual situation. However, the thickness values at each meridional plane position are consistent with those obtained from the tabular calculation method, making this approach both simple and accurate.

Using the calculated s_m values at each computation point, locate the corresponding points on the suction side along each meridional streamline (as shown in Figure 3.26). Connect the corresponding suction-side points on each meridional plane with smooth curves to obtain the suction-side meridional intersection lines (shown as dashed lines in Figure 3.26) that correspond to the pressure-side meridional intersection lines.

5) Drawing the Blade Pattern Layout

As shown in Figure 3.26, use a set of equally (or unequally) spaced axis-normal planes to section the blade. Each plane intersects the blade along two curves (pressure surface and suction surface). Plotting these intersection curves for each section in the planar view yields the template curves (also called mold sections). The specific steps are as follows:

- Draw a set of axis-normal planes and number them 0, 1, 2, 3, 4, ...
- In the planar view, draw meridional lines (a set of rays) corresponding to the angles of the meridional intersection lines from the meridional projection, and number them correspondingly (0, I, II, ...).

- Determine the position of the pressure and suction surfaces in the planar projection based on two principles:
  1) With the blade rotating toward its convex side, removing the rear shroud (viewing from the rear) should reveal the pressure surface.
  2) Removing the front shroud (viewing from the front) should reveal the suction surface. For example, if the rear shroud intersection (which has a smaller inlet radius) is drawn on the right side of the diagram, the right-side diagram should show the template curves for the pressure surface, and the left-side diagram should show those for the suction surface. When viewed from the rear (pressure surface), this blade rotates clockwise; when viewed from the front (suction surface), it rotates counterclockwise.
- Construct the blade's planar projection outline:
  1) Line ① is the intersection of the pressure surface and the front shroud. It is formed by transferring the intersection points of each pressure surface meridional intersection line and the front shroud streamline (from the meridional projection) to their corresponding meridional rays in the planar view using their respective radii, and connecting these points.
  2) Similarly, construct:
     - ② Intersection of the suction surface and the rear shroud.
     - ③ Intersection of the suction surface and the front shroud.
     - ④ Intersection of the pressure surface and the rear shroud.
- Draw the template curves for the pressure and suction surfaces: Transfer the intersection points between the template section planes (axis-normal planes) and the pressure surface (or suction surface) meridional intersection lines to their corresponding meridional rays in the planar view using their respective radii. Connect the resulting points to obtain the pressure surface template curves (1, 2, 3) and the suction surface template curves (1′, 2′, 3′). The radii of the points where the template section planes intersect the front and rear shroud streamlines (in the meridional view) must equal the starting/ending radii of the corresponding template curves in the planar view. The template curves must be smooth.
- The template curves serve as the basis for manufacturing the blade. To avoid distortion during tracing, it is advisable to list the coordinates (radial distance from the axis) of each intersection point between the template curves and the meridional lines.

Impeller Hydraulic Performance Check Upon reaching this stage of the impeller design, the blade shape and dimensions have been fully defined. Before finalizing the blade design, it is essential to check its hydraulic performance. If the results do not meet the requirements, the design should be modified accordingly.

1) Check of the flow passage diffusion between blades

The variation in the cross-sectional area of the flow passage between two blades, particularly the ratio of the outlet area to the inlet area of the effective portion of the passage, has a critical impact on performance. It is recommended that:

$$\frac{S_{\text{out}}}{S_{\text{in}}} = 1.0 - 1.3$$

It is recommended that this ratio should not exceed 1.3. If the ratio is greater than 1.3, the flow passage exhibits excessive diffusion, leading to a decrease in efficiency. In such cases, the design should be modified.

The cross-sectional area between the flow passages is calculated using the following formula:

$$A_i = a_i b_i$$

where

a_i—Width between two blades in the planar view;
b_i—Blade width in the meridional view.

For cylindrical blades, the check can be performed using the method shown in Figure 3.27.

2) Check the variation of the relative velocity w and the angular momentum vuR along the length L of the relative flow streamline.

Following the calculation process outlined in Table 3.9, determine the values of w and $v_u R$ at the intersection points between the spatially curved relative flow streamline and the meridional

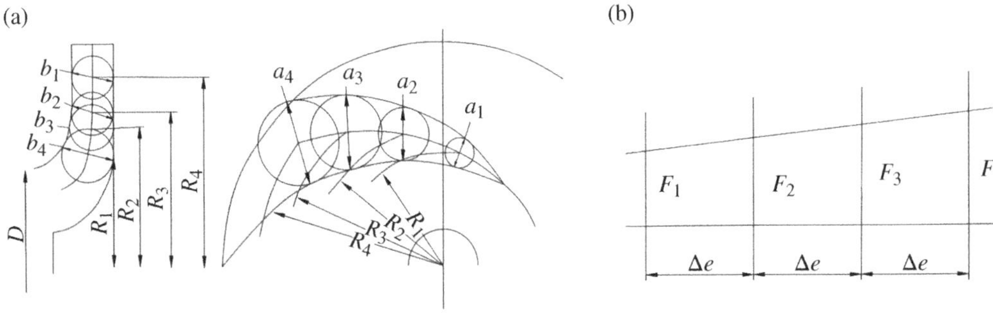

Figure 3.27 Flow passage area check between adjacent blades: (a) check method (b) flow passage area variation curve.

Table 3.9 Calculation of w and $v_u R$.

Parameter	0 (streamline inlet) I II ...... 0 (streamline outlet)
Meridional streamline segment length Δl	
Blade angle β	
Relative flow streamline segment length $\Delta L = \frac{\Delta l}{\sin\beta}$	
Relative flow streamline length $L = \sum \Delta L$	
Actual blade thickness δ	
Angle between meridional streamline and intersection line λ	
Circumferential blade thickness $s_u = \delta\sqrt{1+\left(\frac{\cot\beta}{\sin\lambda}\right)^2}$	
Radius at calculation point R	
Blockage coefficient $\psi = 1 - \frac{Zs_u}{2\pi R}$	
Flow cross-sectional area S	
Meridional velocity $v_m = \frac{Q}{\eta_V S\psi}$	
$w = v_m / \sin\beta$	
$u = \frac{\pi n R}{30}$	
$v_u = u - \frac{v_m}{\tan\beta}$	
$v_u R$	

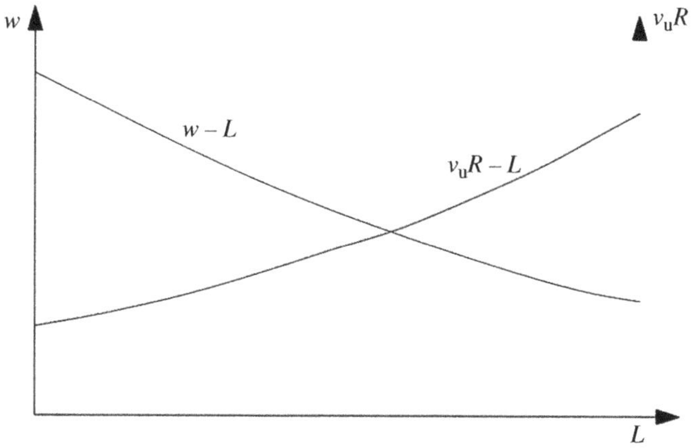

Figure 3.28 Calculation of $w - L$ and $v_uR - L$ curves.

intersection lines 0, I, II, etc. Plot the $w - L$ and $v_uR - L$ curves, as shown in Figure 3.28. These curves should be smooth and monotonic, without irregular fluctuations or the presence of local maxima or minima.

Cavitation Performance Estimation The required NPSH of the designed pump can be estimated using Eq. (3.38):

$$(\mathrm{NPSH})_{\mathrm{r,est}} = \frac{v_0^2}{2g} + \lambda\frac{w_0^2}{2g} \tag{3.38}$$

where

v_0—Absolute velocity at the inlet of the front shroud streamline, without considering blade blockage;
w_0—Relative velocity at the inlet of the front shroud streamline, without considering blade blockage;
λ—Cavitation coefficient, which can be taken as $\lambda = 0.2$–0.4. For pumps with $n_s < 120$, its value can be approximated using Eq. (3.39).

$$\lambda = 1.2\frac{v_0}{u_1} + \left(0.07 + 0.42\frac{v_0}{u_1}\right)\left(\frac{\delta_0}{\delta} - 0.615\right) \tag{3.39}$$

where

δ_0—Blade inlet thickness;
δ—Maximum blade thickness;
u_1—Circumferential velocity at the inlet of the front shroud streamline.

The value of $(\mathrm{NPSH})_{\mathrm{r,est}}$ can also be obtained from Figure 3.29. In this figure, the values of u_1 and v_{m1} used are the circumferential velocity and meridional velocity at the inlet of the front shroud streamline, respectively.

The estimated $(\mathrm{NPSH})_{\mathrm{r,est}}$ and the design requirement $(\mathrm{NPSH})_{\mathrm{r}}$ should satisfy the following condition:

$$(\mathrm{NPSH})_{\mathrm{r,est}} \leqslant (\mathrm{NPSH})_{\mathrm{r}}$$

The theoretical check of the pump's cavitation performance is considered acceptable. Otherwise, the design should be modified.

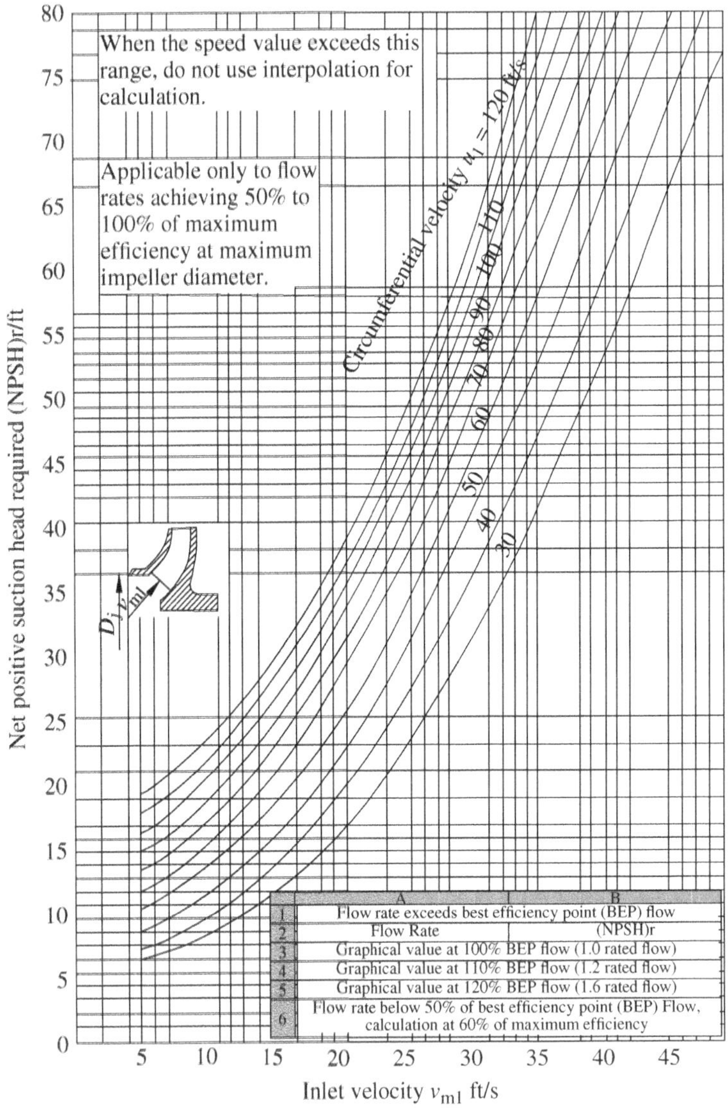

Figure 3.29 $(NPSH)_r$ prediction chart.

3.1.2.4.2 Twisted Triangle Method for Blade Profiling

Profiling Principle A streamline on a flow surface, as shown in Figure 3.30a, is intersected by two sets of mutually perpendicular planes (meridional planes and planes perpendicular to the axis). The intersection lines of these two sets of planes with the flow surface, together with the original streamline, form a series of small right-angled triangles on the flow surface. Since the flow surface is a curved spatial surface, these small triangles are distorted curved triangles.

When the segment length ΔL of the spatial streamline, the corresponding meridional streamline length Δs (the length of the intersection line between the meridional plane and the flow surface), and the circumferential arc length Δu are sufficiently small, the small triangle they form can be approximated as a planar right triangle. Being planar triangles, they can be drawn on a flat surface. These small triangles are drawn end-to-end on a plane while keeping their right-angled sides

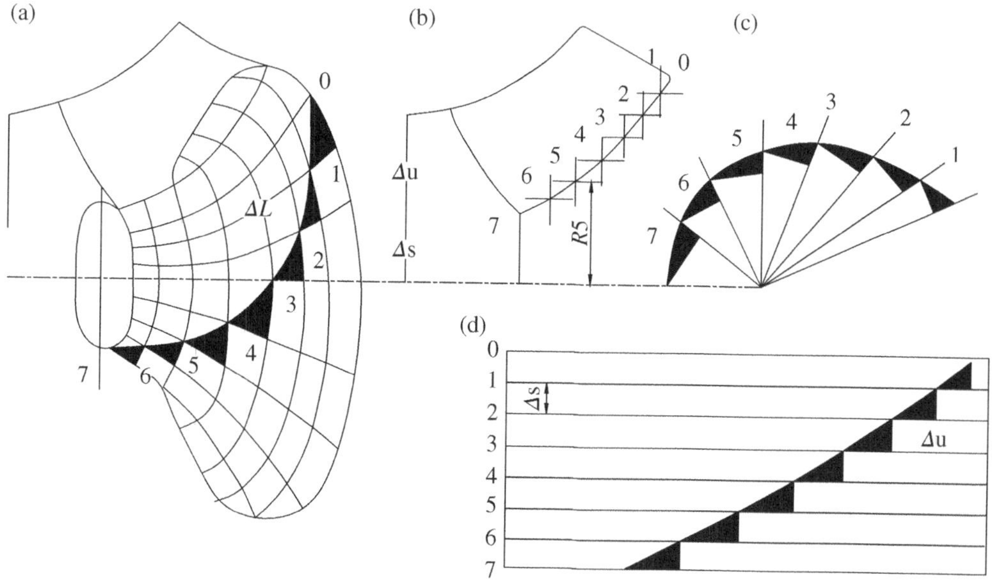

Figure 3.30 Profiling principle of the distorted triangle method: (a) flow surface diagram, (b) meridional view, (c) planar view, (d) development diagram.

parallel, as shown in Figure 3.30(d). This is equivalent to cutting out each small triangle and placing them sequentially on a plane. Clearly, the small triangles on the plane are approximately congruent to their corresponding triangles on the flow surface. Thus, the streamline on the spatial flow surface is represented on the plane using a locally congruent method [10], effectively developing the spatial flow surface into a plane.

When designing the blade, the process is reversed: the streamline is first drawn on the developed planar layout according to the design requirements, and then mapped back onto the flow surface (represented in the planar and meridional projections). The length of the vertical segments Δs on the development drawing equals the predivided segment length of the meridional streamline. Since the corresponding arc lengths Δu are equal on the flow surface diagram, the development drawing, and the planar view in Figure 3.30(c), and the corresponding radii at each point are equal in the meridional view (Figure 3.30(b)) and the planar view, the planar projection of the streamline can be constructed from the development drawing and the meridional projection. The planar projections of several streamlines are then used to construct the meridional intersection lines of the blade, completing the blade profiling.

Profiling Procedure (Figure 3.27)

1) Divide points along the meridional streamlines in the meridional projection, starting from the outlet of each streamline, resulting in points 0, 1, 2, etc. For convenience, the developed lengths of the segments between points are made equal, allowing all three streamlines to share a single development diagram.
2) Create the planar development diagram. Corresponding to the points divided on the streamlines, draw parallel lines spaced at intervals equal to the developed lengths of the segments between the points on the meridional streamline. Number these lines 0, 1, 2, etc.
3) Draw the streamlines on the development diagram. The inlet and outlet points of the drawn streamline should correspond to the positions on the meridional streamline, and the inlet and outlet angles should match the predetermined values. The wrap angle (horizontal length on

the development) can be adjusted flexibly. To reduce rework, the wrap angle can be roughly estimated in advance: using the radius at a point approximately 2/5 of the total meridional streamline length from the inlet, draw a circle such that the arc length for a certain central angle equals the horizontal length of the developed streamline. This central angle roughly equals the wrap angle in the planar view of the streamline. (The horizontal lengths corresponding to the wrap angles for the three streamlines should be determined separately.) All three streamlines can be drawn on one development diagram, or each streamline can be drawn on its own diagram, with the trailing edge starting from point 0 in all cases.

4) Construct the planar projection of the streamlines. Based on the equality between the horizontal length Δu of each small triangle in the development diagram and the corresponding Δu in the planar view (Δu remains constant), and the equality between the radii at the division points on the meridional streamline and the radii in the planar view (R is equal), draw arcs point-by-point to construct the planar projection of each streamline. The relative positions of the streamlines in the planar view can be arranged as needed (e.g., whether the leading and trailing edges lie on the same meridional plane). Modify the profile until it is satisfactory. Starting from the outlet toward the inlet is more convenient when creating the planar view.
5) Draw the meridional intersection lines. In the planar view, draw rays at specific angular intervals (representing meridional planes) and number them 0, I, II, etc. Transfer the intersection points of these rays with the planar streamlines to the corresponding meridional streamlines using their identical radii. The curve smoothly connecting these transferred points is the meridional intersection line.
6) Blade thickening: Once the meridional intersection lines are defined, thickening can be performed similarly to the conformal mapping method using a grid of squares. However, since the distorted triangle method is based on local congruence, the developed streamline is not only similar but also approximately equal in length to the spatial streamline. Therefore, thickening can be applied directly along the developed streamline according to the specified variation of the flow surface thickness s. The thickness perpendicular to the streamline is the flow surface thickness s, the horizontal length on the development is the circumferential thickness s_u, and the vertical length is the meridional thickness s_m. This allows two thickening approaches:
   First method: Transfer s_m to the corresponding points on the meridional streamline; the curve connecting these points is the suction side meridional intersection line.
   Second method: Transfer s_u circumferentially to the corresponding points in the planar view; the curve connecting these points is the planar projection of the suction side streamline, which is then transferred back to define the suction side meridional intersection line.
7) Creating the blade template layout and checking the profiling quality are performed in the same manner as in the conformal mapping method using a grid of squares (Figure 3.31).

3.1.2.4.3 Profiling of Cylindrical Blades For cylindrical blades, which are commonly used in low-specific-speed (low n_s) impellers, profiling can also be performed using either the conformal mapping method (grid of squares) or the distorted triangle method.

In this case, since the planar projection of the entire blade surface collapses into a single curve, profiling can be carried out by dividing points only along the rear shroud meridional streamline. Then, based on the predetermined blade angles β_1 and β_2, only the rear shroud streamline is drawn on the grid or development diagram.

- **Using the grid of squares method:** The intersection points between the streamline on the grid and the vertical grid lines are directly plotted on the planar view at the corresponding radii (from the meridional streamline division points) along the respective meridional rays. Connecting these points smoothly yields the planar projection of the blade profile.

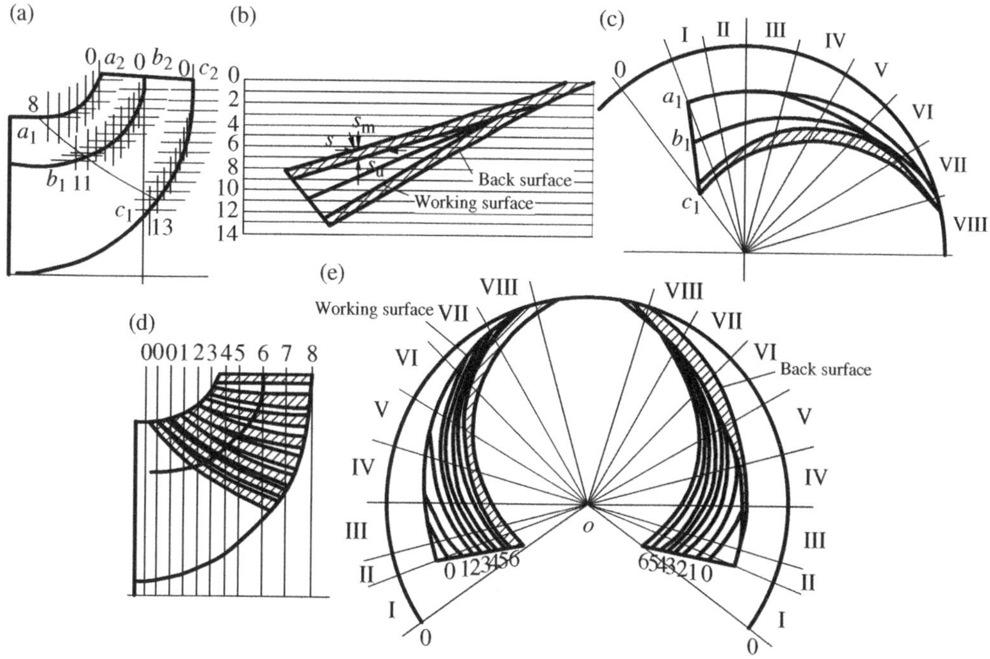

Figure 3.31 Blade profiling using the distorted triangle method: (a) dividing points on the meridional streamline, (b) drawing streamlines on the development diagram, (c) constructing the planar projection of the streamline, (d) drawing the meridional intersection line, (e) creating the template curve.

- **Using the distorted triangle method:** The planar projection of the blade profile is constructed based on the principle that corresponding Δu values and radii are identical between the development diagram and the planar view.

Furthermore, cylindrical blades can also be profiled using circular arc profiles or variable-angle logarithmic spirals.

Circular Arc Profile A circular arc profile can be formed by a single arc or multiple connected arcs. A blade constructed with a single arc tends to be shorter, and the flow passage diffusion characteristics are generally less favorable compared to those achieved with multiple arcs. Using two connected arcs as an example, the drawing method for a circular arc profile is described below and illustrated in Figure 3.32.

1) Draw the arc representing the blade leading edge between diameters D_1 and D_2, with its center at point O.
2) Draw an intermediate circle with a diameter:

$$D_{\mathrm{i}} = \sqrt{\frac{D_1^2 + D_2^2}{2}}$$

The blade angle at this circumference is:

$$\beta_{\mathrm{i}} = \frac{(\beta_1 + \beta_2)(D_{\mathrm{i}} - D_1)}{D_2 + D_1}$$

3) Construct $\angle BAO = \beta_2$, where point A lies on the D_2 circle; construct $\angle AOC = \beta_2 + \beta_{\mathrm{i}}$, where point C lies on the D_{i} circle.

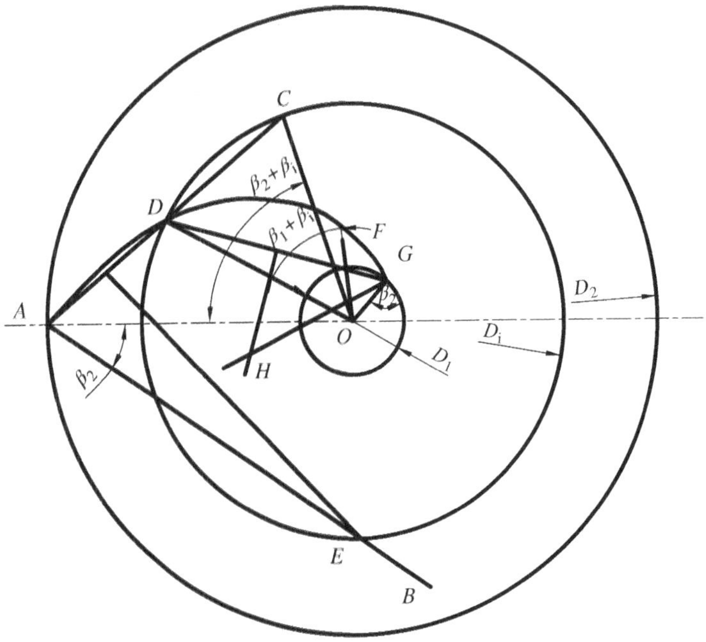

Figure 3.32 Drawing of a circular arc profile.

4) Connect points A and C with a straight line. This line (or its extension) intersects the D_i circle at point D. Construct the perpendicular bisector of segment AD, and find its intersection point E with line AB.
5) Using E as the center and EA as the radius, draw arc AD, which must pass through point D. Construct $\angle DOF = \beta_1 + \beta_i$, where point F lies on the D_1 circle.
6) Connect points D and F with a straight line. This line (or its extension) intersects the D_1 circle at point G. Construct the perpendicular bisector of segment DG.
7) Construct $\angle OGH = \beta_1$, where point H lies on the perpendicular bisector of DG.
8) Using H as the center and DH as the radius, draw arc DG, which must pass through point G.
9) Using points E and H as centers, draw arcs with radii $EA-\delta$ and HD$-\delta$ (where δ is the actual blade thickness), respectively. Appropriately thin the leading-edge region and round the blade head to obtain the planar projection of the thickened cylindrical blade.

From the above profiling process, it can be seen that once D_1, β_1, D_2, and β_2 are determined, the wrap angle of the cylindrical blade using a circular arc profile is also fixed. This limits the ability to adjust the diffusion angle of the flow passage between the blades.

Variable-Angle Logarithmic Spiral Profile As will be discussed later, the outer contour of the volute flow passage should ideally follow the trajectory of free-flowing liquid—a logarithmic spiral. Similarly, when the relative flow streamlines defining the blade surface shape are logarithmic spirals, the hydraulic losses are minimized, leading to higher pump efficiency. Experimental results also indicate that solid–liquid two-phase flow pumps employing logarithmic spiral profiles achieve higher efficiency than those using circular arc profiles.

In the polar coordinate system (r,θ), the general mathematical expression for a logarithmic spiral is:

$$r = ae^{c\theta}$$

Substituting the conditions $\theta = 0$, $r = R_1$ and $\theta = \varphi$, $r = R_2$ into the equation to solve for the coefficients a and c, and then substituting these coefficients back into the original equation yields:

$$r = R_1 e^{\frac{1}{\varphi} \ln \frac{R_2}{R_1} \theta} \tag{3.40}$$

where

R_1—Inlet radius of the rear shroud streamline;
R_2—Outlet radius of the rear shroud streamline;
φ—Wrap angle of the streamline.

From differential geometry, the angle β between the tangent at any point on the spiral and the circumferential direction is given by:

$$\beta = \arctan \frac{r'(\theta)}{r(\theta)} \tag{3.41}$$

From this equation, the angle β at any point on the logarithmic spiral described by Eq. (3.40) can be derived as:

$$\beta = \arctan \left(\frac{1}{\varphi} \ln \frac{R_2}{R_1} \right) \tag{3.42}$$

When R_1 and R_2 are fixed and the wrap angle ψ is specified, Eq. (3.40) represents an equiangular spiral. If the blade inlet and outlet angles are equal ($\beta_1 = \beta_2$), the wrap angle can be determined using Eq. (3.42). Then, the coordinates (r, θ) of points along the equiangular logarithmic spiral profile can be calculated from Eq. (3.40), allowing the blade profile to be drawn. Clearly, when $\beta_1 = \beta_2$, the wrap angle of the equiangular logarithmic spiral profile is also fixed and cannot be adjusted.

For most impellers, $\beta_1 \neq \beta_2$, meaning the blade angle β at various points along the profile is not constant. By specifying different patterns of variation for β with respect to θ, different forms of variable-angle logarithmic spirals can be obtained. The most commonly used functional form for $\beta = f(\theta)$ in impeller design is:

$$\tan \beta = A\theta^K + B \tag{3.43}$$

Substituting the boundary conditions $\theta = 0$, $\beta = \beta_1$ and $\theta = \varphi$, $\beta = \beta_2$ into the Eq. (3.43) to solve for coefficients A and B, and then substituting these coefficients back into the original Eq. (3.43) yields:

$$\tan \beta = \frac{\tan \beta_2 - \tan \beta_1}{\varphi^K} \theta^K + \tan \beta_1 \tag{3.44}$$

Substituting this expression into Eq. (3.41) and applying the outlet boundary condition $\theta = \varphi$, $r = R_2$, obtain:

$$r = R_1 e^{\theta} \left[\frac{\tan \beta_2 - \tan \beta_1}{K + 1} \right] \left(\frac{\theta}{\varphi} \right)^K + \tan \beta_1] \tag{3.45}$$

where the value of K can be determined by the following formula:

$$K = \frac{\varphi(\tan \beta_2 - \tan \beta_1)}{\ln \frac{R_2}{R_1} - \varphi \tan \beta_1} - 1 \tag{3.46}$$

After selecting ψ, the constant K is calculated from Eq. (3.46) based on the predetermined values of R_1, R_2, β_1, and β_2. Then, using Eq. (3.45), the variable-angle logarithmic spiral profile can be determined. Therefore, the characteristics of the profile are highly dependent on the chosen value of ψ. The selected ψ should ensure that the values of β calculated from Eq. (3.44) at various points along the profile change evenly and reasonably from β_1 to β_2. Otherwise, the value of ψ should be adjusted.

3.1.3 Hydraulic Design of the Suction Chamber

The suction chamber is located upstream of the impeller and serves to guide the liquid into the impeller under specified conditions. The flow velocity within the suction chamber is relatively low, resulting in significantly lower hydraulic losses compared to the discharge chamber. However, the flow state in the suction chamber directly affects the flow patterns within the impeller and has a certain impact on pump efficiency, particularly a substantial influence on the pump's cavitation performance. For low-head pumps, the absolute value of the losses may be small, but they constitute a larger proportion of the total head, thus relatively affecting efficiency much more significantly than in high-head pumps.

The requirements for a suction chamber are as follows: ensure the desired velocity field at the impeller inlet, such as uniform velocity distribution, appropriate velocity magnitude, required direction (circulation), and minimized hydraulic losses. Suction chambers can be classified into structural types such as conical, annular, and semi-spiral. In practical applications, these can further derive into spiral-symmetric, asymmetric, and quasi-symmetric forms, as shown in Figure 3.33.

3.1.3.1 Conical Suction Chamber

The conical suction chamber, as shown in Figure 3.34, features a simple structure and excellent performance. The liquid flows within a conical contraction tube where the velocity gradually increases, the distribution remains uniform, and hydraulic losses are low, ensuring a uniform velocity field at the impeller inlet. Due to structural constraints, this type of suction chamber is primarily used in single-stage overhung pumps.

The inlet diameter of the suction chamber, D_s, is typically taken as:

$$D_s = (1.0 - 1.5)\, D_j$$

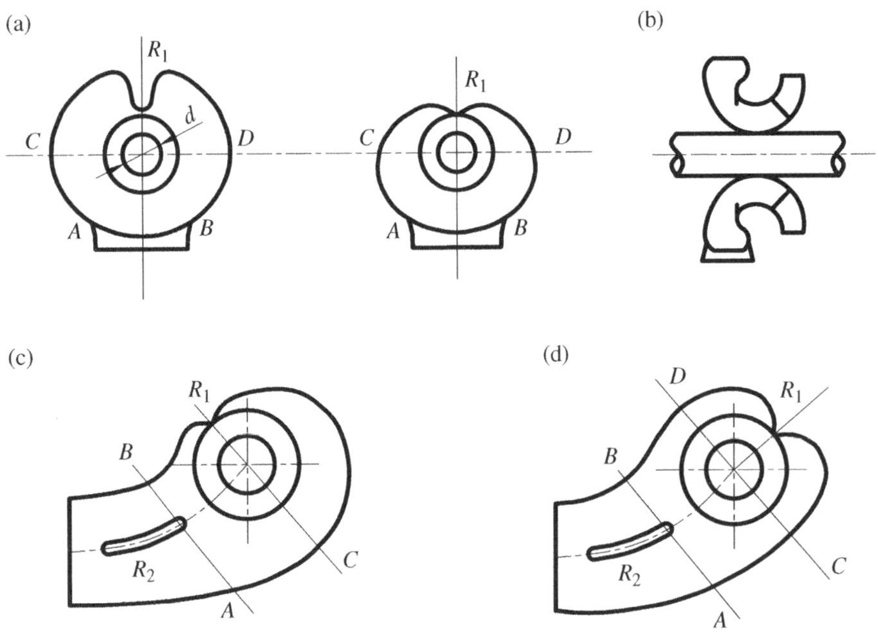

Figure 3.33 Suction chamber configurations: (a) Annular (b) spiral-symmetric (c) asymmetric (d) quasi-symmetric.

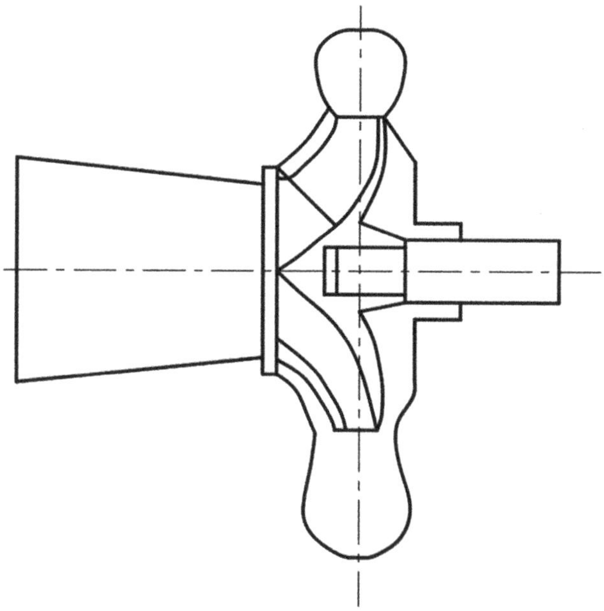

Figure 3.34 Conical suction chamber.

where

D_j is impeller inlet diameter.

The inlet diameter of the suction chamber, which corresponds to the inner diameter at the pump inlet flange, should adopt a standard diameter. Typically, the flow velocity at the suction inlet is approximately 3 m/s. The length of the conical suction chamber depends on the overall pump structure and has no specific requirement.

Adding guide vanes (with uniform height and radius) along the flow direction in the suction pipe, shaping the top into an arc, and streamlining both ends can slightly improve the pump's efficiency.

3.1.3.2 Annular Suction Chamber

The annular suction chamber features a uniform shape and cross-sectional area. Due to its simple and symmetrical structure, it is commonly used in slurry pumps and multistage pumps.

However, the annular suction chamber cannot ensure an axisymmetric and uniform velocity field at the impeller inlet. As shown in Figure 3.35, the liquid enters the annular space through a sudden expansion, then turns axially into the impeller via a sudden contraction. This process results in significant losses and highly nonuniform flow. Clearly, the flow velocity entering the impeller is higher in the lower half than in the upper half. Additionally, liquid in the upper half converges from both sides toward the center, inducing rotational motion, leading to considerable velocity nonuniformity.

Key Design Points for Annular Suction Chambers

1) The inlet of the annular suction chamber is also the pump inlet. The inlet diameter of the suction chamber should adopt a standard diameter.
2) Cross-section of the annular chamber: The 0–0 cross-section of the annular suction chamber is assumed to carry half of the total flow rate. The sizes of other cross-sections should be identical to the 0–0 section. The specific structural shape is determined based on the overall pump layout.

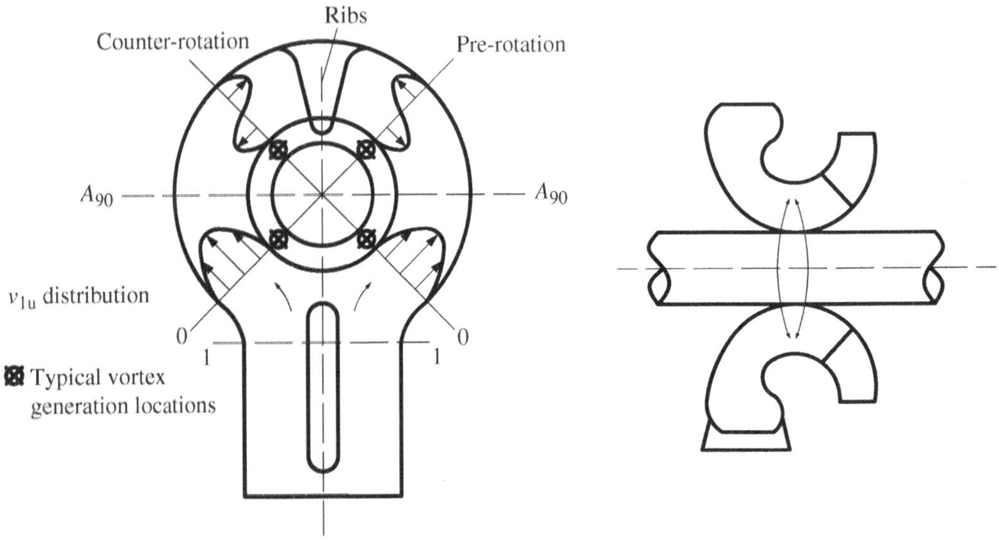

Figure 3.35 Annular suction chamber.

Adding a splitter tongue in the upper part of the annular suction chamber, as shown in Figure 3.35, allows the creation of a quasi-spiral suction chamber (intermediate between annular and semi-spiral types, see Figure 3.35), which can improve flow uniformity.

3.1.3.3 Semi-Spiral Suction Chamber

In a semi-spiral suction chamber, while liquid flows through the chamber's cross-sections, a portion of it simultaneously enters the impeller. Consequently, the cross-sectional area gradually decreases from large to small, and the outer wall forms a spiral shape. Compared to an annular suction chamber, the semi-spiral design helps improve flow conditions and ensures a uniform velocity field at the impeller inlet.

Figure 3.36a,b shows the planar view and meridional projection of a semi-spiral suction chamber, respectively. Typically, the section where the splitter tongue is located is designated as the 0-section. The shape and size of each section are illustrated in the meridional view. The outer contour in the planar view is constructed by transferring the distances from the outer wall to the axis in the meridional view to the corresponding radial lines in the planar view. The general variation of the suction chamber's cross-sectional area is shown in Figure 3.36c.

The splitter tongue is usually positioned at a 45° angle from the horizontal line. Increasing this angle to 90° generally has little impact on pump performance, but further increasing the angle degrades performance.

Figure 3.37 shows the velocity variation at different positions around the semi-spiral suction chamber, starting from the splitter tongue (0°) and moving clockwise to 360°. It can be observed that the meridional velocity v_m remains relatively uniform from 0° to 180° (Section 8). From 180° to 360°, as liquid enters directly from the suction chamber, v_m is higher but still fairly uniform. The swirl velocity v_u is blocked by the splitter tongue at 0°, resulting in a value of zero. From 0° to 180°, the influence of the splitter tongue diminishes, and v_u gradually increases. Beyond 180°, as liquid flows directly into the impeller, v_u gradually decreases. After 270°, due to the reverse rotation direction of the liquid, a negative v_u motion occurs.

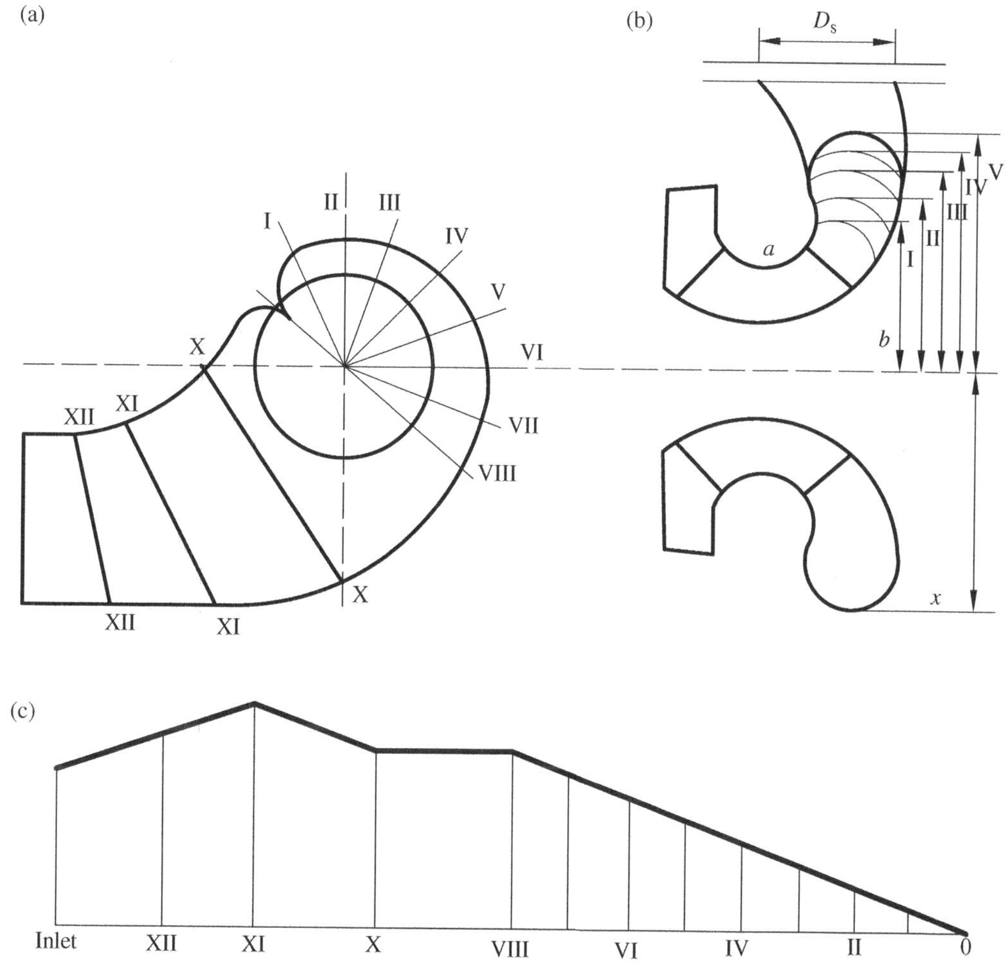

Figure 3.36 Semi-spiral suction chamber: (a) planar projection, (b) meridional projection, (c) area variation diagram.

Although v_u is not very uniform, the semi-spiral suction chamber—with its flow-stabilizing splitter tongue and ability to generate a certain degree of circulation—can achieve hydraulic performance comparable to that of a conical suction chamber if properly designed.

The design method for a semi-spiral suction chamber is as follows:

1) Determine the inlet diameter D_s.
2) Determine the average flow velocity for each section. This velocity is calculated by:

$$v = \left(0.7 - 0.85\right) v_j$$

where v_j is the flow velocity at the impeller inlet.
3) Determine the cross-sectional area of each section. Assuming half of the total flow rate $Q/2$ passes through Section VIII, the area of Section VIII is:

$$S_8 = \frac{Q}{2v}$$

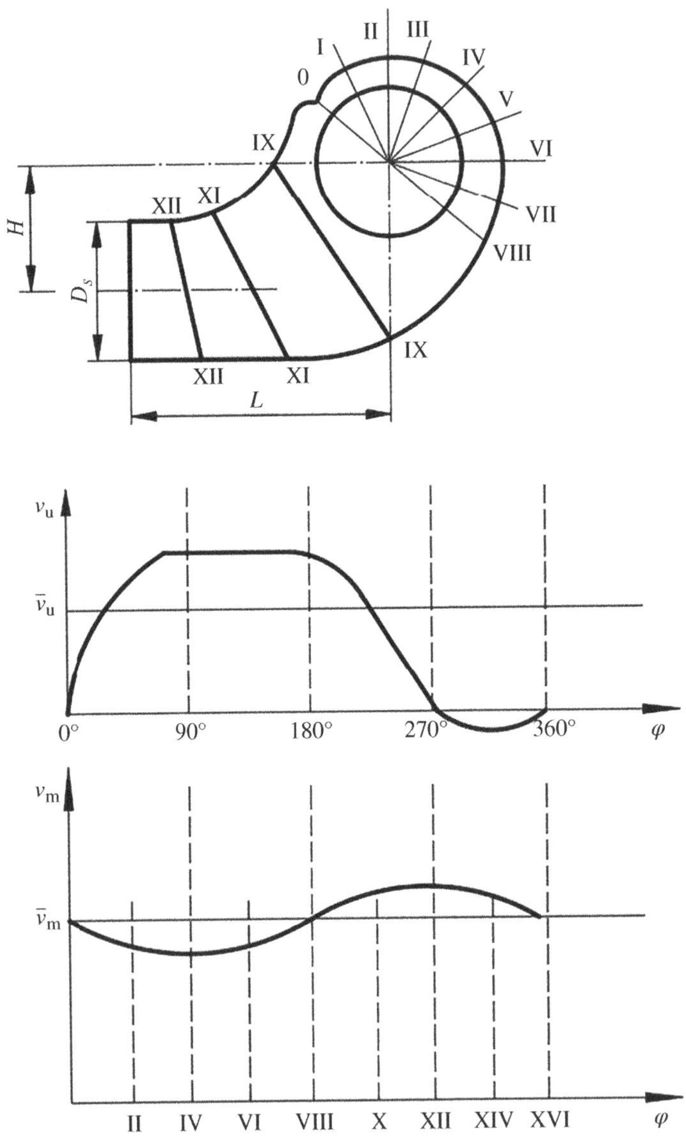

Figure 3.37 Velocity distribution in a semi-spiral suction chamber.

The areas of the other sections are scaled proportionally relative to Section VIII:

$$S_7 = \frac{7}{8}S_8 \; S_6 = \frac{6}{8}S_8 \; S_5 = \frac{5}{8}S_8 \; S_4 = \frac{4}{8}S_8$$

$$S_3 = \frac{3}{8}S_8 \; S_2 = \frac{2}{8}S_8 \; S_1 = \frac{1}{8}S_8 \text{ or } S_\varphi = \frac{\varphi}{\varphi_8}S_8$$

4) Profiling:
① Draw the meridional projection of each section. The shape is determined based on structural constraints (e.g., a wider axial dimension for single-stage pumps, or increased radial dimensions to reduce axial size in multistage pumps). Typically, Section VIII is drawn first, followed by the others. The area of each section must match the calculated value, and the shape and dimensions should vary regularly.

- The area should be measured starting from the splitter tongue (e.g., line ab in Figure 3.36), as the liquid in the annular space outside the tongue undergoes rotational motion and may be excluded from the effective area.
- The shape of the splitter tongue can be designed flexibly but should not interfere with pump manufacturing or assembly. If possible, extending it downward can help reduce the overall pump size.

② Draw the planar view. Transfer the outermost points of each section to the corresponding radial rays in the planar view. Connect these points with a smooth curve to form the contour of the spiral section.

③ Design the transition from the inlet to Section VIII. When drawing the contour, consider both the overall pump structure and flow smoothness. The intermediate sections should transition gradually from the shape of Section VIII to the circular inlet section. The drawing and verification methods are similar to those used for the discharge chamber.

3.2 Hydraulic Design of the Axial Flow Pump

3.2.1 Introduction to Axial-Flow Pump

An axial-flow pump is a kind of vane pump, whose vane unit is a series of airfoils, which form a cylindrical cascade around the hub. Since the trace of the fluid micro mass flowing through the impeller is theoretically located on the cylindrical surface concentric with the rotating axis, the outflow is considered to be along the axial direction after the guide vane racemization, so it is called an axial-flow pump. The flow rate of axial-flow pumps is relatively large, reaching up to 60 m^3/s. The single-stage head is relatively low, usually ranging from 1 to 25 m. Therefore, the specific speed of axial-flow pumps is relatively high, and their common range is 500–1600.

The advantages of axial-flow pumps are (1) The structure is simple, and the cross-sectional area (the plane perpendicular to the rotating shaft) and weight are smaller than those of other types of vane pumps under the conditions of given working parameters. (2) Whether in stop condition or running state, the flow can be easily changed by adjusting the blade angle. (3) Axial-flow pumps are usually of vertical structure, so they occupy a small area and can also be installed outdoors.

The disadvantages of axial-flow pump are (1) Limited self-priming capacity. (2) Low single-stage head. (3) The efficiency curve is steep and the high-efficiency zone is relatively narrow. If the impeller does not have a blade placement Angle adjustment device, the economic efficiency of the mixed-flow pump will be poor when it operates deviated from the design working conditions.

Axial-flow pumps are widely used in irrigation, water supply and drainage, river dredging, power engineering, and other fields where large amounts of liquid need to be transported but the head requirement is not high. They are also applied in nuclear power engineering and ship propulsion.

3.2.2 Structural Parameters and Design Theory

Axial-flow pumps are almost all single-stage. To reduce the floor space, most of them are installed vertically, but there are also horizontal and inclined shaft installation forms, which are used in situations with particularly large flow rates and very low heads. Regardless of the installation form, the flow-through components of an axial-flow pump are generally composed of the suction inlet, impeller, guide vanes, and outlet elbow, as shown in Figure 3.38.

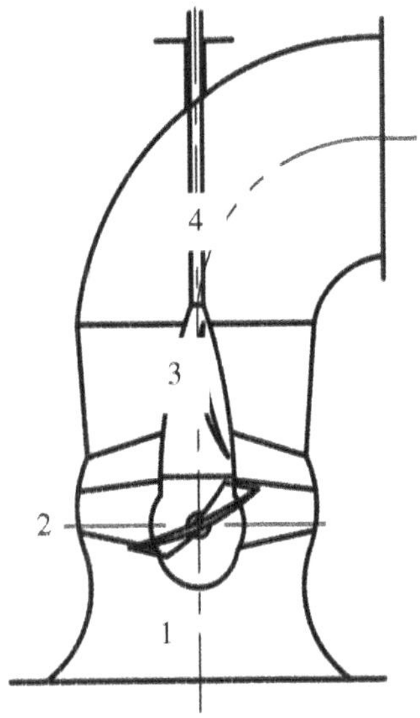

Figure 3.38 Basic parts of an axial-flow pump: 1. Suction, 2. Impeller, 3. Guide vane, 4. Elbow pipe.

3.2.2.1 Cylindrical Layer Independence Hypothesis

The flow of liquid within the impeller of an axial-flow pump is a complex spatial motion. Any kind of spatial motion can be considered as the composition of three mutually perpendicular motions. When studying the motion of water flow in an axial-flow impeller, for convenience, we adopt a cylindrical coordinate system (r, u, z), where r represents the radius direction; u represents the circumferential direction; z represents the axis direction of the pump.

The axial-flow impeller of the liquid velocity component in three axes. In the analysis and design of axial-flow pump blades, the cylindrical layer independence hypothesis is put forward, that is, liquid particles flow on the cylinder surface with the pump axis as the center line, and the movement of liquid particles on the corresponding cylinder surface is not correlated. In other words, within the impeller basin, there is no radial partial velocity. Clearly, the cylindrical surface is the flow surface. According to this assumption, the complex motion within the impeller can be simplified to the study of the flow on the cylindrical surface. Many such cylindrical flow surfaces can be made inside the impeller. The flow on each surface may vary, but the research methods are the same. Therefore, as long as the flow on one surface is thoroughly studied, the flow on the other surfaces can also be solved in a similar way.

A cylindrical surface can be unfolded on a plane after being cut along the generatrix. The cylindrical surface intersects with each blade, and its cross-section (airfoil section or airfoil) forms a set of blade cascades on the plane, as shown in Figure 3.39.

It is easy to see that this group of cascades has the following characteristics: (1) It can be expanded in the plane, that is, it belongs to the plane cascade. (2) If the cascade lines on the plane (the lines connecting the corresponding points of the airfoil in the cascade) are straight lines, it belongs to a linear cascade. (3) If the spacing between each airfoil in the cascade is equal, and the function of each airfoil in the liquid flow is the same, and no boundary airfoil can be distinguished, it belongs to an infinite cascade.

In conclusion, this set of cascades is an infinite planar linear cascade. As long as the flow of one airfoil in the flow cascade is studied, it represents the flow of the entire cascade. Therefore, the study of the flow in the impeller of the axial-flow pump is simplified to the study of the flow of the airfoil in the cascade corresponding to several cylindrical flow surfaces. When several airfoils on cylindrical flow surfaces are combined, an axial-flow pump blade is obtained.

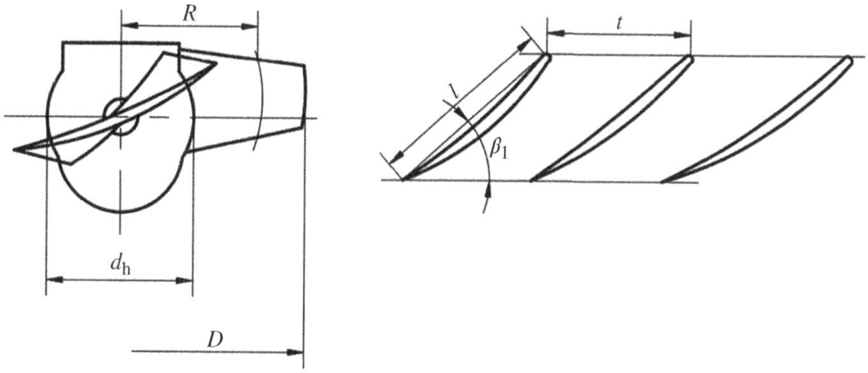

Figure 3.39 Blade cascade of axial-flow pump.

The main parameters of the cascade include: impeller diameter D, airfoil chord length l, impeller hub diameter d_h, chord placement Angle β_1, number of blades z, cascade distance t, cylindrical laminar flow surface radius R, and cascade density s (l/t).

3.2.2.2 Structural Parameters

The pump design process begins by clearly defining the design flow rate Q and head H, and determining the rotational speed n.

3.2.2.2.1 *Impeller Diameter,* D The circumferential velocity at the impeller outlet shall be calculated using Eq. (3.47).

$$u_2 = K_{u2}\sqrt{2gH} \tag{3.47}$$

where K_{u2} is the empirical coefficient, which can be calculated by the formula $K_{u2} = \frac{n_s}{584} + 0.8$. Therefore, the outer diameter $D = \frac{60u_2}{n\pi}$ of the impeller is determined.

3.2.2.2.2 *Hub-to-Tip Ratio,* R_d Moreover, the hub ratio is defined as $R_d = d_h/D$, where R_d ranges from 0.3 to 0.6 and depends on the specific speed denoted as n_s. When the specific speed is relatively high, a smaller hub ratio should be adopted to obtain a larger flow area. As the specific speed increases, a smaller hub ratio should be selected to obtain a larger flow area. For impellers with low specific speed, the hub diameter can be made larger to limit the blade angle at the hub. If the hub diameter is too small, the blade angle may exceed 45°, leading to the occurrence of flow separation.

The hub secures the blades and must satisfy structural and strength requirements for blade installation and adjustment. Hydraulically, reducing the hub ratio decreases hydraulic friction losses and increases the flow area, thereby improving suction performance. However, excessive reduction of the hub ratio increases blade distortion. During off-design operation, this can cause flow disorder, leading to secondary backflow at the impeller outlet. This backflow reduces pump efficiency and narrows the high-efficiency operating range.

The hub ratio is defined as follows:

$$R_d = -3.289\,\psi^2 + 2.522\,\psi + 0.22$$

ψ is the head coefficient, which can be calculated by the formula $\psi = gH/u^2$. The fitting formula is given by

$$\psi = -5.0162 \times 10^{-11} \times n_s^3 + 3.0466 \times 10^{-7} \times n_s^2 - 6.3231 \times 10^{-4} \times n_s + 0.4808$$

3.2.2.2.3 Cascade Density Cascade density is a key geometric parameter of an axial-flow pump impeller, directly influencing pump efficiency and governing suction performance. Reducing s decreases the total blade area and increases the blade loading (pressure difference across blades), impairing suction performance. However, it also reduces the frictional surface area, which can improve efficiency. Consequently, optimizing blade s is critical in axial-flow pump design.

When determining blade s, the following factors must be considered:

1) To ensure effective energy conversion and suction performance, blades require sufficient length to establish a well-defined flow channel, irrespective of blade count. Consequently, when determining s, the number of blades must be considered. Empirical data indicates blade tip solidity values can be guided by Table 3.10.
2) Strategically reduce s at the blade tip while increasing it at the root. This minimizes chord length disparity between inner and outer blade sections, thereby balancing outlet head distribution.

Only by ensuring constant $v_u R$ (velocity moment) at the blade outlet can radial flow be eliminated, losses reduced, and efficiency improved. However, in conventional axial-flow pump blades, the chord length near the hub is significantly shorter than at the tip, typically differing by approximately 1.7 times. The generally recommended relationship between hub cascade density and tip cascade density is $s_h = (1.20-1.25)s_t$. However, the blade sections near the hub inherently operate under adverse conditions. Combined with their short chord length, they fail to achieve energy conversion equivalent to the tip sections. This results in a nonuniform velocity moment ($v_u R$) distribution at the blade outlet, inducing radial flow. Consequently, a linear variation of s along the span is recommended, governed by $s_h = (1.3-1.4)s_t$.

3) Create radius at the leading edge of the blade tip to improve cavitation performance.

Based on the pump's NPSH formula:

$$(\text{NPSH})_r = \frac{v_1^2}{2g} + \lambda\frac{w_1^2}{2g}$$

the blade tip leading edge experiences high relative velocity (w_1), making it a primary cavitation hazard zone. Experimental observations confirm cavitation bubbles consistently initiate at this location. By radiusing the tip leading edge, incoming flow first contacts the root section. A portion of this energized fluid then migrates toward the tip, increasing the rotational velocity component and pressure of fluid approaching the tip section. This reduces the local relative velocity (w_1) and delays bubble formation, thereby enhancing cavitation resistance. Figure 3.40 compares this modified profile with conventional impeller blades. Additionally, minor trailing-edge rounding at the tip reduces chord length disparity between hub and tip sections.

3.2.2.2.4 Blade Quantity and Thickness For axial-flow pump impellers, blade count typically ranges from 2 to 6 blades, depending on specific speed. Lower specific speed designs benefit from increased blade quantity, which expands hydraulic surface area and distributes loading more evenly. Refer to Table 3.11 for empirical blade count selection guidance.

Test results confirmed the axial-flow pump's maximum head remained below the specified limit in Table 3.12.

Table 3.10 Selection of cascade density at the rim of an axial-flow pump.

z	5	4	3
s	0.84–0.94	0.75–0.85	0.65–0.75

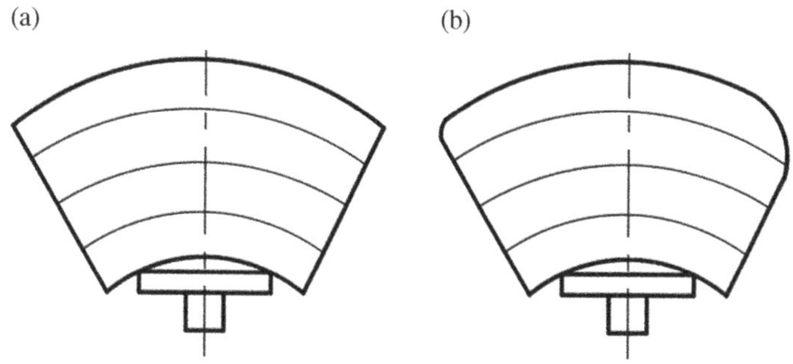

Figure 3.40 Comparison of trailing-edge profile at blade tip sections: (a) Conventional blade, (b) blade radiusing.

Table 3.11 Relationship between blade number and specific speed.

n_s	385	600	800	1000
z	6	5	4	3

Table 3.12 Number of blades vs. maximum head.

z	2	3	4	5	6
H_{max}/m	3	6	10	15	22

3.2.2.3 Design Theory

Based on the independent cylindrical layer assumption, axial-flow pump design methodologies vary according to analytical approaches for infinite cascades.

The airfoil lift design theory posits that fluid flow through an impeller cascade approximates the behavior around an isolated airfoil [11, 12], with minimal interference effects between adjacent blades. This approach treats individual airfoils within axial-flow impeller cascades as independent entities, enabling blade design based on isolated airfoil wind tunnel data. However, practical implementation requires corrections for deviations between cascade flow patterns and isolated airfoil behavior. Consequently, this methodology relies substantially on experimental validation and constitutes a semi-empirical approach.

Rooted in aerodynamic principles, the lift method necessitates empirical corrections due to its inherent limitations. Subsequently, the streamline method—an adaptation of centrifugal pumps' unidimensional design theory for axial-flow applications—evolved. This technique determines blade inlet and outlet setting angles through velocity triangle analysis. Both methodologies will be examined in detail below.

3.2.3 Design of the Axial-Flow Pump Impeller using the Airfoil Lift Method

3.2.3.1 Airfoil and Its Characteristics

As the earliest blade design methodology for axial-flow pumps, the Airfoil Lift Method incorporates fundamental airfoil aerodynamics. Key concepts are briefly introduced below.

As illustrated in Figure 3.41, the primary geometric parameters of an airfoil include:

1) Camber Line—The locus of midpoints between upper and lower surfaces.

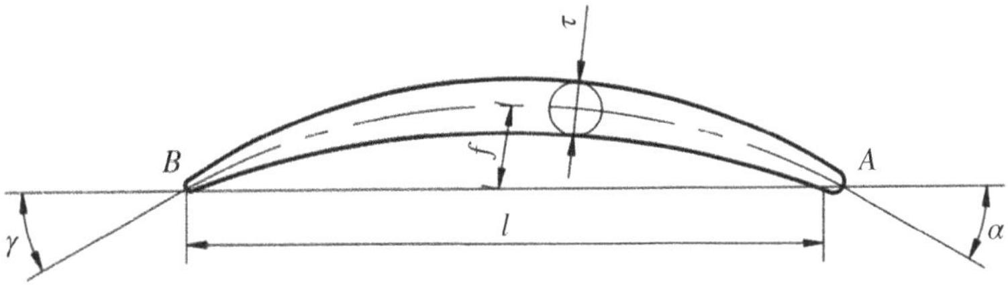

Figure 3.41 Airfoil geometric parameters.

2) Chord—The straight line connecting the leading edge (A) and trailing edge (B), with length l.
3) Thickness(τ)—Maximum distance between upper and lower surfaces measured perpendicular to the chord.
4) Camber (f)—Maximum distance between camber line and chord line.
5) Angle of Attack (α)—Angle between freestream flow direction and chord line.
6) Camber Angle (γ)—Angle between tangents to the camber line at endpoints and chord line.

When a viscous fluid flows past an isolated airfoil, resultant aerodynamic forces act upon it. As depicted in Figure 3.42, these forces can be divided into two orthogonal components: lift L (perpendicular to freestream direction) and drag D (parallel to freestream direction).

The lift L and drag D are computed via Eqs. (3.48) and (3.49).

$$L = C_{\mathrm{L}}\rho\frac{v_\infty^2}{2}A \tag{3.48}$$

$$D = C_{\mathrm{D}}\rho\frac{v_\infty^2}{2}A \tag{3.49}$$

where

v_∞—Freestream velocity;
ρ—Fluid density;
A—Maximum projected area (chord length × span);
C_{L}—Airfoil lift coefficient;
C_{D}—Airfoil drag coefficient.

The magnitudes of C_{L} and C_{D} depend on airfoil geometry, angle of attack α, and Reynolds number. These coefficients can be determined through wind tunnel testing, with results typically presented as C_{L} and C_{D} vs. α curves—known as the airfoil polar curves.

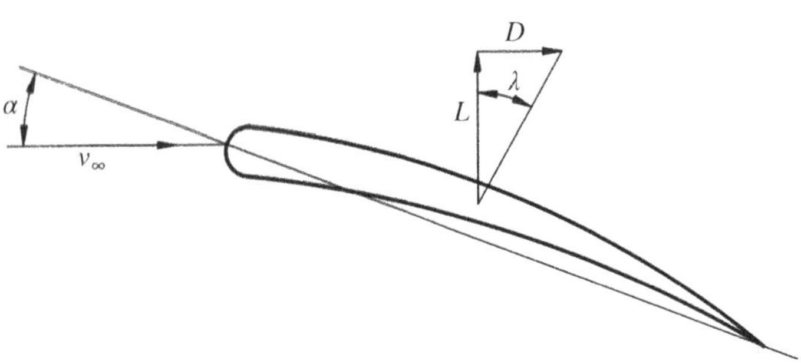

Figure 3.42 Aerodynamic forces on an airfoil.

3.2.3.2 Fundamental Governing Equation for Axial-Flow Pump Blade Design using the Airfoil Lift Method

To ensure efficient energy conversion, the pump's performance parameters—flow rate Q, head H, and rotational speed n—must exhibit deterministic relationships with blade geometric parameters.

Cascade interference effects cause airfoil characteristics in blade rows to differ significantly from isolated airfoil behavior. However, lift and drag forces within cascades remain analogous to single airfoils and are characterized by Eqs. (3.50) and (3.51).

$$L = C_L \rho \frac{w_\infty^2}{2} A \tag{3.50}$$

$$D = C_D \rho \frac{w_\infty^2}{2} A \tag{3.51}$$

where

w_∞—Freestream velocity;
C_L—Airfoil lift coefficient;
C_D—Airfoil drag coefficient.

As depicted in Figure 3.43, the flow surface between cylindrical sections at radii R and $R + dR$ is unrolled into a differential blade row of thickness dR. The resultant fluid dynamic force F acting on the cascade airfoil decomposes into lift L and drag D components. This force equals the reaction force imparted by the airfoil onto the fluid. Hence, the cascade's work done on the fluid is calculated using F.

The power transferred by the cascade to the fluid must equal the hydraulic energy gained by the fluid per unit time. For the differential blade row of thickness dR, the work imparted to the fluid

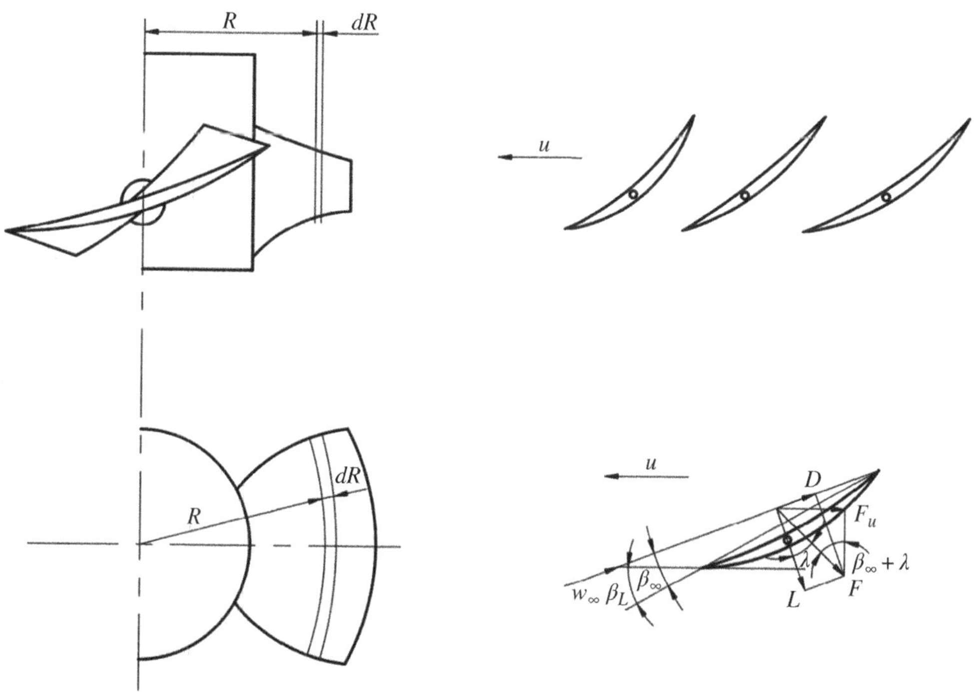

Figure 3.43 Force analysis of the cascade.

within a unit time interval is given by:

$$dP = ZdF_u u$$

The hydraulic energy acquired by fluid traversing the differential blade row per unit time is expressed as:

$$dP = \rho g H_t dQ$$

From the above two equations:

$$H_t = \frac{ZdF_u u}{\rho g dQ} \tag{3.52}$$

As illustrated in Figure 3.43, F_u represents the tangential component of the resultant force F in the circumferential direction:

$$F_u = F\cos(90^\circ - \beta_\infty - \lambda) = F\sin(\beta_\infty + \lambda)$$

$$F = \frac{L}{\cos\lambda}$$

Therefore

$$dF_u = \frac{\sin(\beta_\infty + \lambda)dL}{\cos\lambda}$$

The lift differential dL acting on the differential blade row can be expressed as:

$$dL = C_L \rho \frac{w_\infty}{2} ldR$$

The flow rate differential dQ traversing the differential blade row can be expressed as:

$$Q = 2\pi R dR v_m = Z v_m t dR$$

$$t = \frac{2\pi R}{Z}$$

Substituting the differential tangential force dF_u and flow rate dQ into Eq. (3.52) yields:

$$H_t = C_L \frac{l}{t}\frac{u}{v_m}\frac{\sin(\beta_\infty + \lambda)}{\cos\lambda}$$

Since $v_m = w_\infty \sin\beta_\infty$, the fundamental pump equation becomes:

$$H_t = \frac{u(v_{u2} - v_{u1})}{g} = \frac{u}{g}\Delta v_u$$

Substituting this relationship into the above equation yields:

$$C_L \frac{l}{t} = \frac{2\Delta v_u}{w_\infty}\frac{\sin\beta_\infty \cos\lambda}{\sin(\beta_\infty + \lambda)} \tag{3.53}$$

or

$$C_L \frac{l}{t} = \frac{2\Delta v_u}{w_\infty}\frac{1}{1 + \tan\lambda \tan\beta_\infty} \tag{3.54}$$

Equations (3.53) and (3.54) constitute the fundamental equations for designing axial-flow pump blades using the airfoil lift method. These equations establish the relationship between cascade characteristics (C_L, l/t, λ) and hydrodynamic parameters (β_∞, w_∞, Δv_u). As the pump's performance parameters are external manifestations of these hydrodynamic parameters, the essence of the fundamental equations is to relate pump performance parameters with cascade geometric parameters [3, 13].

In summary, the general procedure for blade design via the lift method comprises: Select l/t ratio and compute C_L using the fundamental equations. Modify the obtained C_L to match isolated airfoil conditions (or approximate as equal without modification). Select airfoil profile; Determine angle of attack α based on the selected airfoil's characteristics; Calculate blade setting angle $\beta_L = \beta_\infty + \alpha$. This procedure can also be executed in reverse sequence.

3.2.4 Axial-Flow Pump Impeller Design using the Streamline Theory

3.2.4.1 Governing Differential Equation for Flow at Impeller Discharge

For axial-flow pump impeller inlet and outlet, energy conservation yields (neglecting blade surface friction):

$$\frac{p_1}{\rho g} + \frac{v_1^2}{2g} + \frac{u_2 v_{u2}}{g} = \frac{p_2}{\rho g} + \frac{v_2^2}{2g}$$

From the velocity triangles, $v_1^2 = v_{m1}^2$, $v_2^2 = v_{m2}^2 + v_{u2}^2$, substitute into the preceding equation.

$$\frac{p_1}{\rho g} + \frac{v_{m1}^2}{2g} = \frac{p_2}{\rho g} + \frac{v_{m2}^2}{2g} - \frac{v_{u2}(2r_2\omega - v_{u2})}{2g}$$

Radial differentiation of the governing equation yields:

$$0 = \frac{1}{\rho}\frac{dp_2}{dr_2} + v_{m2}\frac{dv_{m2}}{dr_2} - \omega v_{u2} - \frac{dv_{u2}}{dr_2}(r_2\omega - v_{u2}) \tag{3.55}$$

The centrifugal force and fluid pressure at the impeller discharge are mutually balanced, that is:

$$\frac{dp_2}{dr_2} = \rho\frac{v_{u2}^2}{r_2} \tag{3.56}$$

Eliminating p_2 from Eqs. (3.55) and (3.56) yield:

$$v_{m2}\frac{dv_{m2}}{dr_2} = \left(\frac{dv_{u2}}{dr_2} + \frac{v_{u2}}{r_2}\right)(r_2\omega - v_{u2}) \tag{3.57}$$

Equation (3.57) constitutes the governing differential equation for the flow field at the pump discharge. Given the relationship between meridional velocity v_{m2} and tangential velocity v_{u2}, this equation can be solved through numerical integration.

3.2.4.2 Free Vortex and Forced Vortex

1) The discharge flow exhibits free vortex characteristics.

$$v_{u2}r_2 = \text{constant}$$

Apply differentiation to r_2 on both sides of this equation:

$$dv_{u2}r_2 + v_{u2}dr_2 = 0$$

Substituting this relationship into Eq. (3.55) yields:

$$v_{m2}\frac{dv_{m2}}{dr_2} = \left(-\frac{v_{u2}}{r_2} + \frac{v_{u2}}{r_2}\right)(r_2\omega - v_{u2}) = 0$$

Since $v_{m2} \neq 0$, the meridional velocity gradient $\frac{dv_{m2}}{dr_2}$ exists. Therefore:

$$v_{m2} = \text{constant}$$

Since $v_u r$ = constant, therefore

$$H_{t\infty} = \frac{u_2 v_{u2}}{g} = \frac{\omega v_{u2} r_2}{g} = \text{constant}$$

Under free vortex conditions, the Euler head remains radially invariant at the discharge while meridional velocity maintains uniform distribution.

2) The discharge flow exhibits forced vortex characteristics.

$$\frac{v_{u2}}{r_2} = \text{constant}$$

Apply differentiation to r_2 on both sides of this equation:

$$\frac{dv_{u2}}{r_2} - \frac{v_{u2}}{r_2^2} dr_2 = 0$$

Substituting this relationship into Eq. (3.57) yields

$$v_{m2}\frac{dv_{m2}}{dr_2} = 2C_1(r_2\omega - v_{u2}) = 2C_1 r_2\left(\omega - \frac{v_{u2}}{r_2}\right) = C_2 r_2$$

By defining $2C_1\left(\omega - \frac{v_{u2}}{r_2}\right) = C_2$, then

$$\int v_{m2}\, dc_{m2} = C_2 \int r_2 dr_2 + C_3$$

$$\frac{v_{m2}^2}{2} = \frac{C_2}{2} r_2^2 + C_3$$

$$c_{m2} = \sqrt{C_2 r_2 + C_3}$$

$$H_{t\infty} = \frac{u_2 v_{u2}}{g} = \frac{\omega v_{u2} r_2}{g} = \frac{C_1 \omega r_2^2}{g}$$

Under forced vortex conditions, the Euler head exhibits a nonuniform radial distribution at the impeller discharge, scaling proportionally to $r_2$2. This results in a progressive increase from hub to tip. Concurrently, meridional velocity rises with r_2, yielding lower values at the hub and higher values at the periphery.

Compared with free vortex designs, forced vortex blades demonstrate: larger blade angles at the hub; Smaller blade angles at the tip. This configuration induces significantly greater blade twist deformation.

3.2.4.3 Circulation Correction and Streamline-Based Design Methodology

Both vortex patterns governing impeller discharge flow exhibit inherent limitations. For free vortex designs: relative flow angles decrease radially outward (minimum at tip); blade twist angles $\Delta\beta = \beta_h - \beta_t$ show a significant gradient; pump efficiency deteriorates rapidly under off-design conditions; high-efficiency operating range narrows substantially. Conversely, forced vortex designs yield: elevated tangential velocity v_u at blade tip; reduced v_u at hub; resulting meridional velocity v_m maldistribution: accelerated flow at periphery—decelerated flow near hub. This nonuniform velocity profile fundamentally constrains achievable pump performance.

Given the limitations inherent in both vortex theories, a modified circulation distribution law is proposed based on statistical analysis of high-performance pump prototypes. This refined methodology employs streamline-based design for axial-flow pump blades. The circulation correction principle follows:

1) Effects of tip clearance and hub interface: The blade hub rotates with the impeller, imparting swirl to adjacent fluid and augmenting tangential velocity components near the hub. Consequently, reduced tangential velocity (v_u) must be enforced in this region during design.

The stationary pump casing induces tip clearance flows that reduce liquid swirl. Additionally, leakage reflux toward the inlet and suction surface backflows diminish the pressure differential between the pressure and suction surfaces of blades. This collectively reduces the outlet head. Consequently, blade design must incorporate increased circulation distribution at the tip region.

2) Tip-region airfoils typically exhibit thin profiles, minimal curvature, and near-straight camber lines. Conventional design approaches yield extremely low -occasionally negative-angles of attack at this location, resulting in deficient lift generation capacity. This aerodynamic underperformance fundamentally contradicts the critical role of tip sections. Conversely, hub-region airfoils feature substantial thickness and pronounced camber. Standard design practice imposes high incidence angles, causing severe blade twist. While performance remains acceptable at design conditions, efficiency deteriorates rapidly during off-design operation, yielding an impractically narrow high-efficiency operating range.
3) Within axial-flow pump impellers, hydraulic losses at constant meridional velocity (v_m) can be minimized by increasing the relative flow angle β_∞ and consequently reducing relative velocity ω_∞. This dual effect enhances hydraulic efficiency because the majority of throughflow passes through the tip region, and a strategic increase of β_∞ at the blade periphery delivers significant efficiency gains.
4) Experimental evidence indicates that increasing the stagger angle of tip-region airfoils does not significantly impair impeller cavitation performance. This phenomenon arises from interdependent mechanisms: tip prerotation necessitates larger blade inlet angles for flow accommodation; enhanced inlet angles: increase blade inlet flow area; reduce inlet meridional velocity (v_{m1}); consequently, decrease the relative inlet velocity (ω_0); the NPSH equation $(\text{NPSH})_r = \frac{v_1^2}{2g} + \lambda\frac{w_1^2}{2g}$ demonstrates that increasing either the blade inlet angle or specifically at the tip region improves cavitation resistance performance.

In conclusion, statistical analysis and optimization of test data from various specific-speed pumps with differing outlet circulation distributions yield a refined correction methodology. This approach modifies the free-vortex-based circulation distribution by progressively increasing circulation at the tip while reducing it at the hub. The correction coefficient varies linearly according to:

$$v_{u2} = \xi v'_{u2}$$

where

v'_{u2} —calculated by tangential velocity component $v'_{u2}\,r = \text{constant}$
v_{u2}—the revised tangential velocity component
ξ—the correction coefficient of v'_{u2}, adopts a coefficient of 0.9 at the hub and 1.1 at the tip, with a linearly distributed variation spanning these endpoints.

3.2.4.4 Blade Inlet Incidence Angle

1) Blade Inlet Prerotation
Experimental measurements confirm significant prerotation at the inlet of axial-flow pump blades. This phenomenon arises due to blade-induced flow effects, where the rotating blades impart angular momentum to the inlet fluid. The degree of prerotation increases radially, peaking at the tip region, thereby influencing the pump's inlet flow conditions. As illustrated in

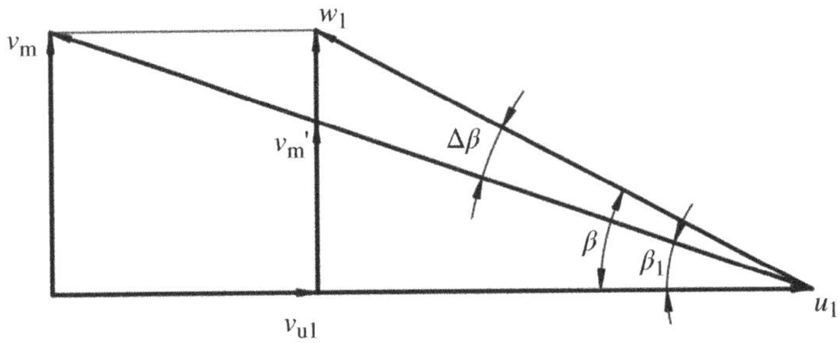

Figure 3.44 Inlet velocity triangle at the blade tip section.

Figure 3.44, the inlet velocity triangle at the blade tip region demonstrates the following relationships: The presence of tangential velocity v_{u1} causes the fluid to follow the path of least resistance (aligned with the blade inlet angle), thereby reducing the meridional velocity from v_m to v'_m. To maintain the original v_m (Q) while preserving the prerotation v_{u1}, the blade angle must be increased from β_1 to β, introducing an incidence angle increment $\Delta\beta$. The adjusted blade angle equals the vector sum of the flow angle and incidence angle: $\beta = \beta_1 + \Delta\beta$.

2) Principles for Incidence Angle Selection

The pump's final operating condition is governed by blade geometry, rendering incidence angle determination critically significant. Selection follows these principles:

1) Address limitations in free-vortex and forced-vortex designs by reducing blade angles at the hub and increasing blade angles at the tip
2) Tip clearance counteracts reduced tangential velocity (v_u drop) from tip leakage flows by: Increasing tip blade angles to augment work input and implementing spanwise-varying incidence with recommended values from 0° to 3°. Higher specific speeds require lower values within this range.

3.2.4.5 Evolution of Profile Radius and Airfoil Thickness Distribution

3.2.4.5.1 Profile Radius **R** Profile geometry specifications: The blade profile contour requires a smooth continuous curve. Common configurations include a circular arc or parabolic curve, with the single-arc geometry illustrated in Figure 3.45.

For circular-arc blade profiles, the following geometric angular relationships:

$$\beta_L = \beta_1 + \gamma,\ \ \beta_L = \beta_2 - \gamma,\ \ \gamma = \theta/2,\ \ \beta_L = (\beta_1 + \beta_2)/2,\ \ \beta_2 - \beta_1 = \theta$$

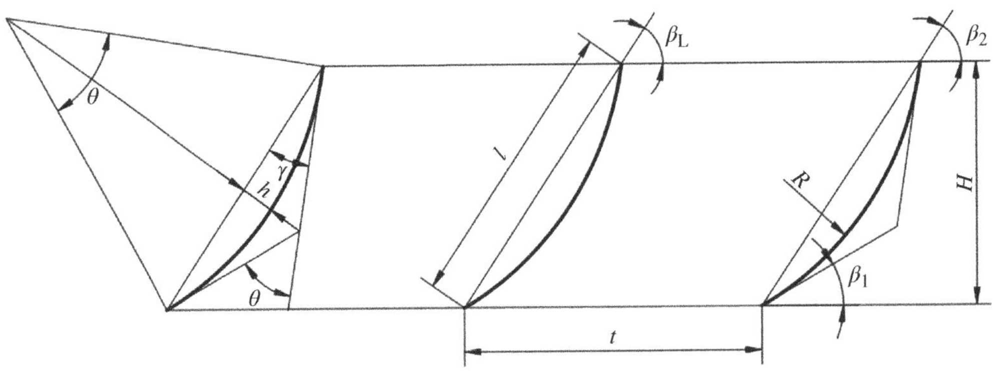

Figure 3.45 Profile geometry specifications.

Height of the profile H:

$$H = l \sin \beta_L = l \sin \frac{\beta_1 + \beta_2}{2}$$

Camber of the profile h:

$$h = R - (R - h) = \frac{l}{2\sin\gamma} - \frac{l}{2\tan\gamma} = \frac{l}{2}\left(\frac{1-\cos\gamma}{\sin\gamma}\right)$$

Radius of the profile R:

$$R = \frac{l}{2\sin\frac{\theta}{2}} = \frac{l}{2\sin\left(\frac{\beta_2-\beta_1}{2}\right)} = \frac{H}{\cos\beta_1 - \cos\beta_2}$$

As Figure 3.45 shows:

$$R^2 = \left(\frac{l}{2}\right)^2 + (R-h)^2, R = \frac{1}{8}\left(\frac{l^2}{h}\right) + \frac{h}{2}$$

The blade profile geometry may alternatively be constructed along any smooth continuous curve, provided the specified inlet and outlet blade angles are maintained.

3.2.4.5.2 Spanwise Thickness Distribution Characteristics Blade thickening may employ any aerodynamic thickness distribution profile. This section introduces a hydrodynamic airfoil design, with its thickness distribution characteristics detailed in Figure 3.46 and Table 3.13.

During blade thickening operations, maintain the pressure-side profile as the datum surface while adding material exclusively to the suction side.

3.2.4.6 Blade Contouring Procedure

1) Establish the chord line of length l at the calculated placement angle β_L for each blade section. Construct the camber line using the specified profile radius (geometric parameters detailed in Figure 3.47). Generate the complete airfoil profile by applying the thickness distribution normal to the camber line.
2) Determination of the airfoil pivot point location
   1) Draw the outer edge airfoil profile.
   2) Locate the pivot axis projection by constructing a normal line to the chord at 40–50% chord length from the leading edge within the blade development diagram.
   3) At the intersection of the pressure-side contour and the chord-normal line, construct a transverse axis representing the blade pivot axis in the meridional development plane.

Figure 3.46 A certain airfoil thickness distribution.

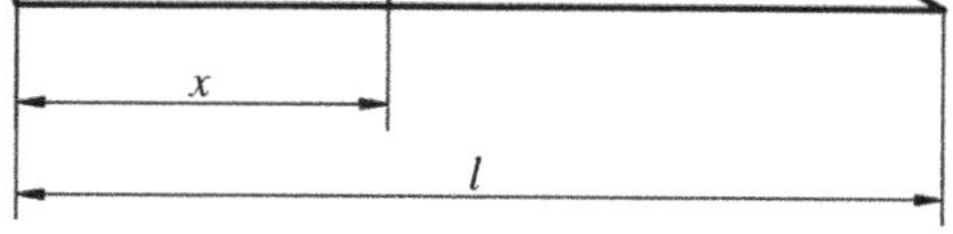

Table 3.13 Characteristics of a certain airfoil thickness distribution.

x/l	0	0.1	0.2	0.3	0.4	0.5	0.6	0.7	0.8	0.9	1
δ/δ_{max}	0	0.489	0.778	0.920	0.978	1	0.883	0.756	0.544	0.356	0

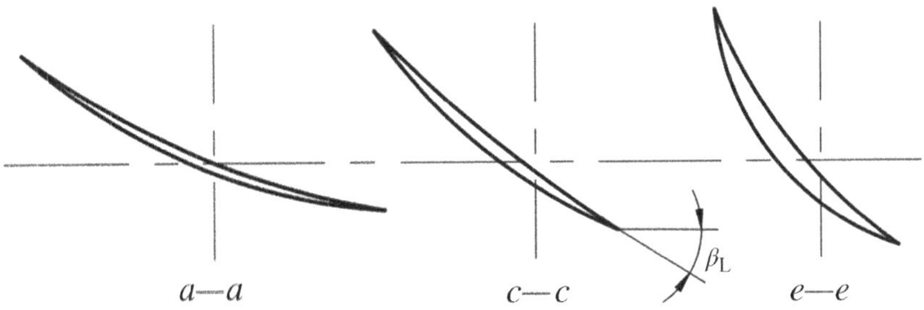

Figure 3.47 Pivot point location for airfoil profiling.

The profiling methodology for nontip blade sections replicates the procedure defined for the tip section (refer to Figure 3.47). However, the pivot axis position exhibits radial progression: its axial offset increases systematically from tip to hub, which induces a blade surface inclination from tip toward hub, suppressing radial flow.

3) Construct circumferential development and meridional projections of the blade geometry

As shown in Figure 3.48, Plot arcs representing hub, tip, and intermediate cross-sections and construct radial rays at uniform angular increments $\Delta\theta$. Then, draw chord-normal reference lines parallel to the meridional axis and set inter-line spacing equal to circumferential arc length l. Establish horizontal lines at uniform axial intervals ΔZ relative to the meridional plane axis

Construct axial-direction reference lines parallel to the meridional datum at uniform intervals ΔZ, symmetrically distributed on both sides of the axial projection axis.

4) Construct the leading (inlet) and trailing (outlet) edges

Map leading/trailing edge endpoints from each airfoil development diagram to corresponding angular positions in the circumferential view. Construct continuous edge contours through all mapped points.

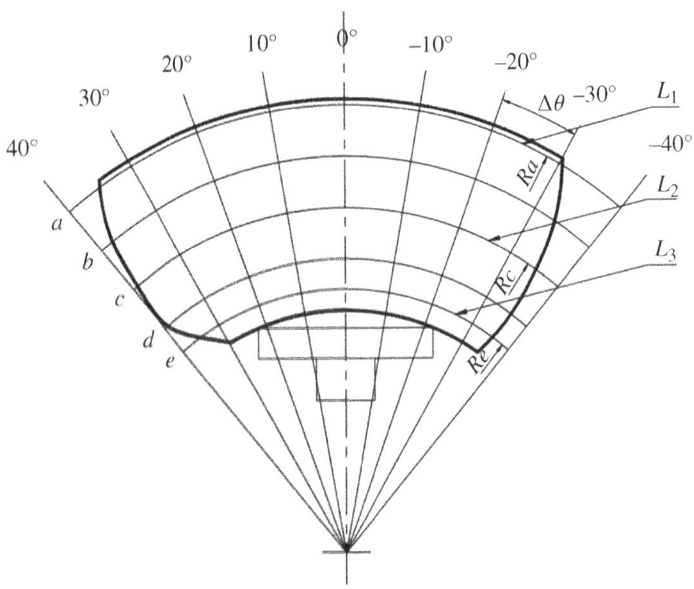

Figure 3.48 Circumferential and meridional projections layout.

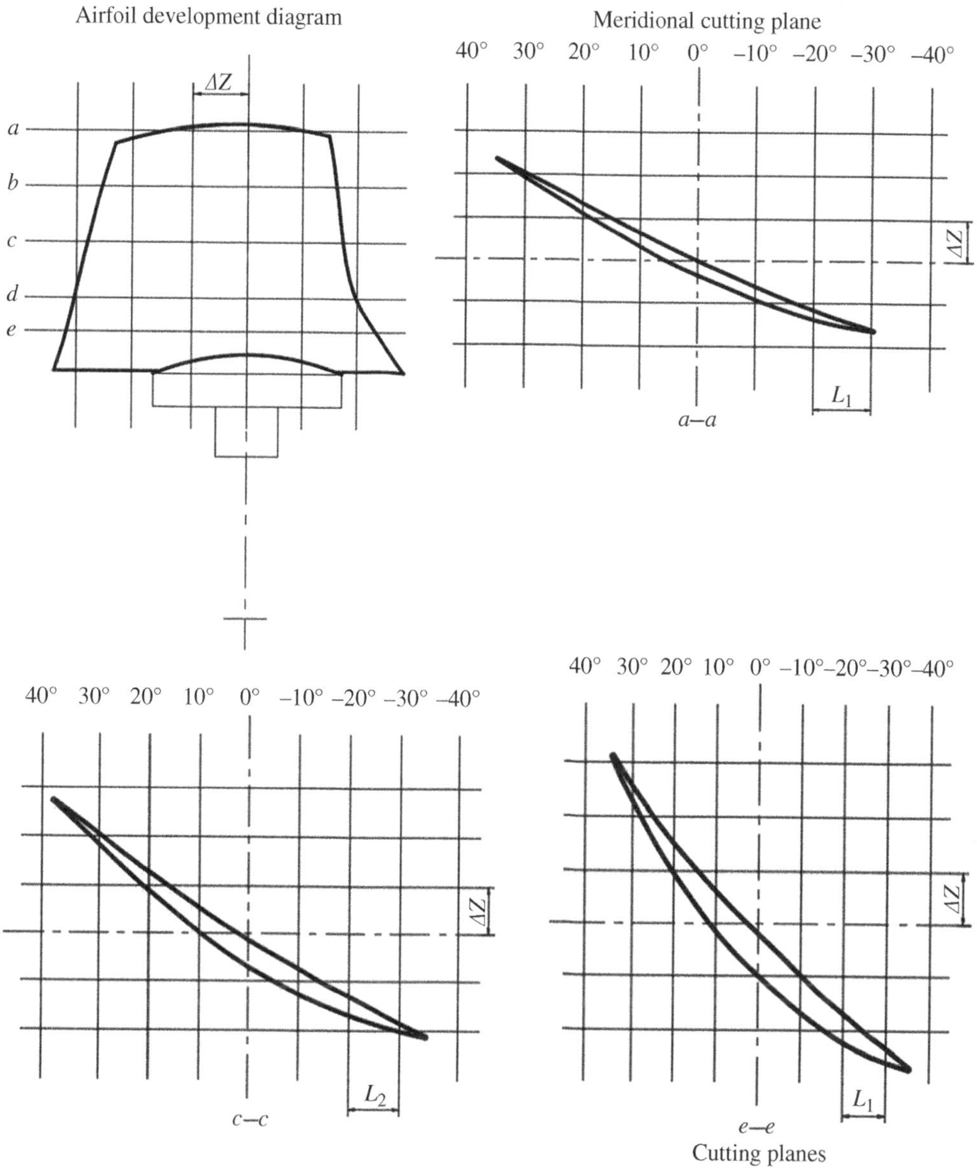

Figure 3.48 *(Continued)*

Project axial distances from the airfoil development diagram to the meridional projection. Generate continuous axial projection lines for both edges.

If the inlet and outlet edges in the plane projection diagram and the axial projection diagram are not smooth, the chord length l of the airfoil or the position of the rotation center should be adjusted to make the adjustment.

5) Surface smoothness validation

In the airfoil development diagram, extract intersection points between designated horizontal cutting planes (denoted 1, 2, 3, …) and the pressure surface. Map these points onto their corresponding angular positions in the circumferential projection view. The resulting curves (labeled 1–1, 2–2, 3–3, …) must exhibit continuity across all sections shown in Figure 3.49.

For each meridional cutting plane (at specified angles −10°, 0°, 10°...), trace intersection points between these vertical planes and the suction surface. Project these points onto the meridional plane while maintaining strict axial position references. The generated contours (demonstrated for 30°, 0°, −30° cases) are shown in Figure 3.49, the meridional cutting plane.

6) Determination of impeller rotation direction

In the circumferential blade layout, the conventional orientation positions the blade crown upward with the axis positioned downward. The configuration presents the inlet edge on the right side and outlet edge on the left, where the left-side outlet edge maintains a higher elevation than the right-side inlet edge. When observing from the outlet perspective, the visible blade surface corresponds to the pressure side (concave surface). This geometric arrangement generates

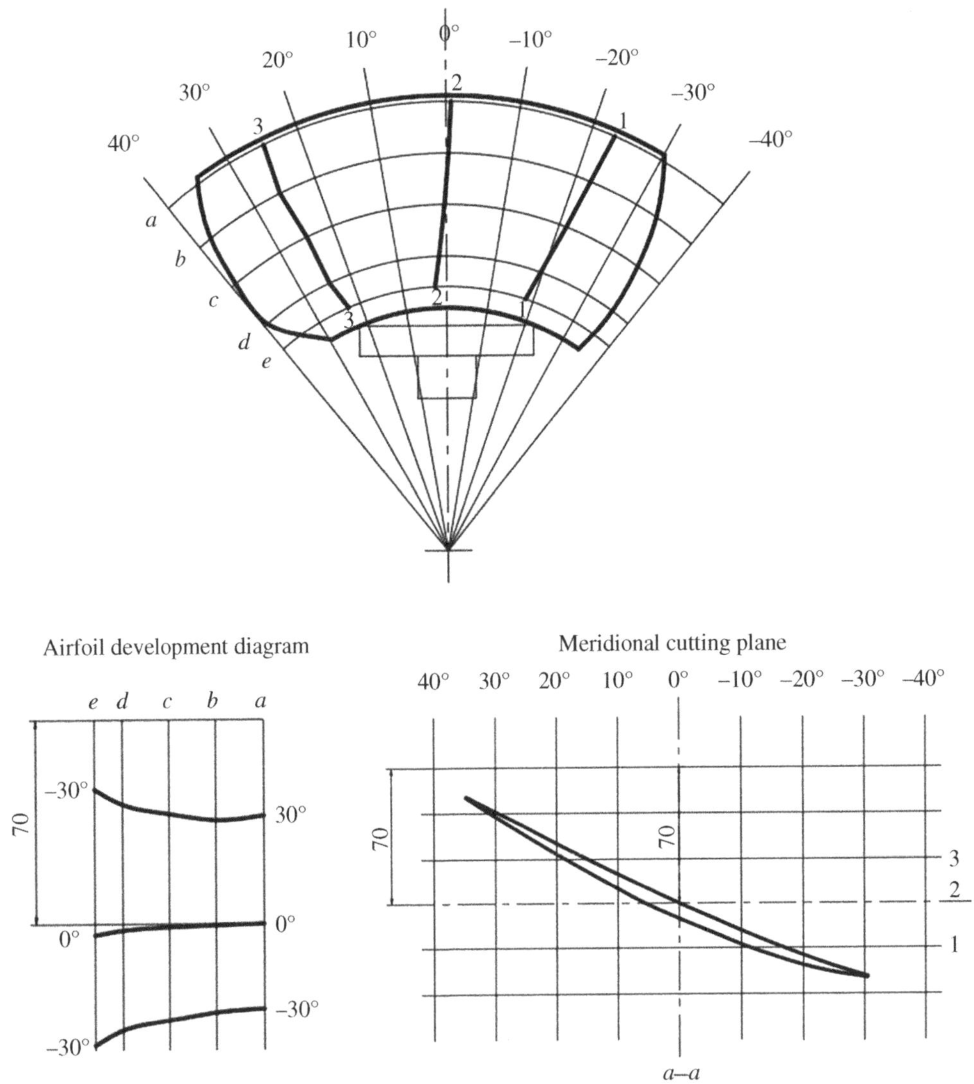

Figure 3.49 Layout check of circumferential and meridional projections.

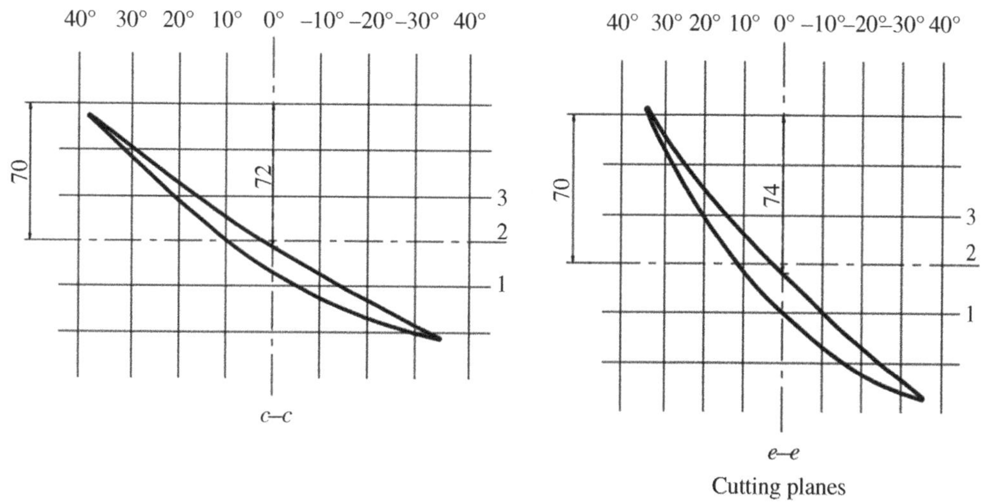

Figure 3.49 (*Continued*)

hydrodynamic forces that induce clockwise rotation when viewed from the outlet, resulting in counterclockwise motion from the inlet viewpoint. Reversing this blade positioning configuration will conversely produce counterclockwise rotation when observed from the outlet direction. The rotational behavior directly correlates with the relative elevation between the inlet and outlet edges and the exposure of the pressure surface.

3.2.5 Axial-Flow Pump Guide Vane Design

The guide vane system in axial-flow pumps serves dual functions: eliminating fluid rotation and converting kinetic energy into pressure energy [4, 14]. The design of guide vanes can be approached using the streamline method, lift method, or arc method [11, 15].

3.2.5.1 Guide Vane Structural Parameter Selection

Standard practice integrates guide vanes with the diffuser tube to minimize pump axial length, forming a composite guide vane body. As shown in Figure 3.50, key structural parameters must

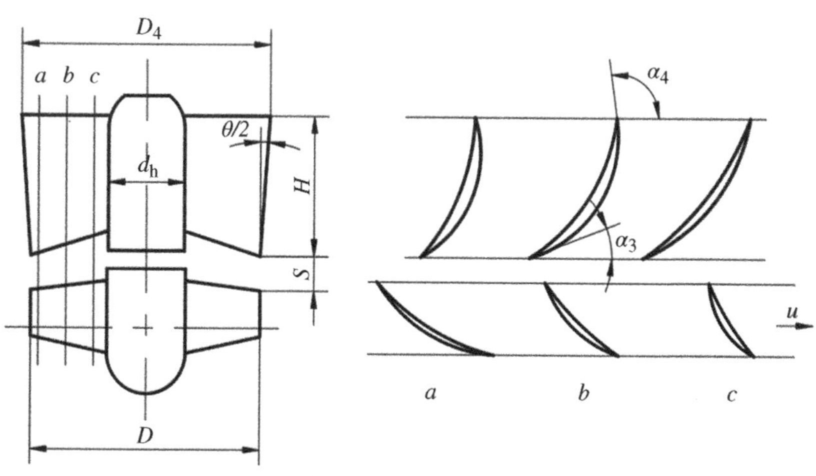

Figure 3.50 Guide vane structural parameters.

coordinate with both impeller chamber and discharge pipe configurations. The guide vane body typically features a 6°–10° diffusion angle, with its inlet edge aligned parallel to the impeller blade outlet edge, maintaining a spacing S between (0.1–0.15) D.

The blade count Z generally ranges from 5 to 10, with higher specific speed pumps requiring fewer blades. To mitigate pressure pulsations, the guide vane quantity should avoid integer multiples of the impeller blade count. The discharge section connects to an elbow pipe, preferably with a constant cross-section or variable-curvature contracting design to minimize hydraulic losses. Standard pipe diameters should be selected for the outlet diameter D_4.

Guide vane axial height H correlates with cascade solidity (l/t) and blade count. Engineering experience demonstrates that optimal performance can be achieved through increased blade count with correspondingly reduced guide vane length.

3.2.5.2 Streamline Method for Axial-Flow Pump Guide Vane Design

Three primary methodologies exist for guide vane design: the lift method, integration method (arc approach), and the more widely adopted streamline method.

The streamline-based design process involves the following systematic approach:

1) Meridional view construction using predetermined structural parameters, develop the axial projection of the guide vane configuration.
2) Stream surface discretization

Accounting for the tapered geometry of guide vanes, the actual flow surface forms a conical section, with its development representing a sector plane. For computational efficiency, this is approximated as a cylindrical surface with acceptable engineering tolerance. Standard practice maintains consistent flow surface radii between guide vanes and corresponding impeller blades.

3) Flow angle determination: Calculate the installation angles (α_3 and α_4) at the guide vane inlet and outlet for each flow surface:

3.2.5.2.1 Inlet Flow Angle Analysis Through velocity triangle construction (Figure 3.51), the inlet liquid flow angle is given by:

$$\tan \alpha_3' = \frac{v_{m3}}{v_{u3}\psi_3}$$

$$v_{m3} = \frac{4Q}{\pi\left(D_3^2 - d_h^2\right)}$$

$$\psi_3 = \frac{t_3 - S_{u3}}{t_3} = 1 - \frac{ZS_{u3}}{\pi D_3}$$

$$S_{u3} = \frac{S_3}{\sin \alpha_3}$$

where

v_{m3}—Meridional velocity at guide vane inlet;
D_3—Inlet diameter of flow surface;
d_h—Hub diameter;
ψ_3—Inlet blade blockage coefficient;
S_{u3}—Circumferential thickness at inlet;
S_3—Normal thickness at inlet (≈ actual blade thickness);
α_3—Inlet installation angle.

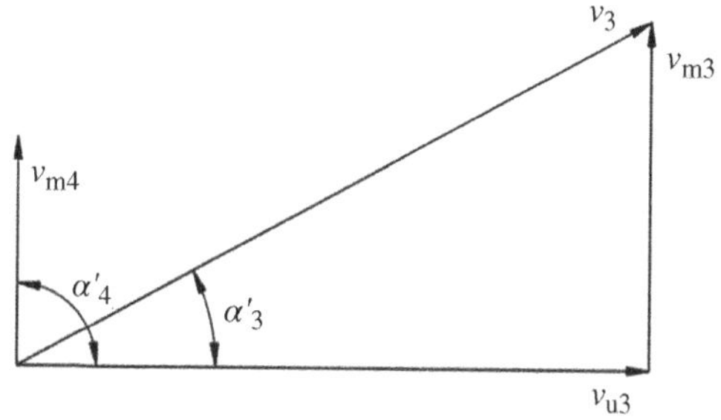

Figure 3.51 Guide vane inlet velocity triangle.

3.2.5.2.2 ***Inlet Installation Angle α_3*** Inlet installation angle

$$\alpha_3 = \alpha_3' + \Delta\alpha$$

The deviation angle $\Delta\alpha$ (impulse angle) should be maintained within 0°–5° for optimal performance. Standard engineering practice recommends eliminating this angular deviation ($\Delta\alpha = 0°$) in most design configurations.

The near-uniform meridional velocity (v_m) at guide vane inlet permits initial determination of the blade inlet angle (α_3) using the mean streamline, with subsequent flow lines derived from $D\times\tan\alpha$ = constant condition.

An iterative computational approach is required since α_3 is initially unknown: first, assume a blade blockage coefficient (ψ_3) to compute preliminary angle α_3', then recalculate ψ_3 using the derived α_3. This iteration continues until convergence between assumed and calculated ψ_3 values is achieved. For most engineering applications, the exact calculation of the inlet angle proves unnecessary given its negligible influence on overall hydraulic performance.

3.2.5.2.3 ***Blade Outlet Angle α_4*** Accounting for finite blade number effects, the outlet angle α_4 must exceed 90° to ensure proper fluid discharge. Current engineering practice adopts values between 80° and 90°, though conventional designs typically maintain α_4 at or slightly below 90°.

3.2.5.2.4 ***Cascade Solidity (l/t)*** The cascade solidity parameter l/t governs the inter-blade flow passage diffusion characteristics. As illustrated in Figure 3.52, the inlet passage width measures $t\sin\alpha_3$ while the outlet passage width equals t, with the flow channel axial length approximated by l. The resulting passage diffusion angle θ thus satisfies the geometric relationship:

$$\tan\frac{\varepsilon}{2} = \frac{t - t\sin\alpha_3}{2l}$$

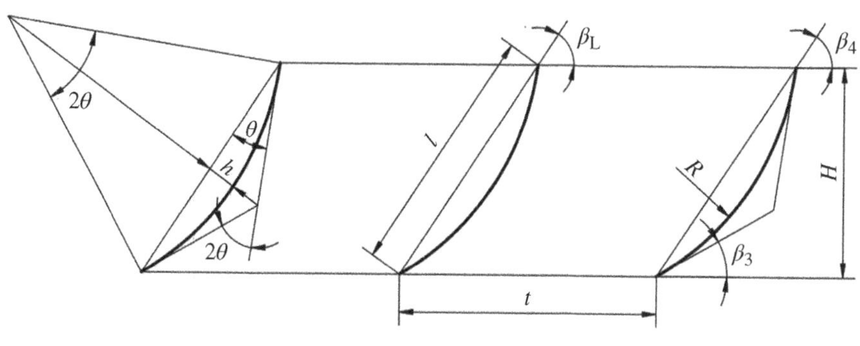

Figure 3.52 Profile of the blade of an axial flow pump.

The diffusion angle ε must be maintained within 6° to 10°, with an absolute maximum limit of 12°. Standard design practice involves first selecting the l/t ratio from relevant technical references, followed by mandatory verification that the resulting diffusion angle falls within this specified optimal range.

3.2.5.2.5 *Vane Height and Profile Radius* Vane height H

$$H = l \sin \beta_L = l \sin \left(\frac{\alpha_3 + \alpha_4}{2} \right)$$

Profile radius R

$$R = \frac{l}{2 \sin \theta} = \frac{l}{2 \sin \left(\frac{\alpha_4 - \alpha_3}{2} \right)} = \frac{H}{\cos \alpha_3 - \cos \alpha_4}$$

or

$$R = \frac{1}{8} \left(\frac{l^2}{h} \right) + \frac{h}{2}$$

The blade profile may alternatively be constructed using any smooth curvature, provided the specified inlet (α_3) and outlet (α_4) angles are maintained.

3.2.5.2.6 *Airfoil Thickness Optimization* The thickness distribution pattern should be uniformly applied across all vanes, with the profile serving as the pressure surface (concave side) while thickening the suction surface. For compact pump designs, constant-thickness blades represent a viable alternative. Blade thickness should be minimized to the greatest extent permitted by manufacturing capabilities and structural integrity requirements. Special attention must be given to aerodynamic refinement—the leading edge should maintain streamlined contours while the trailing edge terminates in a sharp profile.

3.2.5.2.7 *Guide Vane Drawing Conventions* These follow identical drafting methodologies and procedural steps as employed for impeller blade documentation, thus requiring no additional elaboration. The drawings must clearly illustrate the three-dimensional blade geometry through orthogonal projections while maintaining dimensional consistency across all views.

3.3 Hydraulic Design of the Volute Casing

The volute casing, positioned at the outlet of the impeller or between successive impeller stages, serves as a highly critical flow passage component. The efficiency of the volute casing itself and its compatibility with the impeller directly influence the overall efficiency of the water pump. Its primary functions include:

1) Collect the liquid flowing out from the impeller and convey it to the discharge outlet or the suction inlet of the next-stage impeller;
2) Reduce the velocity of the liquid flow, enabling the conversion of kinetic energy into pressure energy;
3) Eliminate the rotational motion of the liquid exiting the impeller to prevent hydraulic losses caused thereby.

During the design process, it is essential to ensure that the geometric profile of the volute casing conforms to the liquid's flow patterns, enabling the volute casing to achieve high efficiency. Meanwhile, it is crucial to guarantee that the flow exiting the impeller is axially symmetric, thereby ensuring stable relative motion within the impeller and minimizing hydraulic losses inside it. The structural forms of volute casings can be classified into spiral volute casings, annular volute casings, diffuser vanes, and so on.

1) The spiral volute casing is often referred to as a scroll casing or volute chamber. From a hydraulic perspective, the flow within a spiral volute casing is relatively ideal, demonstrating strong adaptability and a wide range of high efficiency. However, the flow passages cannot be machined mechanically, and their dimensions, shapes, and surface roughness are directly ensured through casting.
2) The annular volute casing is a special case of the spiral volute casing, featuring a circular (annular) shape. It exhibits relatively poor hydraulic performance and is generally used in pumps with lower specific speeds or for special applications, such as sewage pumps and slurry pumps.
3) Diffuser vanes include radial diffuser vanes, channel-type diffuser vanes, and spatial diffuser vanes, among others. They are primarily employed in multistage pumps and pumps with specific requirements for the outer diameter, such as well pumps.

3.3.1 Hydraulic Design Principles of the Volute Casing

If, after flowing out of the impeller, the liquid enters the space between two parallel flat plates and the viscous frictional force between the liquid and the plates is neglected, this liquid flow is not subjected to any external forces and adheres to the principle of conservation of velocity moment, namely, $v_u R = K_2$ (a constant). To maintain favorable hydraulic performance, the contour profile of the volute casing should align with the trajectory of the water flow [16, 17].

From Figure 3.53, we obtain:

$$\tan\alpha = \frac{v_m}{v_u} = \frac{\dfrac{Q}{2\pi R b}}{\dfrac{K_2}{R}} = \frac{Q}{2\pi b K_2} = \text{constant}$$

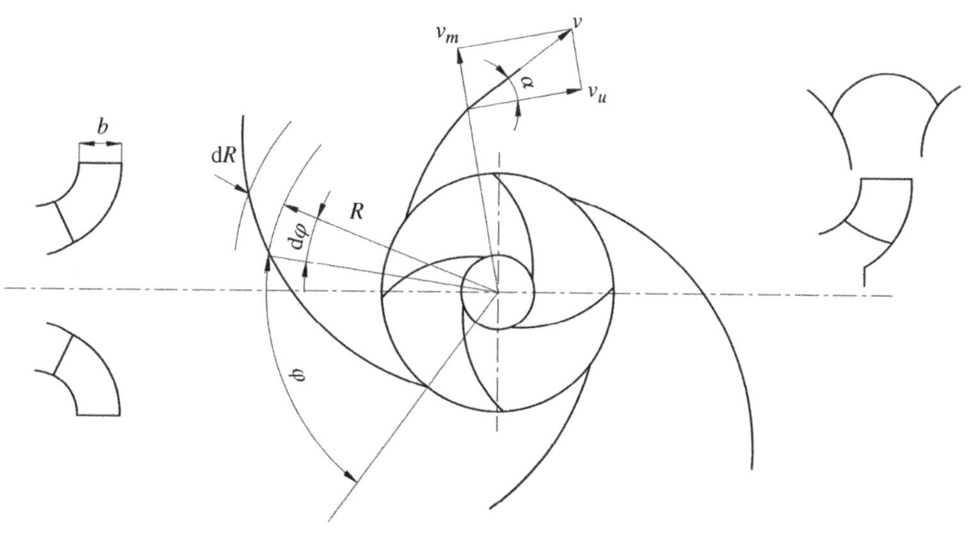

Figure 3.53 Operating principle of the volute casing.

Since Q, b, K_2 are constants, the flow angle of the liquid remains unchanged. That is, the trajectory of the liquid after it flows out of the impeller forms a logarithmic spiral, with the angle between the liquid's flow direction and the circumferential direction remaining constant. This is the origin of the name "volute casing."

Referring to Figure 3.53, the coordinates of any point on the spiral can be expressed as

$$\tan \alpha = \frac{dR}{Rd\varphi'} \frac{dR}{R} = \tan \alpha d\varphi$$

Assuming that $R = R_3$, $\varphi = 0$, $D = D_3$, integrating yields

$$\int_{R_3}^{R} \frac{dR}{R} = \tan \alpha_3 \int_{0}^{\varphi} d\varphi,$$

$$\ln R - \ln R_3 = \tan \alpha_3 \varphi,$$

$$\ln \frac{R}{R_3} = \tan \alpha_3 \varphi,$$

$$R = R_3 e^{\tan \alpha_3 \varphi} \tag{3.58}$$

By utilizing the velocity triangle shortly after the impeller outlet to determine $\tan \alpha_3$, and given different values of the angle φ (in radians), the corresponding radius R can be calculated. Consequently, this logarithmic spiral can be plotted accordingly.

In practical applications of volute casings (spiral-shaped pressure water chambers), to reduce the radial dimensions, the width b of the casing is mostly designed to diverge. This approach allows for the reduction of both v_m and the angle α, thereby achieving the goal of minimizing radial size. Consequently, the actual design of volute casings does not strictly adhere to the aforementioned logarithmic spiral for calculations. Although the flow within the volute casing is highly complex, it generally conforms to the flow pattern where $v_u R = \text{constant}$.

3.3.2 Hydraulic Design of the Volute Chamber

1) Planar Design of the Volute Chamber

   The planar design of the volute chamber can be tailored to specific application requirements, allowing for either central discharge or tangential discharge, as illustrated in Figure 3.54.

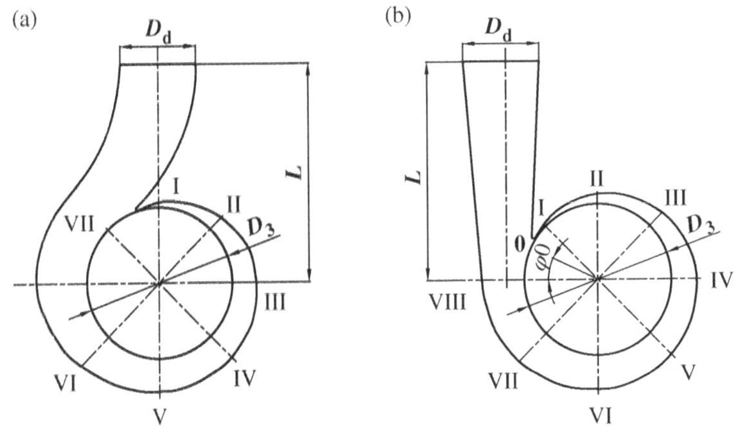

Figure 3.54 Planar layout options for the volute chamber: (a) central discharge; (b) tangential discharge.

2) Calculation and Shape Drawing of the Volute Chamber

1) Base circle diameter D_3

The circle that is tangent to the tip of the divider tongue (i.e., the circle passing through the starting point of the spiral) is referred to as the base circle, denoted by D_3, as shown in Figure 3.55. The diameter D_3 should be slightly larger than the outer diameter D_2 of the impeller to ensure an appropriate clearance between the divider tongue and the impeller. If this clearance is too small, it can easily cause noise and vibration due to fluid flow blockage, and cavitation may also occur at the divider tongue. Increasing the clearance can reduce the nonuniformity of flow around the impeller periphery, lower vibration and noise, and slightly improve efficiency. However, if the clearance is excessively large, in addition to increasing the radial dimensions, the presence of a rotating fluid ring in the clearance consumes a certain amount of energy, leading to a decline in pump efficiency (especially in the low-flow range). Typically, an appropriate value is selected as

$$D_3 = (1.03 - 1.08)\,D_2$$

Pumps with a high specific speed and smaller dimensions should adopt larger clearance values, while those with the opposite characteristics should adopt smaller values (refer to Table 3.14).

2) Inlet width of the volute chamber, denoted as b_3

The width b_3 of the volute chamber inlet is typically larger than the outlet width B_2 of the impeller (including both the front and rear cover plates). There should be at least a certain clearance to compensate for rotor axial displacement and manufacturing tolerances. Currently, in some volute designs, b_3 is made quite wide, allowing the liquid rotated by the

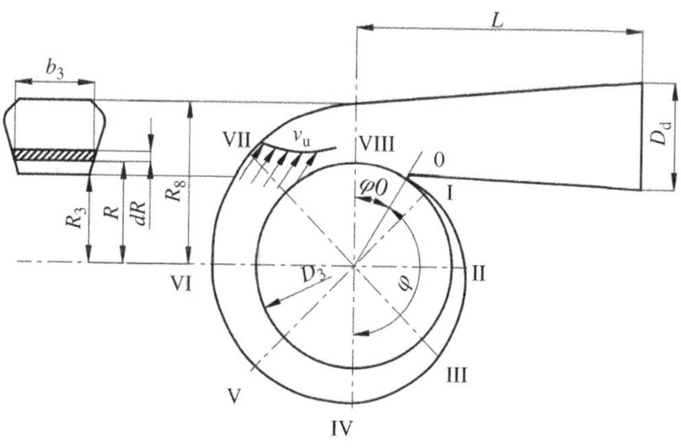

Figure 3.55 Geometric parameters of the volute chamber.

Table 3.14 Reference values for base circle diameter.

Specific speed n_s	Base circle diameter D_3
42–70	$1.05D_2$
70–106	$1.06D_2$
106–177	$1.07D_2$
177–282	$1.08D_2$

Table 3.15 Reference values for the inlet width of the volute chamber.

Specific speed n_s	Width of volute chamber inlet b_3
<70	2.0 b_2
70–212	1.75 b_2
> 212	1.6 b_2

front and rear cover plates of the impeller to flow smoothly into the pressure chamber. This recovers a portion of the disc friction power and improves pump efficiency. Additionally, it accommodates impellers of different widths, enhancing product versatility (see Table 3.15). Typically,

$$b_3 = B_2 + (5 - 10)\text{mm or } b_3 = B_2 + 0.05\, D_2$$

3) Installation angle φ_0 of the volute chamber tongue
   The tongue is located at the starting point of the spiral section of the volute chamber, separating the spiral section from the diffuser tube. By convention, the cross-section passing through the tip of the tongue is referred to as cross-section 0. The angle between the tongue and cross-section VIII (the cross-section passing through the end of the spiral) is termed the installation angle of the tongue, denoted by φ_0.The magnitude of. φ_0 should ensure a smooth connection between the spiral section and the diffuser tube while minimizing the radial dimensions as much as possible. For pumps with a high specific speed, both v_m and α are large, the outer wall of the volute extends significantly in the radial direction. Therefore, a larger angle φ_0 is adopted to maintain a harmonious shape. Table 3.16 presents the relationship between φ_0 and n_s.
4) Helical angle α_0 at the volute tongue
   The helical angle α_0 at the volute tongue refers to the angle between the tangent to the spiral curve and the tangent to the base circle at the tip of the tongue (i.e., the intersection points of the spiral curve and the base circle). Alternatively, it can be approximately considered as the angle between the inner wall of the tongue and the circumferential direction. To conform to the flow pattern and minimize the impact of the liquid flow, the helical angle α_0 at the tongue is generally taken to be equal to the absolute liquid flow angle α_2 at the impeller outlet, that is,

$$\tan \alpha_2 = \frac{v_{m2}}{v_{u2}} \tag{3.59}$$

5) Cross-sectional shape and cross-sectional area of the volute chamber [2, 18]
   The cross-sectional shapes of volute chambers can be categorized into symmetrical and asymmetrical types. For impellers with radial outlet flow, a symmetrical volute cross-section is typically employed, as seen in centrifugal pumps (as illustrated in Figure 3.56a–c). Conversely, for impellers with oblique outlet flow, an asymmetrical volute cross-section is usually

Table 3.16 Relationship between the installation angle of the volute chamber tongue and the specific speed.

n_s	40–60	60–130	130–220	220–360
φ_0	0°–15°	15°–25°	25°–38°	38°–45°

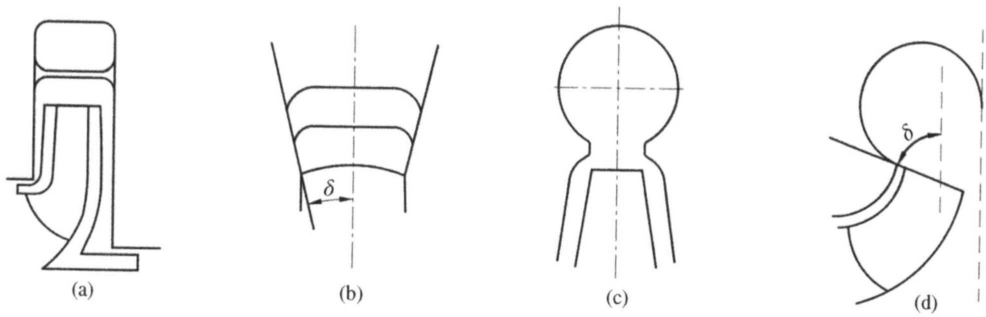

Figure 3.56 Different cross-sectional shapes of volute chambers: (a) rectangular, (b) trapezoidal, (c) circular, (d) asymmetrical (for mixed-flow impellers).

adopted, such as in mixed-flow pumps (as shown in Figure 3.56d). The cross-sectional shape of the volute chamber should conform to the trajectory of the liquid flow.

The cross-sectional shapes of volute chambers can be tailored into rectangular, trapezoidal, or circular forms based on different requirements, as illustrated in Figure 3.56. Tests conducted by some individuals using the same impeller in both rectangular and pear-shaped volute chambers, which shared identical values for D_3, b_3 cross-sectional area, and maximum outer wall dimensions, demonstrated that the rectangular volute chamber exhibited 0.8% lower efficiency at the design point compared to the pear-shaped one, yet it achieved 1.2% higher peak efficiency. This suggests that the cross-sectional shape of the volute chamber has a limited impact on performance, allowing for selection based on structural and manufacturing considerations.

For ease of calculation and drawing, the volute chamber is typically divided into eight cross-sections, each spaced 45° apart from one another, meaning the volute chamber is sectioned by eight meridional planes. During the design process, the eighth cross-section (cross-section VIII) is calculated first, and the other cross-sections are then determined based on it. The shapes of cross-section VIII in various configurations can be determined using analytical methods, but in practical design, the velocity coefficient method is predominantly employed.

The velocity coefficient method is a generalized similarity calculation approach. Similar to the velocity coefficient method for impellers, it involves design based on statistically derived velocity coefficients from well-performing equipment.

$$v_3 = k_3\sqrt{2gH} \tag{3.60}$$

where

v_3—average velocity at the cross-section of the volute chamber;
H—single-stage head of the pump;
k_3—velocity coefficient, whose value can be obtained from Figure 3.57.

The flow rate q passing through cross-section VIII is given by:

$$Q_8 = \frac{\varphi_8}{360}Q = \frac{360 - \varphi_0}{360}Q \tag{3.61}$$

Generally, the flow rate passing through cross-section VIII does not differ significantly from the pump's total flow rate, and there is no harm in adopting a slightly larger volute chamber area. Therefore, the area of cross-section VIII can be calculated using the pump's total flow rate, as follows:

$$S_8 = \frac{Q}{v_3} \tag{3.62}$$

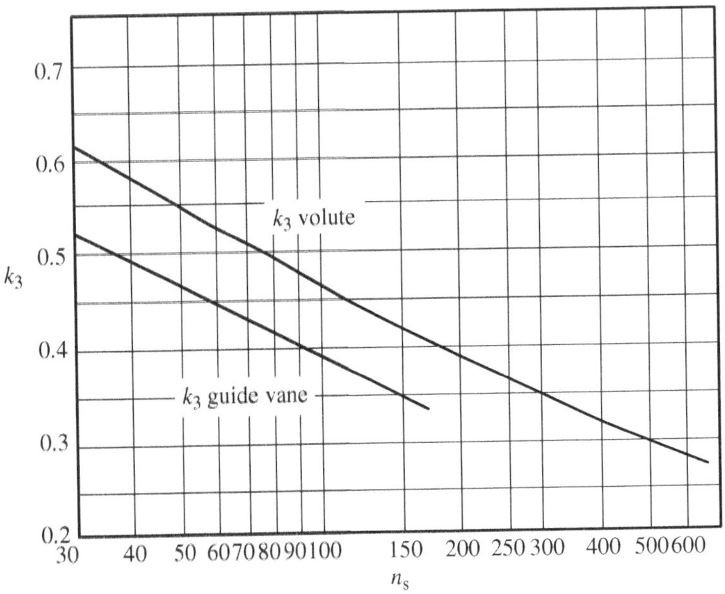

Figure 3.57 Velocity coefficients for the volute casing.

For the areas of other cross-sections, the calculation is no longer based on the assumption that $v_u r = \text{constant}$; instead, it is carried out with the assumption that $v_u = \text{constant}$. Specifically, the area s_φ of each cross-section is directly proportional to the flow rate Q_φ passing through that cross-section, and the flow rate Q_φ is, in turn, directly proportional to the angle φ (the angle between the calculation cross-section and the reference cross-section 0). Therefore, the area s_φ is directly proportional to the angle φ, that is:

$$S_\varphi = \frac{\varphi}{\varphi_8} S_8 \tag{3.63}$$

Based on the calculated cross-sectional area, the shape of cross-section VIII is first sketched, as illustrated in Figure 3.58. The cross-sectional area drawn with reference to $\frac{h}{H} = 0.35 - 0.5$, $\gamma = 15° - 25°$ should be equal to the calculated area of cross-section VIII.
The other cross-sections are generally drawn on the same plane, representing the meridional projections of each cross-section. In other words, the cross-sections of the volute chamber, as intersected by meridional planes, are projected onto a single meridional plane, with their circular arcs flattened accordingly. The area of each cross-section should match the calculated value, and various dimensions, such as the radii of the circular arcs, should vary in a regular and systematic manner.

6) Draw the planar projection view of the volute chamber (Figure 3.59a).
Based on the heights H_i depicted in the meridional cross-sectional views of each section, mark corresponding points on the respective rays in the planar view, and then smoothly connect these points to form the spiral curve on the planar view of the volute chamber. It should be noted that these points should be connected with smooth circular arcs; specifically, the center of the circular arc for the subsequent segment should lie on the extended radius line from the endpoint of the previous arc. Alternatively, every three points can be connected with a single circular arc. Meanwhile, the outline of the diffuser tube should also be drawn. If the volute chamber has a central outlet, cross-section VIII should be rotated by 45° or

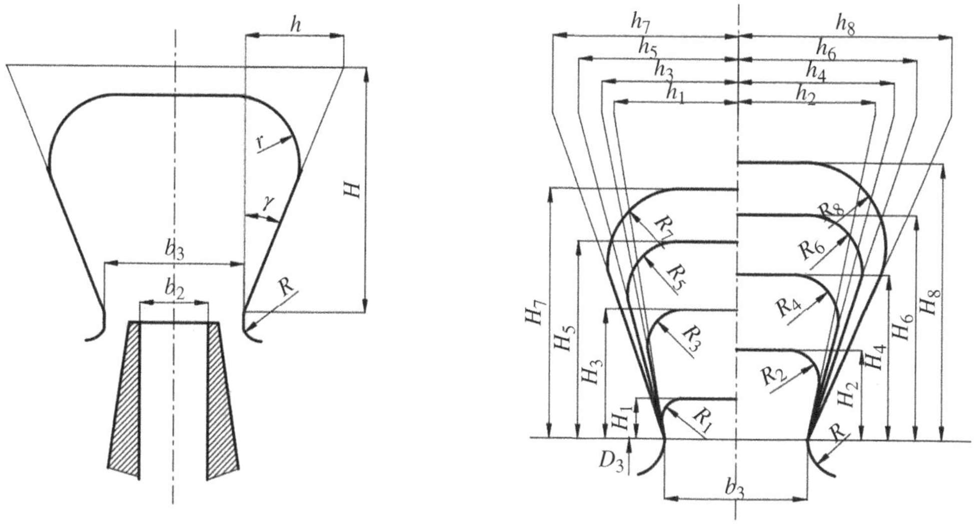

Figure 3.58 Drawing of the cross-sections of a volute chamber.

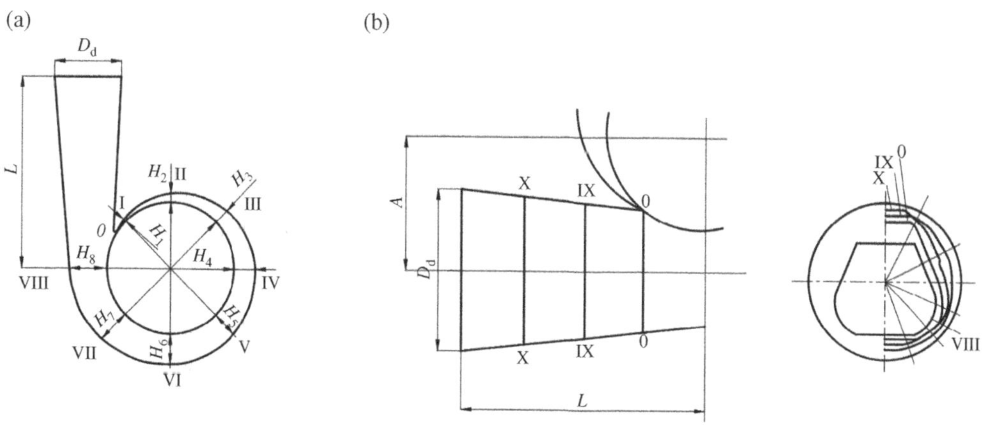

Figure 3.59 The plan view of a volute chamber: (a) Plan view (with the base circle tangent to the tongue), (b) diffuser tube.

any arbitrary angle (Figure 3.54a), and smooth circular arcs should be used to connect from cross-section VIII and the tongue to the outlet.

7) Calculation and shaping of the volute chamber's diffuser tube (Figure 3.59b)
The function of the diffuser tube is to reduce the flow velocity, convert kinetic energy into pressure energy, and simultaneously minimize losses in the discharge piping. The inlet of the diffuser tube can be considered as cross-section VIII of the volute chamber, while its outlet is the pump's discharge port. The main structural parameters of the diffuser tube are as follows:
   1) The discharge diameter D_d should conform to the economic flow velocity and standard diameters, as detailed in Section 3.1.1.3 of this chapter.
   2) The length of the diffusion tube L should be minimized under the conditions that ensure the diffusion angle, as well as facilitate processing and bolt connection, so as to reduce the size of the pump.

3) The commonly used range for the diffusion angle $\theta = 8°–12°$.
Since the inlet S_8 of the diffusion tube is not circular, it is necessary to convert S_8 into an equivalent circular area and then calculate the equivalent angle:

$$\tan\frac{\theta}{2} = \frac{D_{\text{out}} - D_{\text{in}}D_{\text{out}} - \sqrt{\frac{4S_8}{\pi}}}{2L} = \frac{D_{\text{out}} - \sqrt{\frac{4S_8}{\pi}}}{2L} \tag{3.64}$$

where D_{in}—the equivalent diameter at the inlet of the diffusion tube, $S_8 = \frac{\pi}{4}D_{\text{in}}^2$.
The outlet of the diffusion tube features a circular cross-section, while the inlet S_8 has an irregular cross-sectional shape. To ensure a smooth transition from the inlet to the outlet, with a continuously smooth wall surface throughout, the cross-section should gradually change. The following method can be employed for drafting and inspection:
Draw the inlet cross-section S_8 within the outlet circular cross-section, then draw several radial lines. Divide the length of the diffusion tube into equal segments, and correspondingly divide the radial segments between the inlet and outlet cross-sections into equal parts. Connect the resulting points smoothly to obtain the shape of the intermediate cross-sections. This approach ensures a smooth wall surface throughout.

3.3.3 Hydraulic Design of Double Volute Chambers

For large-flow single-stage centrifugal pumps, a double volute chamber design is often adopted for the volute to reduce or eliminate the inherent radial forces present in a single volute chamber. Studies have shown that the radial forces in a double volute chamber are significantly reduced.

The double volute chamber design can essentially be viewed as a composite design where two single volute chambers are arranged in opposite positions and separated by a partition. As shown in Figures 3.60 and 3.61, the throat areas of the two volute chambers are identical, and the specific design method can refer to that of a single volute chamber. The shape of the partition in the double volute chamber not only affects the manufacturing process of the pump but also exerts varying impacts on the radial forces, as illustrated in Figure 3.61.

3.3.4 Design of the Annular Discharge Chamber

The annular discharge chamber is a type of chamber where the cross-sectional areas are equal throughout. In an annular discharge chamber, the flow velocity gradually increases toward the outlet and collides with the liquid flowing out from the impeller, resulting in significant losses. Moreover, under the designed operating conditions, it generates substantial radial forces. However,

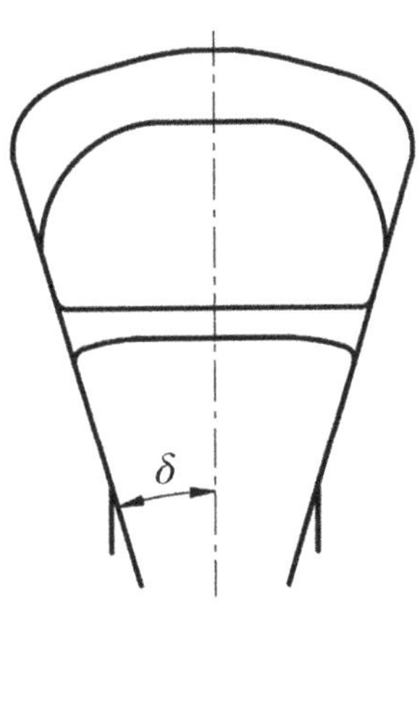

Figure 3.60 Cross-sectional shapes of the double volute chamber.

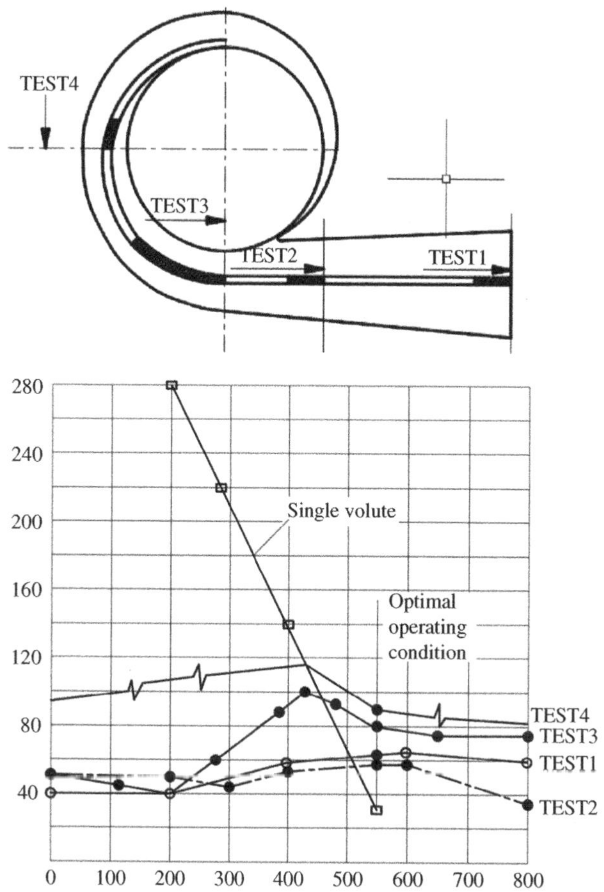

Figure 3.61 Influence of the shape of the volute chamber's partition on radial forces.

its structure is symmetrical and simple. The cross-sections of an annular discharge chamber are typically semicircular or rectangular, as shown in Figure 3.62.

The annular discharge chamber is primarily used in slurry pumps because its large gap at the tongue makes it less prone to clogging by impurities. Additionally, in terms of manufacturing, it facilitates the application of measures such as rubber lining. After the last-stage diffuser of multi-stage pumps, an annular discharge chamber is also commonly employed due to its symmetrical structure, which facilitates the arrangement of tie rods and ensures uniform thermal deformation. Depending on specific circumstances, a quasi-spiral discharge chamber (also known as an intermediate-type discharge chamber), which lies between the spiral and annular discharge chambers, may sometimes be adopted.

The calculation of Section VIII for both the annular discharge chamber and the quasi-spiral discharge chamber is the same as that for the spiral discharge chamber. In the case of an annular discharge chamber, the cross-sectional areas of all sections are identical and equal to the area of Section VIII. For a quasi-spiral discharge chamber, the area of Section VIII equals the calculated area, while the areas of other sections are appropriately reduced.

Typically, the diffuser tube is perpendicular to Section VIII and positioned at a distance of $R/2$ from the central axis. The intersection point A, where the diffuser tube meets the annular discharge chamber, serves as the tongue (or cutwater), as illustrated in Figure 3.62.

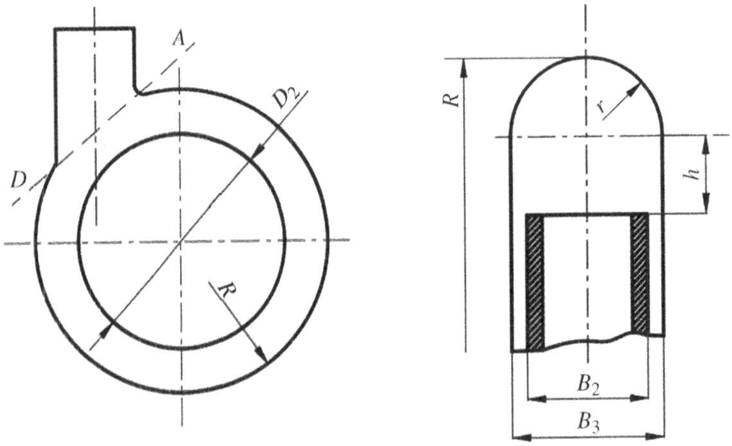

Figure 3.62 Annular discharge chamber.

3.3.5 Design of Radial and Guide Vanes

Diffusers are generally employed between two stages of impellers in multistage pumps or at the pump outlet. The operating principle of a diffuser is the same as that of a spiral discharge chamber, and the design methods are similar. The flow passages between the diffuser vanes can be regarded as several small spiral discharge chambers evenly distributed around the periphery of the impeller.

The structure of diffusers can be classified into radial diffusers, guide vanes, and three-dimensional diffusers.

The flow passage of a radial diffuser consists of an inlet vane, an annular space (the turning section), and an outlet vane. The inlet vane functions as a discharge chamber, while the outlet vane not only serves as a discharge chamber but also guides the liquid into the suction chamber of the next-stage impeller. Therefore, this type of diffuser is a fixed flow-guiding component that fulfills the dual functions of both a suction chamber and a discharge chamber.

The guide vane is fundamentally similar to the radial diffuser, with the key difference being that in a radial diffuser, the liquid exiting the inlet vanes mixes together in the annular space before entering the outlet vanes, as shown in Figure 3.63. In contrast, for a guide vane, as depicted in Figure 3.64, the inlet and outlet vanes form a continuous, integral structure, creating individual small flow passages from the inlet of the positive guide vanes to the outlet of the negative guide vanes, where the liquids in different passages do not mix. The guide vane exhibits superior hydraulic performance compared to the radial diffuser. However, its drawback lies in its complex structure, which makes casting and machining more challenging.

Figure 3.63 Three-dimensional view of radial guide vane. *Source:* Zibo Yinxuan Carbon Technology Co., Ltd.

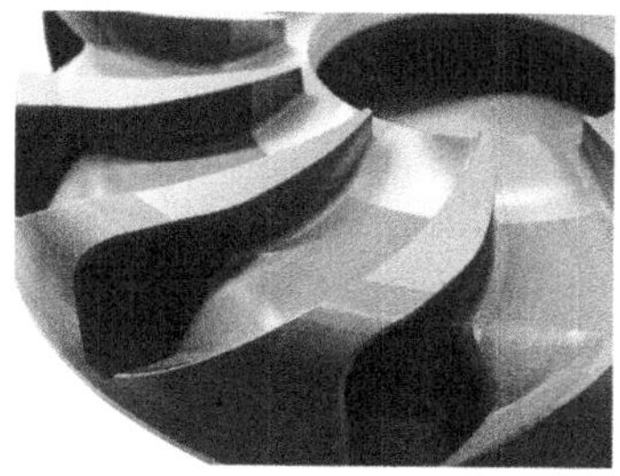

Figure 3.64 Three-dimensional view of channel-type (or flow-passage-type) guide vane. *Source:* Yong Hua Heng Import Co., Ltd.

The three-dimensional diffuser is similar to the guide vane in that its inlet and outlet vanes are integrated into a single unit. Due to the spatial twisting of the vanes, it is referred to as a three-dimensional diffuser. It is primarily used in various well pumps (such as deep-well pumps and submersible pumps) and diffuser-type mixed-flow pumps.

Since the design calculations for the inlet vane and outlet vane sections of both radial diffusers and guide vanes are identical, the following discussion will focus solely on the design calculations for radial diffusers. When designing a guide vane, attention should simply be paid to ensuring a uniform change in the cross-sectional flow area from the outlet of the diffuser tube to the outlet of the negative guide vane flow passage.

3.3.5.1.1 Design and Calculation of Radial Guide Vane Figure 3.65 shows a radial guide vane. In this structure, to reduce the radial dimensions, a triangular window HFG is opened at the tail of the positive guide vane. The liquid passes through this window, changes its direction to axial flow, mixes in the annular space above the negative guide vane, and then enters the negative guide vane. The flow passage of the radial guide vane consists of the following parts: the helical section (AB)—which serves to collect the liquid; the diffuser section (BC)—which converts kinetic energy into pressure energy; the transition section (CD)—which redirects the liquid flow; and the negative guide vane (DE)—which enables the liquid to enter the next-stage impeller at the required velocity and circulation.

1) Base circle diameter D_3

   D_3 refers to the diameter of the circle tangent to the starting point of the helical curve at the inlet of the guide vane. It is selected in the same way as the D_3 of the volute chamber:

   $$D_3 = \left(1.02 - 1.05\right) D_2$$

   For pumps with a high specific speed n_s or small dimensions, the larger value should be chosen. Typically, the maximum radial clearance does not exceed 5 mm. Increasing the clearance will reduce the pump's efficiency.

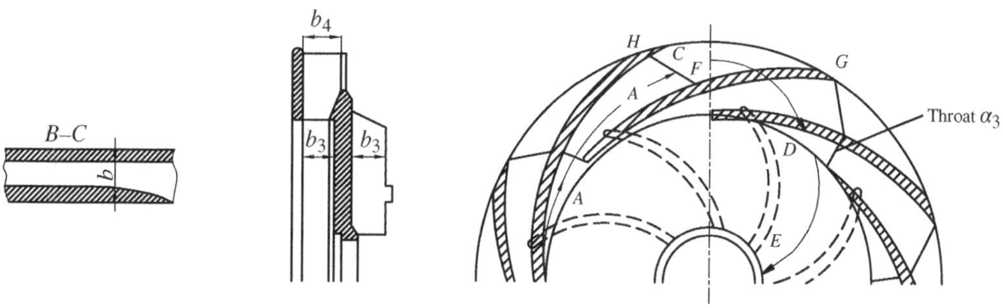

Figure 3.65 Radial guide vane.

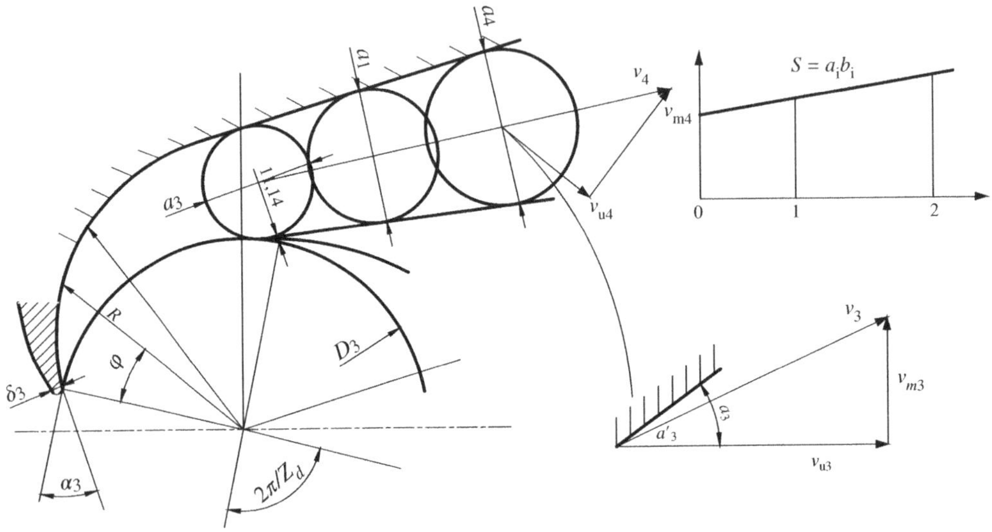

Figure 3.66 Design of the helical section and diffusion section of the inlet guide vane.

2) Width of guide vane inlet (axial width) b_3

$$b_3 = b_2 + (2 - 5)\text{ mm}$$

In the formula, b_2 represents the blade outlet width. For multistage pumps, appropriately reducing b_3 can decrease the axial dimension.

3) Angle and Line Type of the Helical Curve (Figure 3.66): The liquid enters the helical section from the impeller outlet. Here, similar to its flow in the volute chamber, the liquid follows the law of $v_u R = K_2$ (a constant).

1) Setting the angle at the inlet of the guide vane α_3

The meridional velocity at the guide vane inlet v_{m3}, and the blockage coefficient, ψ_3 are given as follows:

$$\psi_3 = 1 - \frac{\delta_3 Z_d}{D_3 \pi \sin \alpha_3} v_{m3} = \frac{Q}{2\pi R_3 b_3 \psi_3}$$

The circumferential component of velocity at the guide vane inlet:

$$v_{u3} = \frac{v_{u2} R_2}{R_3}$$

The inlet fluid flow angle of the guide vane α_3': $\tan \alpha_3' = \frac{v_{m3}}{v_{u3}}$

To improve the shape of the guide vane, it is advisable to make the guide vane inlet angle greater than the fluid flow angle, that is:

$$\alpha_3 = \alpha_3' + \Delta\alpha \qquad \Delta\alpha = 3^\circ - 8^\circ$$

The guide vane inlet angle can also be approximately determined according to Eq. (3.65).

$$\tan \alpha_3 = (1.1 - 1.3) \tan \alpha_2' \tag{3.65}$$

where α_2'—the absolute fluid flow angle at the impeller outlet.

2) The line type of the helical section is the same as that of the volute spiral represented by Eq. (3.58), that is,

$$R = R_3 e^{\varphi \tan \alpha_3}$$

For different given φ angles (in radians), the corresponding radii at φ angles can be calculated using the above formula, thereby enabling the plan view of the helical section to be drawn. Practical experience has demonstrated that the key factor determining the performance of the guide vane is the throat area described below, while the shape of the helical segment has a relatively minor impact on performance. Once the throat area is determined, this segment can be appropriately drawn using arcs without the need for calculating each point on the helix individually.

4) Determination of the throat area and the number of guide vanes

The velocity coefficient method is widely used due to its simplicity in determining the throat area.

Throat area $S = \frac{Q}{Z_d v_3}$

Throat velocity $v_3 = k_3 \sqrt{2gH}$

$$S = a_3 b_3$$

If we assume $a_3 = b_3$, then the number of guide vane blades Z_d can be calculated as

$$Z_d = \frac{Q}{b_3^2 v_3}$$

where

H—Single-stage head;

k_3—Velocity coefficient, which is a function of specific speed and can be determined from the curve in Figure 3.57.

After calculating the value of Z_d, round it to the nearest integer to use as the determined number of guide vanes. The commonly used number of guide vane blades Z_d ranges from 6 to 8 and should be coprime with the number of impeller blades. Table 3.17 provides blade number combinations for reference.

5) Design and calculation of the diffuser section
   1) The inlet area of the diffuser section is equal to the throat area $S = a_3 b_3$
   2) Recommended reference values are as follows:

$$\frac{S_4}{S_3} = 1.5 - 1.8 \qquad D_4/D_3 = 1.3 - 1.5$$

   $L/a_3 = 3 - 4$ (L—Length of the diffuser section) $v_4/v_3 = 0.4 - 0.5$

   3) Straight diffusion offers better performance than curved diffusion, but it has a larger radial dimension.

The divergence angle is determined based on the equivalent circular area:

$$\theta = 2 \arctan \frac{\sqrt{S_4/\pi} - \sqrt{S/\pi}}{L}$$

Table 3.17 Combinations of impeller and guide vane blade numbers.

Number of impeller blades	5			6	7				
Number of guide vane blades	7	8	12	10	9	10	11	12	(15)

Note: Using a coprime number of blades for the impeller and guide vanes can reduce pump vibration.

The selected range for the angle θ is 7°–10°.
The outlet area of the diffuser section is $S_4 = a_4 b_4$ the outlet velocity $v_4 = \frac{Q}{S_4 Z_d}$
The outlet circumferential velocity is $v_{u4} = v_4 \cos \alpha_4$

where α_4 is the included angle between the middle streamline at the outlet of the diffuser section and the circumferential direction (measured from the diagram).

6) Design of the transition section
The inlet area of the annular space should be equal to or slightly larger than the outlet area of the diffuser section. It is beneficial to have a moderately larger area during the turning process.
7) Design of the reverse guide vanes
When designing reverse guide vanes, reference should be made to existing structures with good performance. Consideration should be given to the connection with the forward guide vanes and the inlet of the downstream impeller, striving for smooth and unobstructed flow. This ensures the required changes in velocity magnitude and direction are achieved, as illustrated in Figure 3.67. The specific design steps are as follows:
   1) Determination of key parameters: The number of blades generally matches that of the forward guide vanes. Their circumferential positioning should ensure that the fluid flow between every two forward guide vanes continuously enters the channel between two adjacent reverse guide vanes.
   The inlet diameter D_5 of the reverse guide vanes is generally slightly smaller than D_4, while the outlet diameter D_6 is typically comparable to the impeller inlet diameter D_j or may extend slightly toward the axis as appropriate. The axial width b_5 is determined based on the required velocity change, generally:

$$b_5 = \frac{b_4}{1.05 - 1.15}$$

   Determination of the inlet setting angle α_5 and outlet setting angle α_6 of the reverse guide vanes: After the liquid flows out of the diffuser section of the forward guide vanes and before entering the reverse guide vanes, the velocity moment remains constant.

$$v_{u4} R_4 = v_{u_5} R_5 \qquad v_{u5} = v_{u4} \frac{R_4}{R_5}$$

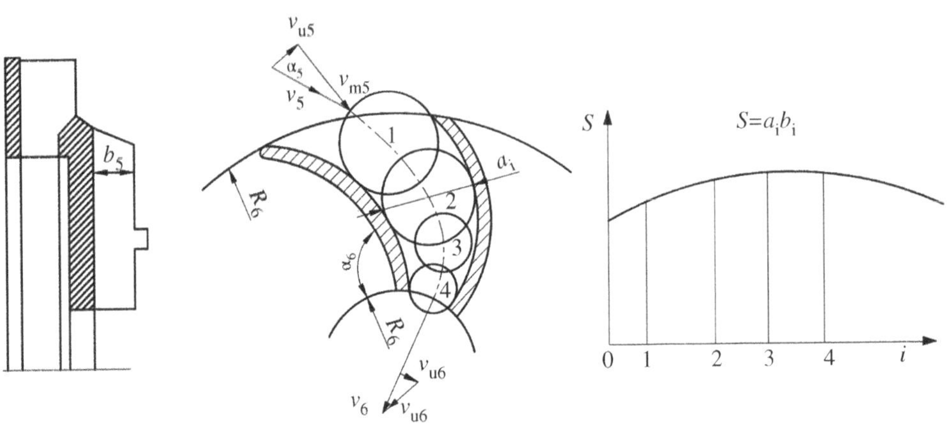

Figure 3.67 Design of reverse guide vanes.

Meridional velocity $v_{m5} = \frac{Q}{D_5 b_5 \pi \psi_5}$

where

ψ_5—blockage coefficient;

δ_5—actual thickness at the inlet of the reverse guide vane;

α_5—inlet setting angle of the reverse guide vane. For the first approximate calculation, can initially assume that $\alpha_5 = \alpha_4$, followed by subsequent verification.

$$\tan \alpha_5' = \frac{v_{m5}}{v_{u5}}$$

To minimize the curvature of the reverse guide vanes, the setting angle α_5 should be made larger than the liquid flow angle, that is, a positive incidence angle is adopted.

$\alpha_5 = \alpha_5' + \Delta\alpha$, where $\Delta\alpha$ is taken as 3°–10°, and a larger value is chosen for pumps with a lower n_s.

The calculated value of α_5 should be equal to or close to the initially assumed value; otherwise, recalculations must be performed.

2) Outlet setting angle α_6 of the reverse guide vanes: To enhance the pump head, α_6 is often set between 90° and 95°. For certain pumps with low specific speeds, to improve flow conditions at small flow rates and achieve a declining performance curve, α_6 is typically set between 60° and 80°. In this case, there is a rotational component v_u at the impeller inlet, which reduces the relative velocity w_1 at the impeller inlet and increases the liquid flow angle β_1 at the blade inlet. This is the primary reason for the impact on performance.

3.3.5.1.2 *Blade Profiling of Guide Vanes*

Drawing of the Forward Guide Vanes (Figure 3.68)

1) Draw the base circle D_3 on the plan view and divide it equally according to the number of guide vane blades to obtain the positions of each division point, labeled as *A*, *B*, and so on.
2) From division point *B*, draw a ray such that the angle between the ray and the circumference of the circle is equal to α_3. Draw a line perpendicular to the ray, and on this line, mark off the inlet thickness δ_3 of the guide vane to obtain point *N*. Then, mark off a segment corresponding to the angle α_3 to obtain point *H*. The line segment *NH* represents the planar width of the guide vane inlet, with the midpoint of *NH* labeled as point *O*.
3) Connect *AH* with one or several circular arcs, ensuring that the angle between the connecting line at point *A* and the circumference of the circle is also equal to the calculated α_3.
4) Thicken the guide vane blade and construct the diffusion section.
5) With point *O* as the center and the length of the diffusion section *L* as the radius, draw an arc. Select a point *O′* on this arc. With *O′* as the center and the outlet cross-sectional height α_4 of the diffusion section as the diameter, draw a circle that is tangent to the back surface of the forward guide vane at point *P* and intersects the extension of line *PO′* at point *P′*. Use one or several circular arcs to connect and form arc *HG*, which is tangent to circle *O* and circle *O′*. After the plan view of the diffusion section is completed, the outer diameter D_4 of the guide vane is also determined, but D_4 should be rounded to an integer value.
6) Once the outer diameter D_4 of the guide vane is determined, the outlet edge *GT* of the diffusion section can be drawn. To fully utilize the function of the diffusion section, the lengths of the two side surfaces *HG* and *BT* of the diffusion section should be approximately equal.
7) Construct the transition section, ensuring that the inlet area of the transition section is equal to or slightly larger than the outlet area of the diffusion section, and it smoothly connects with the inlet of the reverse guide vane.

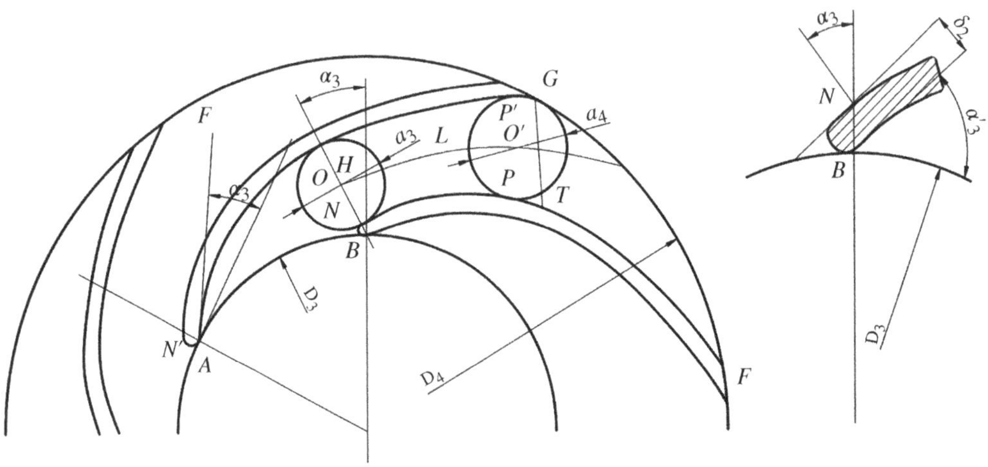

Figure 3.68 Drawing method for guide vanes.

Blade Profiling of Reverse Guide Vanes The profiling of reverse guide vanes can be accomplished using the streamline method or the single circular arc method. The procedure for drawing the plan view of reverse guide vanes using the single circular arc method is illustrated in Figure 3.69.

1) Draw the inlet and outlet circles of the reverse guide vanes with diameters D_5 and D_6, respectively.
2) Draw radii OA and OB such that $\angle AOB = \alpha_5 + \alpha_6$.
3) Connect points A and B, and find the intersection point C of line AB with the outlet circle.
4) Construct $\angle OAM = \alpha_5$, and find the intersection point M of this angle's bisector with the perpendicular bisector of line segment AC.
5) With M as the center and MA as the radius, draw a circular arc AC, which represents the working surface of the reverse guide vane.

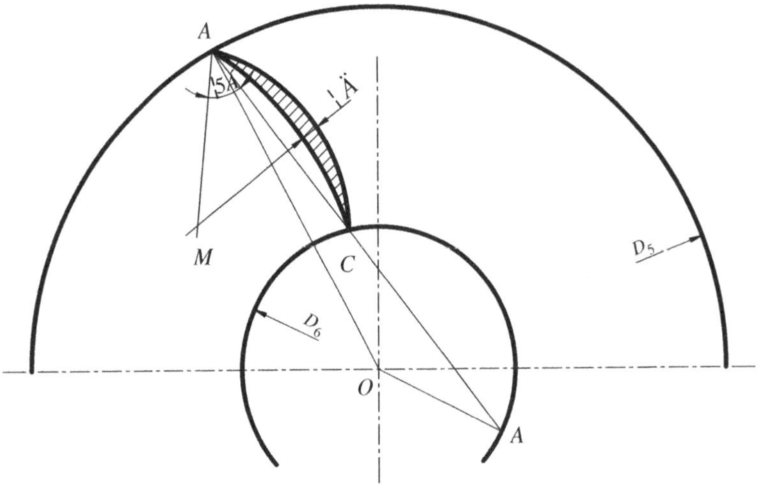

Figure 3.69 Drawing of reverse guide vanes.

6) Determine the thickness δ of the guide vane blade according to technological requirements, and thicken the blade accordingly. The inlet should be designed with a streamlined rounded head, while the outlet should be made as thin as possible.

After completing the blade profiling of the reverse guide vanes, it is essential to check whether the variation in the cross-sectional flow area of the passage is regular, as shown in Figure 3.67. The outlet velocity v_6 should be slightly less than the inlet velocity of the next-stage impeller. If these requirements are not met, improvements can be made by adjusting the axial width or the variation pattern of the blade thickness.

3.3.6 Design of Spatial Guide Vanes

Spatial guide vanes are a commonly utilized type of pressure casing, characterized primarily by their long axial length and short radial dimension, enabling them to perform the functions of an entire pressure casing. Specifically, they collect the liquid exiting the impeller and transport it to the inlet of the next-stage impeller or to the outlet pipeline. They also convert kinetic energy into pressure energy, with v_m showing little change from inlet to outlet, while v_u decreases from a large value at the inlet to zero or a very small value at the outlet. Consequently, v within the guide vane casing gradually decreases, achieving the goal of energy conversion. Additionally, they eliminate the rotational component of the flow; typically, the outlet angle of the guide vanes $\alpha_4 \approx 90°$, resulting in $v_{u4} \approx 0$.

The design of spatial guide vanes, like that of other types of guide vanes, should take into account the following principles:

① The wetted perimeter of the flow passage cross-section should be minimized, making a circular or square shape preferable; ② The shape of the flow passage should change smoothly; ③ The angles at various locations should conform to the laws of fluid flow; ④ The changes in various velocities should be uniform; ⑤ The velocity at the throat section should be controlled to a specific value; ⑥ The divergence angle of the flow passage should be controlled to a specific value.

3.3.6.1.1 Design Calculation of Spatial Guide Vanes

1) Determination of the Meridional Projection

Spatial guide vanes are adaptable to a wide range of n_s, resulting in significant variations in the shape of their meridional projections. When drawing the meridional projection, it is advisable to use the graphical representation of existing products with similar n_s values and good performance as references. Additionally, the following conditions should be considered: smooth connection with the impeller outlet and the inlet of the next-stage impeller (or outlet pipe), ensuring that the position of maximum bulge is not too close to the impeller outlet, giving the outlet section a tendency to guide the flow axially, achieving uniform changes in the flow passage shape, and reasonably determining the inlet and outlet edges of the guide vanes. There are three shapes for arranging the inlet edge of the guide vanes, as shown in Figure 3.70.

In Figure 3.70, the inlet edge of the guide vane in form (a) is essentially parallel to the outlet edge of the impeller, with a small gap between them. This type of guide vane exhibits a high degree of twist, making casting and sand cleaning difficult, and is mostly used in large, high n_s mixed-flow pumps where the guide vane flow passage has minimal curvature. In form (b) shown in Figure 3.70, the inlet edge of the guide vane is located farther away from the outlet edge of the impeller, within a less curved linear section, creating a larger annular space between the impeller outlet and the guide vane inlet. This type of inlet edge is sometimes used in small pumps with very low specific speeds. Form (c) in Figure 3.70 falls somewhere between the

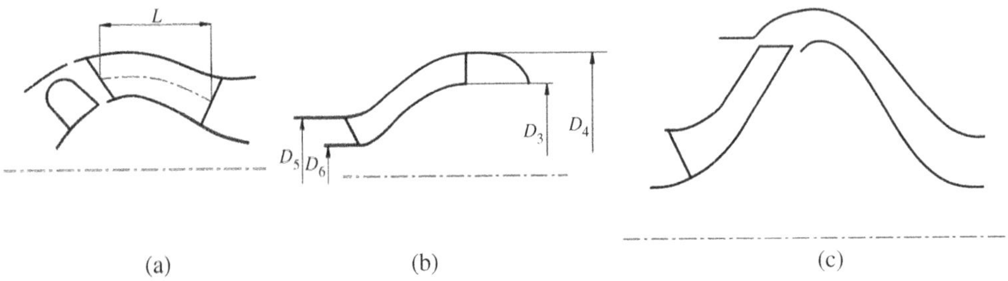

Figure 3.70 Meridional projection shapes of spatial guide vanes: (a) the inlet edge of the guide vane is parallel to the outlet edge of the impeller, (b) the inlet edge of the guide vane is away from the outlet edge of the impeller, (c) the inlet edge of the guide vane is intermediate between the two.

two aforementioned forms and is generally applicable to most pumps. The shape of the meridional projection is crucial, primarily ensuring smooth flow. The following data can serve as a reference:

1) Maximum diameter D_3 of the streamline within the guide vane

$$D_3 = D_2 + (2 - 5)(\text{mm})$$

2) Maximum diameter D_4 of the outer streamline of the guide vane

$$D_4 = D_3 + (0.7 - 1.2)D_2$$

For pumps with low specific speed, a smaller value is adopted; for pumps with high specific speed, a larger value is taken.

3) Axial length L of the guide vane

$$L = (0.5 - 0.7)D_2$$

4) Number of guide vane blades Z_d
$Z_d = Z \pm 1$ or refer to Table 3.17.
5) Wrap angle φ of the guide vane blade

$$\varphi = 60^\circ - 90^\circ$$

2) Examine the changes in the cross-sectional area of the axial-flow passage (similar to the impeller). Typically, the ratio of the maximum area of the guide vane to the outlet area of the impeller ranges from 0.8 to 1.2.
3) Streamline splitting: Usually, only one middle streamline is split, and the design is based on three streamlines. The method for splitting streamlines is the same as that used in impeller design. For small pumps, streamline splitting may be omitted, and the design can be based on two streamlines.
4) Determination of the setting angles at the inlet and outlet of the guide vane
Determination of the inlet setting angle α_3

$$\tan \alpha_3 = \frac{v_{m3}}{v_{u3}}$$

where

v_{m3}—Meridional velocity at the calculation point on the inlet edge of the guide vane;

v_{u3}—Tangential (circumferential) component of velocity at the calculation point on the inlet edge of the guide vane.

$$v_{m3} = \frac{Q}{S_3\psi_3}$$

where

Q—Flow rate of the pump;

S_3—Cross-sectional area of the meridional flow passage at the calculation point on the inlet edge of the guide vane, $S_3 = 2\pi R_c b$;

ψ_3—Blockage coefficient at the inlet of the guide vane.

$$\psi_3 = 1 - \frac{Z_d s_{u3}}{D_3 \pi}$$

where

Z_d—Number of guide vane blades;

D_3—Diameter at the calculation point on the inlet edge of the guide vane;

s_{u3}—Circumferential thickness at the calculation point on the inlet edge of the guide vane.

$$s_{u3} = \delta_3 \sqrt{1 + \cot^2 \alpha_3 / \sin^2 \lambda_3}$$

where δ_3—Actual thickness, typically $\delta_3 = 3 - 6$ mm.

v_{u3} is determined based on the condition that $v_u R$ = constant, that is

$$v_{u3} = \frac{R_2}{R_3} v_{u2}$$

The inlet angle of the guide vane α_3, is generally set to be larger than the flow angle α_3', that is, an incidence angle is taken into account to reduce blade distortion. In other words, $\alpha_3 = \alpha_3' + \Delta\alpha$, where $\Delta\alpha$ is taken as 3°–10° (a larger value is chosen for smaller n_s).

During specific calculations, one can initially assume a value for ψ_3 or α_3, and then perform verification calculations until the finally obtained value matches or closely approximates the assumed value.

The outlet angle of the guide vane α_4 is typically taken as 80°–95°. If a descending performance curve is required, α_4 can be set within the range of 60°–80°.

3.3.6.1.2 Blade Profiling The profiling of spatial guide vanes typically employs the twisted triangle method, with the fundamental principles and steps being essentially the same as those for impeller blade profiling.

The profiling steps (as shown in Figure 3.71) are as follows:

1) Construct wooden mold cross-sections on the meridional projection drawing. The cross-sections are generally spaced at equal intervals, although smaller intervals may be used in areas with significant curvature. Numbering starts from the inlet, as shown in Figure 3.71a with labels 0, 1, 2, 3, …, 9.

   It should be noted that the meridional projection shape of guide vanes differs from that of impellers, extending axially instead. The wooden mold cross-sections are integrated with the division points along the meridional streamlines, with the streamlines intersected by the cross-sections also serving as division points.

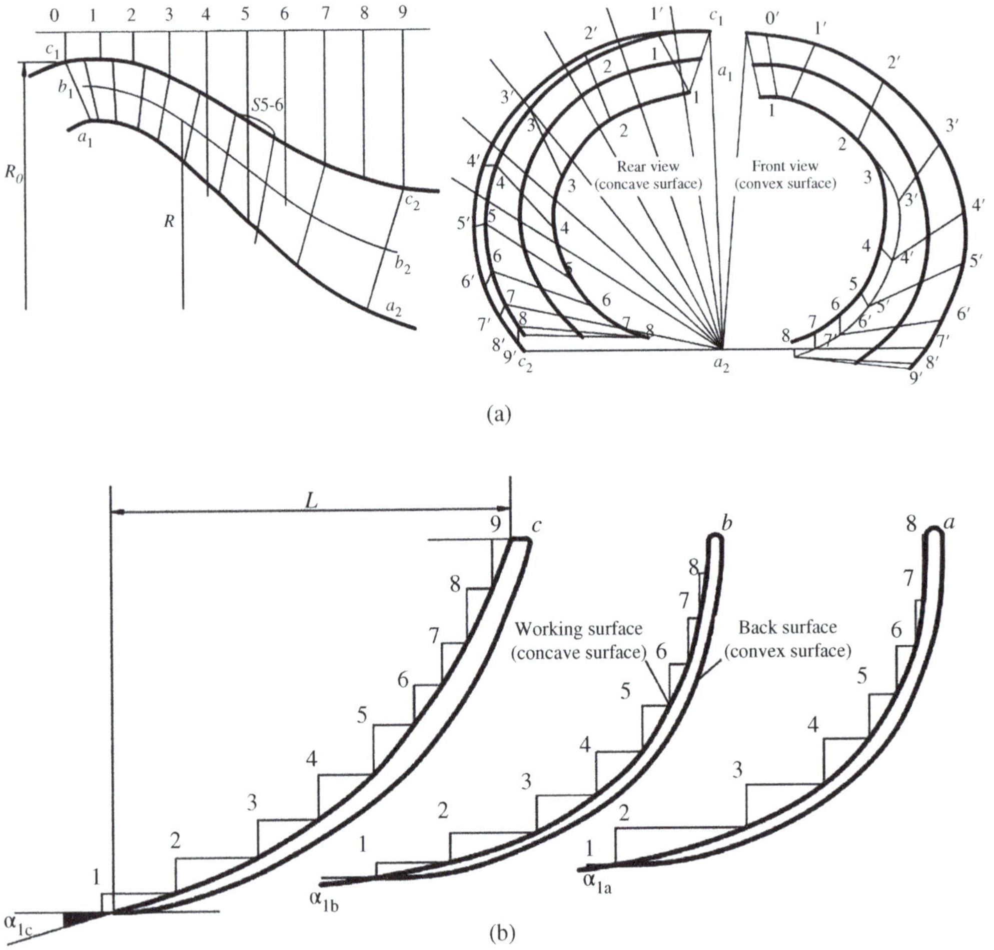

Figure 3.71 Profiling of spatial guide vane blades: (a) meridional projection view, (b) developed streamline view.

2) Create a developed view of the streamlines (Figure 3.71b).

The vertical distance between parallel lines in the developed view should equal the curved length of the corresponding meridional streamline intersected by the wooden mold cross-section. For example, the length s_{5-6} along the meridional streamline should match s_{5-6} in the developed view. Three sets of parallel lines are drawn accordingly for the three streamlines, numbered sequentially as 0, 1, 2, …

When sketching the streamlines in the developed view, ensure that the developed streamlines change uniformly and conform to the calculated inlet and outlet angles, α_3 and α_4. Based on these angles, draw the streamlines for the working surfaces of the three blades. The wrap angle for each streamline is estimated as follows: the horizontal length L of the developed streamline should equal the arc length corresponding to the central angle of a circle drawn with a radius R, where R is the radius at a point approximately one-third of the total length from the inlet along the meridional streamline in the meridional view. That is

$$\varphi = 360^\circ \times \frac{L}{2\pi R}$$

3) Create the planar projection drawing.
Begin by constructing the planar projection of the working surface. The three inlet points, a_1, b_1, and c_1 (as shown in Figure 3.71), may either be arranged on the same meridional plane or on different planes, depending primarily on the requirement that the meridional section lines (obtained by transferring the intersection points of radial lines in the planar view with the three streamlines to corresponding positions along the meridional streamlines at equal radii) should intersect the meridional streamlines at approximately 90°. Additionally, consideration should be given to ensuring uniform variation in the subsequent meridional section lines.
The true included angle γ between the guide vane and the wall surface is given by

$$\cot \gamma = \cot \lambda \cos \alpha$$

Due to the small value of the inlet angle α_3 of the guide vane, the angle λ is approximately equal to the angle γ. If the angle λ is close to 90°, then the angle γ will also be approximately 90°. This arrangement results in minimal blockage, favorable flow conditions within the passage, and ease of casting.
After preliminarily determining the starting positions of the three streamlines on the planar view, the planar projections of the working surfaces of the three streamlines are constructed based on the principles that the radius of corresponding division points on the meridional view equals the radius of the corresponding points on the planar view, and that the horizontal length Δu of corresponding segments on the developed view equals the length Δu of the corresponding segments on the planar view. Taking the streamline c_1–c_2 as an example, the construction begins at the inlet: starting from point c_1, an arc is drawn with a radius R_0 (equal to R_0 on the meridional view). On this arc, a segment length Δu_1 equal to Δu_1 on the developed view is marked. A radial line is drawn through the marked point, and another arc with radius R_1 (equal to R_1 on the meridional view) is drawn to intersect this radial line; the intersection point is the desired point 1. Subsequently, on the arc with radius R_1, a length equal to Δu_2 on the developed view is marked… and the process is repeated accordingly.
The connecting lines 1–1, 2–2, …, between corresponding points represent the wooden mold section lines, which should be straight or smoothly curved and change in a regular manner. If they do not meet these requirements, the shape of the developed streamlines should be adjusted at any time, and the drawing should be redone accordingly.
4) Blade thickening.
Thickening is applied on the developed view based on the flow-surface thickness s. The relationship between the flow-surface thickness and the actual thickness is as follows:

$$s = \delta\sqrt{1 + \cot\ \lambda^2 \cos^2\ \alpha}$$

In the formula, the angles λ and α are measured from the meridional view and the developed view, respectively. Generally, the variation pattern of the flow-surface thickness s can be directly specified for thickening. Subsequently, the circumferential thickness (horizontal segment Δu) at each point on the developed view is transferred to the corresponding point on the planar view. For instance, a length equal to s_{u2} is measured along the arc passing through point 2, thereby obtaining the intersection points 0′, 1, 2′, 3′, …, 9′ on the planar view (left side) for the back (convex) surface streamline c_1–c_2 of the blade and its intersection with the outer wall.
The points 0′, 1′, 2′, 3′, …, 9′ on the c_1–c_2 streamline and the points 0, 1, 2, 3, …, 8 on the a_1–a_2 streamline (the intersection line between the working surface streamline a_1–a_2 and the inner wall) are symmetrically transferred to the right side. Thickening is then applied along the corresponding arcs of the inner streamline a_1–a_2 according to the circumferential thickness

from the developed view, yielding points 0′, 1′, 2′, …, 8′. Similarly, thickness is added to the middle streamline, and corresponding points are connected to obtain the wooden mold section lines for the back surface. As seen in the figure, on the same radial line, the radius of the convex surface is greater than that of the concave surface.

5) Quality inspection of blade profiling.
   The wooden mold section lines should ideally be straight or smoothly curved with regular variations. Similarly, the meridional section lines should exhibit regular changes, and the angle between the meridional section lines and the streamlines at the inlet should be close to 90°. In Figure 3.71, the meridional section lines (I, II, …, VII) on the concave surface of the blade are illustrated.

3.4 Hydraulic Design of the Vortex Pump

3.4.1 Characteristics of the Vortex Pump

The structural design of the vortex pump is simple, consisting mainly of a pump cover, an impeller (a disk with multiple radial blades on its outer edge), and a pump body. A schematic diagram of its structure is shown in Figure 3.72. An annular flow passage is formed between the pump body and the impeller, allowing liquid to enter through the suction inlet, gain kinetic energy from the rotating impeller, and then be discharged through the outlet. The characteristics of a vortex pump include high head, low flow rate, low efficiency, poor cavitation resistance, and a significant drop in efficiency as the viscosity of the pumped liquid increases. However, due to its simple structure, compact size, and lightweight design, the vortex pump offers advantages in specific applications.

Vortex pumps are widely used in situations requiring high head but low flow rates. They are commonly found in fountain systems, small-scale water supply systems, laboratories, and industrial applications, where their compact design and excellent head capabilities make them suitable choices. Additionally, they are frequently used in cooling liquid circulation systems and certain chemical processing applications.

The vortex pump transfers energy to the liquid in the flow passage through the impeller blades. This process, where energy is transferred via momentum exchange in three-dimensional flow, is repeated multiple times within the pump's flow passage [1, 19]. As a result, the vortex pump

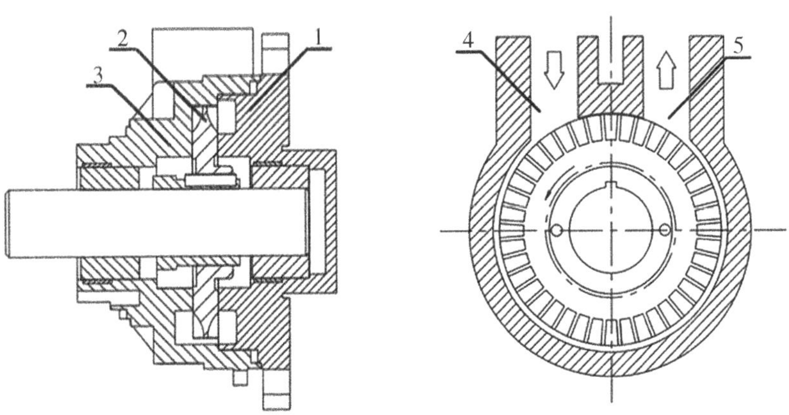

Figure 3.72 Schematic diagram of the vortex pump: 1. Pump cover, 2. Impeller, 3. Pump body, 4. Suction inlet, 5. Discharge outlet.

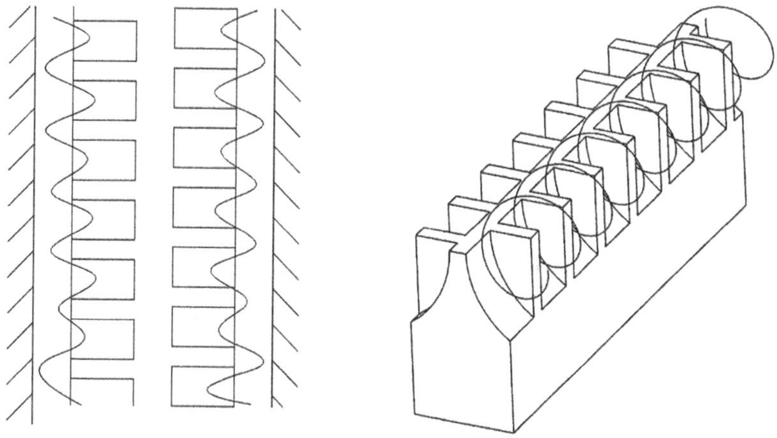

Figure 3.73 Schematic diagram of liquid motion in a vortex pump.

achieves a higher head than other types of blade pumps, which would otherwise be impossible. Figure 3.73 illustrates the motion of liquid inside a vortex pump.

As the impeller rotates, the liquid inside both the impeller and the flow passage undergoes circumferential motion. The circumferential velocity of the liquid within the impeller is greater than that within the flow passage, meaning the centrifugal force acting on the liquid inside the impeller is stronger. This results in an annular flow from the impeller to the flow passage, as shown in Figure 3.73. This motion resembles a vortex, which is how the vortex pump gets its name. The vector of this vortex is perpendicular to the axial plane and points along the longitudinal length of the flow passage; thus, it is also referred to as a longitudinal vortex.

As the liquid enters the flow passage from the impeller blades, it transfers part of the kinetic energy obtained from the blades to the liquid inside the pump passage, imparting an impulse in the direction of rotation. At the same time, some lower-energy liquid re-enters the impeller. Due to the longitudinal vortex, each time the liquid passes through the impeller within the flow passage, it gains additional energy. This repeated energy transfer process is the reason why the vortex pump achieves a higher head than conventional blade pumps.

In addition to the longitudinal vortex, a radial vortex also exists at the inlet region of the vortex pump blades. Figure 3.74 illustrates the velocity triangles of the liquid as it enters and exits the impeller.

As shown in the figure, the incidence angle of the liquid upon entry is relatively large. Due to the deviation of the relative velocity from the blade direction at the inlet, a vortex is formed. The vector of this vortex is parallel to the blade leading edge, which means it is also parallel to the radial direction of the impeller, hence it is called a radial vortex.

The radial vortex is sometimes carried into the space between the blades, but it does not contribute to any additional energy transfer. In some cases, it is carried into the flow passage, transferring part of its energy to the liquid inside the passage.

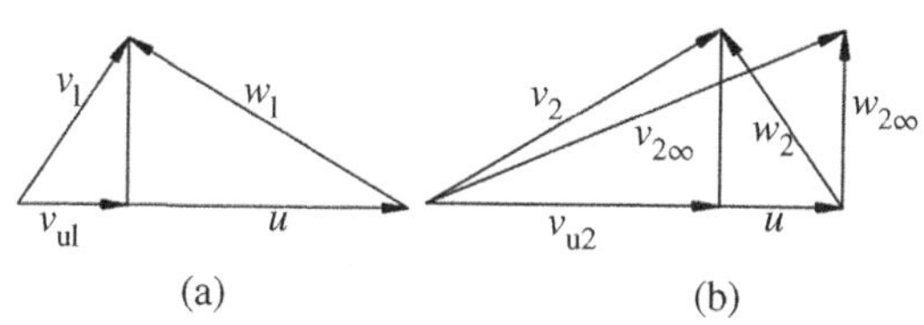

Figure 3.74 Velocity triangles of the vortex pump: (a) Velocity triangle of fluid inflow into the impeller, (b) velocity triangle of fluid outflow from the impeller.

When both longitudinal and radial vortices exist and transfer energy simultaneously, which one dominates depends on the ratio and shape of the impeller and the flow passage geometry, as well as the operating conditions of the pump. For a typical vortex pump, the possibility of the liquid carrying radial vortices into the flow passage is small, so the effect of radial vortices is negligible and can be ignored.

The strength of the longitudinal vortex is directly related to the liquid velocity in the flow passage, which is also related to the flow rate. As the flow rate increases, the longitudinal vortex weakens. Let the liquid velocity in the flow passage be v and the cross-sectional area of the passage be A. When $v = u$ (i.e., $Q = A_u$), the centrifugal force of the liquid in the impeller and flow passage is equal, and no longitudinal vortex is generated. However, when $v = 0$ (i.e., $Q = 0$), the centrifugal force difference is maximized, and the longitudinal vortex is the strongest. Since the vortex pump transfers energy through the exchange (impact) of momentum between the liquid exiting the impeller and the liquid in the flow passage, there is a significant impact loss, resulting in lower efficiency for vortex pumps.

The theoretical head generated by the liquid after passing through a single impeller.

$$H_{t1} = \frac{1}{g}(u_2 v_{u2} - u_1 v_{u1}) \tag{3.66}$$

The theoretical head generated by the liquid after passing through the impeller once.

$$H_t = i \times H_{t1} = i\frac{1}{g}(u_2 v_{u2} - u_1 v_{u1}) \tag{3.67}$$

The velocities v_{u2}, u_1, and v_{u1} are all proportional to u_2. Therefore, the basic equation for the vortex pump can be derived as:

$$H = \psi \frac{u_2^2}{2g} \tag{3.68}$$

The coefficient ψ in the equation represents the head coefficient, which cannot be calculated precisely at this stage and can only be determined using statistical methods based on specific speed [6, 20].

According to the impeller suction and pressure outlet relative position, the vortex pump can be divided into closed impeller and open impeller types. A closed impeller refers to one where the flow passage completely isolates the blades from the pump's suction inlet and discharge outlet. In a vortex pump with a closed impeller, the liquid enters the flow passage through the pump's suction inlet, then flows into the outer edge of the impeller, and is finally discharged through the flow passage to the discharge outlet. In contrast, an open impeller has blades that are directly connected to both the suction inlet and discharge outlet of the pump. In an open impeller, the liquid directly enters between the blades, then flows through the passage and exits the discharge outlet. Open impellers have better cavitation resistance, making them suitable for self-priming pumps and pump systems handling mixtures of gas and liquid. However, they are larger in size and have lower efficiency.

3.4.2 Hydraulic Design of Vortex Pump

The working principle of a vortex pump is unique, and the internal flow process is complex. Currently, there is no precise theoretical design method. Therefore, the model substitution method (similar to centrifugal pumps) and empirical coefficient design methods are commonly used. Below are the steps for the widely used empirical coefficient design method.

3.4.2.1 Empirical Coefficient Design Method

3.4.2.1.1 Design Parameters Flow rate Q, head H, NPSH, temperature, density, viscosity of the fluid, etc.

3.4.2.1.2 Rotational Speed At high rotational speeds, the pump efficiency is relatively higher, with a smaller volume and lighter weight. However, as the speed increases, the gas bubbles inside the pump become smaller, and the gas–liquid mixture becomes more uniform, making it difficult to separate, which negatively affects the self-priming performance.

3.4.2.1.3 Specific Speed The calculation formula for the specific speed of a vortex pump is as follows:

$$n_s = \frac{3.65n\sqrt{Q}}{H^{3/4}} \tag{3.69}$$

The specific speed formula for a multistage vortex pump is as follows:

$$n_s = \frac{3.65n\sqrt{Q}}{(H/i)^{3/4}} \tag{3.70}$$

In the formula, n represents the rotational speed (r/min), and i denotes the number of stages.

The specific speed range for a vortex pump is $n_s = 5$–40. If the specific speed is too low, a multistage vortex pump can be designed. However, the number of stages should not be excessive. For an open-type multistage pump, the maximum number of stages should be limited to six, while a closed-type pump is typically designed with two stages.

3.4.2.1.4 Structural Form When selecting the structural form, factors such as cavitation performance, self-priming capability, and gas-liquid mixed transportation should be considered. Additionally, the impeller and flow channel design, as well as the choice between single-stage and multistage configurations, should be taken into account.

3.4.2.1.5 Impeller Diameter D

$$D = \frac{60}{\pi n}\sqrt{\frac{2gH}{\psi}} \tag{3.71}$$

In Eq. (3.71), D represents the centroid diameter of the flow passage.

For a closed impeller, D is taken as the impeller's outer diameter; for an open impeller, it is taken as the centroid diameter of the flow passage. The head coefficient ψ is related to the specific speed and the impeller configuration, and can be determined according to Figure 3.75. For open impellers, the impeller outer diameter D_2 is related to parameter a and the impeller thickness b; when $a = 0.5b$, $D_2 \approx D + b$, and when $a = 0.7b$, $D_2 \approx D + 0.6b$.

3.4.2.1.6 Flow Passage Cross-Sectional Area A The average velocity v of the liquid in the flow passage is related to the tangential speed. Under typical design operating conditions, the average velocity in the flow passage is taken as 5–6 m/s, and it can be calculated using the following formula:

$$v = k_v u \tag{3.72}$$

In the equation, for closed impellers, u represents the tangential speed at the outer circle of the impeller; for open impellers, u refers to the tangential speed at the centroid of the flow passage. k_v represents the velocity coefficient within the flow passage; for closed impellers, $k_v = 0.5$–0.6; for open impellers, $k_v = 0.55$–0.65.

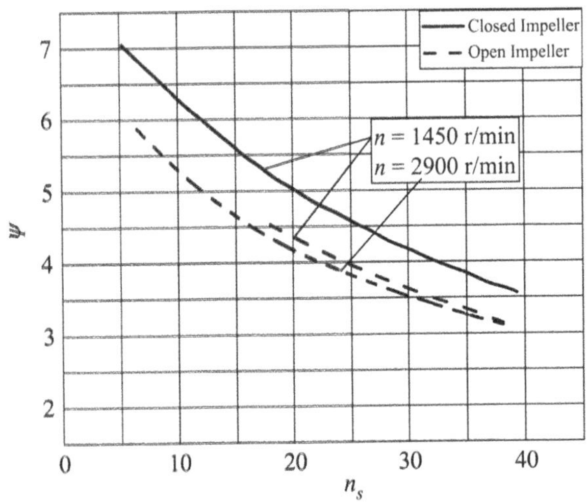

Figure 3.75 Head coefficient vs. specific speed.

The flow rate at the maximum efficiency point of a vortex pump is determined by the flow passage. When the flow passage area increases, the optimal operating point shifts toward higher flow rates and the head curve becomes flatter. Conversely, if the flow passage area is insufficient, the flow rate will not meet the design requirements. The flow passage area A is calculated using the following formula:

$$A = \frac{Q}{\eta_v v} \tag{3.73}$$

In the formula, Q represents the design flow rate; A is the cross-sectional area of the flow passage (excluding the area occupied by the impeller); and η_v is the volumetric efficiency, with $\eta_v = 70\text{–}80\%$ (use the lower value).

3.4.2.1.7 *Impeller Width* b The impeller width, b, is calculated using the following formula:

$$b = k\left(\frac{Q}{\eta_v k_v}\right)^{1/2}\left(\frac{\psi}{H}\right)^{1/4} \tag{3.74}$$

In the formula, k is the impeller width coefficient, which is related to the form of the flow passage cross-sectional area and is selected according to Table 3.18.

Table 3.18 Selection of impeller width coefficient k.

	Closed-type				Open-type			
Impeller types	**Trapezoidal**			**Rectangular**	**Semicircular**			**Rectangular**
Flow passage dimensions	$a = 0.25b$	$a = 0.3b$	$a = 0.35b$	—	$a = 0.5b$	$a = 0.6b$	$a = 0.7b$	—
Coefficient k	0.461	0.447	0.431	$\dfrac{0.475}{\left[\dfrac{b}{a} + 2\dfrac{c}{b}\left(\dfrac{b+h}{a}\right)\right]^{1/2}}$	0.308	0.296	0.286	0.36

3.4.2.1.8 Flow Passage Hydraulic Dimensions Usually, the hydraulic dimensions are determined based on the cross-sectional area of the flow passage and the optimal dimension ratios of the flow passage.

1) Hydraulic dimensions for closed impeller with open flow passage
   1) Trapezoidal flow passage
      A trapezoidal flow passage is shown in Figure 3.76. $c \approx 0.5b$, $h \approx b$, $a = (0.25–0.35)\, b$, $R = 0.5b$, and r is determined by the radius of the milling cutter.
   2) Rectangular flow passage
   A rectangular flow passage is shown in Figure 3.77. Here, $c = (0.4–0.5)\, b$, $h \approx (1.1–1.7)\, b$, $a = (0.35–0.8)\, b$.
2) Hydraulic dimensions for open impeller with closed flow passage
   1) Semi-circular flow passage
      A semi-circular flow passage is shown in Figure 3.78. Here, $c \approx b$, $h \approx 2b$, $a = (0.5–0.7)\, b$, $e = (0.3–0.5)\, b$; $R_1 = b$, $R_2 = R_3 \approx b$, $D_2 \approx D + h - 2a \approx D + (1–0.6)\, b$.
   2) Rectangular flow passage
      A rectangular flow passage is shown in Figure 3.79. Here, $b/c = 1.07$, $h = 2c$, $D_2 \approx D + h$, and e/h is selected according to Table 3.19.

Based on the above dimensions, draw the cross-sectional view of the flow channel. The cross-sectional area of the flow channel should be equal to or slightly larger than the calculated area.

Figure 3.76 Closed impeller trapezoidal flow passage.

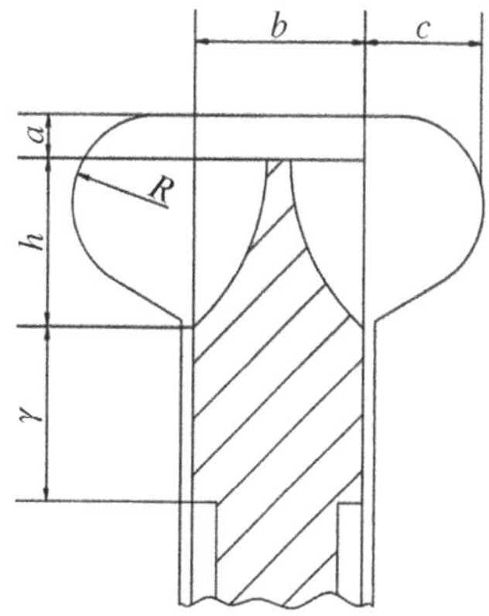

Figure 3.77 Closed impeller rectangular flow passage.

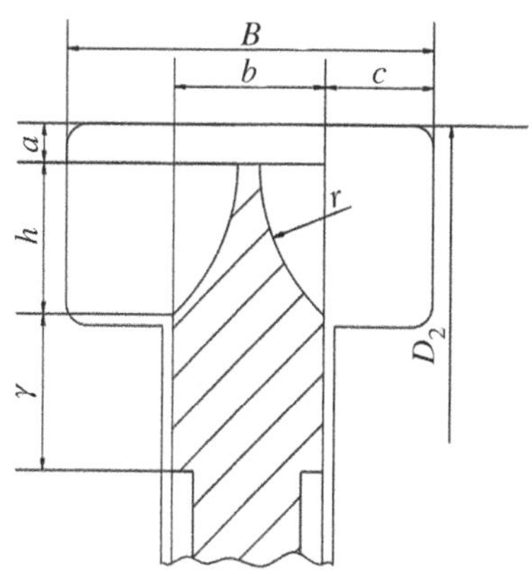

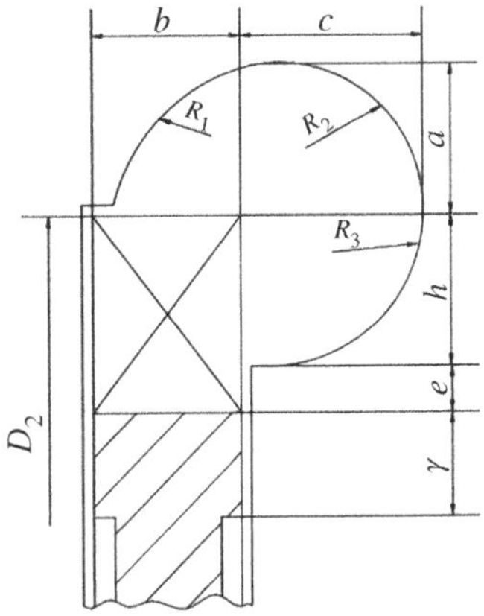

Figure 3.78 Open impeller semi-circular flow passage.

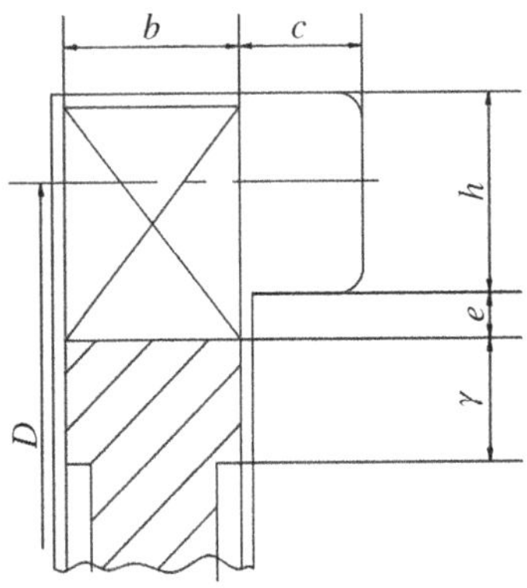

Figure 3.79 Open impeller rectangular flow passage.

Table 3.19 Relationship between e/h and η_v.

η_v	15	20	25	30	35	40	45	50
e/h	0.24	0.29	0.34	0.385	0.435	0.48	0.53	0.58

3.4.2.1.9 Selection of the Number of Blades The number of blades has a significant impact on the pump's performance. As the number of blades increases, the head increases significantly, while the power consumption also slightly rises. However, when the number of blades reaches a certain threshold, the head (H), efficiency (η), and power (P) no longer change. Generally, the number of blades ranges from 24–60. The selection of the number of blades should also consider the impeller diameter and manufacturing method. For closed impellers, the ratio of the blade spacing at the maximum radius to the blade height is typically between 0.6 and 1.0.

3.4.2.1.10 Selection of Blade Cross-Section Shape For milled closed impellers, the blades are typically radial with a uniform thickness. For cast open impellers, trapezoidal cross-section blades can be used.

3.4.2.1.11 *Tongue Wrap Angle θ* The minimum width of the tongue must be greater than twice the pitch of the blades to effectively separate the high-pressure outlet region from the low-pressure inlet region. The tongue wrap angle θ is calculated using the following equation:

$$\theta \geq \frac{2 \times 360}{z} \tag{3.75}$$

3.4.2.1.12 *Sealing Dimension γ at the Impeller End Face Clearance* If the γ value is too small, the sealing performance will be poor, resulting in significant end-face leakage. If γ is too large, the machining workload increases, and at low flow rates, end-face wear is likely to occur. In some cases, structural constraints may also prevent a large γ value. Generally, γ is set between 10 and 20 mm, with larger values chosen for impellers with a larger diameter.

3.4.2.1.13 *Clearances δ_1 and δ_2* The axial clearance δ_1 refers to the gap between the pump casing and cover and the impeller end face, also known as the end-face clearance. The radial clearance δ_2 is the gap between the pump tongue and the impeller outer edge. Both δ_1 and δ_2 have a significant impact on the shape of the pump performance curve. The volumetric loss in vortex pumps is primarily caused by these two clearances.

Typically, δ_1 ranges from 0.1 to 0.25 mm, and δ_2 ranges from 0.1 to 0.3 mm, with larger values chosen for larger pumps.

3.4.2.1.14 *Inlet and Outlet Pipe Diameter* d The flow velocity in the piping of a vortex pump is usually $v = 1\text{–}1.5$ m/s, from which the pipe diameter can be determined. For a closed impeller with a trapezoidal flow passage, the pipe diameter should satisfy: $d \geq b + 2c$, for an open impeller with a single-sided flow passage, the requirement is:

3.4.2.1.15 *Calculation of Maximum Head, Maximum Power, and Radial Force*

1) *Maximum head (power)*

   The relationship between the maximum head, maximum power, and the design head (power) within the operating range is as follows:

$$H_{max} = (1.4 - 1.6)H \tag{3.76}$$

$$P_{max} = (1.2 - 1.6)P \tag{3.77}$$

2) *Impeller radial force*

   The pressure of the liquid in the flow channel gradually increases from the suction inlet to the discharge outlet. The pump casing tongue separates the suction inlet from the discharge outlet, and the pressure along this length can be considered to vary along the diameter. The pressure distribution around the flow channel is asymmetrical, thus generating a radial force on the impeller. This radial force causes shaft deflection, which may lead to end-face wear and other issues. The radial force is calculated using the following formula:

$$F_r = krpb \tag{3.78}$$

   where b is the impeller width, r is the impeller radius, p is the pump pressure, and k is a coefficient with a value ranging from 1.12 to 1.24. A smaller n_s (specific speed) corresponds to a larger k value.

3.5 Hydraulic Design of Multistage Pumps

The centrifugal pump is the most commonly used pump among all pumps. When the actual working condition of the application requires high head, one can choose to increase the head of multiple impellers in series to achieve the desired goal. Multistage centrifugal pumps have the advantages of high head and low ratio of rotational speed [2, 7] compared to single-stage centrifugal pumps. Compared to piston pumps and diaphragm pumps and some other types of pumps, they can convey a larger flow in irrigation and drainage, industrial manufacturing, petroleum industry, urban water supply and drainage, high-rise buildings etc. by increasing the pressure in a variety of large-scale engineering projects.

Multistage centrifugal pumps commonly used in industry are mainly horizontal open-type and sectional multistage pumps. Horizontal open-type structure is characterized by the upper and lower pumps through the axis of the horizontal section of the docking, inlet and outlet pipe, part of the worm and flow casting in the lower pump casing. They usually have symmetrical arrangement, therefore most of the axial force is balanced, and therefore there is no need to install an axial force balancing device. The sectional structure is characterized at each level by the next impeller located in the diffuser shell, and all levels are in series fixed together by a fixed rod. They have low maintenance, the benefits of high-pressure resistance, and are not easy to leak. The working principle of multistage pumps is that the impeller rotates by the drive of the motor to draw liquid into the impeller, which is then transported to the next impeller through the guide vane, and so on until finally discharged from the outlet pipeline. Since more than one impeller works on the liquid, a multistage pump can increase the pump head without changing the flow rate. As shown in Figure 3.80, the structure of multistage centrifugal pumps is very complex, and it is necessary to take into account the impact of various aspects on performance during design. After years of development, the multistage centrifugal pump design method has matured, its hydraulic design and single

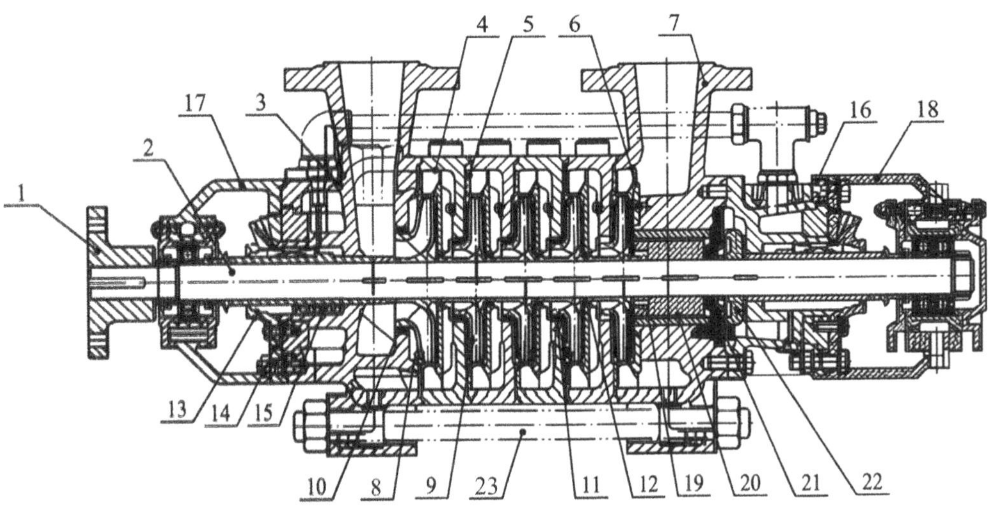

Figure 3.80 Structure of a multistage centrifugal pump: 1. Coupling, 2. Axis, 3. Water inlet section, 4. Middle section, 5. Guide vane, 6. Final guide vane, 7. Outlet section, 8. Head impeller, 9. Impeller, 10. Head seal ring, 11. Sealing ring, 12. Guide vane sleeve, 13. Packing gland, 14. Filler, 15. Packing ring, 16. Tail cover, 17. Bearing component a, 18. Bearing part b, 19. Balance drum, 20. Balance sleeve, 21. Pressure adjusting ring, 22. Throttle ring, 23. Tighten the bolts.

centrifugal pump design method are very similar. The difference lies in that in the multistage pump design, single stage centrifugal pump structures are added in series [5, 18].

Design parameters of multistage pump: Flow rate Q, Single-stage head of multistage pump H, Speed n, Specific speed n_s, Series i.

3.5.1 Determination of Inlet and Outlet Diameters of Multistage Pumps

The multistage pump inlet diameter is the diameter of the inlet flange pipe. Pump inlet diameter is determined based on the inlet for the required flow rate to ensure that the fluid enters the multistage pump at a reasonable speed. Designed for manufacturing convenience, if the flow rate is larger, reduce the volume of the pump and select a smaller inlet diameter to improve the inlet flow rate. Considering anti-cavitation, one may need to choose a large diameter inlet. Inlet diameter D_s is determined by the following formula:

$$D_s = K_s\sqrt[3]{\frac{Q}{n}}, \quad K_s = 4 - 5$$

The diameter of the flange at the pump outlet is the pump outlet diameter. Usually, the low head pump outlet diameter can be equal to the inlet diameter. When the pump head is high, in order to reduce the volume of the pump, the pump head can be lower than the inlet diameter, generally calculated according to the following formula:

$$D_d = K_d\sqrt[3]{\frac{Q}{n}}, \quad K_d = 3.5 - 4.5$$

3.5.2 Hydraulic Design of Impeller

The first stage impeller needs to be designed first, and the main geometric parameters of the impeller are shown in Figure 3.81.

1) Determine the impeller hub diameter:

   First, determine the pump shaft diameter with the following formula:

$$d = \sqrt[3]{\frac{M_n}{0.2[\tau]}}, \quad M_n = 9550\frac{p_c}{n}$$

In the above formula:

M_n—torques, N·m

p_c—Calculated power, can be taken as $p_c = 1.2p$

$[\tau]$—Permissible shear stress of the material, Pa, can be obtained from empirical data

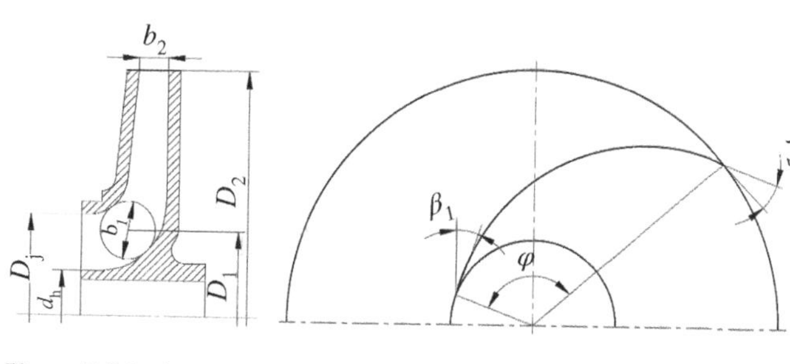

Figure 3.81 Impeller geometry of first-stage pump and main parameters.

2) Two sets of options for designing the impeller:
Design centrifugal pump mainly in accordance with the known design parameters requirements through a variety of options to optimize the choice in order to meet the necessary performance conditions to get the centrifugal pump hydraulic losses smaller and making the pump's external characteristics in line with the design requirements. A wealth of experience in design as well as sufficient reference materials is available to design an excellent hydraulic model for the pump. The design of the impeller includes two sets of programs: one is to select the same or similar model impeller with the same or similar specific rotational speed for a similar design, and the other program is to use the velocity coefficient method for design.

3.5.2.1.1 Similarity Design Method A similarity design method which is simple, reliable, and easy to implement, has become a major pump design method. The principle is to select a model pump similar to the pump to be designed, its specific speed being equal to the pump to be designed and meets the geometric similarity, motion similarity, and power similarity with the full size pump based on the law of similarity to determine its parameters. The design of centrifugal pumps using the similarity design method requires excellent hydraulic models for reference, and if there is no such model similar to the pump to be designed available in the literature, the pump cannot be designed by using the similarity design approach. The design steps include the following:

1) The pump specific speed n_s of the pump to be designed is given by the following equation:

$$n_s = \frac{3.65n\sqrt{Q}}{H^{3/4}}$$

2) Model pumps selection: the requirements for the model pumps are:
① The n_s of the model pump is equal or similar to the n_s of the design pump.
② The model pump has high efficiency (wide range of high efficiency), good anti-cavitation performance, and the shape of the characteristic curve meets the requirements.
③ The technical data of the model pump is reliable and complete.
④ In order not to lose similarity, the ratio of the real and model pump
Reynolds number is in the range of $^{Re}/_{Re_m} = 1.0 - 1.5$

$$\mathrm{Re} = \frac{u_2 D_2}{v_2}$$

In the above formula:
D_2—Impeller outer diameter, m
u_2—Circumferential velocity of impeller outlet, m/s
v—Kinematic viscosity of the conveyed fluid, m^2/s

3) Dimension less variable λ
In case of geometric similarity and kinematic similarity, by the law of similarity, assuming that the volumetric efficiency and the hydraulic efficiency of the model pump and the design pump are equal,

$$\lambda = \frac{D}{D_M} = \sqrt[3]{\frac{Q}{Q_M}\frac{n_M}{n}}, \quad \lambda = \frac{D}{D_M} = \frac{n_M}{n}\sqrt[3]{\frac{H}{H_M}}$$

If the values of λ calculated by the above two formula are slightly different, in that case the larger value or the average of the two can be taken as the value for conversion. In general, the closer λ

is to 1, more accurate is the conversion method. Finally, some important geometric parameters may need to be corrected according to the experience of the designer.

4) Calculate the size of the design pump
Multiply the various dimensions of the model overflow section by the size factor λ to get the corresponding dimensions of the overflow section of the design pump. The vane angle of the design pump should be equal to the corresponding angle of the model pump.

$$D = D_M \lambda \, \beta = \beta_M$$

Parameters such as blade thickness, which sometimes cannot be calculated using the similarity conversion values, can be determined on a case-by-case basis.

5) Conversion of performance curves for the design pump
In the model pump performance curve, take a number of points, then according to similarity conditions, convert them to the corresponding points in the design pump, and plot the corresponding performance curve.

$$\frac{Q}{Q_M} = \frac{n}{n_M}\left(\frac{D}{D_M}\right)^3 \quad Q = \lambda^3 \frac{n}{n_M} Q_M$$

$$\frac{H}{H_M} = \left(\frac{n}{n_M}\right)^2 \left(\frac{D}{D_M}\right)^2 \quad \eta = \frac{\rho g Q H}{1000P}$$

$$\frac{P}{P_M} = \left(\frac{n}{n_M}\right)^3 \left(\frac{D}{D_M}\right)^5 \quad P = \lambda^5 \left(\frac{n}{n_M}\right)^3 P_M$$

$$\eta = \frac{\rho g Q H}{1000P}$$

6) Drawing of the design pump
According to the size and angle of the conversion, taking into account the specific circumstances of the vane thickness, sealing clearance, and other appropriate modifications, draw the design

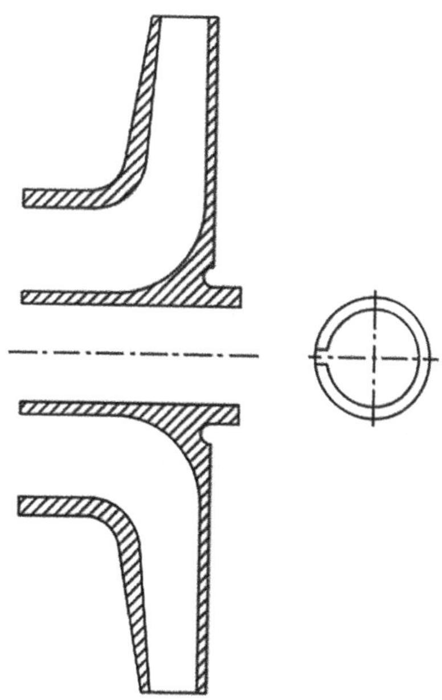

Figure 3.82 2D sketch of impeller.

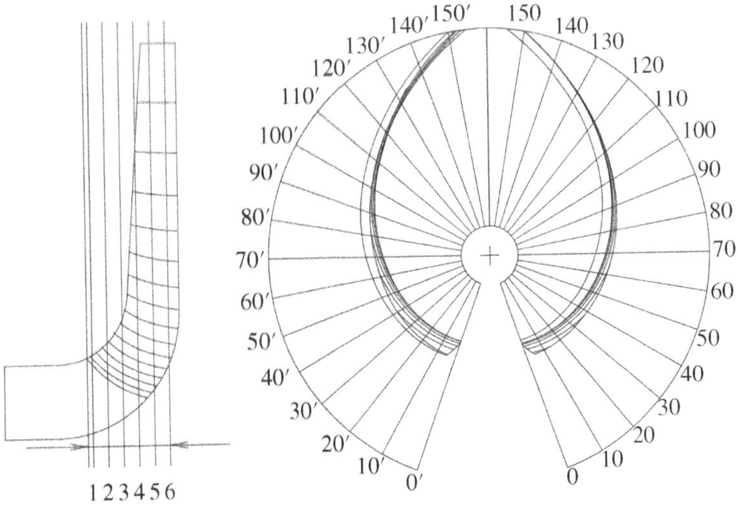

Figure 3.83 Blade axial cutout and wood mold drawing.

pump drawings. A vane cutting diagram can also be obtained in the conversion of the main dimensions, see redrawing (Figure 3.82).

A drawing of the blade wood mold is shown in Figure 3.83.

7) Problems to be aware of with a similarity design method

① Efficiency correction issues

Usually, in the similarity design method, the efficiency of the model pump and the design pump are supposedlly equal. In fact, due to the large pump and small pump (similar pump) flow channel relative roughness, relative clearance and vane relative thickness, etc., the large pump hydraulic efficiency and volumetric efficiency are higher than that of the small pump, the mechanical efficiency is often slightly higher. Therefore, when the size difference between similar pumps is large, the size effect of the correction should be considered.

② Modification of modeling issues

The existing model pump may be very good; however, if the design pump n_s is different, one can modify the model pump by changing its performance parameters such that the n_s of the two pumps is the same. Then, use the modified model pump for the design of the new design pump.

3.5.2.1.2 Velocity Coefficient Method The velocity coefficient method is very similar to the similarity design method described in the previous section, the difference being that in the similarity design method the model conversion is based on a similar model pump, while the velocity coefficient method is based on a series of similar pumps based on the design. In other words, the velocity coefficient method is based on a similar principle as the similarity method but uses statistical coefficients to calculate the size of the various parts of the overflow components. Since the velocity coefficient is a statistical value, the value introduced in different materials may not be exactly the same, therefore the application should be specifically analyzed.

Steps for calculating the main dimensions of an impeller by the velocity coefficient method:

① Impeller inlet diameter D_j

Since there are impellers with hubs (shaft-through impellers) and impellers without hubs (hanging-wall impellers), the concept of the equivalent diameter of the impeller inlet, D_0, is introduced in order to exclude the effect of the hub from the problem, i.e., to consider the general case. D_0 as the diameter of the circle area is equal to the impeller inlet to remove the effective area of the hub, that is:

$$\frac{D_0^2\pi}{4} = \frac{\left(D_j^2 - d_h^2\right)\pi}{4}$$

② Impeller outlet diameter D_2

$$D_2 = k_D\sqrt[3]{\frac{Q}{n}}\ k_D = 9.35k_{D2}\left(\frac{n_s}{100}\right)^{-1/2}$$

In the above formula:

$k_{D2}D_2$ modifying factor (Table 3.20)

③ Impeller outlet width b_2

$$b_2 = k_b\sqrt[3]{\frac{Q}{n}}\ k_b = 0.64k_{b2}\left(\frac{n_s}{100}\right)^{5/6}$$

In the above formula:

$k_{b2}b_2$ modifying factor (Table 3.21 and Figure 3.84)

④ Blade inlet angle β_1

Generally speaking, the blade inlet angle is larger than the liquid flow angle, i.e., $\beta_1 > \beta_1'$. Positive impulse angle is $\Delta\beta = \beta_1 - \beta_1'$, taking the value range of $\Delta\beta = 3°–15°$, the positive impulse angle is taken to improve the cavitation performance of the impeller, and it will not affect the efficiency.

Table 3.20 Relationship between k_{D2} and specific speed of multistage pumps.

n_s	20	30	40	50	60	70	80	90	100	110	120	130
k_{D2}	1.145	1.112	1.085	1.065	1.052	1.04	1.032	1.025	1.022	1.022	1.022	1.022
n_s	140	150	160	170	180	190	200	210	220	230	240	250
k_{D2}	1.022	1.024	1.027	1.029	1.032	0.038	1.047	1.058	1.074	1.093	1.117	1.146

Table 3.21 Relationship between k_{b2} and specific speed for multistage pumps.

n_s	20	30	40	50	60	70	80	90	100	110	120	130
k_{b2}	2.6	2.315	2.033	1.765	1.556	1.41	1.31	1.246	1.2	1.161	1.128	1.101
n_s	140	150	160	170	180	190	200	210	220	230	240	250
k_{b2}	1.078	1.055	1.03	1.003	0.976	0.95	0.924	0.898	0.87	0.842	0.814	0.787

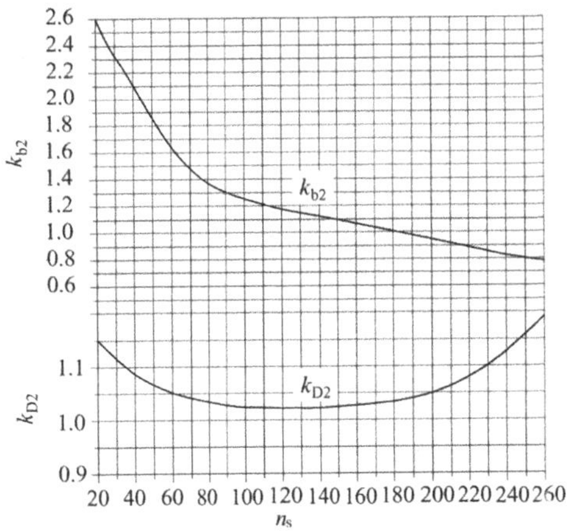

Figure 3.84 Relationship between k_{D2}, k_{b2}, and specific speed for multistage pumps.

The blade inlet angle β_1 is obtained by adding the liquid flow angle and impulse angle of each flow line. The blade angle is used to calculate the crowding coefficient, which is used to calculate the liquid flow angle. After obtaining the liquid flow angle, the blade angle can be derived.

⑤ Blade setting angle β_2

The outlet setting angle of the vanes is also one of the key factors in determining the hydraulic performance of the impeller of a centrifugal pump. For the same flow condition, with increase in β_2, the head increases. Because the outlet setting angle increases, the speed v_{2u} also increase and head increases as shown in Figure 3.85.

⑥ Number of vanes Z

The number of vanes should be determined in such a way so as to minimize fluid congestion in the pump and friction on the surface of the vanes, and the vanes should have sufficient length to adequately ensure smooth fluid flow. Based on the past experience, the number of vanes is generally selected as Z: 5–7.

After completing the above design, a two-dimensional impeller drawing similar to Figures 3.82 and 3.83 for the velocity coefficient method can be drawn.

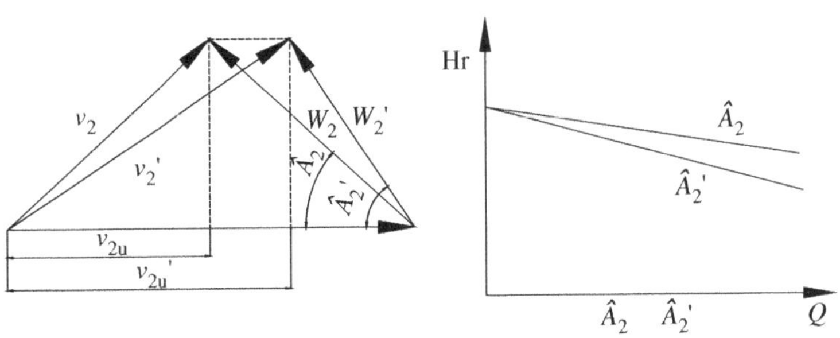

Figure 3.85 Relationship between head and outlet angle of repose.

3.5.3 Design of the Main Parameters of the Guide Vane

Guide vanes are generally used in multistage pumps between two neighboring impellers, or at the pump outlet, for collecting the liquid flowing from the impeller and transporting it to the inlet of the next impeller, and at the same time, they can make the fluid move relatively smoothly reducing the hydraulic loss, and reducing the fluid flow rate to increase the potential energy of the fluid. The flow path of the radial guide vane is shown in Figure 3.86, which consists of a helix section, a diffusion section, a transition section, and a counter-guide vane. The shape of the guide vane is complex, and its internal flow velocity is very high, accounting for a large proportion of the hydraulic loss of the pump. Reasonable selection of geometric parameters has an important impact on the performance of the pump.

3.5.3.1 Positive Guide Vane Design

1) Guide vane base circle diameter D_3

$$D_3 = D_2 \times (1.02 - 1.05) \quad \text{or} \quad D_3 = D_2 + (1 - 3)$$

2) Guide vane inlet width (axial width) b_3

$$b_3 = b_2 + (3 - 5)$$

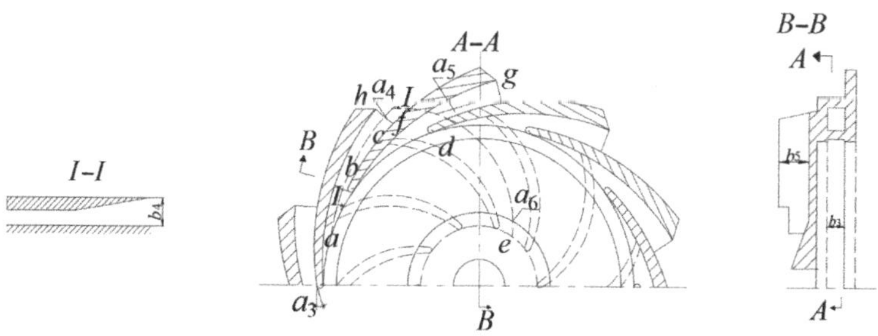

Figure 3.86 Radial guide vane structure.

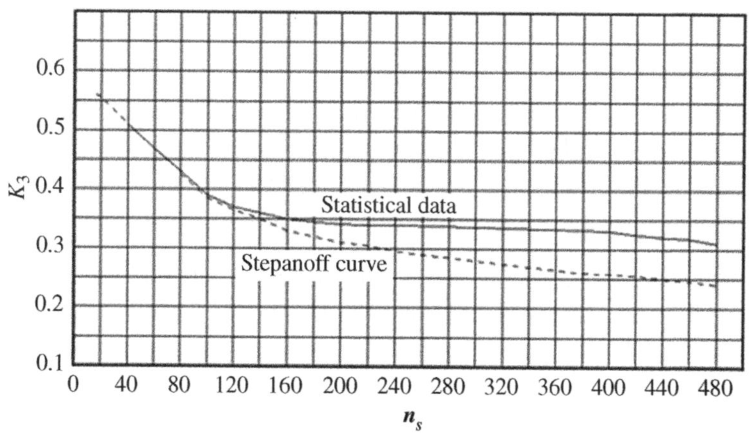

Figure 3.87 Velocity coefficient K_3 vs. specific rotation speed.

3) Guide vane inlet placement angle α_3
The guide vane inlet axial surface velocity v_{m3} and the blockage coefficient ψ_3 are:

$$v_{m3} = \frac{Q}{2\pi R_3 b_3 \psi_3} \quad \psi_3 = 1 - \frac{\delta_3 z}{D_3 \pi \sin\theta}$$

Guide vane inlet circumferential velocity: $v_{u3} = \frac{v_{u2} R_2}{R_3}$
Inlet flow angle of guide vane α_3': $\tan\alpha_3' = \frac{v_{m3}}{v_{u3}}$

4) Guide vane throat area and shape
Throat area $A_3 = a_3 b_3$ (a_3 is the width of the throat plane, b_3 is the width of the throat axis). It is generally recommended that $a_3 = (0.6\text{–}1)\, b_3$.
Velocity coefficient method calculation

Throat speed: $v_3 = K_3\sqrt{2gH}$

where K_3 is the velocity coefficient, as shown in Figure 3.87.

Throat area $F = \frac{Q}{z v_3}$,

where z is the number of guide vanes, usually the number of guide vane blades is 6–8.

5) Guide vane outer diameter D_4

$$D_4/D_3 = 1.3 - 1.5$$

6) Design and calculation of the diffusion section
The inlet area of the diffusion section is the throat area $F_3 = a_3 b_3$, reference value:
$F_4/F_3 = 1.5\text{–}1.8 \quad D_4/D_3 = 1.3\text{–}1.5$
$L/a_3 = 3\text{–}4 \quad v_4/v_3 = 0.4\text{–}0.5$
Determination of diffusion angle by equivalent circular area

$$\varphi = 2\arctan\frac{\sqrt{F_4/\pi} - \sqrt{F_3/\pi}}{L}$$

The range of the φ angle is 6°–8°.

7) Transition section design
The area of the transition section should be equal to or slightly larger than the outlet area of the diffusion section, and the area of the turning process should be appropriately enlarged or kept unchanged.

3.5.3.2 Return Guide Vane Design

The design of the return guide vane refers to the existing structure of good performance, consider a positive guide vane and the lower impeller inlet connection, and strive to flow smoothly and unobstructed, to ensure that the required speed size and direction of change. The specific steps are:

1) Determination of the main dimensions

The number of blades of the return guide vanes is generally the same as that of the positive guide vanes, and they should be positioned in a circumferential orientation so that the liquid flow between each of the two positive guide vanes enters the flow channel between the two return guide vanes in a continuous manner.

The inlet diameter D_5 of the return guide vane is generally slightly smaller than the outer diameter D_4 of the positive guide vane; the outlet diameter D_6 is generally equal to the inlet diameter D_j of the impeller, or appropriately extended a little in the axial direction; the axial width b is determined in accordance with the required speed changes, generally $b_5 = (1.5\text{–}1.8)\, b_2$.

2) Angle of placement of the inlet of the return guide vane α_5

The fluid flows out of the diffusion section of the positive guide vane and the velocity momentum remains constant until it reaches the return guide vane.

$$v_{u4}R_4 = v_{u5}R_5$$

Axial surface velocity: $v_{m5} = \frac{Q}{D_5 b_5 \pi \psi_5}$, blockage factor: $\psi_5 = 1 - \frac{\delta_5 z}{D_5 \pi \sin \alpha_5}$.

where

δ_5—True thickness of the inlet of the return guide vane;

α_5—The angle of placement of the inlet of the return guide vane, it can be calculated in the first approximation by assuming that $\alpha_5 = \alpha_4$ and then checking the calculation.

$$\tan \alpha_5' = \frac{v_{m5}}{v_{u5}}$$

In order to minimize the bending of the return guide vane, take the placement angle as large as a thousand times the liquid flow angles, i.e., take a positive impulse angle:

$$\alpha_5 = \alpha_5' + \Delta\alpha$$

$\Delta\alpha$ ranges from 3° to 10°, with pumps having a small n_s taking the maximum value.

The calculated α_5 should be equal or similar to the preassumed value, otherwise it should be recalculated.

3) Angle of placement of return guide vane α_6

Under small flow conditions, in order to improve the anti-cavitation performance of the pump, usually take $\alpha_6 = 55°–80°$.

References

1 Xingfan, G. (1995). *Modern Pump Technology Manual*. Beijing: Aerospace Press.
2 Shouqi, Y. (2014). *Pump Theory and Technology*. Beijing: Mechanical Engineering Press.
3 Stepanoff, A.J. (1993). *Centrifugal and axial flow pumps: theory, design, and application*. Krieger Publishing Company.
4 Ding, C.W. (1981). *Centrifugal and Axial Flow Pumps*. Beijing: China Machine Press.
5 Zha, S. (ed.) (1988). *Theory and Hydraulic Design of Vane Pumps*. Beijing: China Machine Press.
6 Shenyang Pump Research Institute (1983). *Vane Pump Design Manual*. Beijing: China Machine Press.
7 Lobanoff, V.S. and Ross, R.R. (1992). *Centrifugal Pumps Design & Application*, 2e. Houston: Gulf Publishing Company.
8 Li, S.H. and Wu, Y.L. (1990). *A Course in Pump Design*. Beijing: China Machine Press.
9 Fuchslocher and Schulz, H. (1963). *Die Pumpen: Arbeitsweise, Berechnung, Konstruktion fur Studierende des Maschinenbaus und zum Selbststudium*. Berlin: Springer.
10 Pfleiderer, C. (1959). *Die Kreiselpumpen: fur Flussigkeiten und Gase: Wasserpumpen, Ventilatoren, Turbogeblase, Turbokompressoren*. Berlin: Springer.
11 Li, S.H. (2006). *Off-Design Conditions and Optimal Design of Vane Pumps*. Beijing: China Machine Press.
12 Sitaram, N., Lakshminarayana, B., and Ravindranath, A. (1981). Conventional probes for the relative flow measurement in a rotor blade passage. *ASME Journal of Engineering for Power* 103: 406–414.

13 Zierke, W.C., Farrell, K.J., and Straka, W.A. (1995). Measurements of the tip clearance flow for a high-reynolds-number axial-flow rotor. *ASME Journal of Tubomachinery* 117: 522–532.

14 Chen, N.X. and Wu, Y.L. (2003). *Centrifugal Pumps*. Beijing: China Machine Press.

15 Tuzson, J. (2000). *Centrifugal Pump Design*, 3e. New York: A Wiley-Insterscience Publication.

16 Hui, Q. and Rennian, L. (2021). *Theory and Design of Solid-Liquid Two-Phase Flow Pumps*. Beijing: China Machine Press.

17 The Institution Industries Division of the Institution of Mechanical Engineers (1982). *Centrifugal Pumps—Hydraulic Design*. London: The Institution of Mechanical Engineers London.

18 Turton, R.K. (1994). *Rotodynamic Pump Design*, 2e. New York: Cambridge University Press.

19 Troskolański, A.T. and Łazarkiewicz, S. (1976). *Kreiselpumpen: Berechnung und Konstruktion. Kreiselpumpen*. Berlin: Springer.

20 Zierke, W.C., Straka, W.A., and Taylor, P.D. (1995). An experimental investigation of the flow through an axial-flow pump. *ASME Journal of Fluids Engineering* 117: 485–490.

4

Computational Fluid Dynamics Methodology for Impeller Pumps

4.1 Introduction to Fundamental Concepts of CFD

Computational fluid dynamics (CFD) is a technology that utilizes computers to simulate the fluid motion; it is capable of modeling the behavior of various types of fluids including gases, liquids, and multiphase fluids. The application scope of CFD technology is very broad, encompassing aerospace, automotive, shipbuilding, agriculture, architecture, energy, environment, and other industries. For example, in the aerospace sector, CFD technology can be used to design and optimize the aerodynamic shapes of aircraft, rockets, missiles, and other aerial vehicles, as well as gas turbines and the combustion chambers of engines. In the automotive industry, it aids in the design and optimization of vehicle exteriors, chassis, engines, and improving the aerodynamic efficiency and performance of the vehicle. In shipbuilding, CFD technology is applied to the design and optimization of hull shapes, and propellers for improving the hydrodynamic performance of ships. In architectural engineering, it assists in the design and optimization of building ventilation, air conditioning, and thermal performance. In the energy industry, CFD technology is utilized for the design and optimization of the hydrodynamic performance of water pumps, hydro-turbines, tidal turbines and wind-turbines. In the field of environmental protection, CFD technology can provide simulations to address issues related to air pollution, water pollution, and noise pollution, facilitating the implementation of effective environmental protection measures.

The development history of CFD technology can be traced back to the 1940s, with its initial applications primarily in the aerospace field. With the continuous development of computer technology, CFD technology has advanced rapidly, and its application scope has continuously expanded. Currently, CFD technology has become one of the important tools for engineering design and optimization, significantly shortening product development cycles, reducing development costs, and enhancing the product performance and reliability. In summary, CFD technology has become a crucial engineering computation method that can simulate the motion and performance of various fluid systems, providing strong support for engineering design and optimization. With the continuous development of computer hardware and numerical methods, the application prospects of CFD technology will continue to become broader.

The fundamental principle behind the CFD technology is to discretize the basic partial differential equations (PDEs) of fluid motion (conservation of mass, conservation of momentum, and conservation of energy) using the appropriate stable numerical algorithms and then solve the resulting algebraic equations applying the well established techniques from linear algebra and matrix theory. The main building blocks of CFD technology include establishing a geometric model, defining a computational domain around or inside the model, surface and volume grid generation in the computational domain, and then solving the governing equations on the generated mesh with appropriate boundary conditions, and finally postprocessing the numerically generated solution data.

Among these building blocks, establishing a geometric model is the first step, which requires converting the geometric shape of the fluid system into a mathematical model (CAD model) that can be processed by a computer. For this purpose, the well known software e.g. Ansys SpaceClaim exists. The next step is to generate a structured (hexahedral) or unstructured (tetrahedral) or hybrid mesh inside the computational domain including the mesh on its surface using software such as Ansys ICEM CFD. The discretization of the governing equations involves transforming the continuous partial differential equations of the fluid system into discrete algebraic equations, typically using numerical methods such as the finite volume method, finite element method, or spectral method. The solution of these equations involves using numerical methods to solve the discretized system of equations, usually through iterative or direct methods. For this purpose, one can employ the well known Ansys Fluent/CFX flow solver. As a final step, the postprocessing involves analyzing and visualizing the CFD simulation results to evaluate and optimize the performance of the fluid system.

In CFD technology, the Navier–Stokes equations are the fundamental governing equations for describing the fluid motion, which are highly nonlinear momentum equations whose numerical solution has been a challenge for more than a century. With the availability of increasingly powerful computers, last fifty years have seen rapid progress in the development of efficient and stable numerical algorithms for the solution of these equations. These equations are solved in conjunction with the conservation of mass equation and the conservation of energy equation. In this chapter, we consider only the incompressible Navier-Stokes equations since the fluid medium is considered mainly incompressible. Equation (4.1) represents the incompressible Navier–Stokes equations in a Cartesian coordinate system (x, y, z). The three velocity component are denoted as (u, v, w), pressure is denoted by p, μ denotes the dynamic viscosity and ρ denotes the density of the fluid medium, and f denotes the body force.

$$\frac{\partial u}{\partial t} + u\frac{\partial u}{\partial x} + v\frac{\partial u}{\partial y} + w\frac{\partial u}{\partial z} = f_x - \frac{1}{\rho}\frac{\partial p}{\partial x} + \frac{\mu}{\rho}\left(\frac{\partial^2 u}{\partial x^2} + \frac{\partial^2 u}{\partial y^2} + \frac{\partial^2 u}{\partial z^2}\right)$$

$$\frac{\partial v}{\partial t} + u\frac{\partial v}{\partial x} + v\frac{\partial v}{\partial y} + w\frac{\partial v}{\partial z} = f_y - \frac{1}{\rho}\frac{\partial p}{\partial y} + \frac{\mu}{\rho}\left(\frac{\partial^2 v}{\partial x^2} + \frac{\partial^2 v}{\partial y^2} + \frac{\partial^2 v}{\partial z^2}\right)$$

$$\frac{\partial w}{\partial t} + u\frac{\partial w}{\partial x} + v\frac{\partial w}{\partial y} + w\frac{\partial w}{\partial z} = f_z - \frac{1}{\rho}\frac{\partial p}{\partial w} + \frac{\mu}{\rho}\left(\frac{\partial^2 w}{\partial x^2} + \frac{\partial^2 w}{\partial y^2} + \frac{\partial^2 w}{\partial z^2}\right) \tag{4.1}$$

Eq. (4.1) along with the continuity equation for conservation of mass is solved for the velocity components (u, v, w) and pressure p. Thus the incompressible fluid motion is described by only the velocity field and pressure field. However, the continuity equation does not contain a pressure term, therefore there is weak coupling between the velocity field and the pressure field. In order to address this problem numerically, the SIMPLE algorithm was developed by Patankar and Spalding in 1972 which is now a part of almost all CFD solvers. Since then, variants of SIMPLE and several other algorithms have been developed and they are part of Ansys Fluent/CFX and other flow solvers. In CFD simulations, the proper imlementation of boundary conditions has a significant impact on the results. Boundary conditions may include velocity boundary conditions and/or pressure boundary conditions at various boundaries of the computational domain. The choice of boundary conditions has a crucial influence on the accuracy and reliability of the simulation results.

4.2 Numerical Methods

The numerical methods employed in various building blocks of CFD are briefly discussed in the following sections.

4.2.1 Geometry Modeling

Using three-dimensional geometry modeling software such as Creo, SolidWorks, Ansys SpaceClaim as needed, the 3D geometric shape of the vane pumps are modeled, and then the computational simulation domains are extracted, including the inlet domain, the conical connecting pipe domain (connecting the impeller domain and the inlet domain), the impeller domain, the guide vane domain, the annular vortex domain, and the outlet domain. Finally, all solid domains of the pump are assembled. The specific 3D geometric modeling process for different vane pumps is provided in detail for five different pumps - centrifugal pump, axial-flow pump, mixed-flow pump, vortex pump and multi-stage pump later in Chapters 5–10. It should be noted that Creo and SolidWorks are featured based parametric CAD systems used for creating complex design-intent-driven models while SpaceClaim is a direct modeling tool designed for fast geometry editing and simulation.

4.2.2 Mesh Generation

Meshing involves discretizing a continuous geometric region (such as a model of a structure or a fluid domain) into multiple small elements or units for computation using numerical methods (like finite difference, finite volume or finite element). In engineering and scientific computing, particularly in topology optimization and finite element method (FEM), meshing plays a crucial role. By meshing, originally complex continuous problems are transformed into multiple smaller problems that can be solved on a computer using numerical methods.

The commonly used mesh types in numerical computation include structured, unstructured, and hybrid (combination of structured and unstructured) meshes:

1) **Structured mesh:**
   The nodes are arranged in an orderly manner with clear relationships among the neighboring nodes, making it suitable for implementation of many algorithms and efficient processing. The 2D mesh elements are quadrilateral and 3D mesh elements are hexahedral. A variety of methods and associated software exist for generating 2D and 3D structured meshes in a computational domain.
2) **Unstructured mesh:**
   The distribution of nodes and elements is arbitrary, allowing for better handling of boundaries. Unstructured mesh generation methods employ certain criteria for optimization and user's judgment during the mesh generation process, enabling the production of high-quality meshes and easy control over mesh size and node density in a specified region of the computational domain. However, storage is more complex, requiring larger memory space and relatively higher computational cost.

For unstructured grid generation in a computational domain, a variety of methods and associated software are available; some of them are listed below:

1) **Delaunay method:**
   This is a commonly used unstructured mesh generation algorithm; it generates triangular meshes in 2D and tetrahedral meshes in 3D. It generates high-quality mesh for a given set of discrete points. The Delaunay method employs the circumcircle criterion to generate 2D triangular meshes and a circumsphere criterion for generating 3D tetrahedral meshes.
2) **Advancing-front method (frontal method):**
   The key idea behind this method is to generate a mesh on the original boundary surface and then progressively advance until the entire domain is filled with the mesh. The frontal method can produce high-quality meshes with controllable gradients, ensuring the integrity of target boundaries. However, in a 3D domain, some areas may be difficult to fully fill with mesh.

3) **Spatial decomposition-based method:**
This method primarily uses spatial tree structures, such as 2D quadtrees and 3D octrees. It creates a tree structure based on the distribution of geometric objects and points, and then generates a mesh within each tree node. This method is relatively simple and can dynamically generate unbalanced trees to adapt to mesh requirements in different regions.

Mesh Quality

The quality of the mesh directly determines the accuracy and convergence of computations. The quality of the mesh is mainly reflected in its following attributes:

1) **Mesh density:**
Mesh density directly affects the accuracy and computational load of the results. A coarser mesh may miss some geometric and physical details, affecting the accuracy of the results; while a denser mesh improves accuracy but also increases the computational load.
2) **Mesh distribution:**
A reasonable mesh distribution ensures the accuracy and stability of the results. In practical applications, mesh distribution may need to be adjusted based on the geometric characteristics and physical properties of the model.
3) **Mesh shape:**
The shape of the mesh also affects the accuracy of the results. Irregularly shaped mesh elements may produce larger errors in computations. Therefore, it is necessary to generate regularly shaped mesh elements as much as possible.

The evaluation parameters for mesh quality include maximum edge length, minimum edge length, maximum angle, minimum angle, aspect ratio, collapse ratio, dihedral angle, and skewness. These parameters help users assess mesh quality and make corresponding changes in the mesh.

4.2.3 Governing Equations of Fluid Dynamics

The conservation equations of fluid dynamics, namely the continuity equation, momentum equations, and energy equation, are the basic equations governing the fluid motion based on the law of conservation of mass, Newton's second law, and the law of conservation of energy. These equations are numerically solved by CFD [1].

In an inertial frame of reference in Cartesian coordinate system, the basic governing equations describing the internal instantaneous flow state in hydraulic machinery include the following two equations [2].

Continuity equation

$$\frac{\partial \rho}{\partial t} + \frac{\partial(\rho u_{\mathrm{i}})}{\partial x_{\mathrm{i}}} = 0 \tag{4.2}$$

and the Navier–Stokes equations

$$\frac{\partial(\rho u_{\mathrm{i}} u_{\mathrm{j}})}{\partial x_{\mathrm{j}}} + \frac{\partial(\rho u_{\mathrm{i}})}{\partial t} = -\frac{\partial p}{\partial x_{\mathrm{i}}} + \frac{\partial}{\partial x_{\mathrm{j}}}\mu\left(\frac{\partial u_{\mathrm{i}}}{\partial x_{\mathrm{j}}}\right) + S_{\mathrm{i}} \tag{4.3}$$

where ρ denotes the fluid density, u_{i} is the velocity component in the i th direction ($i = x, y$, or z), p is the pressure, t is the time, x_{i} is the spatial coordinate (x, y, or z), μ is the dynamic viscosity, and S_{i} is the source term in the i th direction ($i = x, y$, or z). In Eqs. (4.2) and (4.3), i and j represent the directional component (i or $j = x, y$, or z) of a variable and follow the summation convention in tensor algebra. Eq. (4.2) reduces to Eq. (4.1) for incompressible flow.

4.2.4 Simulation of Turbulent Flows

The internal flow of hydraulic components such as an impeller and a guide vane in hydraulic machinery can be considered as a 3D incompressible unsteady turbulent flow.

The numerical simulation of turbulent flow remains the pacing item and the most difficult problem in CFD. The main turbulent flow simulation methods can be categorized as (1) the direct numerical simulation (DNS) and (2) the large eddy simulation (LES) of the Navier–Stokes equations, and (3) the simulation based on the Reynolds-averaged Navier–Stokes (RANS) equations in conjunction with a turbulence model needed to accomplish the closure of RANS equations by modeling the Reynolds or turbulent stresses. This classification of turbulent flow solution methods is shown schematically in Figure 4.1 [3–8].

DNS refers to the direct numerical solution of instantaneous or unsteady Navier–Stokes equations [9–12] without any approximation or modeling. All other (non-DNS) methods aim to analyze the turbulent flow by employing some degree of approximation and simplification to instantaneous or unsteady Navier–Stokes equations, which in principle include all length and time scales of turbulence. Based on different approximation and simplification methods, they can be divided into LES, Reynolds time-averaging, and statistical averaging methods. In order to use DNS to calculate some highly complex turbulent fluid motions at high Reynolds numbers (HRNs), very small time and spatial discretization steps must be employed in solving the Navier–Stokes equations in order to obtain the detailed local spatial structure and temporal characteristics of turbulent fluid motion. For the complete 3D complex geometric model, like those in hydraulic machinery, the existing supercomputing power is not sufficient to realize such calculations using DNS in the foreseeable future, since the DNS of turbulence requires very large computer memory and operational speed. At present, DNS cannot be employed for practical engineering calculations; however, a great deal of exploratory work is underway as more and more computing power of exascale computers is becoming available. The LES method, on the other hand, employs the unsteady Navier–Stokes equation to directly simulate the large-scale eddies of turbulent flow, but the small-scale eddies are not directly simulated. Thus, in the LES conceptually, the turbulent fluid motion is considered to be composed of large- and small-scale eddies. The influence of small eddies on large eddies is considered by an approximate model. The standard method is to filter the small-scale eddies with smaller scales of turbulence compared to the specified filter function from the instantaneous Navier–Stokes equations so that the filtered equation then describes the

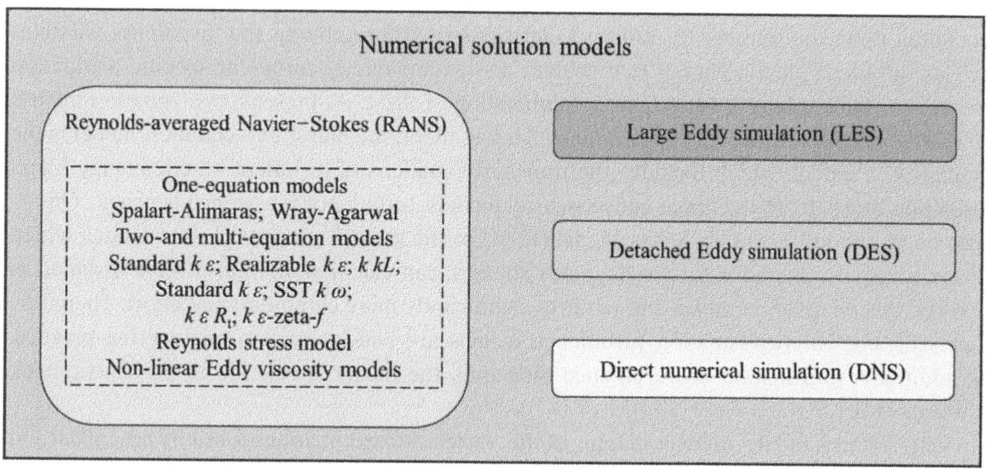

Figure 4.1 Numerical solution models for turbulent flow simulations.

flow field with large eddies. In this process, an additional stress term is added to the equations describing the motion of large eddies, which reflects the influence of filtered small eddies on large eddy motion. The stress term is similar to the Reynolds stress term in the RANS equations; this stress term is called the subgrid scale stress in LES. Most subgrid Reynolds stress models (RSMs) in the literature since the original model of Smagorinsky are based on an eddy viscosity concept, which relates the subgrid scale stress to strain tensor using the Boussinesq-type approximation. Although the requirements of LES for computer memory and operational speed are still relatively high for many routine 3D industrial applications, they are, however, much lower than those for DNS. As a result, LES is being increasingly used in 3D simulations at HRNs using the currently available high-end computing hardware.

Instead of solving the Navier–Stokes equations by DNS or LES, the Reynolds time-averaging method is employed to obtain the RANS equations for incompressible flows or the mass-weighted RANS equations for compressible flows (also called the Favre equations) are solved in conjunction with a turbulence model in most commercial CFD codes for industrial applications;, these methods give results of acceptable engineering accuracy on currently available computing hardware. Thus the RANS method has become the most widely used approach for turbulent flow simulations in hydraulic machinery.

Finally, the statistical averaging method is based on the statistical theory of turbulent fluid motion extended from the turbulence correlation function. It mainly uses the correlation function and spectral analysis method to analyze the turbulence structure, which primarily involves the flow of small-scale eddies. It is hardly used in engineering practice.

Now, the most widely used RANS method, which requires modeling of Reynolds stresses in order to achieve closure of the governing equations, is described in detail. The problem of closure in RANS equations is referred to in the literature as "turbulence modeling." Over a century, a wide variety of turbulence models of different levels of complexity have been developed for use with RANS equations; they vary in accuracy and efficiency depending upon the application, but none of them can be considered to approach universality. Most of the turbulence models are based on the concept of "eddy viscosity" in the Boussinesq approximation, which postulates that the turbulent stress tensor is proportional to the strain-rate tensor with a proportionality constant defined as eddy viscosity, analogous to Stokes' hypothesis used in the derivation of Navier–Stokes equations from Newton's second law. Depending upon the complexity, most linear eddy-viscosity models are classified as the zero-equation, one-equation, and two-equation models depending upon the number of transport equations used for defining the turbulence variables such as turbulent kinetic energy k, turbulent dissipation rate ε, turbulent specific dissipation rate ω, or turbulent length scale L, or a combination of these. At present, two-equation models are widely used in engineering applications. Among them, the oldest two-equation model is the standard k–ε model, which includes the transport equations of turbulent kinetic energy k and dissipation rate ε. In all the linear eddy-viscosity models, turbulence is assumed isotropic. On the other hand, the full Reynolds Stress Model (RSM) or the second-moment closure model, which has six transport equations for six Reynolds stresses, can calculate the anisotropic turbulence; however, it is relatively complex and requires significantly more computational effort. Therefore, at present, one- or two-equation turbulence models are widely used in engineering practice. For additional information on turbulence modeling, the reader is referred to the classic book by Wilcox [13].

In what follows, the Reynolds-averaging of the Navier–Stokes equations is briefly described, and the derivation of the RANS equations is presented. In Reynolds-averaging, the instantaneous flow variables in the Navier–Stokes equations are decomposed into two parts: the average value over a certain time period and the time-dependent fluctuation. For example, the velocity components can be decomposed as:

$$u_i = \overline{u}_i + u'_i \tag{4.4}$$

where $\overline{u}_i$ and u'_i represent the average and fluctuating velocity components, respectively ($i = 1, 2, 3$)

Similarly, for scalar variables ϕ such as pressure and energy, one can write:

$$\phi = \overline{\phi} + \phi' \tag{4.5}$$

By introducing the decomposition of variables given in Eqs. (4.4) and (4.5) into the continuity equation (4.2) and Navier–Stokes equations (4.3) and performing the Reynolds averaging on the equations over a time period T, the continuity equation and the momentum equations can be written in a Cartesian coordinate system (by dropping the overbar on the mean velocity $\overline{u}$) in tensor notation as:

$$\frac{\partial \rho}{\partial t} + \frac{\partial}{\partial x_i}(\rho u_i) = 0 \tag{4.6}$$

$$\frac{\partial(\rho u_i)}{\partial t} + \frac{\partial}{\partial x_j}(\rho u_i u_j) = -\frac{\partial p}{\partial x_i} + \frac{\partial}{\partial x_j}\left[\mu\left(\frac{\partial u_i}{\partial x_j} + \frac{\partial u_j}{\partial x_i} - \frac{2}{3}\delta_{ij}\frac{\partial u_k}{\partial x_k}\right)\right] + \frac{\partial}{\partial x_j}\left(-\rho\overline{u'_i u'_j}\right) \tag{4.7}$$

Equations (4.6) and (4.7) are called the RANS equations. They have basically the same form as the unsteady Navier–Stokes equations except that the flow variables including the velocity components are replaced by their time averages, and there is an additional term $-\rho\overline{u'_i u'_j}$ in Eq. (4.6) which represents the Reynolds or turbulent stress due to the influence of turbulent fluctuations.

In Eq. (4.6), the Reynolds stress term is unknown and is not defined in terms of the other flow variables such as velocity and pressure. This presents the so-called "closure problem" for the solution of Eqs. (4.6) and (4.7). In order to close the equations, the Reynolds stress term $-\rho\overline{u'_i u'_j}$ needs to be modeled in terms of the flow variables. The modeling of Reynolds stress term in Eq. (4.7) is referred to in the literature as "turbulence modeling." The most used method to model the Reynolds stress is to employ the Boussinesq hypothesis, which relates the Reynolds stress tensor with the strain-rate tensor via an eddy viscosity μ_t, as shown below:

$$-\rho\overline{u'_i u'_j} = \mu_t\left(\frac{\partial u_i}{\partial x_j} + \frac{\partial u_j}{\partial x_i}\right) - \frac{2}{3}\left(\rho k + \mu_t\frac{\partial u_k}{\partial x_k}\right)\delta_{ij} \tag{4.8}$$

Boussinesq hypothesis is applied in the derivation of all one- and two-equation turbulence models described in the next section.

Turbulence models are needed in a wide range of applications in engineering fields, e.g. in aircraft design, automotive design, and design of rotating machinery to name a few. By using turbulence models in conjunction with RANS equations, turbulent behavior in fluids can be predicted with acceptable engineering accuracy, then the design schemes can be optimized, and product performance and efficiency can be improved. In industrial applications, the selection of turbulence models needs to consider several factors, such as fluid properties (single or multiphase flow), flow properties (boundary layers, shear layers, wakes, separated flow and stall etc.), and computational resources (simple algebraic to multi- transport equation models), etc. For fluid problems with low Reynolds numbers, RANS models can be used, for example, to predict hydrodynamic performance in ship design. For fluid problems with high Reynolds numbers, LES models preffered, e.g. in predicting the aerodynamic performance in aircraft design. Sometimes, hybrid models that combine RANS and LES models such as Detached Eddy Simulation (DES) can be used to balance the cost relationship between computational accuracy and computational load. Furthermore, the research in turbulence modeling faces some fundamental challenges and difficulties.

For example, turbulence models need to consider multiscale turbulent structures, but the computational resources are generally limited, therefore the challenge is how to accurately and efficiently handle the multiscale problems. Additionally, in some applications the turbulence models may need to account for the instability and transitional behavior of flow, which adds substantially more complexity in modeling.

Over a century, the turbulence models have been developed as simple algebraic equations or partial differential equations that determine the turbulent eddy viscosity in the Boussinesq approximation for the Reynolds stresses (Eq. (4.8)). Based on the assumptions made for characterization of the turbulent process by key turbulence variables such as turbulent kinetic energy, dissipation or dominant length scale, simple algebraic or complex multi-transport equations based turbulence models are formulated which are described in the existing literature as follows.

The algebraic models, also known as the Zero-equation models are the simplest of all turbulence models. These models are usually based on empirical formulas for eddy viscosity derived from some rationale about the turbulence mixing length scale wherein the empirical constants are determined by comparison with experimental data for simple flows such as the turbulent flow over a flat plate. These models are suitable for computation of simple turbulent flows. The most common zero-equation models include the Cebeci–Smith (C-S) model and the Baldwin–Lomax (B-L) model. Among them, the B-L model is generally preferred for computation of turbulent boundary layers since it does not require explicit determination of the outer edge of the boundary layer. However, the disadvantage of zero-equation models is that they cannot accurately capture the turbulent flow features in multi-dimensional flow since they do not consider the transport and dissipation of turbulent kinetic energy, and therefore may not accurately simulate the flows with separated and recirculating regions and flows involving shear layers and wakes.

One-equation models introduce a partial differential equation to describe the transport of a turbulence variable, which is usually related to turbulent kinetic energy or turbulent eddy viscosity. These models are certainly more complex and computationally more expensive than zero-equation models but provide higher accuracy. There are several one-equation models described in the literature, well-known among them include the Spalart–Allmaras (SA) model, the Wray-Agarwal (WA) model and the Baldwin–Barth (BB) model. Among them, the SA model is most widely used for wall-bounded flows in the aerospace field and has shown good results. In recent years, the WA model is increasingly being used in a wide variety of applications in aerospace, automotive, and wind energy industries among others as well as in hydraulic machinery. The advantage of one-equation models is that they are relatively simple and computationally efficient. However, since they involve only one a transport equation for a turbulence variable, they nevertheless have certain limitations when simulating complex turbulent flows.

Two-equation models on the other hand describe the turbulent flow characteristics by introducing two additional transport equations for two turbulence variables, which are usually the turbulent kinetic energy and turbulent dissipation rate or turbulent specific dissipation rate. These models are more complex than zero-equation and one-equation models but offer higher accuracy and a broader range of applicability. The most common two-equation models include the k–ε model, the k–ω model and the SST k-ω model.

k–ε Model: In this model, k represents turbulent kinetic energy, and ε represents the dissipation rate of turbulent kinetic energy. This model assumes that the flow is fully turbulent, and the influence of molecular viscosity can be neglected. It was first proposal by Launder and Spalding, and over the years the standard k–ε model has become one of the primary turbulence modeling tool in engineering flow field calculations. The RNG k–ε model, derived from rigorous statistical techniques, adds a condition to the ε equation, effectively improving the accuracy by taking into account the turbulent vortices.

k–ω Model: In this model, ω is the specific dissipation rate. It was proposed by Wilcox to overcome some of the weaknesses of the k–ε model, particularly in the near-wall region, where the k–ω model performs significantly better than the k–ε model. However, the k–ω model is highly sensitive to the free-stream value of ω. Menter combined the k–ε and k–ω models using a blending function to design a new Shear Stress Transport (SST) k-ω model, which retains the excellent characteristics of the k–ω model in the near-wall region while overcoming its sensitivity to the free-stream value of ω.

Two-equation models are widely used in engineering applications, especially in the field of hydraulic machinery. However, they still have certain limitations, such as failing to accurately simulate turbulent characteristics in some complex flows with rotating stall.

In addition to the aforementioned three types turbulence models, there are more complex multi-equation models, namely the four-equation and seven-equation models. These models describe turbulent characteristics by introducing more partial differential equations for turbulence variables, thereby presumably providing higher accuracy and a broader range of applicability. However, due to their higher complexity, the computational load also increases accordingly, which may limit their application in engineering.

The following sections describe some of the most widely used one-equation and two-equation turbulence models. It should be noted that since they are rarely used in hydraulic machinery flow simulations, other one-equation (e.g., Baldwin–Barth) and two-equation (e.g., k-kL) models are not described here. Furthermore, there are some more complex models, such as the seven-equation models based on various turbulent quantities and the full Reynolds Stress Model (RSM), which are not discussed here since they have not been found to improve accuracy but rather increase the computational cost of hydraulic machinery flow simulations. All types of turbulence models are listed on the NASA Turbulence Modeling Resource (TMR) (https://turbmodels.larc.nasa.gov/) [14]. For more information on turbulence modeling, please refer to Wilcox's classic book [13].

4.2.4.1 Standard k–ϵ Model

The standard two-equation k–ϵ model employs the transport equation for turbulent kinetic energy k and the transport equation for turbulent dissipation rate ϵ to define the eddy viscosity. The model assumes that the flow is fully turbulent [15].

The equations of turbulent kinetic energy k and dissipation rate ε can be written as:

$$\frac{\partial(\rho k)}{\partial t} + \frac{\partial(\rho k u_i)}{\partial x_i} = \frac{\partial}{\partial x_j}\left[\left(\mu + \frac{\mu_i}{\sigma_k}\right)\frac{\partial k}{\partial x_j}\right] + G_k + G_b - \rho\varepsilon - Y_M + S_k \tag{4.9}$$

$$\frac{\partial(\rho \varepsilon)}{\partial t} + \frac{\partial(\rho \varepsilon u_i)}{\partial x_i} = \frac{\partial}{\partial x_j}\left[\left(\mu + \frac{\mu_t}{\sigma_c}\right)\frac{\partial \varepsilon}{\partial x_j}\right] + C_{1\varepsilon}\frac{\varepsilon}{k}(G_k + C_{3\varepsilon}G_b) - C_{2\varepsilon}\rho\frac{\varepsilon^2}{k} + S_\varepsilon \tag{4.10}$$

where the turbulent (or eddy) viscosity μ_t is computed by combining k and ε as follows:

$$\mu_t = \rho C_\mu \frac{k^2}{\varepsilon} \tag{4.11}$$

where $C_\mu = 0.09$. In Eqs. (4.9) and (4.10), G_k represents the generation of turbulent kinetic energy due to the mean velocity gradients (it is described in the next section), G_b is the generation of turbulence due to buoyancy, which can be defined as:

$$G_b = \beta g_i \frac{\mu_t}{Pr_t}\frac{\partial T}{\partial x_i} \tag{4.12}$$

where Pr_t is the turbulent Prandtl number for the turbulent kinetic energy and g_i is the component of the gravitational vector in the i th direction. For the standard k–ε model, the default value of Pr_t is 0.85.

The coefficient of thermal expansion β is defined as:

$$\beta = -\frac{1}{\rho}\left(\frac{\partial \rho}{\partial T}\right)_p \tag{4.13}$$

In Eq. (4.9), Y_M represents the effect of the expansion of compressible turbulence on the total dissipation rate; this term is modeled based on a proposal by Sarkar [16] as:

$$Y_M = 2\rho\varepsilon M_t^2 \tag{4.14}$$

where M_t is the turbulent Mach number defined as:

$$M_t = \sqrt{\frac{k}{a^2}} \tag{4.15}$$

where a is the speed of sound.

In Eqs. (4.9) and (4.10), $C_{1\varepsilon}$, $C_{2\varepsilon}$, and $C_{3\varepsilon}$ are constants. σ_k and σ_ε are the turbulent Prandtl numbers for k and ε, respectively. S_k and S_ε are user-defined source terms.

In ANSYS software, the default values of the model constants are $C_{1\varepsilon} = 1.44$, $C_{2\varepsilon} = 1.92$, and $C_{3\varepsilon} = 0.09$, and the turbulent Prandtl numbers associated with the turbulent kinetic energy k and dissipation rate ε are $\sigma_k = 1.0$ and $\sigma_\varepsilon = 1.3$, respectively. For $C_{3\varepsilon}$, it becomes 1 for the buoyant shear layers for which the main flow direction is aligned with the direction of gravity [17] and for buoyant shear layers that are perpendicular to the gravitational vector, $C_{3\varepsilon}$ becomes 0. For simulations in the fluid machinery, the buoyant term is not considered, and it is 0 in all the calculations reported in this book.

4.2.4.2 Shear Stress Transport (SST) k–ω Model

The SST k–ω is a widely used two-equation linear eddy viscosity model developed by Menter. It combines the features of the standard k–ε model and Wilcox's k–ω model [18]. It has been established by researchers over four decades that Wilcox's k–ω model is more accurate in the boundary layers near the solid boundaries, while the k–ε model is more accurate in the free-stream and other shear flow regions in the flow field. The switching function F_1, in the model allows the model to switch between the k–ω model in the near-wall region and to the k–ε model in the free-stream region. This characteristic of the SST k–ω model allows the model to be accurate throughout the flow field. Since the first published version of the model in 1994 by Menter [17], various modifications to the model have been reported [19]. Here, the following version of the model is described, which has been recognized as relatively more accurate compared to the other versions for turbulent flow simulations in the field of rotating machinery. The relevant transport equations are as follows [20]:

$$\frac{\partial(\rho k)}{\partial t} + \frac{\partial(\rho k u_i)}{\partial x_i} = \frac{\partial}{\partial x_j}\left(\Gamma_k \frac{\partial k}{\partial x_j}\right) + G_k - Y_k + S_k \tag{4.16}$$

$$\frac{\partial(\rho \omega)}{\partial t} + \frac{\partial(\rho \omega u_i)}{\partial x_i} = \frac{\partial}{\partial x_j}\left(\Gamma_\omega \frac{\partial \omega}{\partial x_j}\right) + G_\omega - Y_\omega + S_\omega \tag{4.17}$$

where G_k represents the production term of turbulent kinetic energy k, G_ω represents the production term of the turbulent dissipation rate ω; Γ_k and Γ_ω represent the coefficients of diffusion term for k and ω, respectively; Y_k and Y_ω represent the dissipation terms of k and ω, respectively; and D_ω represents the cross-diffusion term calculated as described below. S_k and S_ω are user-defined source terms for the k and ω equations, respectively.

The expressions for the diffusion coefficients can be written as:

$$\Gamma_k = \mu + \frac{\mu_t}{\sigma_k} \tag{4.18}$$

$$\Gamma_\omega = \mu + \frac{\mu_t}{\sigma_\omega} \tag{4.19}$$

where σ_k and σ_ω represent the turbulent Prandtl numbers in k and ω equations, respectively. The turbulent viscosity coefficient μ_t is defined as:

$$\mu_t = \frac{\rho k}{\omega}\frac{1}{\max\left[\frac{1}{a^*}, \frac{SF_2}{a_1\omega}\right]} \tag{4.20}$$

where S is the strain-rate magnitude and

$$\sigma_k = \frac{1}{\frac{F_1}{\sigma_{k,1}} + \frac{(1-F_1)}{\sigma_{k,2}}} \tag{4.21}$$

$$\sigma_\omega = \frac{1}{\frac{F_1}{\sigma_{\omega,1}} + \frac{(1-F_1)}{\sigma_{\omega,2}}} \tag{4.22}$$

In Eq. (4.20), α^* is the low Reynolds number correction coefficient (also known as the intermittent coefficient) given by:

$$\alpha^* = \alpha^*_\infty \left(\frac{\alpha^*_0 + \frac{Re_t}{R_k}}{1 + \frac{Re_t}{R_k}}\right) \tag{4.23}$$

where $Re_t = \rho k/\mu\omega$, $R_k = 6$, $\alpha^*_0 = \beta_i/3$, and $\beta_i = 0.072$. Note that, in HRN flows, $\alpha^* = \alpha^*_\infty = 1$.

In Eq. (4.21), $\sigma_{k,1} = 1.176$, $\sigma_{k,2} = 1.0$ and in Eq. (4.17), $\sigma_{\omega,1} = 2.0$ and $\sigma_{\omega,2} = 1.168$.

The functions F_1, and F_2 are defined as follows:

$$F_1 = \tan h\left(\phi_1^4\right) \tag{4.24}$$

$$\phi_1 = \min\left[\max\left(\frac{\sqrt{k}}{0.09\omega y}, \frac{500\mu}{\rho y^2\omega}\right), \frac{4\rho k}{\sigma_{\omega,2}D^+_\omega y^2}\right] \tag{4.25}$$

$$F_2 = \tan h\left(\phi_2^2\right) \tag{4.26}$$

$$\phi_2 = \max\left[2\frac{\sqrt{k}}{0.09\omega y}, \frac{500\mu}{\rho y^2\omega}\right] \tag{4.27}$$

$$D^+_\omega = \max\left[2\rho\frac{1}{\sigma_{\omega,2}}\frac{1}{\omega}\frac{\partial k}{\partial x_j}\frac{\partial \omega}{\partial x_j}, 10^{-10}\right] \tag{4.28}$$

where y is the distance from the wall and D^+_ω is the positive component of the cross-diffusion term.

In Eq. (4.16), the turbulence kinetic energy production term G_k is defined as:

$$G_k = -\rho\overline{u'_i u'_j}\frac{\partial u_j}{\partial x_i} \tag{4.29}$$

G_k is evaluated in a manner consistent with the Boussinesq hypothesis, i.e.,

$$G_k = \mu_t S^2 \tag{4.30}$$

In Eq. (4.17), the turbulence dissipation rate term G_k is defined as:

$$G_k = \alpha\frac{\omega}{k}G_k \tag{4.31}$$

In Eq. (4.31), the coefficient α is given by:

$$\alpha = \frac{\alpha_\infty}{\alpha^*}\left(\frac{\alpha_0 + \frac{Re_t}{R_\omega}}{1 + \frac{Re_t}{R_\omega}}\right) \tag{4.32}$$

In Eq. (4.16), the dissipation term for k is defined as:

$$Y_k = \rho\beta^* f_{\beta^*} k\omega \tag{4.33}$$

where

$$f_{\beta^*} = \begin{cases} 1 \\ 1 + 680\chi_k^2 & \chi_k \leq 0 \\ 1 + 400\chi_k^2 & \chi_k > 0 \end{cases} \tag{4.34}$$

where

$$\chi_k \equiv \frac{1}{\omega^3}\frac{\partial k}{\partial x_i}\frac{\partial \omega}{\partial x_j} \tag{4.35}$$

and

$$\beta^* = \beta_i^*[1 + \xi^* F(M_t)] \tag{4.36}$$

$$\beta_i^* = \beta_\infty^* \left(\alpha_\infty^* \left(\frac{\frac{4}{15} + \left(\frac{Re_t}{R_\beta}\right)^4}{1 + \left(\frac{Re_t}{R_\beta}\right)^4} \right) \right) \tag{4.37}$$

In Eq. (4.17), the dissipation term for ω is defined as

$$Y_\omega = \rho\beta f_\beta \omega^2 \tag{4.38}$$

where

$$f_\beta = \frac{1 + 70\chi_\omega}{1 + 80\chi_\omega} \tag{4.39}$$

$$\chi_\omega = \left| \frac{\Omega_{ij}\Omega_{jk}S_{ki}}{(\beta_\infty^*\omega)^3} \right| \tag{4.40}$$

$$\Omega_{ij} = \frac{1}{2}\left(\frac{\partial u_i}{\partial x_j} - \frac{\partial u_j}{\partial x_i} \right) \tag{4.41}$$

$$\Omega_{ij} = \frac{1}{2}\left(\frac{\partial u_i}{\partial x_j} - \frac{\partial u_j}{\partial x_i} \right) \tag{4.42}$$

$$\beta = \beta_i \left[1 - \frac{\beta_i^*}{\beta_i}\xi^* F(M_t) \right] \tag{4.43}$$

In Eq. (4.38), the compressibility function, $F\,(M_t)$ is given by:

$$F(M_t) = \begin{cases} 0 & M_t \leq M_{t0} \\ M_t^2 - M_{t0}^2 & M_t > M_{t0} \end{cases} \tag{4.44}$$

where

$$M_t^2 \equiv \frac{2k}{a^2} \tag{4.45}$$

$$M_{t0} = 0.25 \tag{4.46}$$

$$a = \sqrt{\gamma RT} \tag{4.47}$$

Note that, in the high Reynolds number form of the SST k–ω model, $\beta_i^* = \beta_\infty^*$. In incompressible flow, $\beta^* = \beta_\infty^*$.

The other constants are $\alpha^*_\infty = 1$, $\alpha_\infty = 1$, $\alpha_0 = 1/9$, $\beta^*_\infty = 0.09$, $R_\beta = 8$, $R_\omega = 2.95$, $\xi^* = 1.5$, $\sigma_k = 2.0$, and $\sigma_\omega = 2.0$.

4.2.4.3 Spalart-Allmaras (SA) Model

The SA turbulence model is an industry-standard one-equation linear eddy-viscosity turbulence model developed for the prediction of aerodynamic flows. It is a low Reynolds number model and is therefore accurate in the wall region of a boundary layer. The SA model is given by the following equation:

$$\frac{\partial \widetilde{v}}{\partial t} + u_j \frac{\partial \widetilde{v}}{\partial x_j} = c_{b1}(1 - f_{t2})\widetilde{S}\widetilde{v} - \left[c_{\omega 1} f_\omega - \frac{c_{b1}}{\kappa^2} f_{t2}\right]\left(\frac{\widetilde{v}}{d}\right)^2 + \frac{1}{\sigma}\left[\frac{\partial}{\partial x_j}\left\{(v + \widetilde{v})\frac{\partial \widetilde{v}}{\partial x_j}\right\} + c_{b2}\frac{\partial \widetilde{v}}{\partial x_l}\frac{\partial \widetilde{v}}{\partial x_l}\right] \tag{4.48}$$

The turbulent eddy viscosity is given by the equation

$$\mu_t = \rho \widetilde{v} f_{v1} \tag{4.49}$$

Near-wall blocking is accounted for by the damping function f_{v1}

$$f_{v1} = \frac{\chi^3}{\chi^3 + c_{v1}^3} \tag{4.50}$$

$$\chi = \frac{\widetilde{v}}{v} \tag{4.51}$$

The remaining functions are defined as follows

$$\widetilde{S} = \Omega + \frac{\widetilde{v}}{\kappa^2 d^2} f_{v2}, f_{v2} = 1 - \frac{\chi}{1 + \chi f_{v1}} \tag{4.52}$$

where $\Omega = \sqrt{2W_{ij}W_{ij}}$ is the magnitude of the vorticity, and d is the distance of the flow field point to the nearest wall. W_{ij} is defined by the following equation:

$$W_{ij} = \frac{1}{2}\left(\frac{\partial u_i}{\partial x_j} - \frac{\partial u_j}{\partial x_i}\right) \tag{4.53}$$

$$f_\omega = g\left[\frac{1 + c_{\omega 3}^6}{g^6 + c_{\omega 3}^6}\right]^{\frac{1}{6}}, \quad g = r + c_{\omega 2}(r^6 - r), \quad r = \min\left[\frac{\widetilde{v}}{\widetilde{S}\kappa^2 d^2}, 10\right] \tag{4.54}$$

$$f_{t2} = c_{t3}\exp(-c_{t4}\chi^2) \tag{4.55}$$

where the constant $c_{b1} = 0.1355$, $\sigma = 2/3$, $c_{b2} = 0.622$, $\kappa = 0.41$, $c_{\omega 2} = 0.3$, $c_{\omega 3} = 2$, $c_{v1} = 7.1$, $c_{t3} = 1.2$, $c_{t4} = 0.5$, and $c_{\omega 1} = c_{b1}/\kappa 2 + (1 + c_{b2})/\sigma$.

4.2.4.4 Wray-Agarwal (WA) Model

The WA turbulence model is a one-equation linear eddy-viscosity model derived from the k–ω closure. It has been shown to have improved accuracy over the SA model and is competitive with the SST k–ω model in a large number of 2D and 3D complex flows. It behaves like the two-equation k–ω model in the near-wall region and like the standard two-equation k–ε model in the free-stream and shear flow regions. The inclusion of the cross-diffusion term in the R (= k/ω) equation and a blending function that allows smooth switching between the two destruction terms are the most innovative features of this model. The transport equation for the variable R in the WA model is as follows:

$$\frac{\partial R}{\partial t} + u_j \frac{\partial R}{\partial x_j} = \frac{\partial}{\partial x_j}\left[(\sigma_R R + \nu)\frac{\partial R}{\partial x_j}\right] + C_1 RS + f_1 C_{2k\omega} \frac{R}{S}\frac{\partial R}{\partial x_j}\frac{\partial S}{\partial x_j} - (1 - f_1) C_{2k\varepsilon} R^2 \left(\frac{\frac{\partial S}{\partial x_j}\frac{\partial S}{\partial x_j}}{S^2}\right) \tag{4.56}$$

In Eq. (4.56), S is the mean strain given by:

$$S = \sqrt{2S_{ij}S_{ij}} \text{ where } S_{ij} = \frac{1}{2}\left(\frac{\partial u_i}{\partial x_j} + \frac{\partial u_j}{\partial x_i}\right) \tag{4.57}$$

The turbulent kinematic eddy viscosity is given by the equation:

$$\nu_T = f_\mu R \tag{4.58}$$

The wall-blocking effect is accounted for by the damping function f_μ as:

$$f_\mu = \frac{\chi^3}{\chi^3 + C_w^3}, \chi = \frac{R}{\nu} \tag{4.59}$$

The switching function f_1 allows the WA model to behave as a one-equation k–ω or one-equation k–ε model and is defined as:

$$f_1 = \min\left(\tan h\left(\arg_1^4\right), 0.9\right) \tag{4.60}$$

$$\arg_1 = \frac{1 + \frac{d\sqrt{RS}}{\nu}}{1 + \left[\frac{\max(d\sqrt{RS}, 1.5R)}{20\nu}\right]^2} \tag{4.61}$$

where d is the minimum distance to the nearest wall.

The various constants in the WA model are given in Table 4.1.

Table 4.1 Model constants of the WA 2017 model.

Constant	Value
$C_{1k\omega}$	0.0829
$C_{1k\varepsilon}$	0.1127
C_1	$f_1(C_{1k\omega} - C_{1k\varepsilon}) + C_{1k\varepsilon}$
$\sigma_{k\omega}$	0.72
$\sigma_{k\varepsilon}$	1.0
σ_R	$f_1(\sigma_{k\omega} - \sigma_{k\varepsilon}) + \sigma_{k\varepsilon}$
$C_{2k\omega}$	$C_{2k\omega} = \frac{C_{1k\omega}}{\kappa^2} + \sigma_{k\omega}$
$C_{2k\varepsilon}$	$C_{2k\varepsilon} = \frac{C_{1k\varepsilon}}{\kappa^2} + \sigma_{k\varepsilon}$
κ	0.41
C_ω	8.54
C_m	8

4.2.4.5 Detached Eddy Simulation (DES) Model

The DES is a hybrid RANS/LES model which can be considered as a modification of the original RANS turbulence model in which the model switches to a subgrid-scale (SGS) formulation in regions fine enough for LES calculations. The regions near the solid boundaries, where the turbulent length scale is smaller than the maximum grid dimension, are assigned the RANS equations for solution, but as the turbulent length scale exceeds the grid dimension, the regions are solved using the LES subgrid scale model. Therefore, the grid resolution is not as demanding as for pure LES, thereby considerably cutting down the cost of the computations with increased accuracy compared to RANS simulation. Although DES was initially formulated for use with the RANS/SA model, it can be implemented with other RANS models described in the sections above by appropriately modifying the length scale, which is explicitly or implicitly involved in the RANS turbulence model. The construction of the DES model for the one-equation SA model is described below.

The transport equation for the kinematic eddy viscosity for the SA model can be rewritten as:

$$\frac{\partial}{\partial t}(\rho\hat{\upsilon}_T) + \frac{\partial}{\partial x_{\rm i}}(\rho\hat{\upsilon}_T u_{\rm i}) = \frac{1}{\sigma}\left\{\frac{\partial}{\partial x_{\rm i}}\left[(\rho\upsilon_l + \rho\hat{\upsilon}_T)\frac{\partial\hat{\upsilon}_T}{\partial x_{\rm i}}\right] + c_{b2}\rho\frac{\partial\hat{\upsilon}_T}{\partial x_{\rm i}}\frac{\partial\hat{\upsilon}_T}{\partial x_{\rm i}}\right\}$$
$$+ c_{b1}\rho\hat{S}\hat{\upsilon}_T - c_{\omega 1}f_\omega\rho\left(\frac{\hat{\upsilon}_T}{d}\right)^2 \tag{4.62}$$

where $x_{\rm i}$, t are space and time coordinates, respectively, $\hat{\upsilon}_{\rm T}$ is the kinematic eddy viscosity coefficient, and d is the distance to the nearest wall. The values of other coefficients in Eq. (4.62) are given in Reference [21].

The DES method modifies the dissipation term in the SA model, that is, d in the dissipation term (last term on RHS of Eq. (4.62)) is replaced by $\widehat{d}_{\rm DES}$, so that the calculated flow field is related to the grid scale. $\widehat{d}_{\rm DES}$ is defined as:

$$\widetilde{d}_{\rm DES} = \min\,(d, C_{\rm DES}\Delta) \tag{4.63}$$

where $C_{\rm DES}$ is a constant = 0.65 and is the filtering scale of the SGS model with the maximum scaling values $\Delta_{\rm i} = \max(\Delta x_{\rm i}, \Delta y_{\rm i}, \Delta z_{\rm i})$. This modification to the SA model, while very simple, changes the interpretation of the model substantially. This modified distance function causes the model to behave like a RANS model in regions close to the walls, and as an LES (e.g., Smagorinsky-like) model away from the walls.

Near the wall, $d_1 < C_{\rm DES}\Delta_1$ and $d_2 < C_{\rm DES}\Delta_2$, therefore, $\tilde{d}_{\rm DES,\ i} = d_{\rm i}$, where the DES method matches the RANS method, while away from the wall, $d_3 > C_{\rm DES}\Delta_3$ and $d_4 > C_{\rm DES}\Delta_4$, therefore, $\tilde{d}_{\rm DES,i} = C_{\rm DES}\Delta_{\rm i}$ and the reduction in the turbulent eddy viscosity coefficient is determined by the local grid-scale Δ. When the production term and dissipation term in the transport equation, Eq. (4.62) reach equilibrium, one obtains,

$$c_{b1}\rho\hat{S}\hat{\upsilon}_{\rm T} = c_{\omega 1}f_\omega\rho\left(\frac{\hat{\upsilon}_{\rm T}}{C_{\rm DES}\Delta}\right)^2 \tag{4.64}$$

which gives $\hat{\upsilon}_{\rm T} \propto \hat{S}(C_{\rm DES}\Delta)^2$. Thus, when the local equilibrium occurs, the eddy viscosity coefficient is proportional to the amplitude of the deformation rate multiplied by the square of the filtering scale, which is the requirement of the Smagorinsky SGS eddy-viscosity model, realizing the effect similar to LES implicit filtering.

It should be noted that while the SA model-based DES acts as LES with a wall model. DES based on other turbulence models (like the two-equation models) behaves as a hybrid RANS-LES model. Grid generation is more complicated than for a simple RANS or LES case due to the switch between RANS and LES. Furthermore, DES is a nonzonal approach and provides a single smooth velocity field across the RANS and the LES regions of the solution.

4.2.4.6 Large Eddy Simulation (LES) Model

The LES is a well-established model for simulating turbulent flows. A theory developed by Kolmogorov in 1941 showed that the large eddies in a turbulent flow are dependent on the geometry while the smaller eddies are universal. This concept is used in the formulation of LES model from the Navier–Stokes equations [22].

In LES, the turbulent motion is assumed to be composed of vortex structures of different scales. Through a filter function, the instantaneous motion of the turbulent flow field is decomposed into two parts: large-scale vortex motion and SGS vortex motion. The large-scale vortex structures (large eddies) are calculated directly, while the SGS vortex structures (small eddies) are modeled. Thus, LES requires much less time and space resolution compared to the DNS method and can calculate the turbulent flow field characteristics at HRN on the currently available computer hardware, accurately reflecting the movement, development, and dissipation process in turbulent flow.

Filtering in LES is defined as the convolution of a function u with a filtering Leonard term G as [23]:

$$\overline{u}_{\mathrm{i}}(\vec{x}) = \int G(\vec{x} - \vec{\xi})u(\vec{\xi})d\vec{\xi} \tag{4.65}$$

It results in

$$u_{\mathrm{i}} = \overline{u}_{\mathrm{i}} + u_{\mathrm{i}}' \tag{4.66}$$

where $\overline{u}_{\mathrm{i}}$ is the resolvable large-scale part and u_{i}' is the SGS part of the variable u. However, most practical implementations of LES utilize the computational grid as the filter (the box filter) and do not employ explicit filtering.

The filtered equations can be obtained from the incompressible Navier–Stokes equations given below:

$$\frac{\partial u_{\mathrm{i}}}{\partial t} + u_{\mathrm{j}}\frac{\partial u_{\mathrm{i}}}{\partial x_{\mathrm{j}}} = -\frac{1}{\rho}\frac{\partial p}{\partial x_{\mathrm{i}}} + \frac{\partial}{\partial x_{\mathrm{i}}}\left(\upsilon\frac{\partial u_{\mathrm{i}}}{\partial x_{\mathrm{j}}}\right) \tag{4.67}$$

Substituting $u_{\mathrm{i}} = \overline{u}_{\mathrm{i}} + u_{\mathrm{i}}'$ and $p = \overline{p} + p'$ in Eq. (4.67) and then filtering the resulting equation using Eq. (4.65) results in the equation for the resolved field:

$$\frac{\partial \overline{u}_{\mathrm{i}}}{\partial t} + \overline{u}_{\mathrm{j}}\frac{\partial \overline{u}_{\mathrm{i}}}{\partial x_{\mathrm{j}}} = -\frac{1}{\rho}\frac{\partial \overline{p}}{\partial x_{\mathrm{i}}} + \frac{\partial}{\partial x_{\mathrm{i}}}\left(\upsilon\frac{\partial \overline{u}_{\mathrm{i}}}{\partial x_{\mathrm{j}}}\right) + \frac{1}{\rho}\frac{\partial \tau_{\mathrm{ij}}}{\partial x_{\mathrm{j}}} \tag{4.68}$$

The new term $\frac{\partial \tau_{\mathrm{ij}}}{\partial x_{\mathrm{j}}}$ in Eq. (4.68) arises from the nonlinear advection terms since

$$\overline{u_{\mathrm{j}}\frac{\partial u_{\mathrm{i}}}{\partial x_{\mathrm{j}}}} \neq \overline{u}_{\mathrm{j}}\frac{\partial \overline{u}_{\mathrm{i}}}{\partial x_{\mathrm{j}}} \tag{4.69}$$

Therefore,

$$\tau_{\mathrm{ij}} = \overline{u}_{\mathrm{i}}\overline{u}_{\mathrm{j}} - \overline{u_{\mathrm{i}}u_{\mathrm{j}}} \tag{4.70}$$

Similar equations can be derived for the SGS field.

Boussinesq hypothesis is used in SGS modeling of turbulence. The SGS turbulent stress can be modeled as:

$$\tau_{\mathrm{ij}} - \frac{1}{3}\tau_{\mathrm{kk}}\delta_{\mathrm{ij}} = -2\mu_{\mathrm{t}}\overline{S}_{\mathrm{ij}} \tag{4.71}$$

where $\overline{S}$ is the strain-rate tensor for the resolved scale defned as:

$$\overline{S}_{\mathrm{ij}} = \frac{1}{2}\left(\frac{\partial \overline{u}_{\mathrm{i}}}{\partial x_{\mathrm{j}}} - \frac{\partial \overline{u}_{\mathrm{j}}}{\partial x_{\mathrm{i}}}\right) \tag{4.72}$$

In Eq. (4.71), μ_t is the SGS turbulent eddy viscosity. Employing Eqs. (4.67)–(4.72), the Navier–Stokes equations become:

$$\frac{\partial \overline{u}_i}{\partial t} + \overline{u}_j \frac{\partial \overline{u}_i}{\partial x_j} = -\frac{1}{\rho}\frac{\partial \overline{p}}{\partial x_i} + \frac{\partial}{\partial x_i}\left([\upsilon + \upsilon_t] \frac{\partial \overline{u}_i}{\partial x_j} \right) \tag{4.73}$$

where the incompressibility constraint has been used to simplify the equation, and the pressure is modified to include the trace term $\frac{1}{3}\tau_{kk}\delta_{ij}$.

In order to close the equations and determine the filtered velocity field $\overline{u}(x,t)$ and the filtered pressure $\overline{p}(x,t)$, one needs to model the anisotropic residual-stress tensor $\tau^r_{ij}(x,t)$. The Smagorinsky model is the simplest model which has been proven to perform reasonably well. In this model, the residual SGS eddy viscosity υ_t, is modeled to represent the motion of SGS eddies υ_t is modeled as [24]:

$$\upsilon_t = l_s^2 (2\overline{S_{lk}S_{lk}})^{\frac{1}{2}} = (C_S\Delta)^2 (2\overline{S_{lk}S_{lk}})^{\frac{1}{2}} \tag{4.74}$$

where the Smagorinsky length scale is defined by $l_s = C_S\Delta$ and C_S is the Smagorinsky coefficient and Δ is the local grid scale. In Ansys Fluent, Δ is computed according to the volume of the computational cell using

$$\Delta = V^{\frac{1}{3}} \tag{4.75}$$

Lilly derived a value of 0.17 for C_S for homogeneous isotropic turbulence in the inertial subrange [25]. However, this value was found to cause excessive damping of large-scale fluctuations in the presence of mean shear and in transitional flows near the solid boundary and needed to be reduced in such regions. In short, C_S is not a universal constant, which is the most serious shortcoming of this simple model. Nonetheless, a C_S value of around 0.1 has been found to yield the best results for a wide range of flows and is used as the default value in Fluent.

In OpenFOAM, the Smagorinsky coefficient C_S is calculated by the two coefficients C_k and C_ε as follows:

$$(C_S)^2 = \left(\frac{C_k}{C_\varepsilon}\right)^{\frac{1}{2}} \tag{4.76}$$

The two coefficients are determined based on the behavior of the model when applied to canonical free shear flows and a turbulent boundary layer. The lower bounds of C_k and C_ε are 0.085 and 0.984, respectively, and the upper bounds of C_k and C_ε are 0.103 and 1.152, respectively. The standard values of C_k and C_ε are 0.094 and 1.048, respectively.

4.2.5 Wall Functions

A wall function is a mathematical model that approximates the velocity profile near the wall in a known functional form (generally includes some form of law of the wall), thereby requiring lesser number of grid points for flow resolution near the wall in CFD simulations. This is employed for computational efficiency especially in computation of 3D flows.

The y^+ is a dimensionless parameter used to describe the relative magnitude of viscous effects in the fluid near the wall. It is the ratio of the wall distance y to the wall friction velocity u, i.e., $y^+ = y\, xu/\upsilon$, where ν is the dynamic viscosity of the fluid. The value of y^+ is crucial for the selection of wall functions and the grid density near the wall. Generally, when the y^+ value is small (e.g., $y^+ < 5$), the wall function is more applicable within the viscous sublayer and can well describe the velocity and shear stress distributions near the wall. When the y^+ value is large (e.g., $y^+ > 30$), the applicability of the wall function decreases, and a more refined wall model is needed to describe the details of the flow.

In CFD software (such as Fluent), commonly used wall functions include the standard wall function, the scalable wall function, and the nonequilibrium wall function. The standard wall function: when used in regions with small y^+ values (typically $y^+ < 15$), it may lead to deteriorated numerical solutions and cannot guarantee solution accuracy. The scalable wall function: when refining the grid along the wall direction, it can avoid the problem of deteriorated solutions caused by the standard wall function when y^+ values are small. For any grid refined along the wall, the scalable wall function can provide consistent numerical solutions. When the y^+ value is greater than 15, it functions the same way as the standard wall function. The nonequilibrium wall function: it takes into account the influence of pressure gradients on velocity field and is therefore suitable for simulating complex flows in presence of pressure gradients, such as flows with recirculation, separation and reattachment.

The applicability of wall functions is also closely related to the grid distribution near the wall. When generating the grid near the wall, it is necessary to ensure that the first grid point away from the wall is placed within the range where the logarithmic law of the wall holds, i.e., in the fully developed turbulent flow region. If the y^+ value exceeds this range, the grid in that area should be refined, and the calculation should be repeated. The y^+ value should then be checked again until it meets the requirements. The wall functions are primarily used in turbulence models for computing high Reynolds numbers flows. For low Reynolds number flows, within the viscous sublayer, wall functions may not be applicable.

In summary, wall functions have significant application value in CFD, but they need to be carefully selected and adjusted based on specific flow problems and computational requirements. By understanding the definition of wall functions, the relationship with the y^+ parameter, types and their characteristics, the relationship with grid distribution near the wall, as well as their application scope and limitations, wall functions can be effectively applied for CFD simulations.

4.2.6 Discretization Methods

Discretization is the process of converting continuous problems into discrete ones; it is primarily achieved by dividing the computational domain into discrete points or regions. For numerical computation, discretization methods are needed to numerically solve the continuous mathematical problems described by ordinary and partial differential equations and integrals equations.

The commonly used methods for discretization can be classified as follows:

1) **Finite difference method:**
   The computational domain is divided into a number of grid points or nodes using a mesh generation software. At each grid node, the first - and second - order partial derivatives in the governing partial differential equations are replaced by first- or second - order differences using the Taylor series expansion at the nodal point resulting in a system of algebraic finite difference equations with a finite number of unknowns at discrete points. This method is suitable for solving partial differential equations and has been widely used; its main drawback is that the conservation property is not preserved for each grid cell.
2) **Finite element method:**
   A continuous structure or medium is divided into a finite number of small elements. By approximating the flow properties of each small element, the flow behavior in the entire computational domain is obtained. This method is widely used in structural mechanics, earthquake engineering, and other fields, especially for the study of fluid-structure interactions.

3) **Finite volume method:**
   It is similar to the finite element method, but with a greater focus on integrating and discretizing the physical quantities within control volumes. In contrast to finite-difference method, it conserves conservation properties of the fluid in each control volume. This method is widely used in fluid mechanics and other area of computational physics.
4) **Numerical integration method:**
   The integration interval is divided into discrete subintervals. The function values on each subinterval are then calculated and summed to obtain the integral value over the entire interval. This includes methods such as the rectangle method, trapezoidal rule, Simpson's rule and quadrature rules. These methods can approximate the integral value of a function to some extent and are important tools for solving the discrete functions.

4.2.7 Interpolation Schemes and Algorithms

Interpolation in numerical computation is an important mathematical method used to construct a function based on known data points to estimate the values at unknown data points. Interpolation is the process of "inserting" some values between known data points, requiring that the interpolation function must pass through all discrete points, and the values at points outside the discrete points are considered as "inserted" values. There are global interpolation methods and local interpolation methods (such as piecewise linear interpolation). Interpolation error is usually a point-wise error or maximum norm error, and the quality of interpolation is often reflected through certain local properties, such as Runge's phenomenon or Gibbs oscillation.

The common interpolation methods include the following:

1) **Linear interpolation:**
   It is one of the simplest interpolation methods, assuming that the function values at two points are known, then the function value at an intermediate point is estimated by the straight line between these two points.
   The linear interpolation formula is: $f(x) = f(x_0) + (x - x_0) * (f(x_1) - f(x_0))/(x_1 - x_0)$. Here, $f(x)$ represents the function value at the desired interpolation point, $f(x_0)$ and $f(x_1)$ are the function values at the two known points, and x_0 and x_0 are the x-coordinates of the two known points.
   Linear interpolation is simple to calculate but has low accuracy, suitable for cases with few data points or low requirements for data accuracy.
2) **Lagrange interpolation:**
   It is a polynomial-based interpolation method that constructs a polynomial using the function values at multiple known points and uses this polynomial to estimate the function values at other points.
   The general form of the Lagrange interpolation polynomial is: $f(x) = \Sigma[f(x_i) * Li(x)]$ $(i = 0, 1, 2, \ldots, n)$. Here, $f(x)$ represents the function value at the desired interpolation point, $f(x_i)$ are the function values at the known multiple points, and $Li(x)$ are the Lagrange basis functions, with the expression for the Lagrange basis function being: $Li(x) = \Pi\ [(x - x_j)/(x_i - x_j)]$ $(i \neq j,\ i, j = 0, 1, 2, \ldots, n)$.
   Lagrange interpolation can construct a polynomial that passes through all known points, but when there are many known points, the degree of the polynomial will be high, potentially leading to unstable interpolation results (such as Runge's phenomenon).
3) **Piecewise linear interpolation:**
   It is a method of approximating functions piecewise linearly by dividing the entire interpolation interval into multiple small segments and using linear interpolation on each segment to estimate

the function values. The process of piecewise linear interpolation is divided into two steps: first, determine the small segment where the point to be interpolated is located, and then perform linear interpolation on the segment to estimate the function value.

Piecewise linear interpolation avoids the drawbacks of high-degree polynomials and maintains the stability of the interpolation function, but discontinuities may occur at the connections between segments.

4) **Newton interpolation:**

   It is also a polynomial-based interpolation method that uses the concept of divided differences to construct the interpolation polynomial.

   The general form of the Newton interpolation polynomial is: $f(x) = f(x_0) + (x - x_0) * f[x_0, x_1] + (x - x_0) * (x - x_1) * f[x_0, x_1, x_2] + \ldots$. Here, $f(x)$ represents the function value at the desired interpolation point, $f(x_0)$ is the function value at a known point, and $f[x_i, x_i + 1, \ldots, x_i + k]$ is the k-th order divided difference.

   Newton interpolation adds only one term to the interpolation polynomial for each additional node, facilitating recursive calculations and having a smaller computational load than Lagrange interpolation.

5) **Hermite interpolation:**

   It is a method that uses function values and derivative values for interpolation, constructing an interpolation polynomial through the function values and derivative values at known points.

   The general form of Hermite interpolation is: $f(x) = f(x_0) + (x - x_0) * f'(x_0) + (x - x_0)\hat{}2 * f'(x_0)/2 + \ldots$.

   Hermite interpolation not only requires the interpolation function to have the same value as the function at the nodes but also requires the interpolation function to have the same first-order, second-order, or even higher-order derivative values as the function, making it suitable for situations where high smoothness of the interpolation function is required.

6) **Spline interpolation:**

   It is a method of piecewise polynomial interpolation that divides the entire interpolation interval into multiple small segments, each of which uses a low-degree polynomial for interpolation. Common spline interpolation methods include linear spline interpolation, quadratic spline interpolation, and cubic spline interpolation, with cubic spline interpolation being the most commonly used.

   Spline interpolation can construct a smooth and continuous interpolation function, particularly suitable for situations where smooth curves or surfaces need to be drawn.

References

1 Hirsch, C. (2007). *Numerical Computation of Internal and External Flows: The Fundamentals of Computational Fluid Dynamics*. Elsevier.

2 Pulliam, T.H. and Zingg, D.W. (2014). *Fundamental Algorithms in Computational Fluid Dynamics*. New York: Springer.

3 Anderson, J.D. and Wendt, J. (1995). *Computational Fluid Dynamics*. New York: McGraw-hill.

4 Glatzel, T., Litterst, C., Cupelli, C., Lindemann, T., Moosmann, C., Niekrawietz, R., Streule, W., Zengerle, R., and Koltay, P. (2008). Computational fluid dynamics (CFD) software tools for microfluidic applications–a case study. *Computers & Fluids* 37 (3): 218–235.

5 Versteeg, H.K. (2007). *An Introduction to Computational Fluid Dynamics the Finite Volume Method*. Pearson Education India.

6 Moser, R.D., Haering, S.W., and Yalla, G.R. (2021). Statistical properties of subgrid-scale turbulence models. *Annual Review of Fluid Mechanics* 53 (1): 255–286.
7 Dick, E. (2009). Introduction to finite element methods in computational fluid dynamics. In: *Computational Fluid Dynamics* (ed. E. Dick), 235–274. Berlin, Heidelberg: Springer Berlin Heidelberg.
8 Wilcox, D.C. (1988). Multiscale model for turbulent flows. *AIAA Journal* 26 (11): 1311–1320.
9 Moin, P. and Mahesh, K. (1998). Direct numerical simulation: a tool in turbulence research. *Annual Review of Fluid Mechanics* 30 (1): 539–578.
10 Ishihara, T., Gotoh, T., and Kaneda, Y. (2009). Study of high-Reynolds number isotropic turbulence by direct numerical simulation. *Annual Review of Fluid Mechanics* 41 (1): 165–180.
11 Lee, M. and Moser, R.D. (2015). Direct numerical simulation of turbulent channel flow up to. *Journal of Fluid Mechanics* 774: 395–415.
12 Eggels, J.G.M., Unger, F., Weiss, M.H., Westerweel, J., Adrian, R.J., Friedrich, R., and Nieuwstadt, F.T.M. (1994). Fully developed turbulent pipe flow: a comparison between direct numerical simulation and experiment. *Journal of Fluid Mechanics* 268: 175–210.
13 Wilcox, D.C. (1998). *Turbulence Modeling for CFD*. La Canada: DCW industries https://turbmodels.larc.nasa.gov/.
14 Launder B E. Lectures in mathematical models of turbulence. 1972.
15 Sarkar and Lakshmanan, B. (1991). Application of a Reynolds stress turbulence model to the compressible shear layer. *AIAA Journal* 29 (5): 743–749.
16 Henkes, R., Van Der Vlugt, F.F., and Hoogendoorn, C.J. (1991). Natural-convection flow in a square cavity calculated with low-Reynolds-number turbulence models. *International Journal of Heat and Mass Transfer* 34 (2): 377–388.
17 Menter, F.R. (1994). Two-equation eddy-viscosity turbulence models for engineering applications. *AIAA Journal* 32 (8): 1598–1605.
18 Menter, F.R., Kuntz, M., and Langtry, R. (2003). Ten years of industrial experience with the SST turbulence model. *Turbulence, Heat and Mass Transfer* 4 (1): 625–632.
19 Fluent, A. (2011). *Fluent 14.0 User's Guide*. Ansys Fluent Inc.
20 Vatsa, V.N., Lockard, D.P., and Spalart, P.R. (2017). Grid sensitivity of SA-based delayed-detached-eddy-simulation model for blunt-body flows. *AIAA Journal* 55 (8): 2842–2847.
21 Bailly, C. and Comte-Bellot, G. (2015). *Turbulence*. Springer.
22 Hou, Z. (2017). Uncertainty quantification of wall-modeled large eddy simulation (WMLES) model in OpenFOAM.
23 Smagorinsky, J. (1963). General circulation experiments with the primitive equations: I. The basic experiment. *Monthly Weather Review* 91 (3): 99–164.
24 Scotti, A., Meneveau, C., and Lilly, D.K. (1993). Generalized Smagorinsky model for anisotropic grids. *Physics of Fluids A: Fluid Dynamics* 5 (9): 2306–2308.
25 Lilly, D.K. (1992). A proposed modification of the Germano subgrid-scale closure method. *Physics of Fluids A* 4: 633–635.

5

3D Geometry Modeling of Impeller Pumps

5.1 Geometry Modeling Software

SolidWorks is a subsidiary of Dassault Systèmes S.A., specializing in the development and sale of mechanical design software for the Windows platform. As a comprehensive software solutions provider, Dassault Systèmes not only offers extensive software support but also provides value-added services with internet integration capabilities for manufacturing enterprises. The company's product portfolio covers the entire product lifecycle, including design, engineering, manufacturing, and product data management, offering a range of industry-leading software systems. Among them, CATIA V5 is a well-known software developed by Dassault Systèmes and is widely used in various complex engineering projects. With its powerful functionality and flexible applications, Dassault's computer-aided design (CAD) products maintain a strong market presence worldwide, making them a trusted choice for many engineers and manufacturing companies [1].

SolidWorks is a powerful and feature-rich 3D CAD software, widely favored for its unique advantages. The software is characterized by three key attributes: robust functionality, ease of learning and use, and continuous technological innovation. These features make SolidWorks a leading mainstream 3D CAD solution. With SolidWorks, users can explore multiple design solutions, effectively reducing design errors and improving overall product quality. Its intuitive user interface and comprehensive toolset allow engineers and designers to get started effortlessly, whether they are beginners or experienced professionals. Additionally, SolidWorks provides extensive learning resources and community support, enabling users to leverage the software more efficiently for design tasks. In summary, SolidWorks not only meets complex design requirements but also enhances work efficiency through its user-friendly operation, making it an indispensable tool for engineers and designers in product development [2].

Therefore, SolidWorks and Creo are adopted for the 3D modeling of all pumps.

5.1.1 Centrifugal Pump

The key components of a centrifugal pump include the impeller and the volute casing [3]. The centrifugal impeller modeling is performed employing the following steps.

5.1.1.1 Impeller Modeling of the Centrifugal Pump

5.1.1.1.1 Creating a PRT File Now, the modeling process for the centrifugal pump impeller begins. Double-click [SolidWorks], then click [Home], and select [Part] to enter the drawing interface, as shown in Figure 5.1.

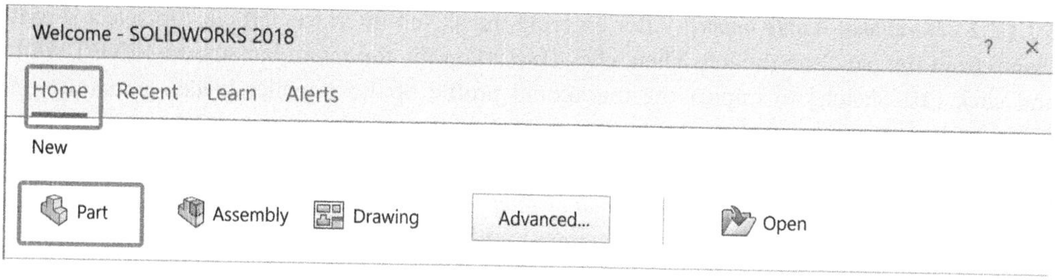

Figure 5.1 Creation of a Part file in SolidWorks.

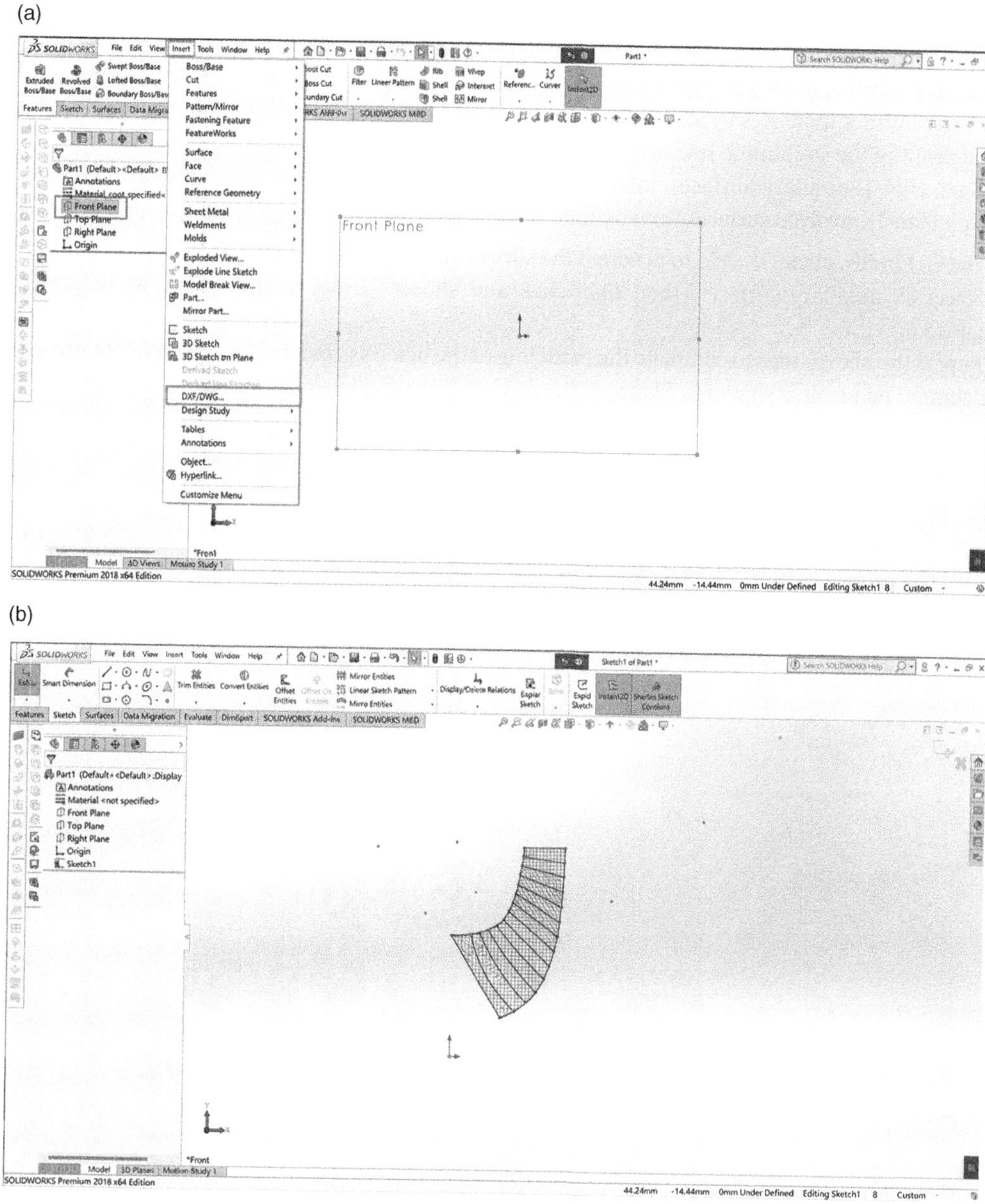

Figure 5.2 Meridional profile import: (a) Sketch interface, (b) import data.

5.1.1.1.2 Meridional Profile Import After entering the sketch interface, left-click to select [Front Plane] from the panel on the left. Then, click [Insert] on the top menu bar, choose [DXF/DWG], and click [2D Sketch] to import the meridional profile of the impeller blades, as shown in Figure 5.2.

5.1.1.1.3 Creating Surfaces Using the Rotation Method Next, on the rotation mold diagram, create surfaces by moving the corresponding flow lines to their respective positions. Click the [Surfaces] menu and then click [Revolved Surface]. In the [Axis of Revolution] field, select the rotation axis. In [Direction 1], determine the rotation direction. Choose the [Blind] rotation method and set the rotation angle according to the sweep angle and the number of flow lines. For example, input the number 7.5 for this case. In [Selected Contours], select the rotation lines, as shown in Figure 5.3, which are the surfaces formed by the rotation of the flow lines.

Each streamline is then revolved sequentially to form surfaces. The rotation angle between adjacent streamlines increases by 7.5°, progressing from the impeller inlet to the impeller outlet. The final result of the streamline revolutions is shown in Figure 5.4.

Next, click [Boundary Surface], then under [Direction 1], sequentially select the boundaries of the previously revolved surfaces from bottom to top (including the inlet edge of the blade's working surface). Finally, click OK ✓ to generate the surface, as shown in Figure 5.5.

Select all auxiliary surfaces, then right-click and choose [Hide] to hide them, as shown in Figure 5.6.

Repeat the above steps to complete the modeling of the blade's pressure side and perform surface stitching. The result is shown in Figure 5.7.

5.1.1.1.4 Creating the Cover Plate After creating the single blade, import the .dwg file for the hub. The import method is the same as the one used for the meridional profile of the blade. Select

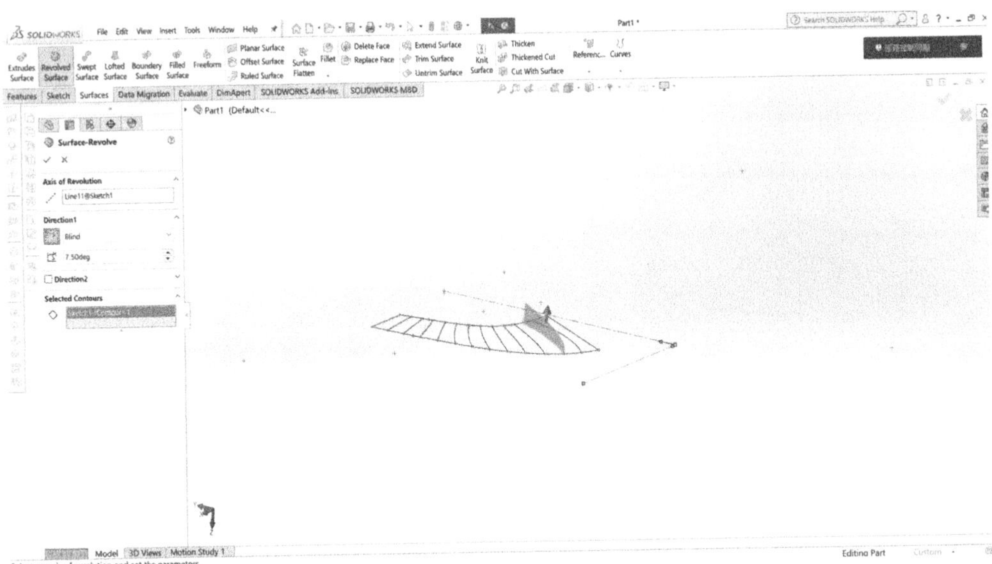

Figure 5.3 Surface rotation of flow lines.

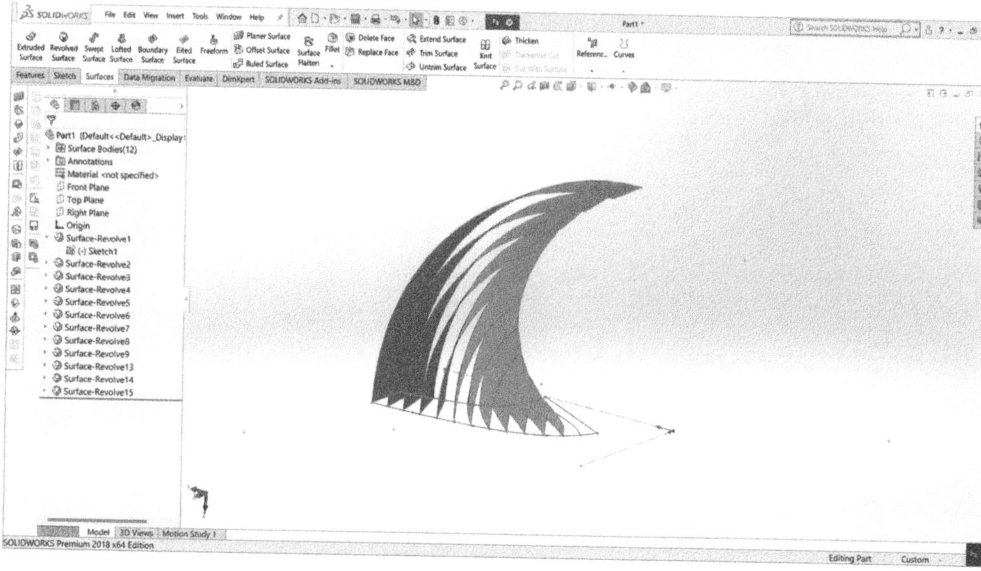

Figure 5.4 Result of streamline revolutions.

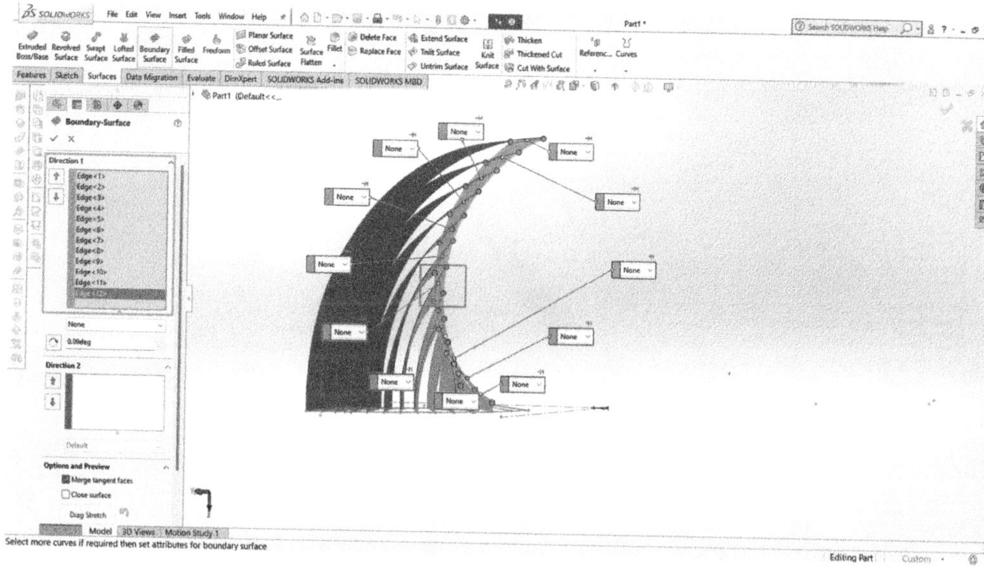

Figure 5.5 Creation of Boundary Surface.

the newly imported hub section, then click [Features] in the status bar and choose [Revolved Boss/Base] to rotate it 360°, forming the cover plate of the impeller, as shown in Figure 5.8.

5.1.1.1.5 Blade Array Finally, click [Circular Pattern] under [Features] to create a circular array of the previously created blades. In [Direction 1], rotate the axis of revolution, and set the rotation angle to 360°. Then, select the array feature in [Features and Faces], as shown in Figure 5.9.

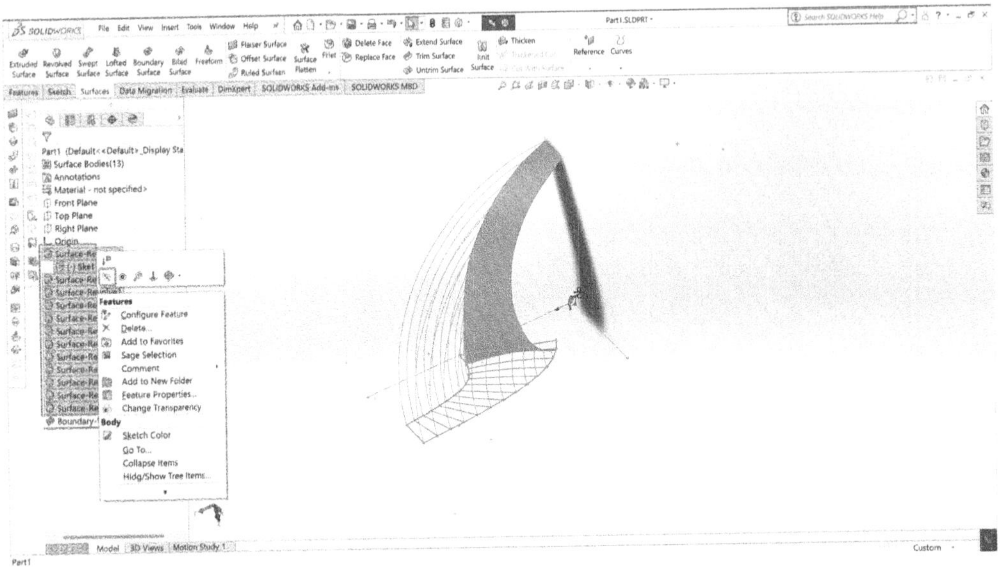

Figure 5.6 Suction side of the blade.

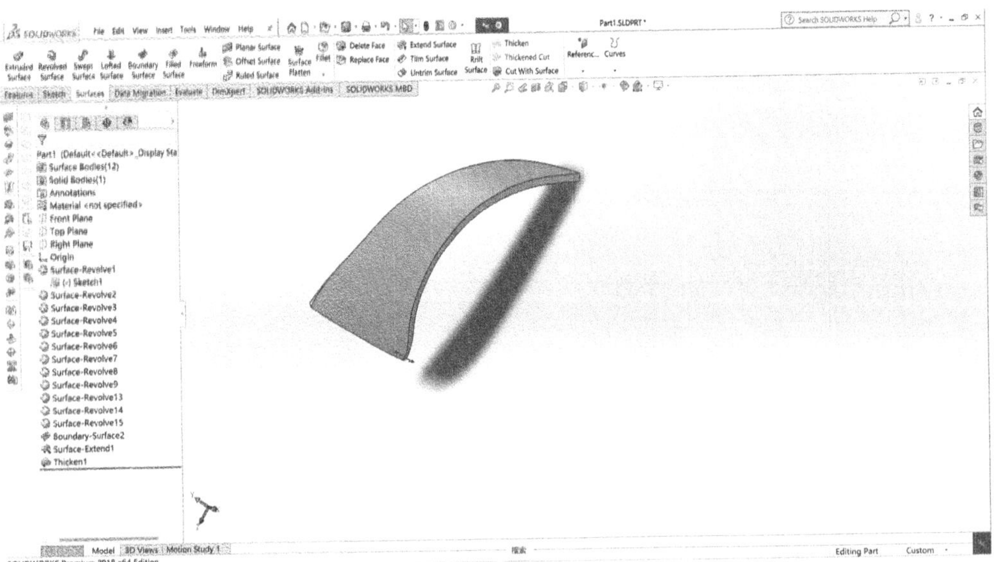

Figure 5.7 Creation of a single blade.

5.2 Axial Flow Pump

5.2.1 Modeling of the Impeller and Inlet Conical Tube

5.2.1.1 Modeling Approach

First, create points on the blade airfoil. Next, connect the points to form the blade airfoil control line. Then, use the blade airfoil control line to create the blade's working surface, back surface, hub surface, and rim surface. Finally, stitch and solidify the blade's working surface, back surface, hub surface, and rim surface into a solid blade. Additionally, create the complete impeller

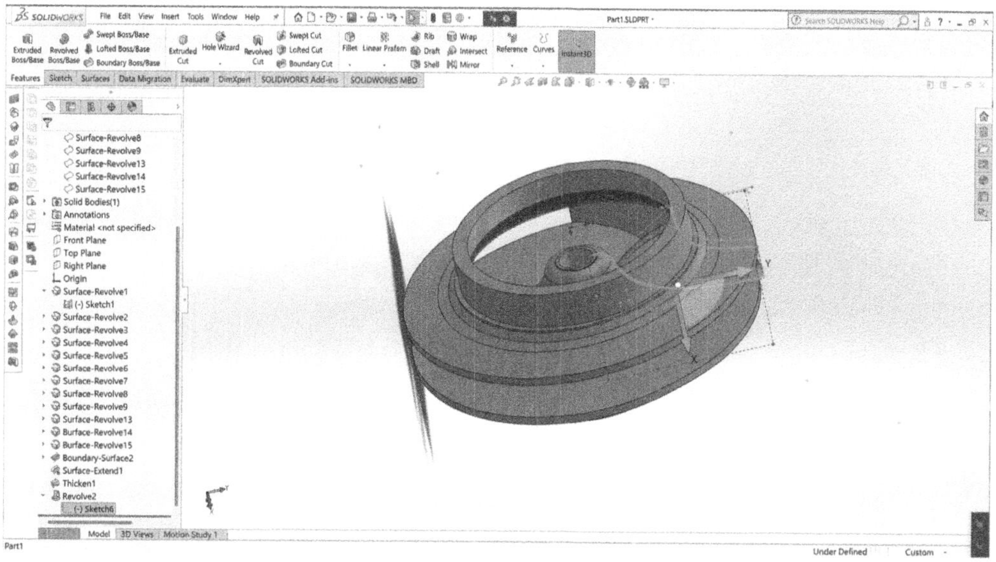

Figure 5.8 Impeller cover plate.

blade model by arraying a single blade. Select the axis of rotation for the impeller, and use the revolution command to create the impeller hub and casing. The impeller hub and casing are then merged with the blades into a single solid, completing the modeling of the impeller and inlet conical tube.

Start Creo, then select [Select Working Directory] from the quick access toolbar. In the pop-up file explorer, specify the location to store the 3D model and name the file, then click [OK] to confirm, as shown in Figure 5.10.

Click [New] in the quick access toolbar to open the new file window. In the [Type] dropdown, select [Part]; in the [Sub-type] dropdown, select [Solid]. Enter the desired file name, such as AP_Impeller, in the [File Name] field. Uncheck the default selection for [Use default template], then click [OK] to confirm.

Next, in the [New File Options] window, select the required [Template], typically [mmns_part_solid_abs], and click [OK] to confirm, as shown in Figure 5.11.

First, based on the cylindrical coordinate system, create coordinates of the points on the blade in Creo. There are two common methods for creating points:

Method 1: Select [Point] from the ribbon menu, then choose [Coordinate System]. The Offset Coordinate System window will pop up. In the [Placement] section under [References], select the coordinate system of the workspace. In the [Type] dropdown, choose Cylindrical Coordinate System. Then, input the coordinates for each point on the airfoil one by one and confirm.

Method 2: Create a new text file (with the .txt extension), input the coordinate information for each point on the airfoil, and save the file. Then, change the file extension to .ibl or .pts. Select [Point] from the ribbon menu, then choose [Offset Coordinate System]. The Offset Coordinate System window will pop up. In the [Placement] section under [Reference], select the workspace's coordinate system. From the [Type] dropdown menu, choose [Cylindrical], then click [Export]. In the file explorer that appears, import the previously created coordinate file, and click [OK] to confirm, as shown in Figure 5.12. (*Note:* In the text file, each point's coordinates must be entered on a new line. The three coordinate values of each point must be separated by 3–10 characters and aligned vertically with the respective axis names.)

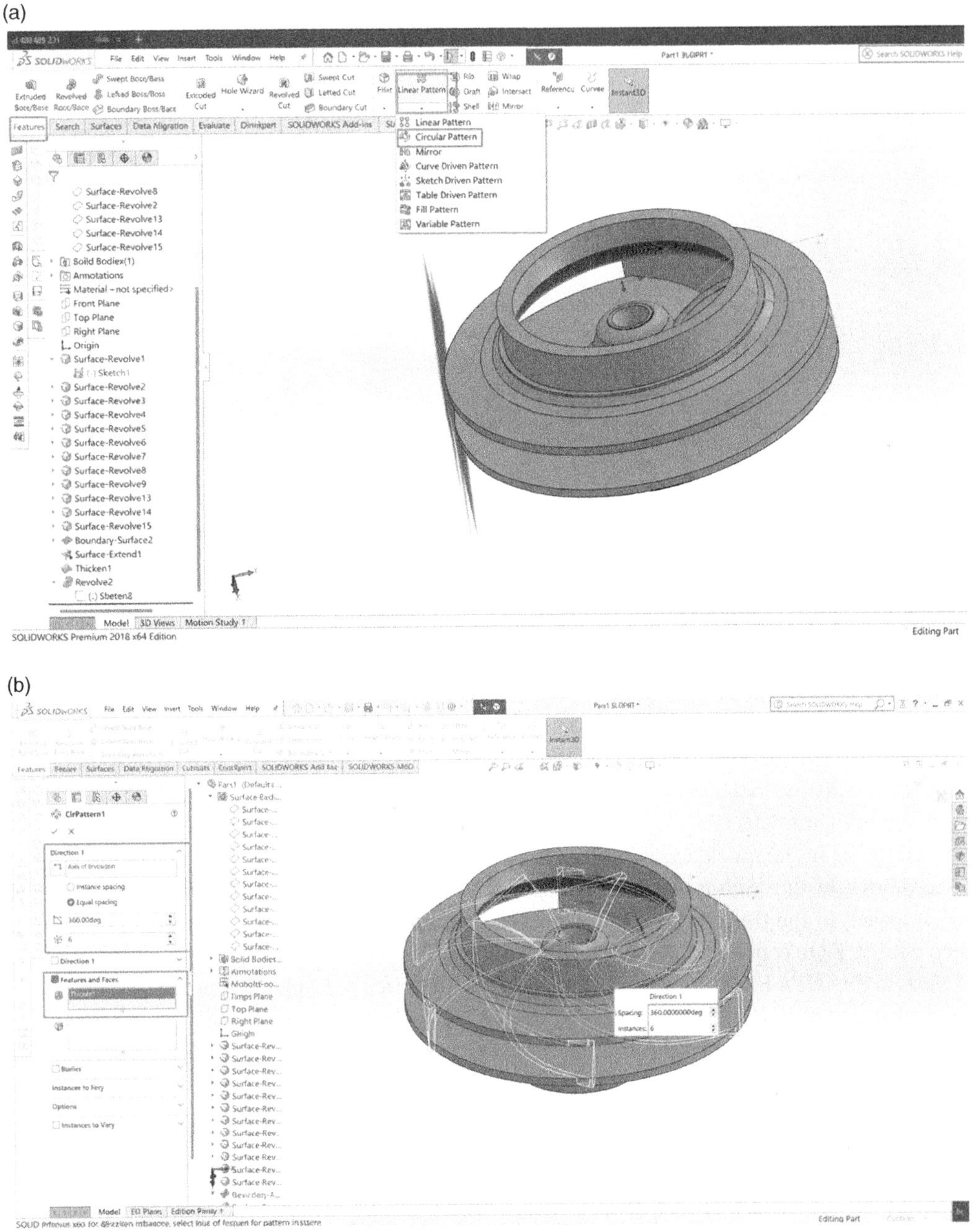

Figure 5.9 Complete impeller: (a) Circular array creation, (b) rotation.

Next, connect the created data points to generate the airfoil control lines. In the [Datum] ribbon, click the [Curve] dropdown menu and select [Curve through Points]. The workspace will switch to curve drawing mode. In the [Type] section of the ribbon, click [Spline], or choose [Spline] under [Connect to previous point by] in the dialog box to create the airfoil curve using a spline. Then, sequentially select the points created on each airfoil section. Once all the points on a section's control line are selected, click the [OK] button in the ribbon to confirm, as shown in Figure 5.13.

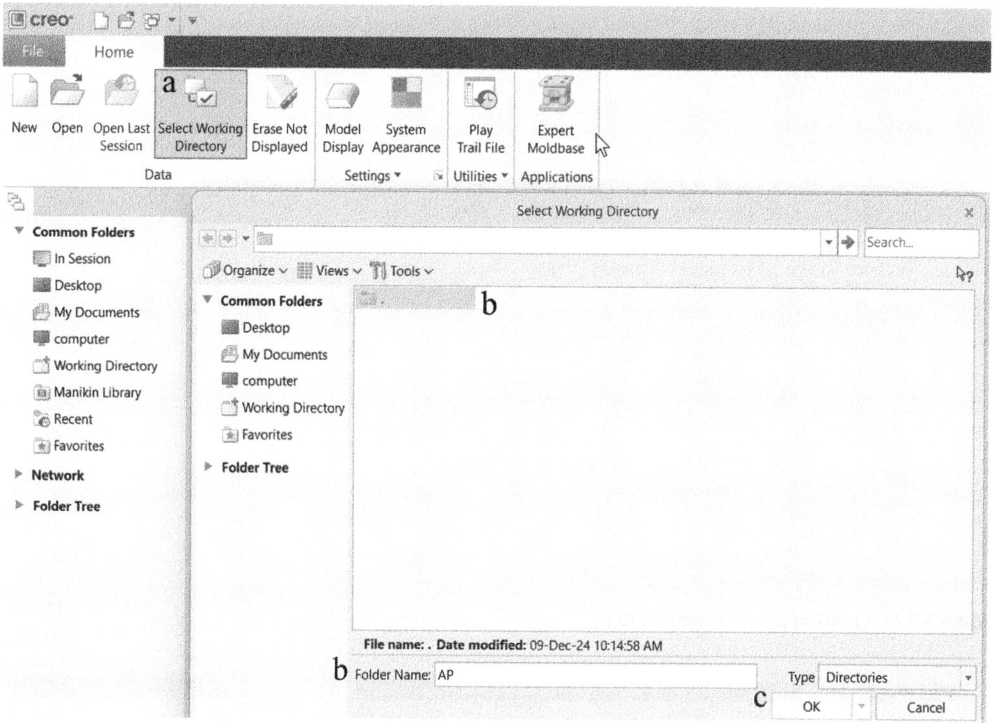

Figure 5.10 Selecting Working Directory.

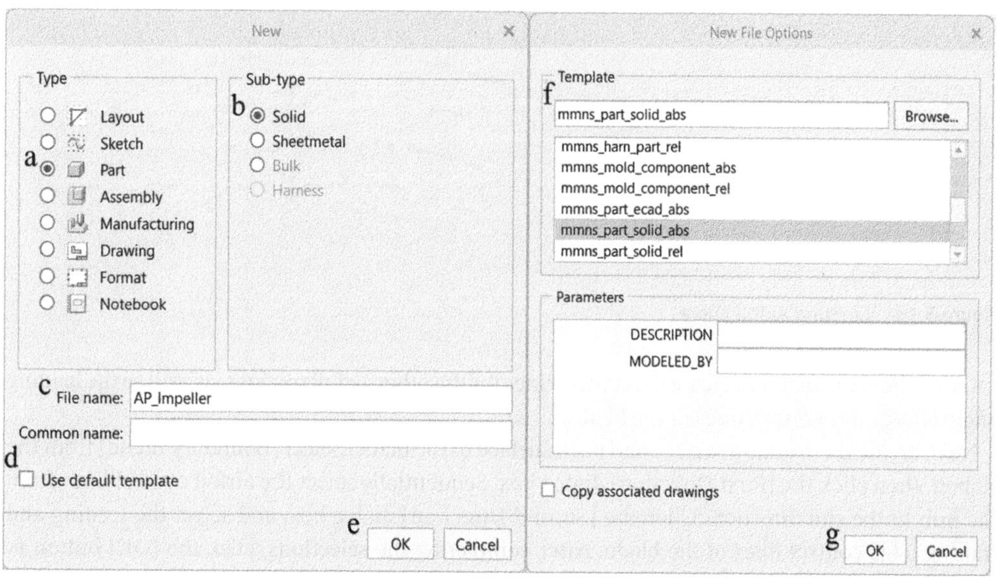

Figure 5.11 New Part creation and unit settings.

Point	R	θ	Z
0	130	-39.2	19.57
1	130	-35	24.85
2	130	-30	32.12
3	130	-15	39.4
4	130	-20	46.47
5	130	-15	53.27
6	130	-10	59.96
7	130	-5	66.2
8	130	0	72.1
9	130	5	77.6
10	130	10	82.9
11	130	15	88.6
12	130	20	93.4
13	130	25	98.5
14	130	30	103.6
15	130	35.25	108.63
16	130	-35	27.05
17	130	-30	36.12
18	130	-25	44.9
19	130	-20	53.47
20	130	-15	61.4
21	130	-10	69.27
22	130	-5	76.2
23	130	0	82.3
24	130	5	87.6
25	130	10	92.6
26	130	15	96.9
27	130	20	100.6
28	130	25	103.8
29	130	30	106.6

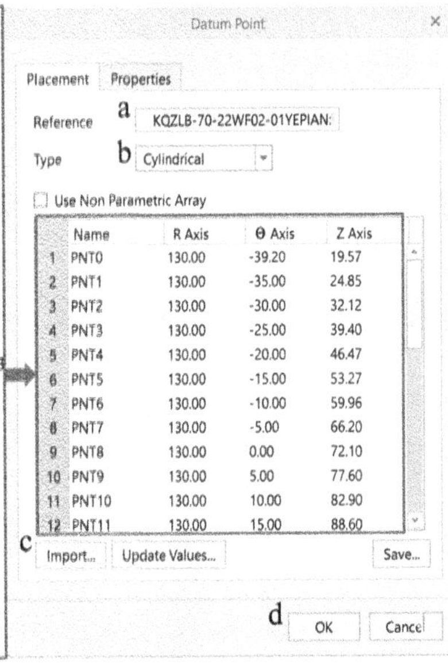

Figure 5.12 Creating points on the airfoil.

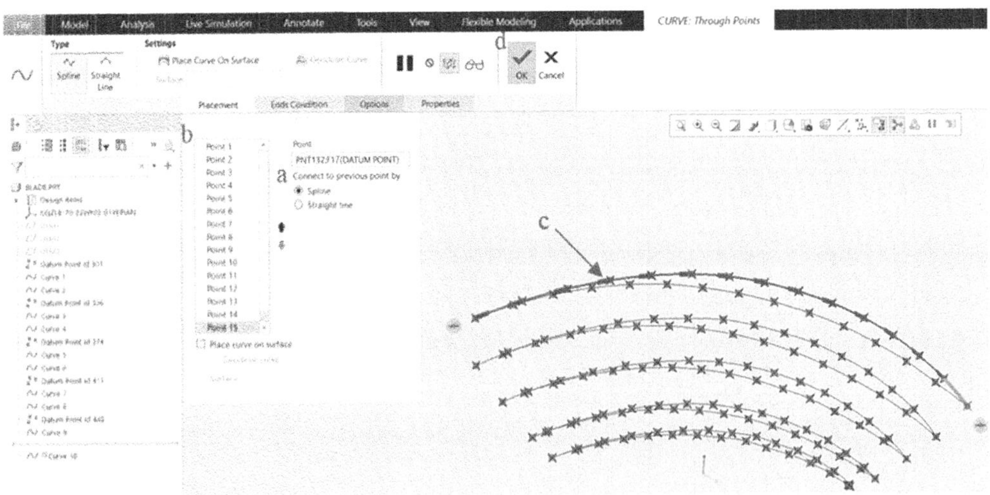

Figure 5.13 Creating airfoil curve.

Using the same method, create the control lines for the other airfoil sections, as well as the leading and trailing edge control lines of the blade.

Next, create the working surface and back surface of the blade. Select [Boundary Blend] from the ribbon, then click the [First Direction] dialog box. Sequentially select the airfoil control lines from the hub to the rim direction. Click the [Second Direction] dialog box, and select the leading and trailing edge control lines of the blade. After completing the selections, click the [OK] button to confirm, as shown in Figure 5.14. The method for creating the blade hub surface and rim surface is similar, so it will not be elaborated here.

Finally, stitch the blade's working surface, back surface, hub surface, and rim surface together to form a solid blade. Select [Merge] from the ribbon, pick the created blade working surface, back

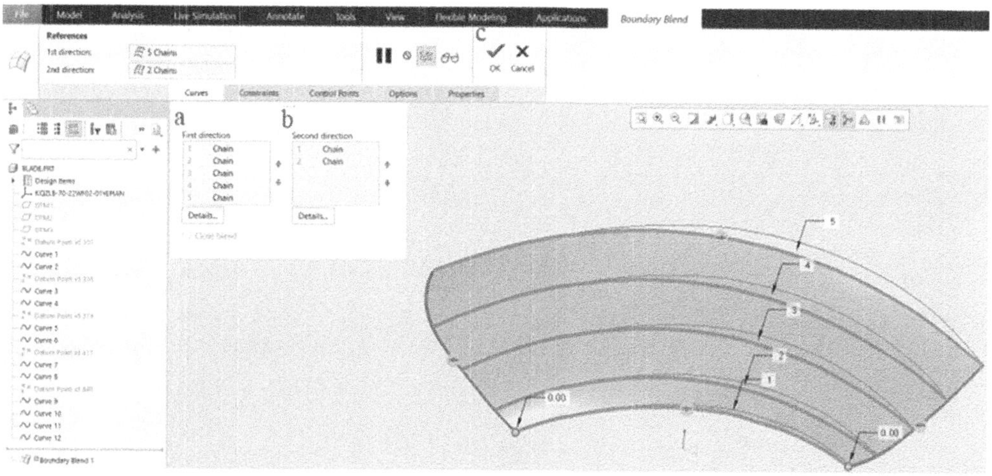

Figure 5.14 Creating blade surface.

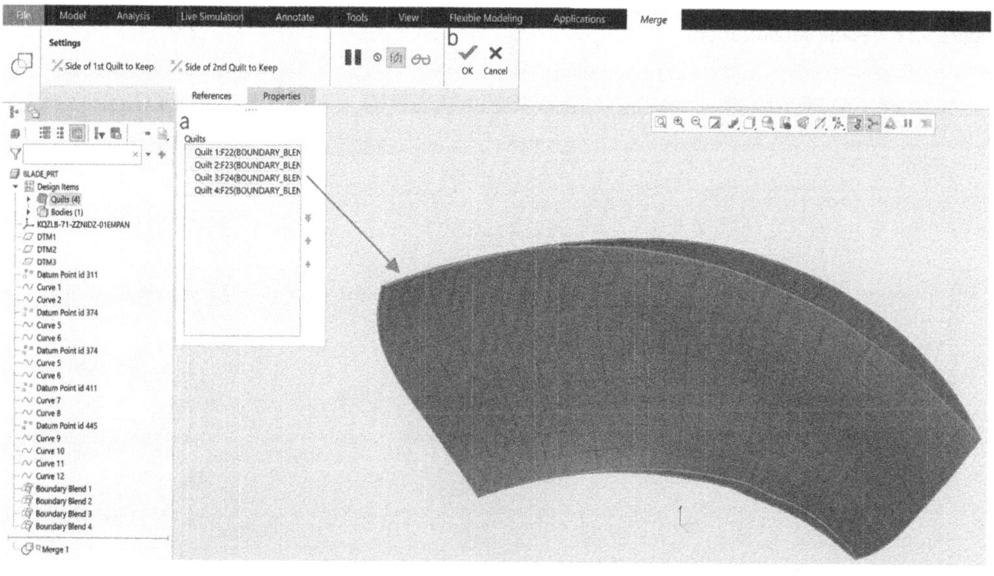

Figure 5.15 Blade surface merging.

surface, hub surface, and rim surface, then click the [OK] button to confirm, completing the surface merging of the blade, as shown in Figure 5.15.

Next, select the merged surface and choose [Solidify] from the ribbon to complete the creation of the solid blade, as shown in Figure 5.16.

In the [Model Tree] on the left side of the workspace, select the series of operations for Boundary Blend, Surface Merge, and Solidify. Right-click and choose [Group] from the pop-up menu to place them into a group. Select the created group, then click the [Pattern] button from the ribbon to create the blade array. In the [Type] dropdown menu, select [Axis] and choose the axis of rotation for the impeller in the workspace. Since the axial flow pump impeller has four blades, set the array number to 4, and the array angle to 90°. Click [OK] to confirm.

After completing the blade array, the full impeller blade is created, as shown in Figure 5.17.

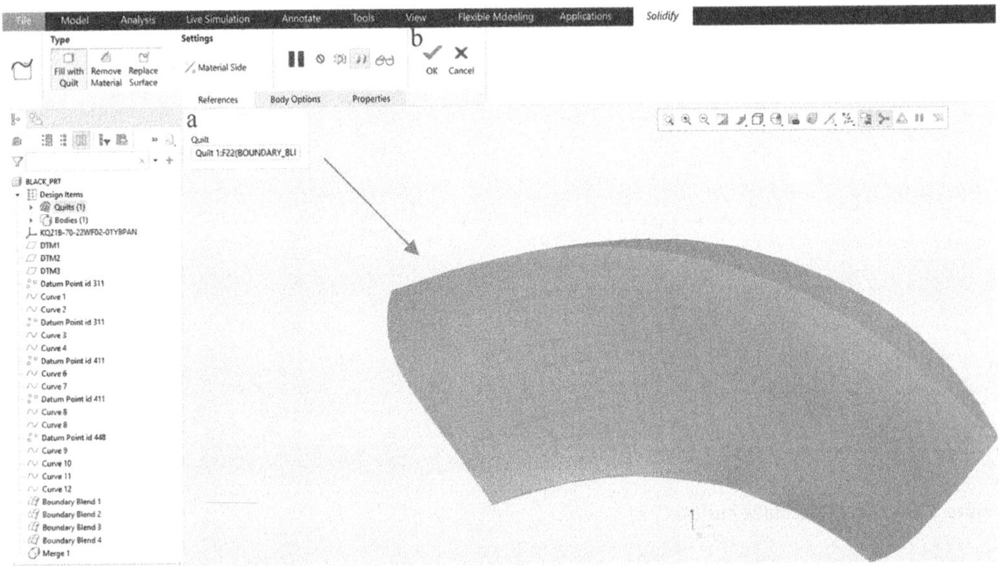

Figure 5.16 Blade solidification.

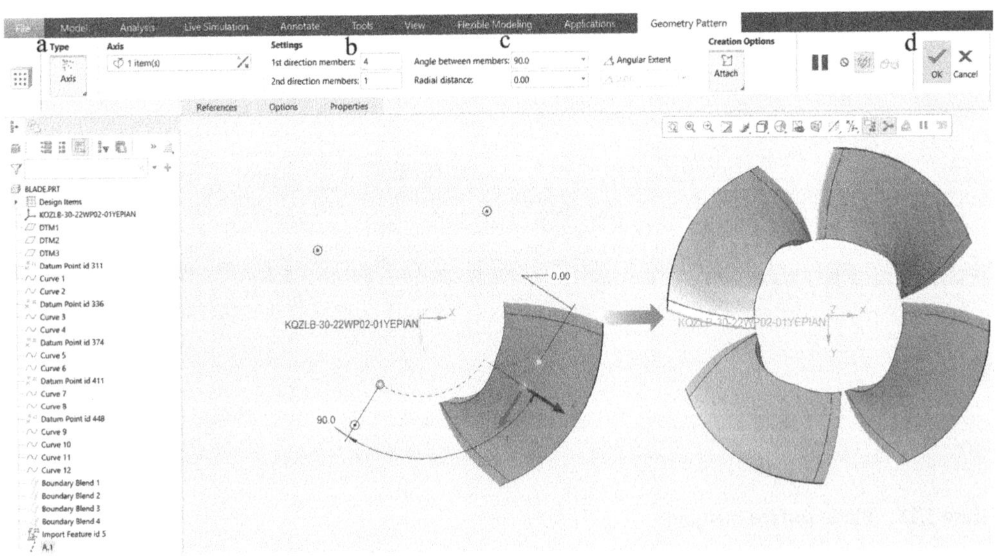

Figure 5.17 Blade array replication.

In the workspace, select the impeller's axis of rotation and use the revolution command to create the impeller hub and casing. Select and click a reference plane in the workspace. In the pop-up dialog box, click [Sketch]. The workspace will switch to sketch mode. Draw a closed sketch on the reference plane and click [OK] to confirm, as shown in Figure 5.18.

Next, click [Revolve] in the ribbon to switch to the revolution command interface. In the [Placement] dialog box, click [Sketch] to select the created revolution sketch, then click [Axis] to select the axis of rotation. The software will default to a rotation angle of 360°. Click [OK] to confirm and complete the modeling of the impeller hub and casing.

Finally, the 3D model of the impeller and inlet conical tube is obtained as shown in Figure 5.19.

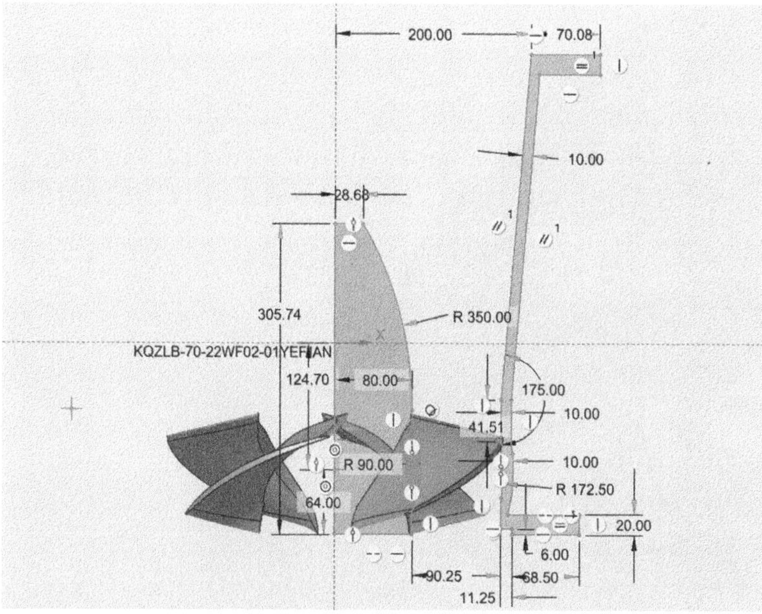

Figure 5.18 Hub and casing modeling sketch.

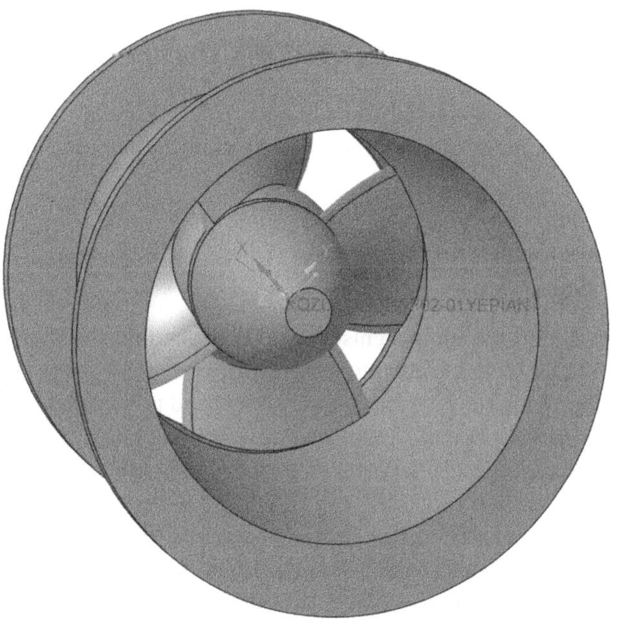

Figure 5.19 3D model of impeller.

5.2.2 Modeling of the Guide Vanes and Outlet Conical Tube

5.2.2.1 Modeling Approach

The modeling process for the guide vanes and outlet conical tube is similar to that of the impeller and inlet conical tube. First, create points on the blade airfoil. Next, connect the points to form the

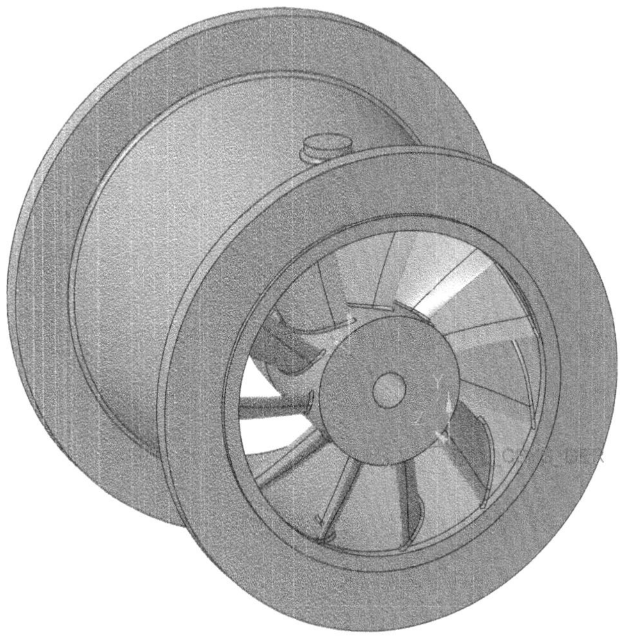

Figure 5.20 3D model of guide vanes.

blade airfoil control line. Then, use the airfoil control line to create the blade's working surface, back surface, hub surface, and rim surface. Finally, stitch and solidify these surfaces into a solid blade.

Further, create the complete guide vane model by arraying a single blade. Select the guide vane's axis of rotation in the workspace and use the revolution command to create the guide vane hub and casing.

The final 3D model of the guide vanes and outlet conical tube is shown in Figure 5.20.

5.2.3 Modeling of the Outlet Bend Tube

To model the outlet bend tube before assembling the components of the axial flow pump, the Sweep command is used to create the outlet bend tube model. Since the outlet bend tube model is relatively simple, one can directly continue modeling from the guide vanes and outlet bend tube model.

First, select and click a reference plane in the workspace. In the pop-up dialog box, click [Sketch] to switch to sketch mode. Draw the center axis of the outlet bend tube on the reference plane, and click [OK] to confirm.

Next, click [Sweep] in the ribbon. In the [Type] section of the pop-up page, select [Solid]. Under the [References] dialog, choose the previously created center axis in the [Trajectories] field. Then, click [Sketch] in the [Section] section to draw the cross-section of the outlet bend tube. Once the section is complete, confirm and return to the [Sweep] page. Finally, click [OK] to complete the outlet bend tube model, as shown in Figure 5.21.

5.2.4 Model Assembly

Based on the models created in the previous three sections, the impeller and inlet conical tube model, along with the guide vanes and outlet bend tube model, are assembled together to form the complete 3D axial flow pump model. The specific assembly process is as follows:

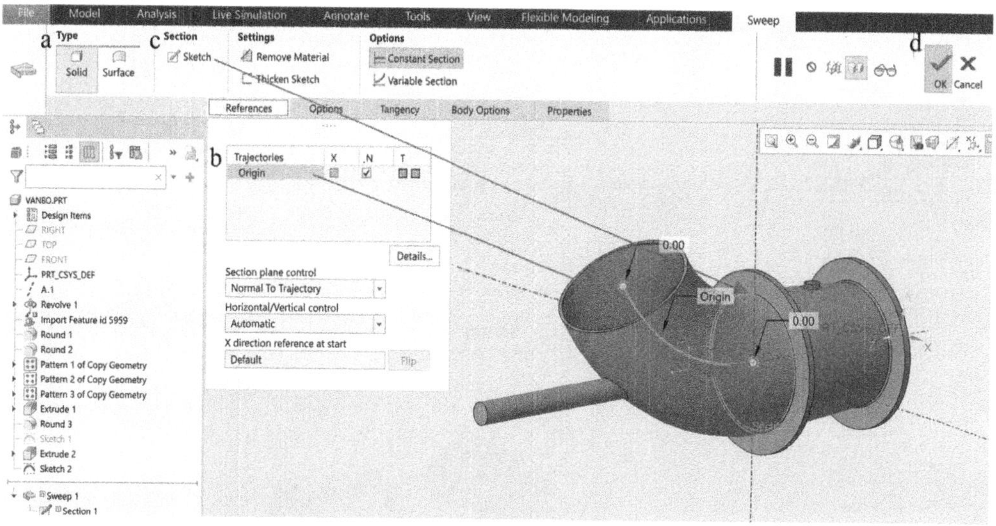

Figure 5.21 Outlet bend tube modeling.

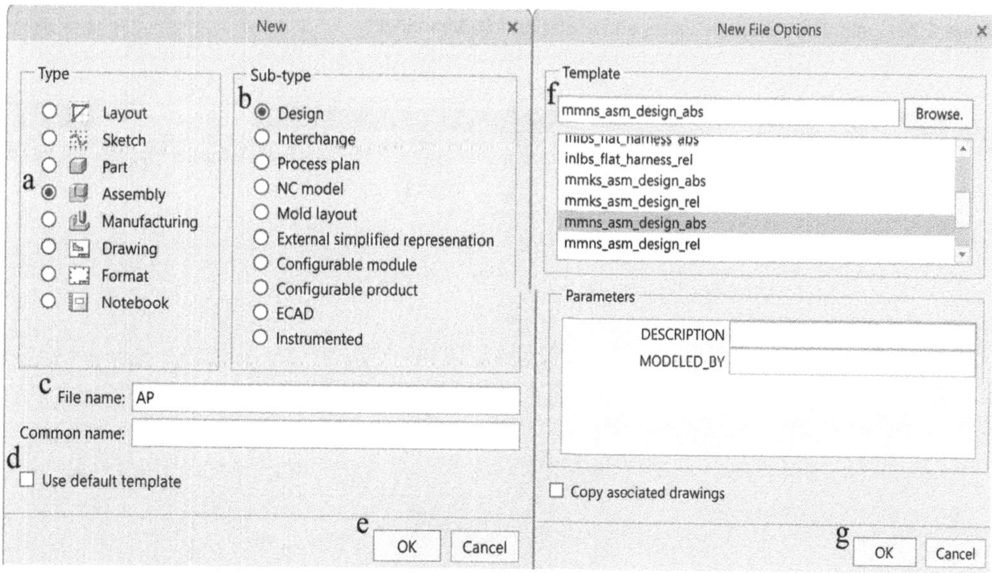

Figure 5.22 Model assembly.

First, create a new assembly. Click [New] on the Quick Access toolbar to open the New window. In the [Type] dropdown, select [Assembly]; in the [Sub-type] dropdown, select [Design]; in the [File name] field, enter the desired file name, such as AP. Uncheck the default [Use default template] option and click [OK] to confirm. Then, in the pop-up [New File Options] window, select the required [Template], typically [mmns_asm_design_abs], and click [OK] to confirm, as shown in Figure 5.22.

Click the [Assemble] tab in the ribbon, then select the guide vanes and outlet bend tube model file from the pop-up file explorer and click the [Open] button to import it. In the [Constraints] section of the ribbon, from the [Current constraint] dropdown menu, select [Default], or in the [Placement]

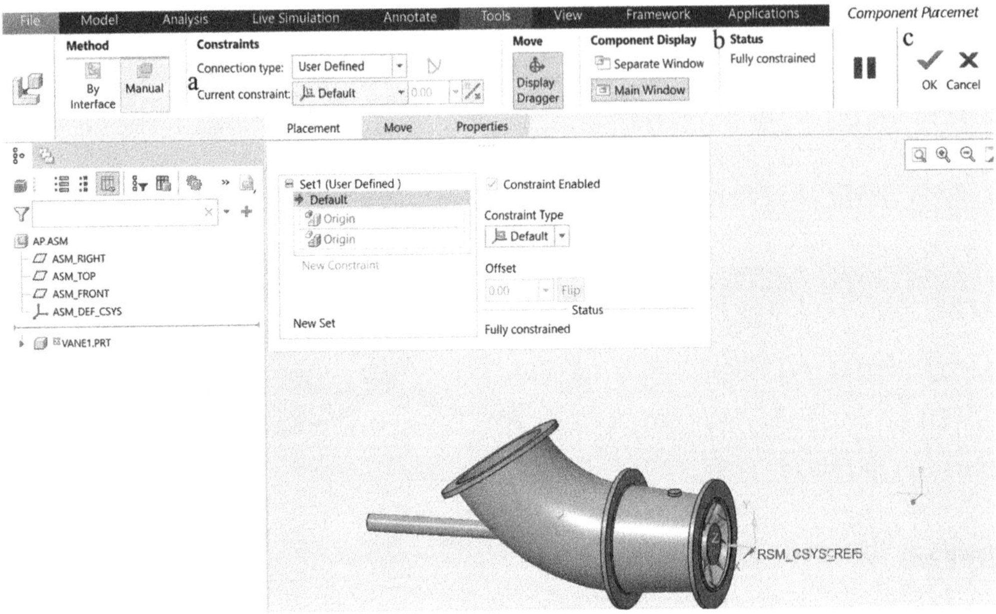

Figure 5.23 Importing of guide vanes and outlet bend tube model.

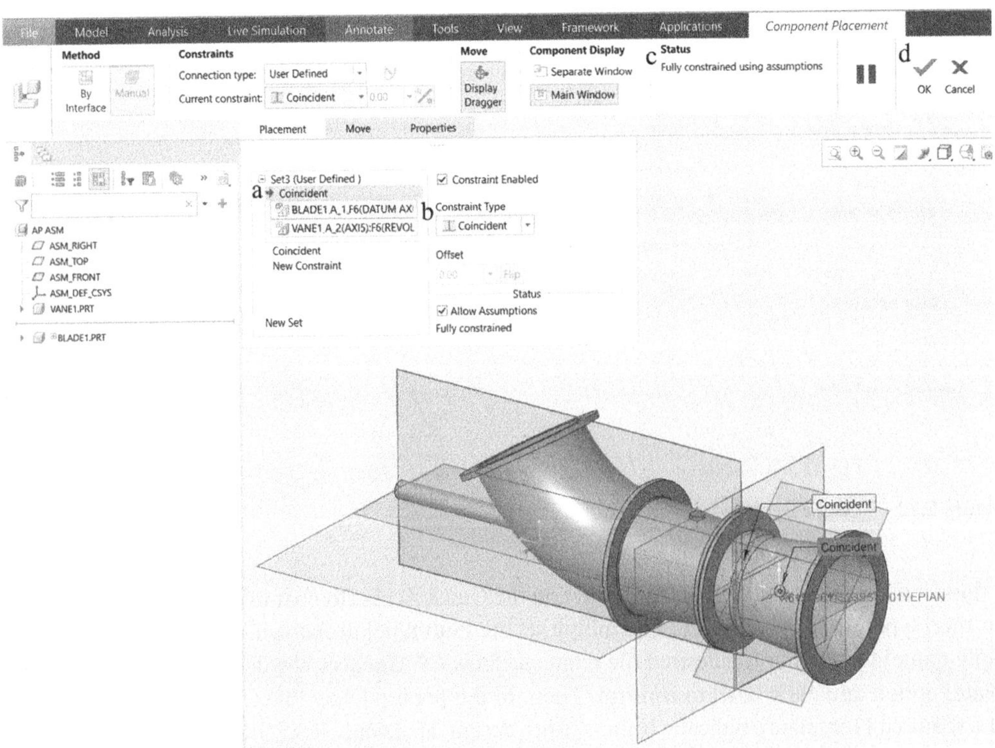

Figure 5.24 Axial flow pump model assembly.

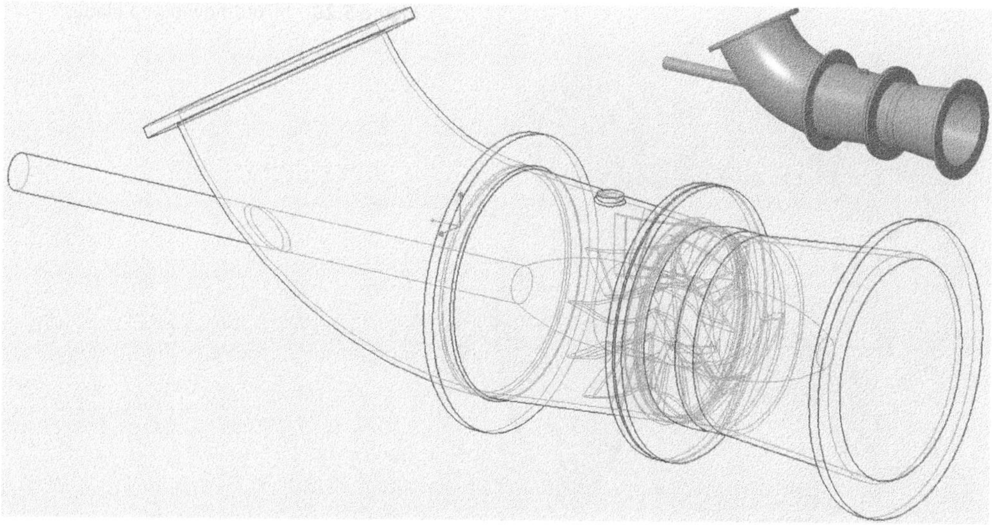

Figure 5.25 3D model of axial flow pump.

dialog box, select [Default] from the [Constraint Type] dropdown menu. The [Status] section of the ribbon will display as [Fully constrained]. Click [OK] to confirm, as shown in Figure 5.23.

Next, click the [Assemble] tab in the ribbon, select the impeller and inlet conical tube model file from the pop-up file explorer, and click the [Open] button to import it. Select the impeller hub and guide vane hub rotation axes, and in the [Placement] dialog box, select [Coincident] from the [Constraint Type] dropdown menu. Then, select the impeller hub and guide vane hub cross-sections, and in the [Placement] dialog box, select [Coincident] from the [Constraint Type] dropdown menu. Subsequently, observe that the [Status] section of the ribbon shows as [Fully constrained using assumptions]. Click [OK] to confirm, as shown in Figure 5.24.

Finally, the 3D axial flow pump model is obtained as shown in Figure 5.25. To facilitate subsequent numerical simulations, the axial flow pump model is saved in commonly used CAD intermediate formats, such as STEP, IGES, and X_T.

5.3 Mixed-Flow Pump

5.3.1 Impeller Modeling

5.3.1.1 Modeling Approach

As shown in Figure 5.26, first create the axial section lines for the blade's working surface and pressure surface. Then, use the [Boundary Blend] command to form the working surface and pressure surface of the blade. Next, use the [Surface Merge] command to solidify the blade and create a complete blade. After that, observe the axial view of the graphic through a 2D image and perform a sectioning operation. Subsequently, use the [Array] command to create an array of blades to form all the blades. Finally, design the impeller hub, completing the model [4].

Figure 5.26 Mixed-flow pump blades.

5.3.1.1.1 Create a New Blade .prt File

[1] Launch Creo. In the menu bar, click [File] → [New] to open the "New" dialog box, or directly click the button in the title bar.

[2] The "New" window pops up as shown in Figure 5.27. In the Type field, select [Part]. In the Name field, enter the desired filename, such as: mix-impeller. Uncheck the [Use default template] option, then click the [OK] button. A new file options window will appear. Select the required template, typically [mmns_part_solid], then click [OK] to confirm.

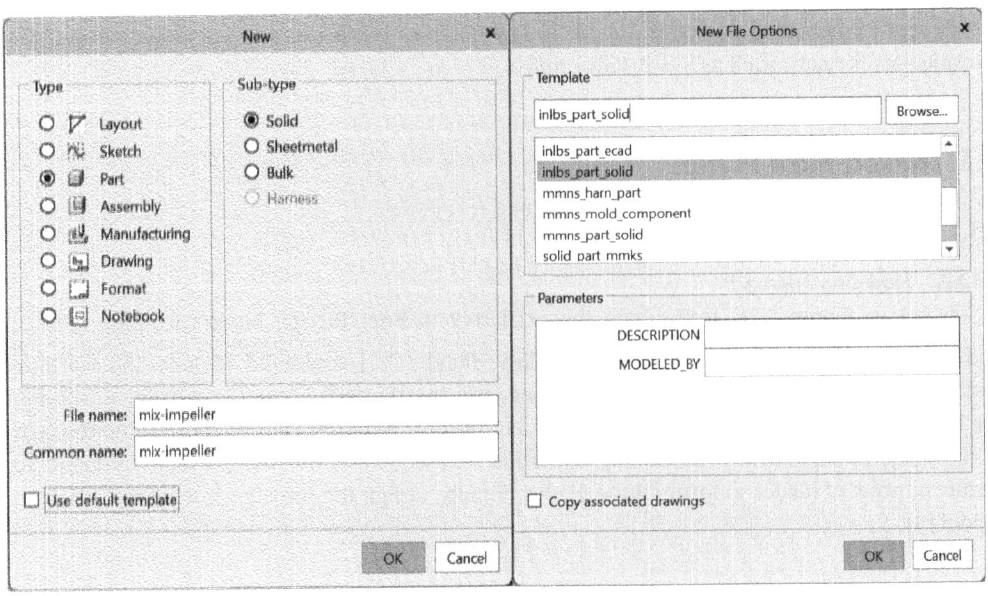

Figure 5.27 New Part file.

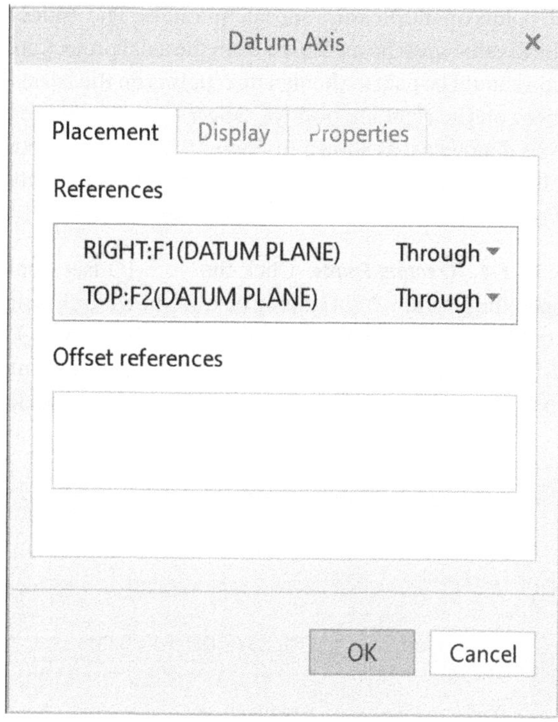

Figure 5.28 Creating the rotation axis.

5.3.1.1.2 Create the Rotation Axis Click the [Axis] button in the features toolbar to open the [Datum Axis] dialog box, as shown in Figure 5.28. In the working area, select the RIGHT and TOP planes. Choose the [Through] option, then click the [OK] button to confirm.

5.3.1.1.3 Determination of Cylindrical Coordinate Points When using the point-plotting method, it is necessary to organize the coordinate values of points on both the pressure side and suction side of the blade. These coordinates are in cylindrical format, consisting of radial coordinate R, angular coordinate θ, and axial coordinate Z. The blade wooden pattern drawing contains detailed data for

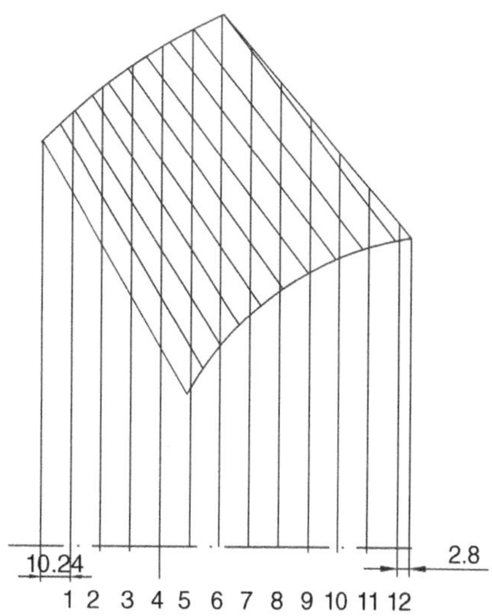

Figure 5.29 Axial projection of the blade.

all points on the pressure and suction sides. The values of R and the angle can be read directly, while the Z-value must be measured from the axial projection of the blade. As shown in Figure 5.29, attention should be paid to the sign of Z: points on the left side of the datum are considered negative, and those on the right are positive. Select the [Axis] button in the features toolbar to open the Datum Axis dialog box, as shown in Figure 5.28. In the working area, select the RIGHT and TOP planes. Choose the [Through] option, then click the [OK] button to confirm.

5.3.1.1.4 Creating Points Click the [Offset Coordinate System] button in the model tab to open the [Datum Point] dialog box, as shown in Figure 5.30. In the [Reference] section, select the coordinate system in the workspace (e.g., PRT_CSYS_DEF:F4). In the [Type] section, choose Cylindrical Coordinate System. One can either manually enter the coordinate values or click [Import...] to import a point file (in .pts or .ibl format). Finally, click the [OK] button to confirm.

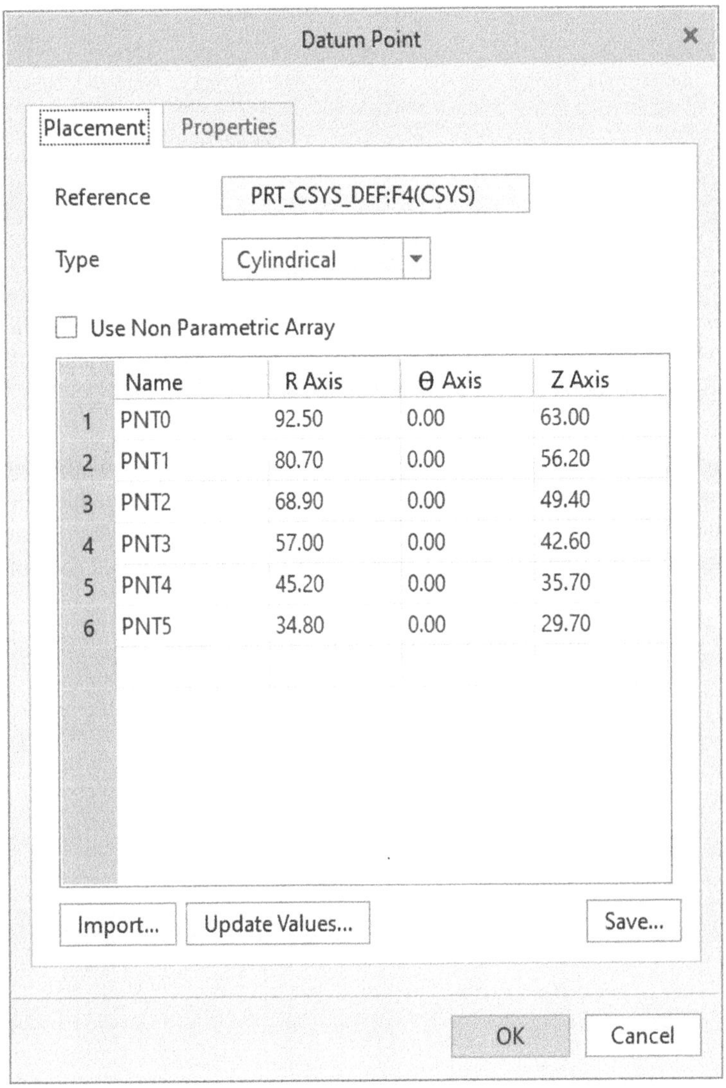

Figure 5.30 Point input settings.

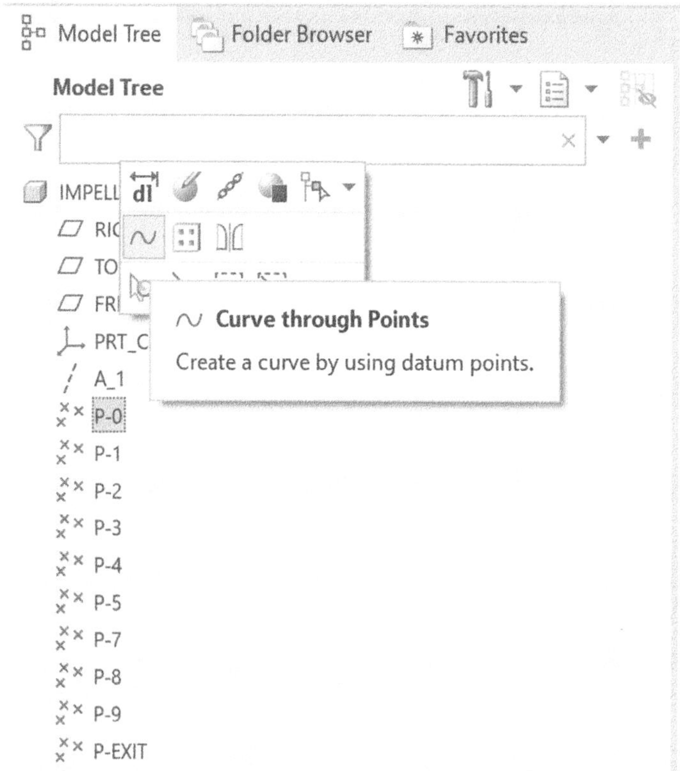

Figure 5.31 Curve through Points.

5.3.1.1.5 Drawing Wooden Mold Section Curves Click on the point on the 0° wooden mold section line in the model tree (Datum Point ID 44). A shortcut bar will appear, as shown in Figure 5.31.

Click [Curve through Points]. In the [Settings] section, select [Use spline to connect to the previous point]. In [Properties], you can customize the curve name. Click [OK] to confirm, resulting in Curve 1 (Figures 5.32 and 5.33).

Repeat this process to complete the drawing of all wooden mold section curves on the working surface, as shown in Figure 5.34.

5.3.1.1.6 Connection of Hub and Shroud Edges Click on [Curve] in the model toolbar and select [Curve through Points], then sequentially select the points in the workspace one by one. After completing the connection, click [OK]. For the settings in the [Settings] panel, refer to the previous step (Figure 5.35).

5.3.1.1.7 Drawing the Blade Pressure Surface Click the [Boundary Blend] button in the model toolbar. In the First Direction chain collector, select the two curves for the shroud and hub. In the Second Direction chain collector, select the inlet and outlet curves of the blade. Under [Options], add influence curves, and under [Properties], name it PRESSURE_SURFACE (Figures 5.36 and 5.37).

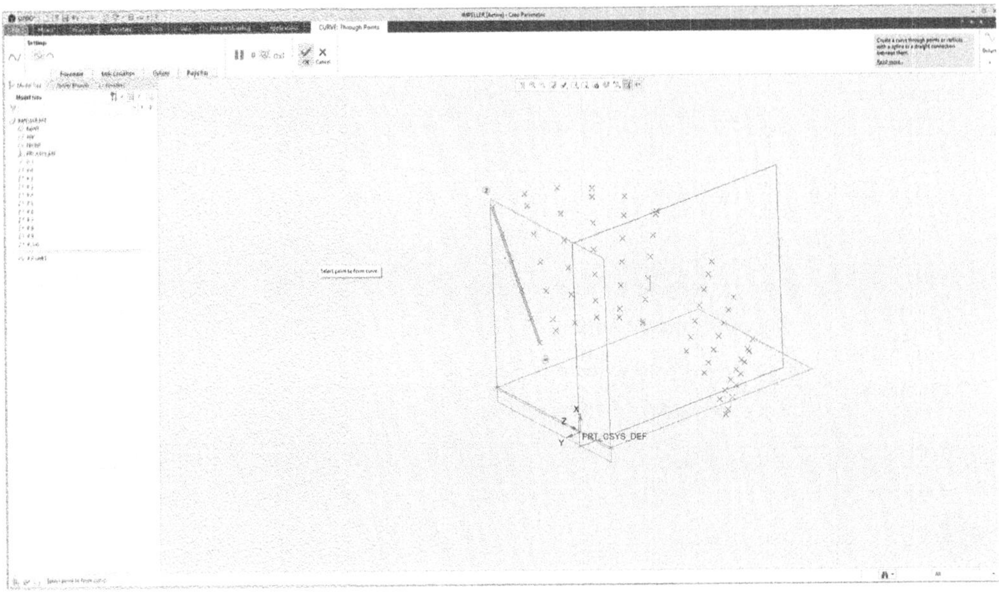

Figure 5.32 Curve through Points—Settings.

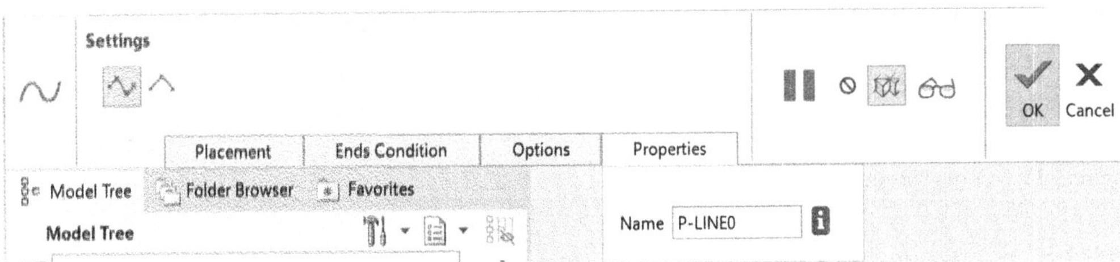

Figure 5.33 Curve through Points—Properties.

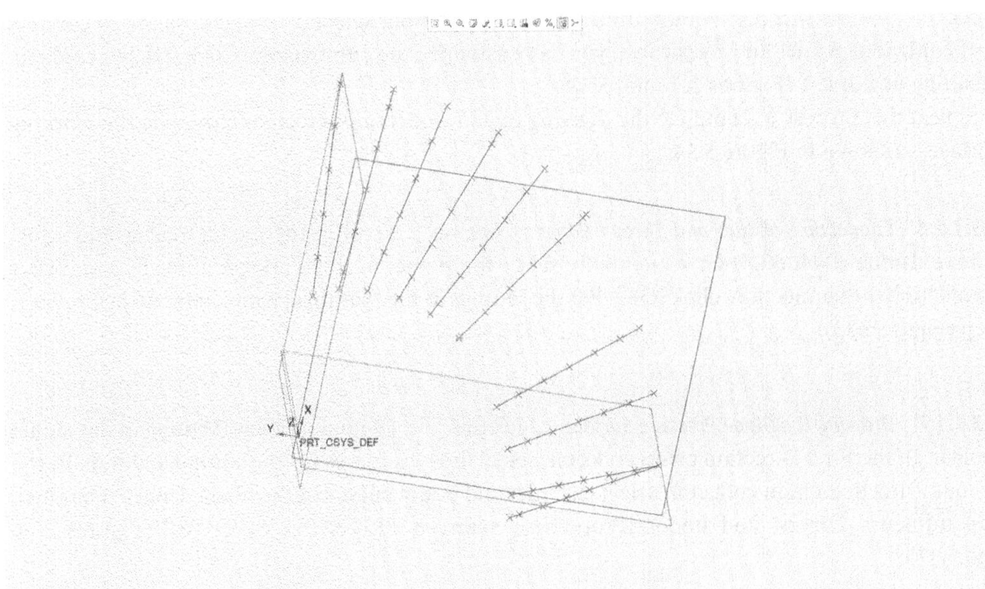

Figure 5.34 Curve diagram.

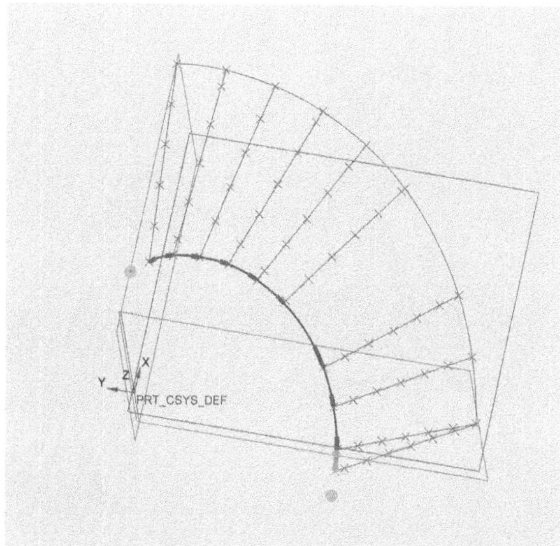

Figure 5.35 Hub and shroud edge connection.

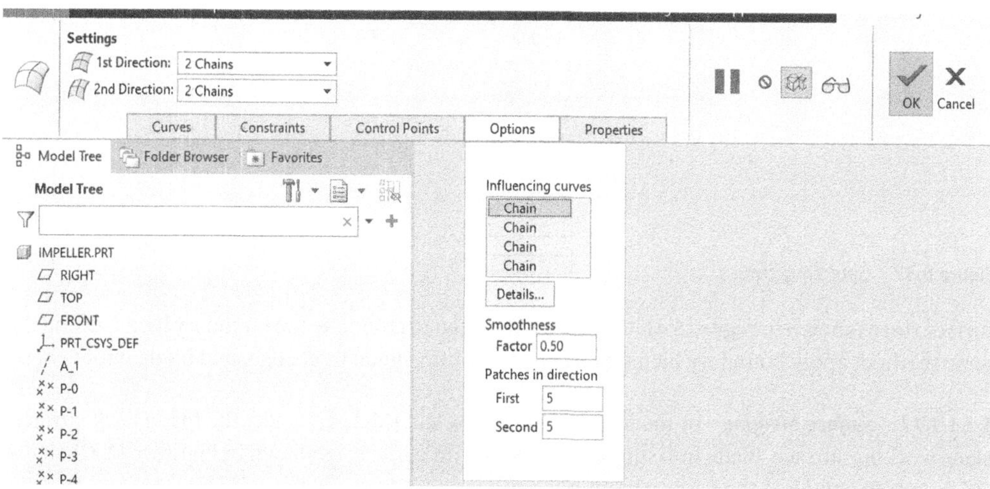

Figure 5.36 Boundary Blend settings panel.

5.3.1.1.8 Drawing the Pressure Surface of the Blade Following steps 3–7, the pressure surface of the mixed-flow pump blade is created by placing points, connecting them, and applying Boundary Surface blending. The result is shown in Figure 5.38.

5.3.1.1.9 Surface Extension Select the working surface's rim boundary line, then go to [Edit] → [Extend], choose [Extend Surface along Original Surface], and input an extension distance of 5 mm, as shown in Figure 5.39. Click OK to complete the surface extension. Follow the same process to extend the hub of the working surface, with an extension distance of 5 mm, as shown in Figure 5.40. Similarly, extend the blade suction surface's rim and hub using the same method.

5.3.1.1.10 Surface Boundary Blending In the features toolbar, select [Boundary Blend]. In the [First Direction Chain Collector], select the two edges of the extended working surface rim and the back

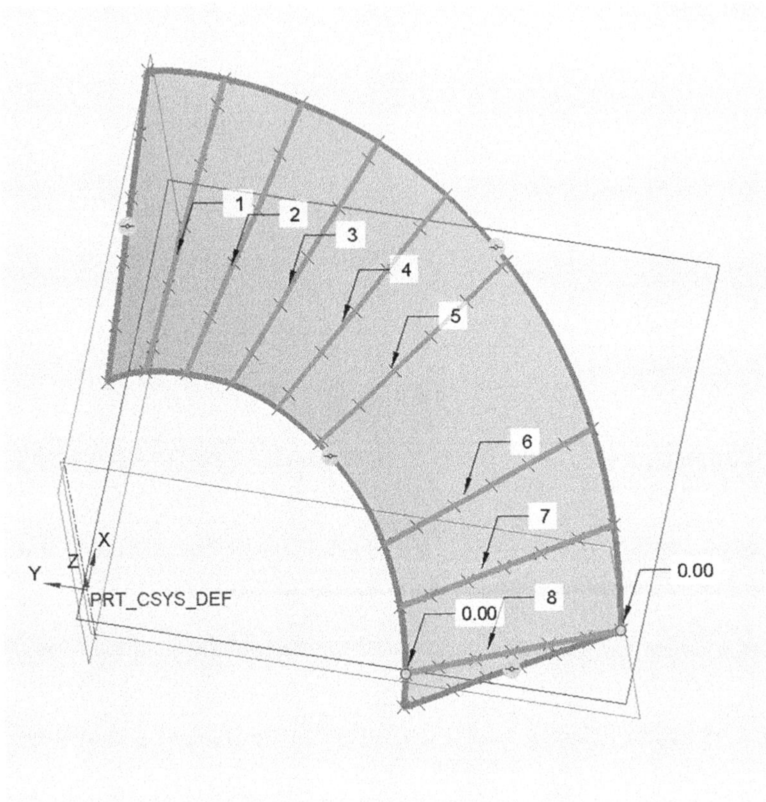

Figure 5.37 Selecting curves.

surface rim as shown in Figure 5.41. Click the OK button to complete the boundary blend. Using the same method, apply boundary blending to the blade hub, blade inlet edge, and blade outlet edge.

5.3.1.1.11 Surface Merging In the features toolbar, select [Merge]. Select the blade back surface, blade working surface, blade hub surface, and blade rim surface, as shown in Figure 5.42. Click the OK button to complete the surface merging.

5.3.1.1.12 Fillet the Blade Inlet and Outlet Edges In the features toolbar, select [Fillet] and choose the [Collect Mode]. Select both the inlet edges of the blade working surface and the back surface, as shown in Figure 5.43. Click the OK button to complete the filleting. Use the same method to fillet the blade outlet edges.

5.3.1.1.13 Create a Group On the left side of the working interface, under the model tree, select the operation icons that you want to group together. Right-click on them and choose [Group] from the pop-up shortcut menu. This will group the key components of the generated body together, as shown in Figure 5.44.

5.3.1.1.14 Blade Array Select the group generated under the model tree, and then choose Pattern from the feature toolbar to array the blades. As shown in the figure, select [Axis] as the array center and choose the axis created in Step 2. Since the model has four blades, set the number of arrays to 4, with an array angle of 90°. Click OK to complete the array, as shown in Figure 5.45.

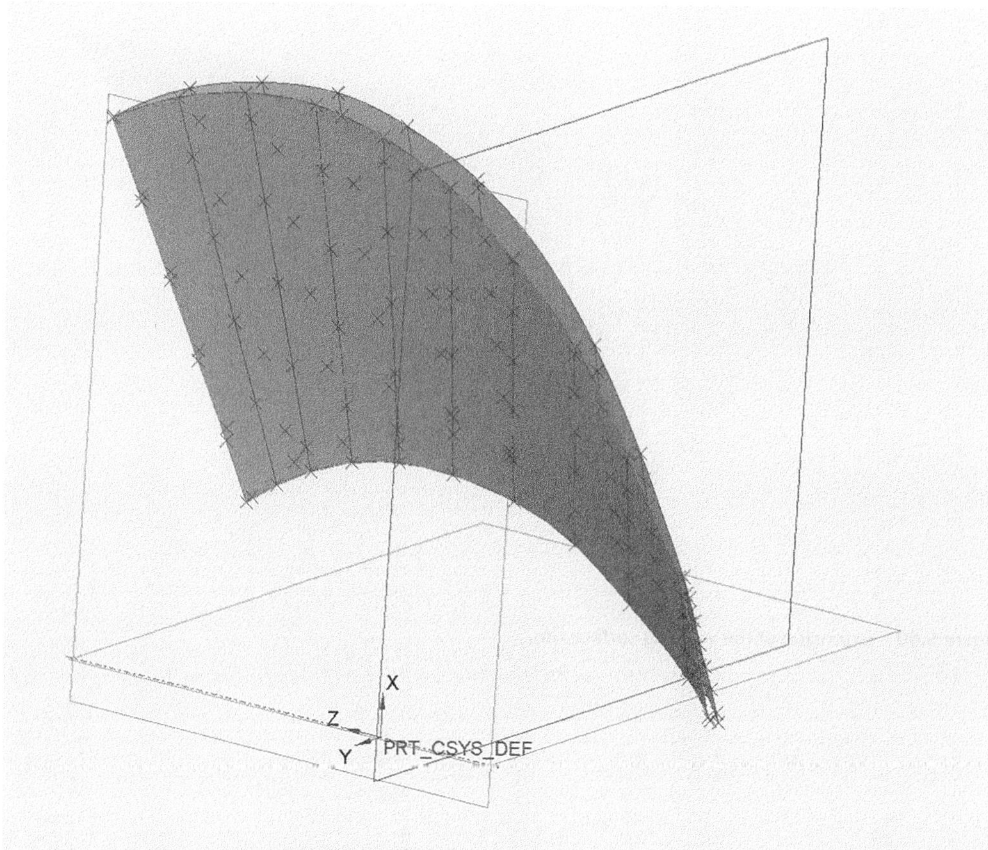

Figure 5.38 Blade suction surface.

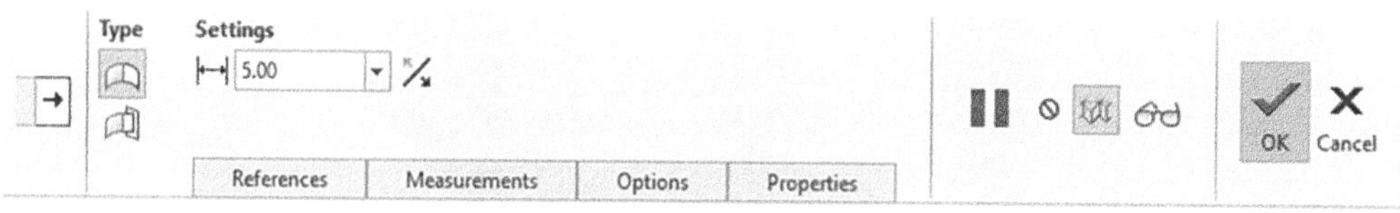

Figure 5.39 Extension distance.

5.3.1.1.15 Blade Solidification Select a single blade after the array, then go to [Edit] → [Solidify], and click OK to complete the solidification. Follow the same steps to solidify the other blades. Note: One can check the model by selecting the icons in the model display toolbar. When the boundaries of the body are shown as white lines, it indicates that the solidification was successful, as shown in Figure 5.46, where the boundary is displayed as a white line. This is the boundary outline before solidification, which can be used to verify if solidification was successful.

5.3.1.1.16 Impeller Hub Drawing In the feature toolbar, select [Sketch] or click the  icon to enter the sketching mode. Draw the section as shown in Figure 5.47 to generate the shape of the hub through rotation. After completing the sketch, click OK to finish the drawing.

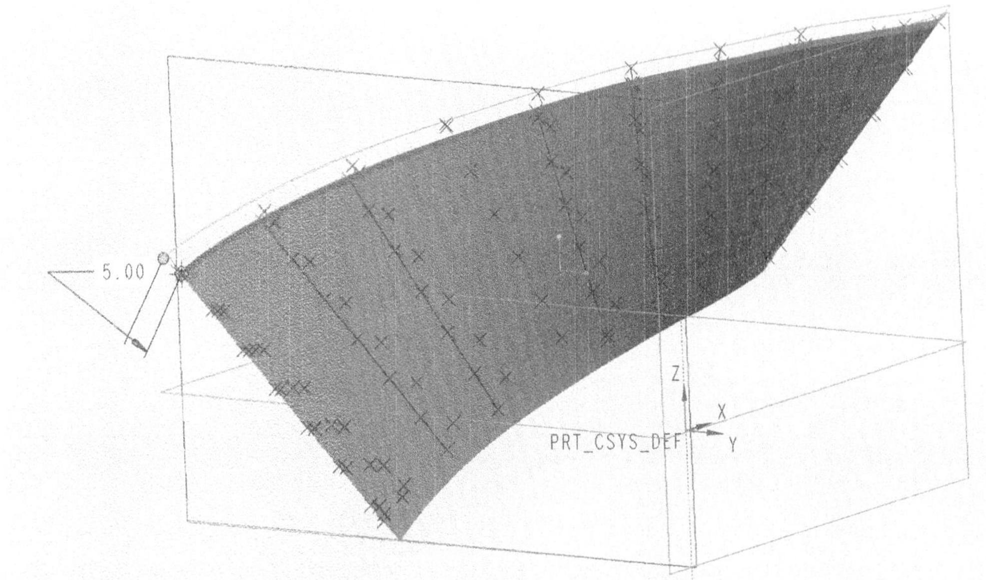

Figure 5.40 Extension of the working surface rim.

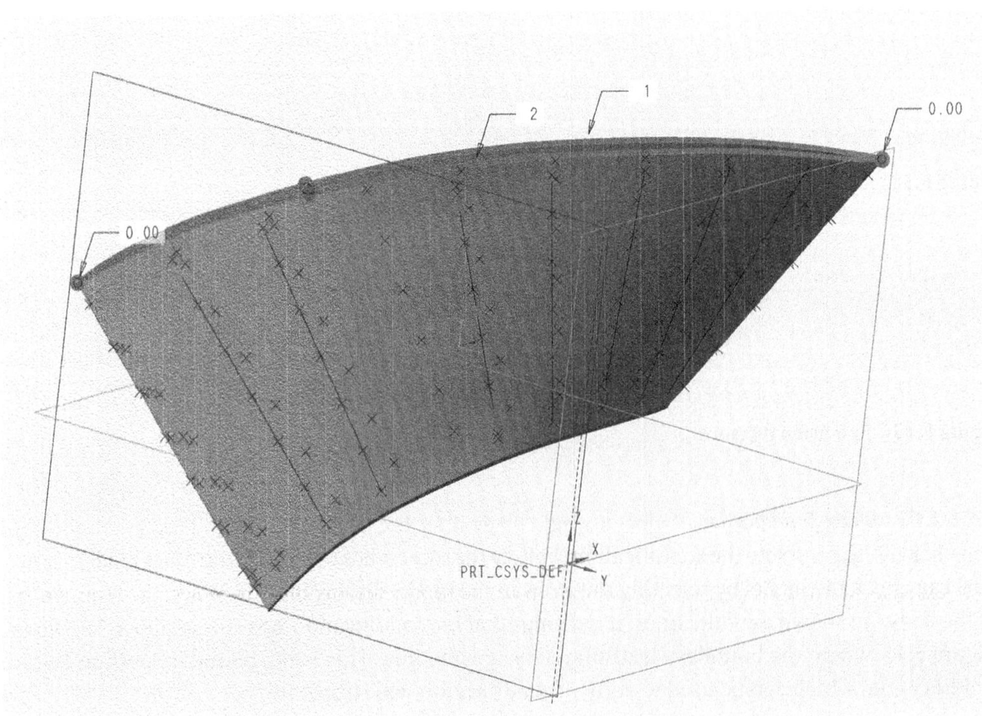

Figure 5.41 Rim surface drawing.

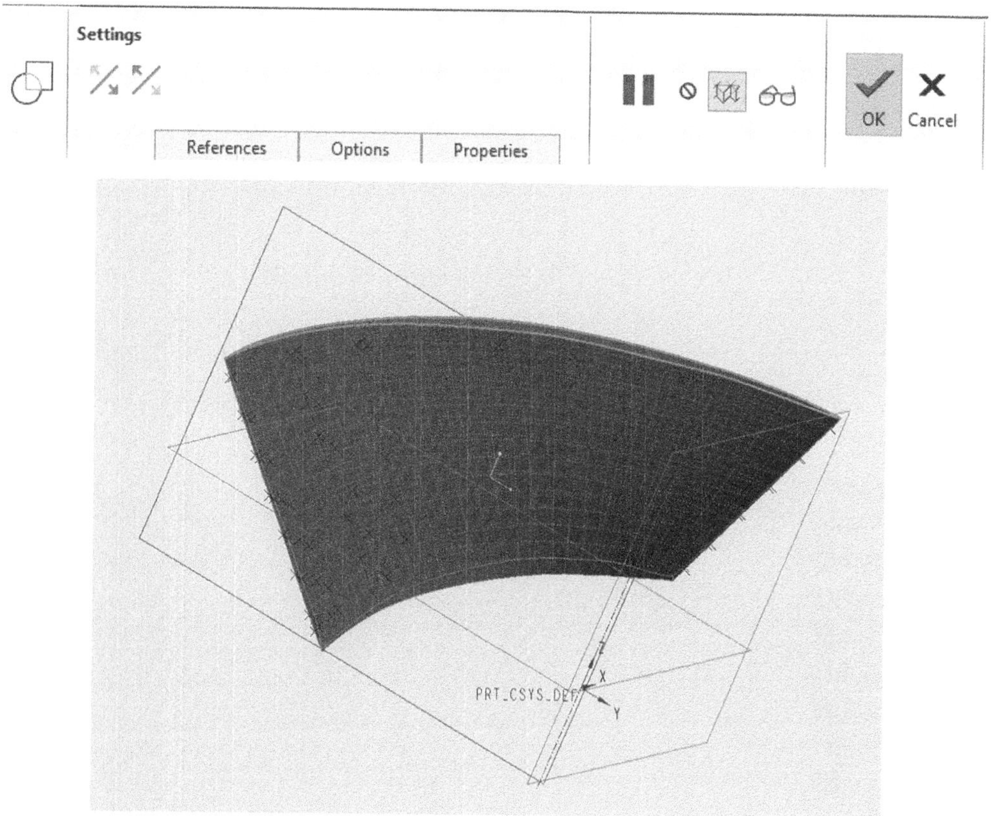

Figure 5.42 Surface merging.

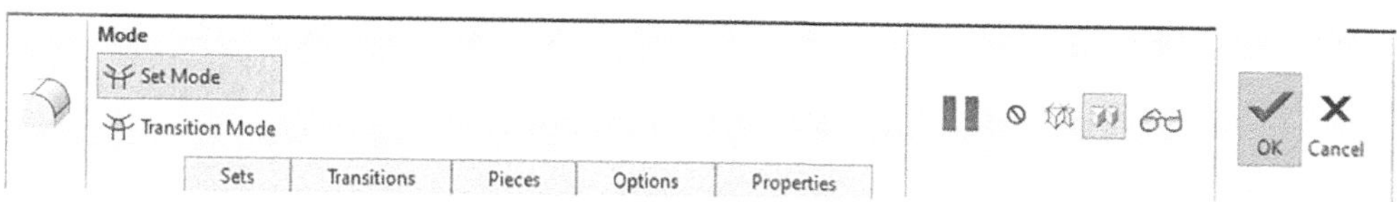

Figure 5.43 Filleting.

Rotating the Sketch (Figure 5.48). In the feature toolbar, select [Revolve]. Choose the sketch as the surface and select the axis A_1 for rotation in the [Settings]. Click the OK button to complete the drawing, as shown in Figure 5.49.

5.3.1.1.17 Complete the Model Design Save the file by clicking [File] → [Save a Copy], choose the file type as stp format, then click the [OK] button. In the pop-up [Export STP] dialog, check the [Solid] option, and click [OK] to finish the saving process.

5.3.2 Guide Vane Modeling

5.3.2.1 Modeling Approach

As shown in Figure 5.50, the guide vane is created by first creating the axial section lines for the guide vane blade's work surface and back surface. Then, use the [Boundary Blend] command to

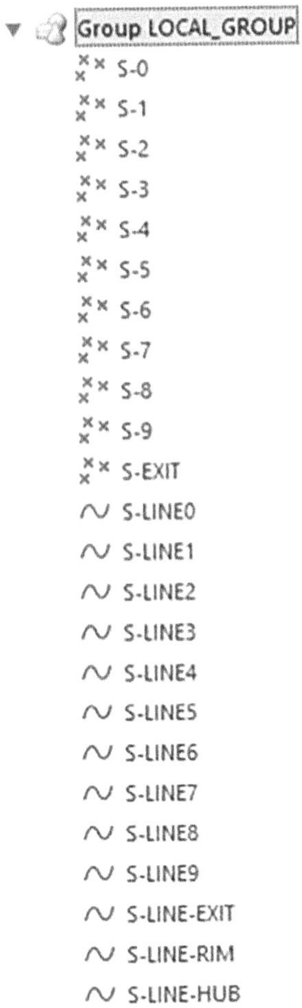

Figure 5.44 Create group.

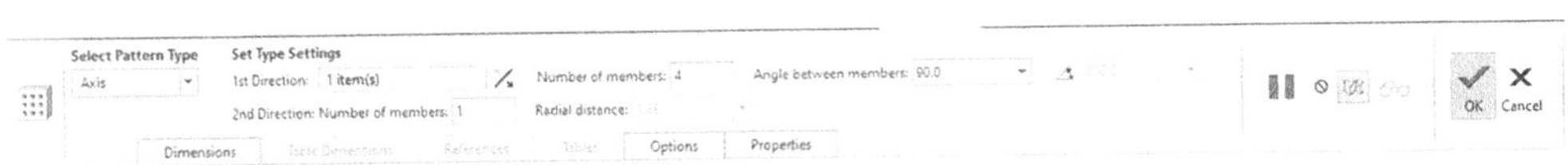

Figure 5.45 Blade array.

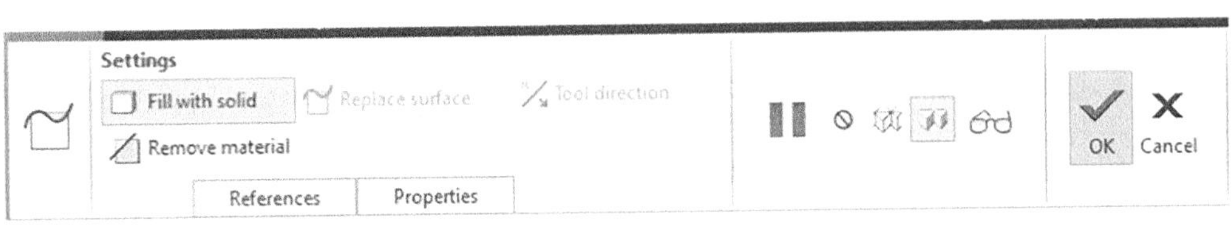

Figure 5.46 Blade solidification.

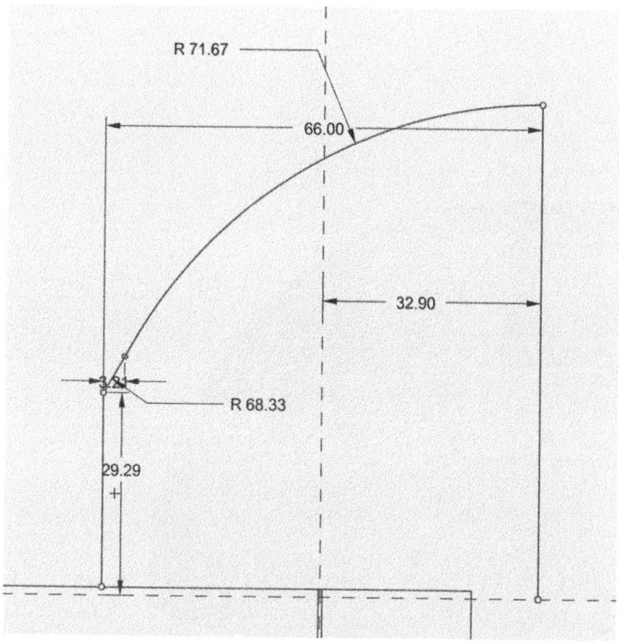

Figure 5.47 Hub Sketch.

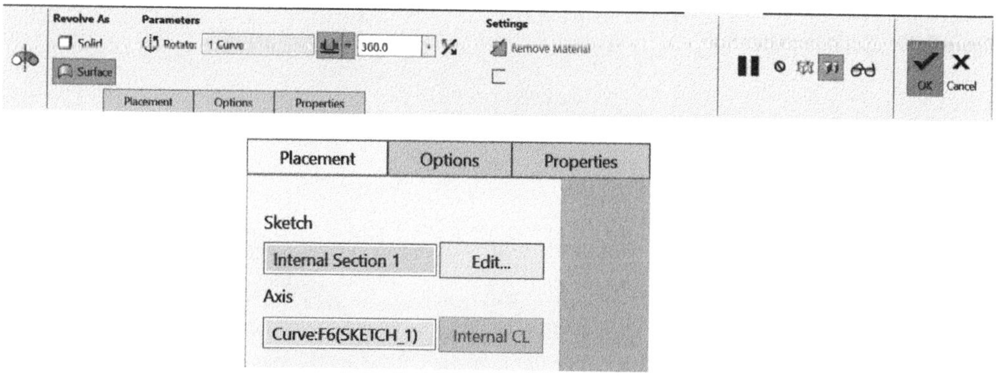

Figure 5.48 Hub section revolution.

form the work and back surfaces of the guide vane blade. After that, use the [Merge] and [Solidify] commands to combine and solidify the guide vane blade into a complete blade. The axial view of the 2D diagram is then used to observe the geometry for necessary cutting. Afterward, use the [Array] command to duplicate the guide vane blades, forming all of them, and finally, the model is completed. (Here, the guide vane hub is simplified, with the focus on the drawing of the guide vane blades.)

5.3.2.1.1 Spatial Guide Vane Drawing and Data Reading Using the point-plotting method, the coordinates of each point on the work surface and back surface of the blade need to be organized. These coordinates are in cylindrical coordinates, specifically the radial coordinate R, angular coordinate θ, and axial coordinate Z. The coordinate point data is obtained from the blade wooden model diagram, as shown in Figure 5.51.

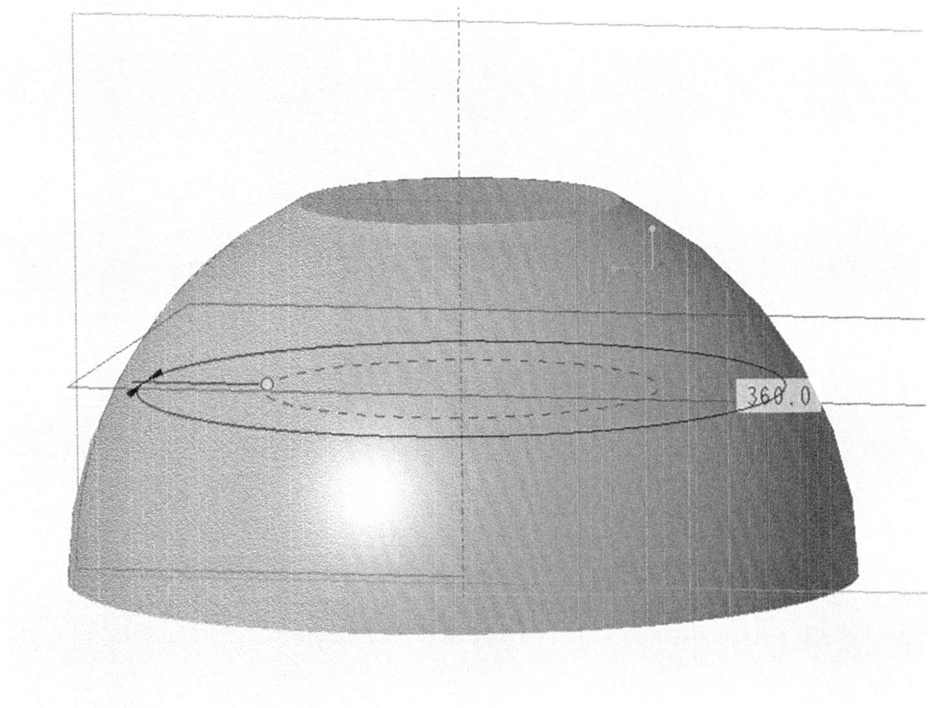

Figure 5.49 Solid impeller hub.

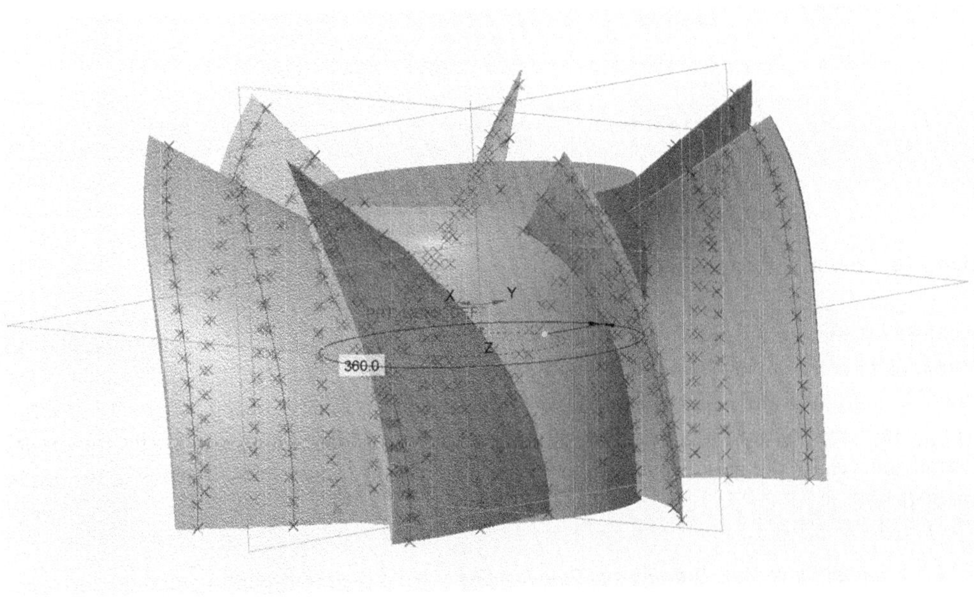

Figure 5.50 Spatial guide vane.

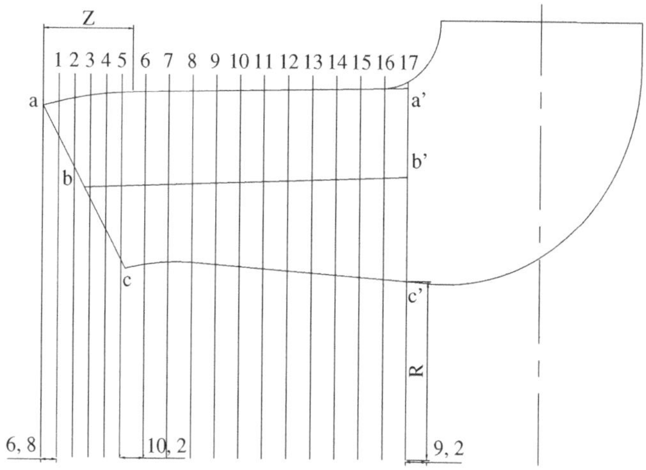

Figure 5.51 Coordinate point reading.

5.3.2.1.2 ***Inputting Coordinate Points*** Click the [Offset Coordinate System] button in the model toolbar to open the [Datum Point]dialog box, as shown in Figure 5.52. In the [Reference] section, select the coordinate system in the workspace, such as PRT_CSYS_DEF:F4 (Coordinate System). In the [Type] section, select cylindrical coordinates, input the coordinate values, or click [Import…] to import the point file (e.g., .pts, .ibl files). Then click the [OK] button.

5.3.2.1.3 ***Curve Connection*** Click on the points in the model tree to open the shortcut toolbar, as shown in Figure 5.53. Click [Curve through Points], and in the [Settings] section, select [Use spline to connect this point to the previous point]. In the [Properties] section, you can customize the curve name. Click [OK] to create [Curve 1] (Figures 5.54 and 5.55).

Repeat this process to complete the drawing of all the wooden model section lines on the working surface, as shown in Figure 5.56.

5.3.2.1.4 ***Boundary Blending*** Click the [Boundary Blend] button in the model toolbar. In the first direction chain collector, select three curves. In the [Options], additional influencing curves can be added, and in the [Properties], the surface can be named according to the requirements (Figures 5.57 and 5.58).

5.3.2.1.5 ***Surface Extension*** Select the boundary line of the working surface rim, then go to [Edit] → [Extend], choose [Extend Surface Along Original Surface], enter an extension distance of 10 mm, as shown in Figure 5.59, and click the [OK] button to complete the surface extension (Figure 5.60). Similarly, extend the working surface hub with an extension distance of 5 mm, as shown in Figure 5.61. Use the same method to extend the rear surface rim and hub of the blade.

5.3.2.1.6 ***Surface Merge*** In the feature toolbar, select [Merge], and simultaneously select the blade back surface, blade working surface, blade hub surface, and blade rim surface, as shown in Figure 5.62. Click the OK button to complete the surface merge.

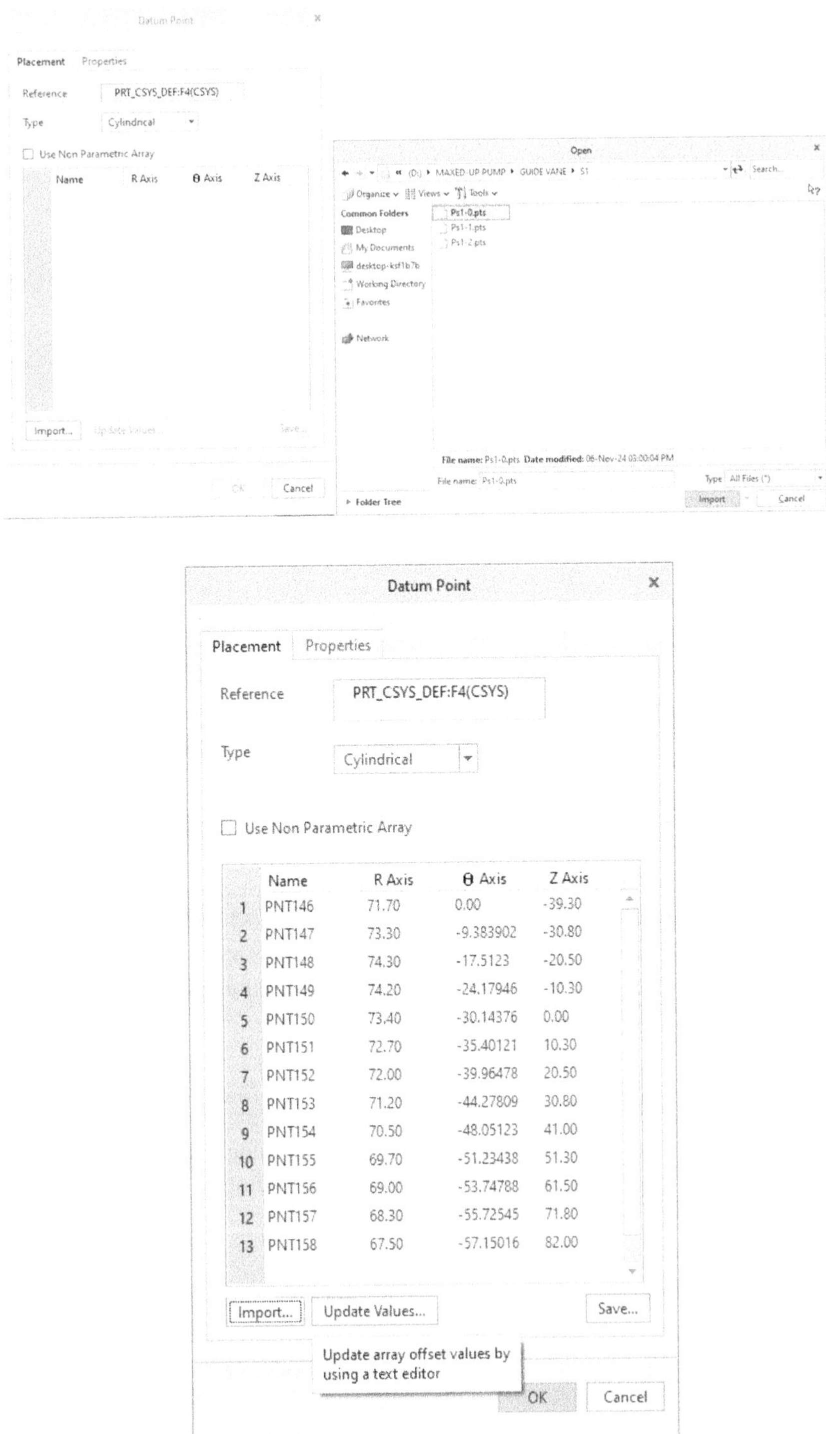

Figure 5.52 Inputting coordinate points.

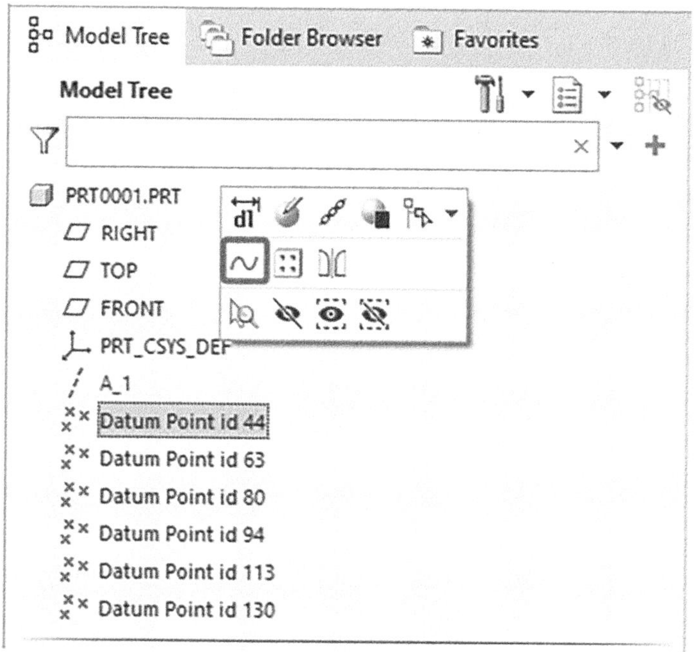

Figure 5.53 Curve through Points.

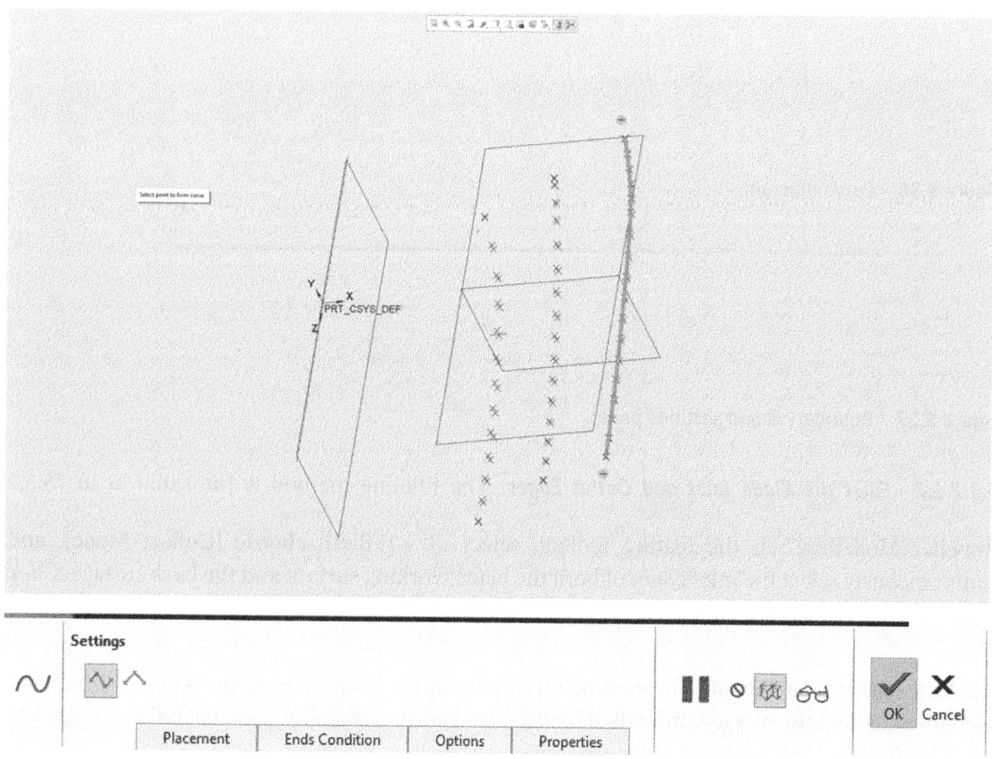

Figure 5.54 Curve through Points–Settings.

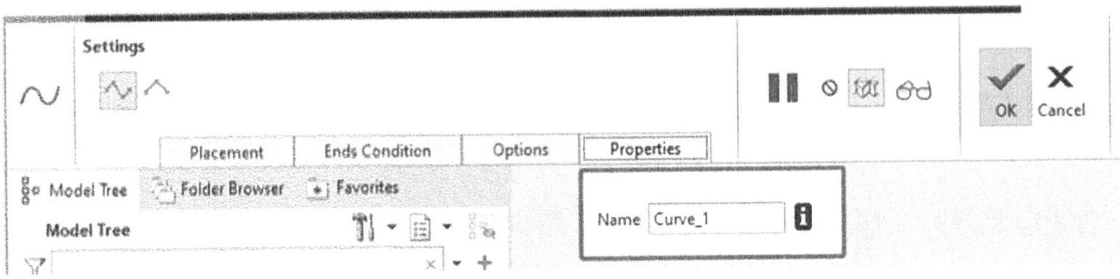

Figure 5.55 Curve through Points—Properties.

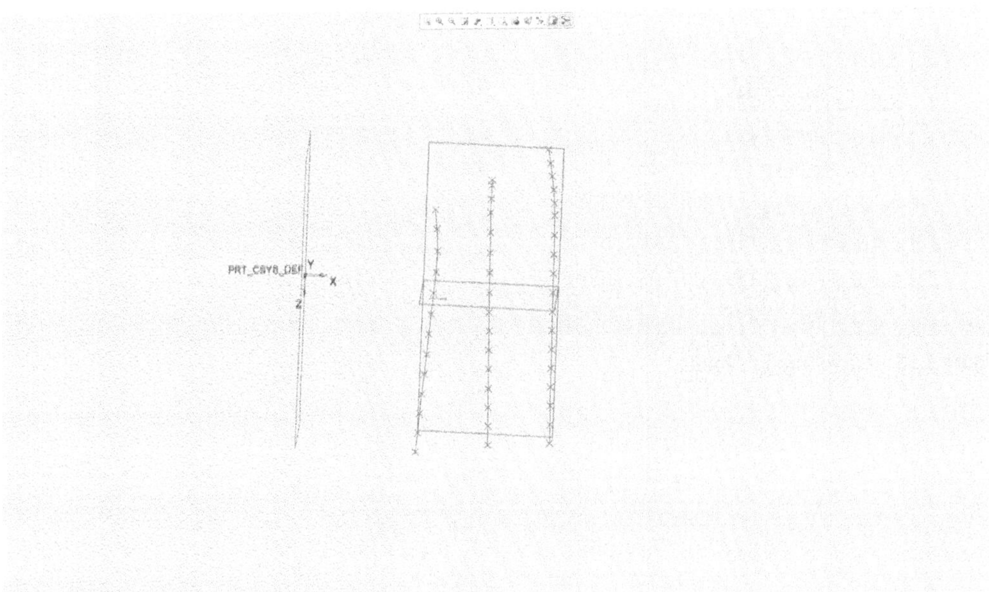

Figure 5.56 Curve diagram.

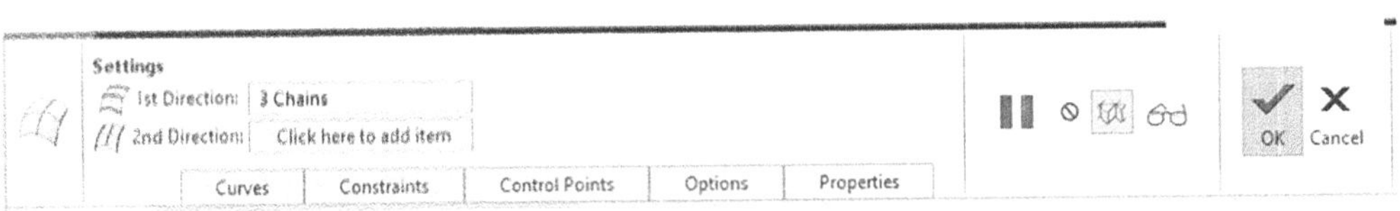

Figure 5.57 Boundary Blend settings panel.

5.3.2.1.7 Fillet the Blade Inlet and Outlet Edges The filleting method is the same as in "5.3.1 Impeller Modeling." In the feature toolbar, select  [Fillet], choose [Collect Mode], and simultaneously select the inlet edges of both the blade working surface and the back surface. Click OK to complete the fillet. Similarly, apply the fillet to the blade outlet edges.

5.3.2.1.8 Blade Solidification The method for solidifying the blade is the same as in "5.3.1 Impeller Modeling." Select the merged individual blade, go to [Edit] → [Solidify], and click OK to complete the solidification. Follow this step for other blades as well.

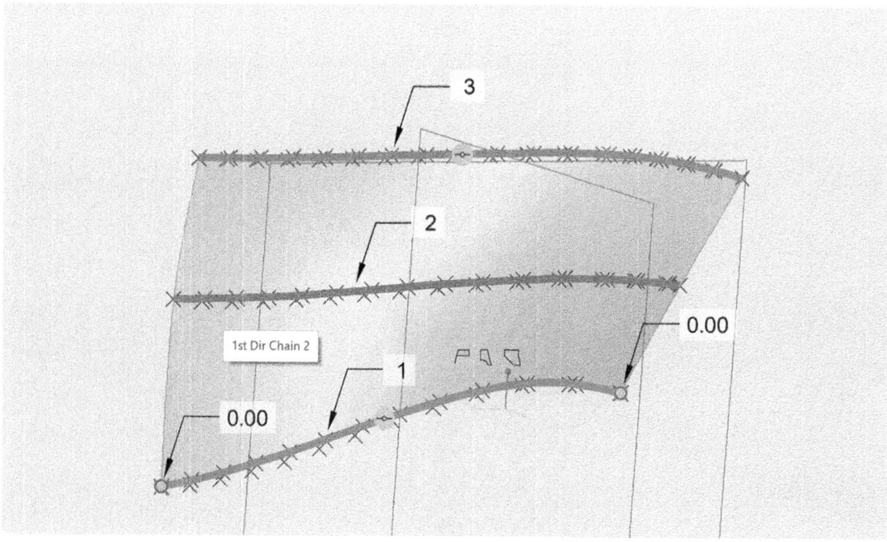

Figure 5.58 Selecting curves.

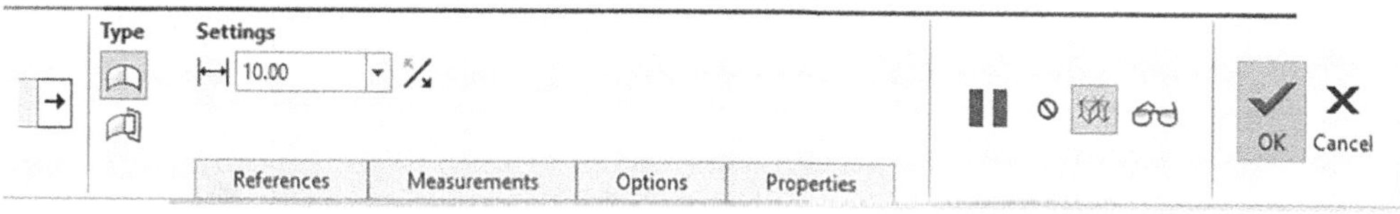

Figure 5.59 Extension distance.

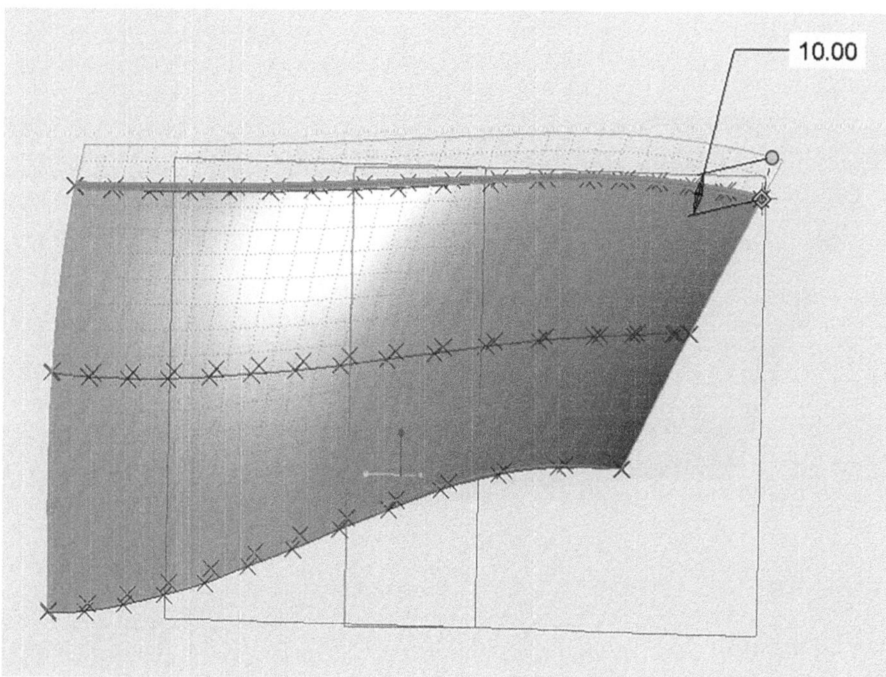

Figure 5.60 Extension of the working surface rim.

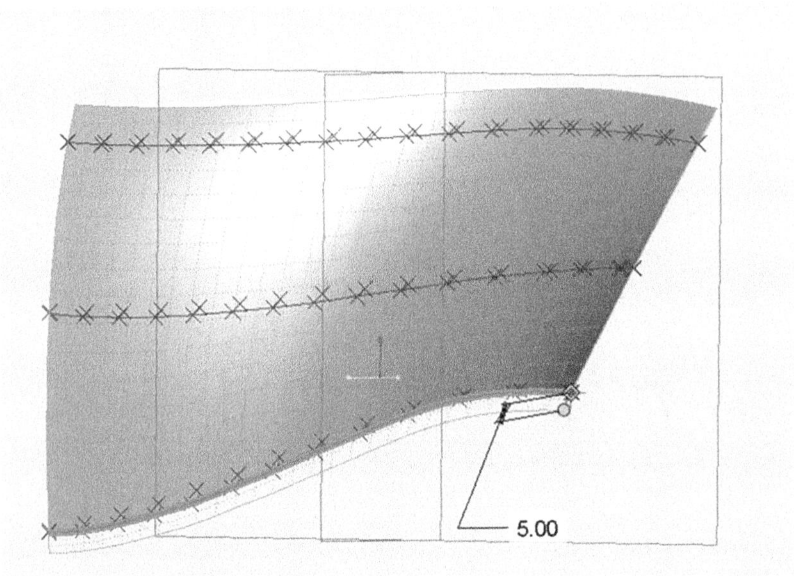

Figure 5.61 Extension of the working surface hub.

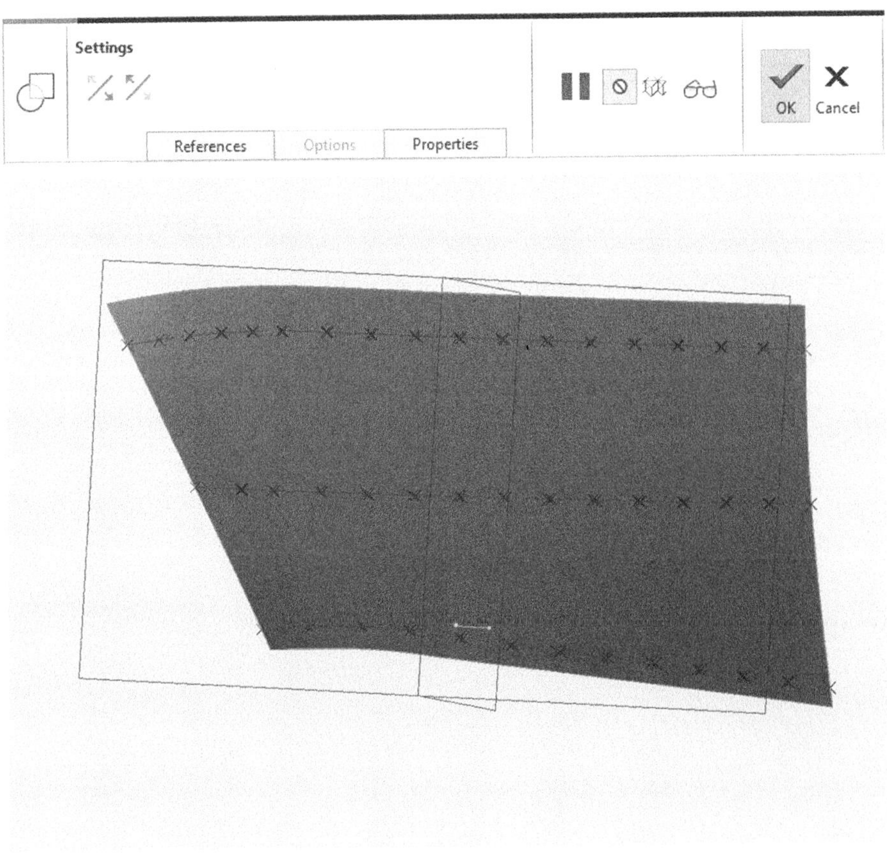

Figure 5.62 Surface merge.

Figure 5.63 Grouping.

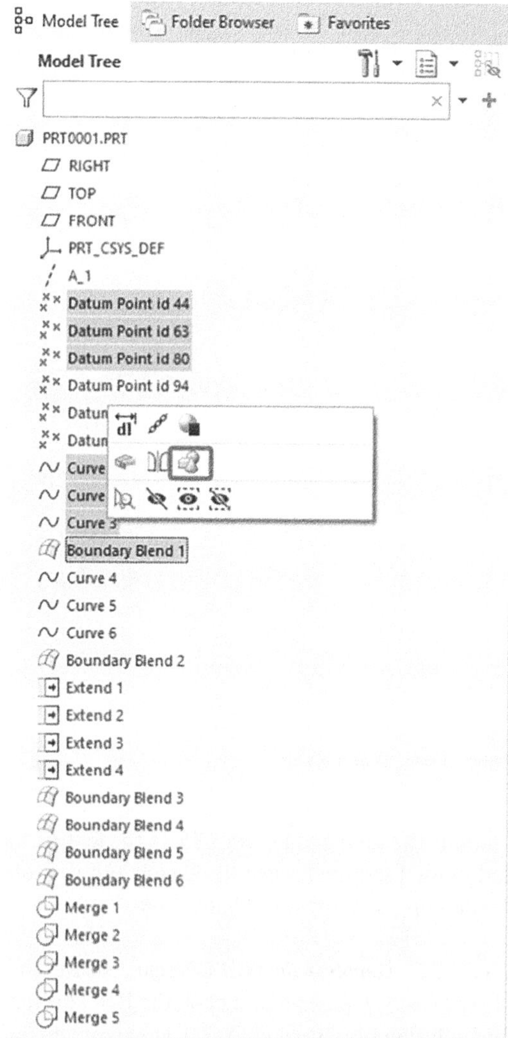

5.3.2.1.9 ***Create Group*** On the left side of the workspace, in the model tree, simultaneously select the operation icons that need to be grouped. Right-click, and from the pop-up context menu, select [Group] to place the key components of the generated body into a group, as shown in Figure 5.63.

5.3.2.1.10 ***Blade Array*** Select the group generated in the model tree. In the feature toolbar, select [Array] to array the blades, as shown in the figure below. Choose the Axis as the array center and select the axis generated in step 2. Since this model has seven blades, set the number of array instances to 7 and the angle to 51.43°. Click OK to complete the array, as shown in Figure 5.64.

5.3.2.1.11 ***Guide Vane Hub Sketching*** In the feature toolbar, select [Sketch] or click the 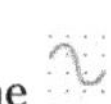 icon to enter sketching mode. Draw the section as shown in Figure 5.65. This shape can then be rotated to generate the hub's form. Once completed, click OK to finish the sketch.

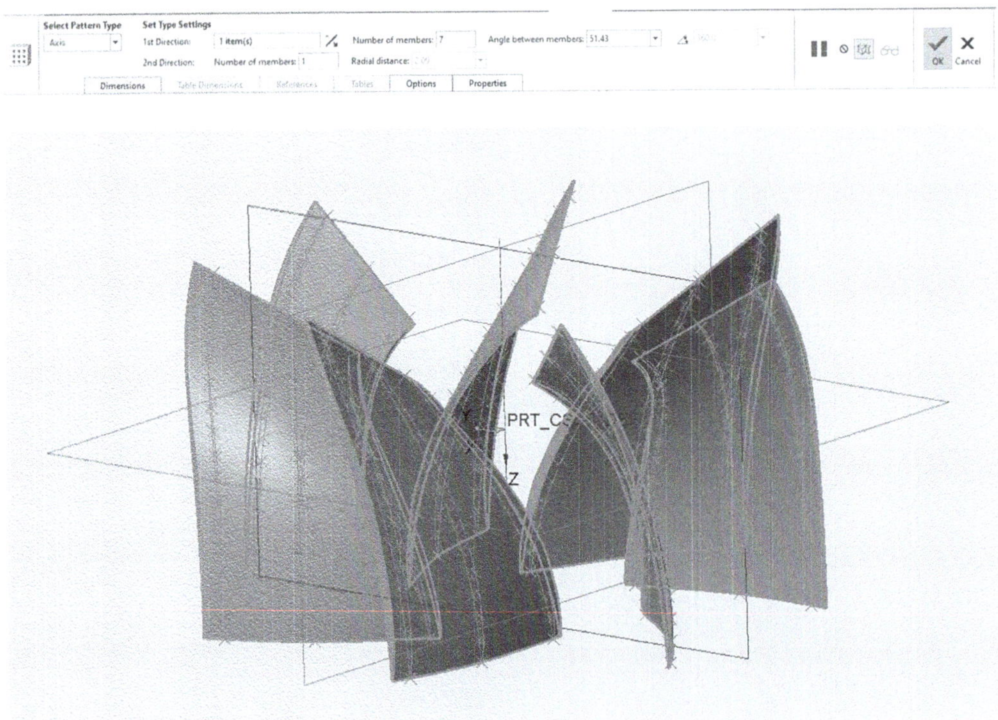

Figure 5.64 Blade array.

Rotate the sketched Figure 5.65. In the feature toolbar, select [Revolve], choose it as a surface, and in the [Settings] select the sketch and the rotation axis A_1. Click the OK button to complete the drawing, as shown in Figure 5.66.

5.3.2.1.12 Complete the Model Design Save the file by clicking [File] → [Save a Copy], choose the file type as STP format, and click the [OK] button. In the [Export STP] dialog that appears, check the [Solid] option, then click [OK] to complete the save.

5.4 Vortex Pump

5.4.1 Impeller Modeling

Launch Creo, and in the menu bar, click [File] → [New] to open the "New" dialog box, or directly click the button on the title bar. The popped-out [New] window is shown in Figure 5.67. In the Type column, select [Solid]; in the Name field, enter the desired file name, for example: VP-impeller, uncheck the [Use default template] option, and click the [OK] button. A new file option window will then appear to select the required template, as shown in Figure 5.67. Generally, choose [mmns_part_solid_abs], then click the [OK] button [5].

Click the [Axis] button under the Datum section in the Feature toolbar. The [Datum Axis] dialog box pops up, as shown in Figure 5.68. Next, click the Z-axis, select the [Through] option, and then click the [OK] button to create a rotational axis. Afterward, proceed to create the rotational axes for the X-axis and Y-axis.

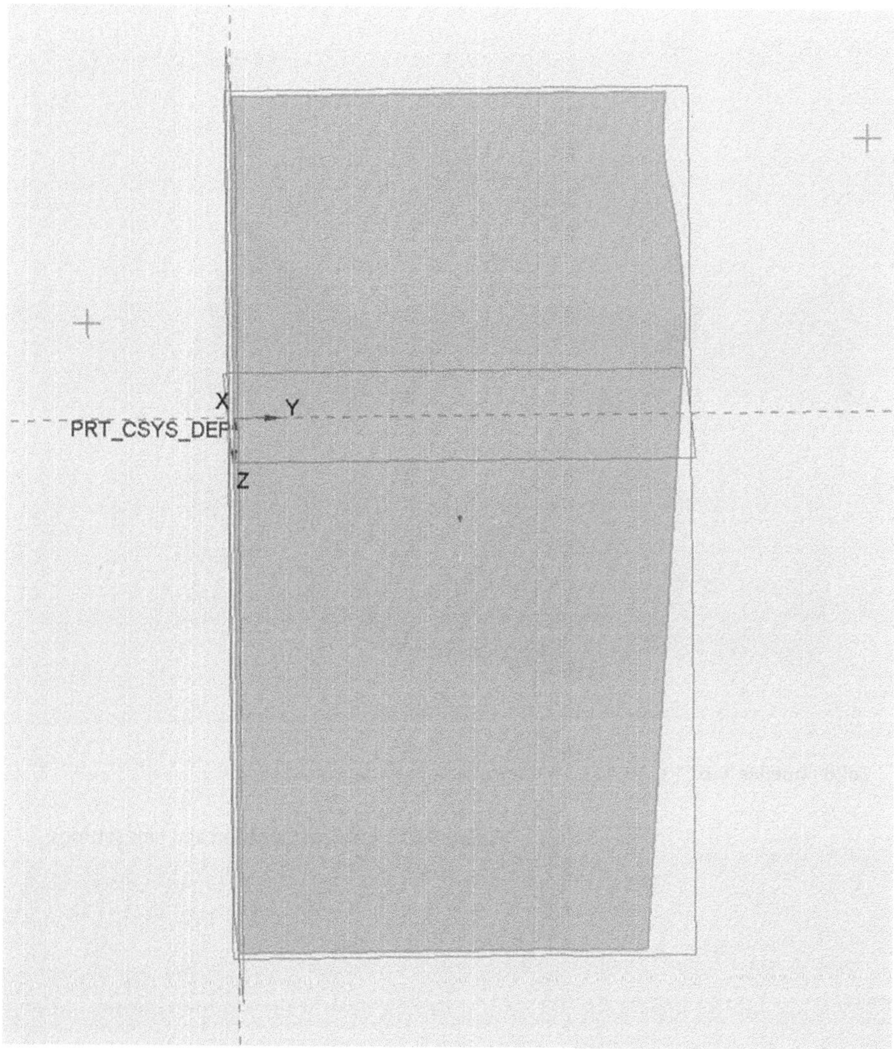

Figure 5.65 Hub sketch.

This book adopts a closed impeller for modeling and numerical simulation. Since the closed impeller of a vortex pump exhibits both central and axial symmetry, the [Revolve] command can be used to create the impeller. First, click the [Revolve] button under the Shapes section in the Feature toolbar. The interface will switch to the [Revolve] function page. Next, determine the sketching plane by selecting the Front Plane to enter the sketching interface. Click [Sketch View], and the perspective will automatically focus on the selected sketch plane (as shown in Figure 5.69).

In the sketching interface, draw the side view of the impeller. Since the [Revolve] command in Creo does not allow self-intersecting geometry, only half of the impeller's side view needs to be drawn. After completing the sketch, click Confirm to enter the [Revolve] command interface, set the rotation angle to 360°, then select the rotation axis. After previewing, click Confirm to generate the basic impeller shape as shown in Figure 5.70.

For closed impellers, the blades on the upper and lower sides of the impeller can either align or be staggered. In this book, the impeller blades are staggered. Therefore, one side of the blade is

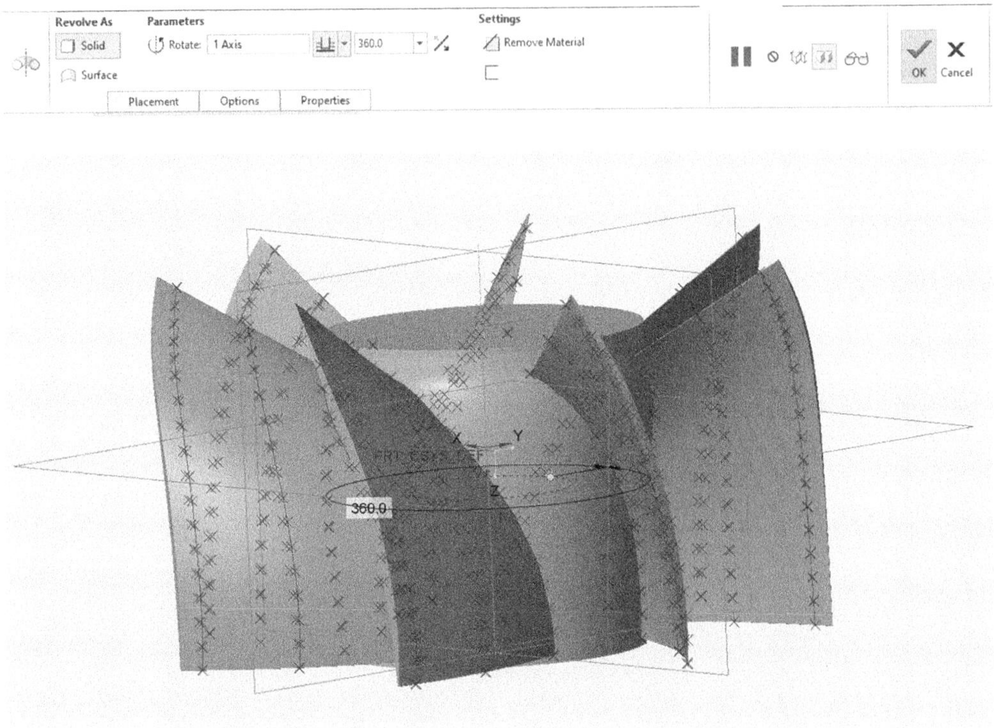

Figure 5.66 Solid impeller hub.

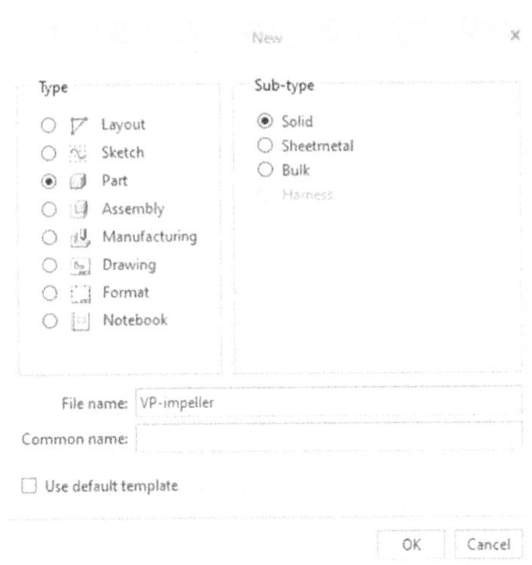

Figure 5.67 New part creation and unit settings.

drawn first, followed by the other side. First, click the [Extrude] button under the Shapes section in the Feature toolbar to enter the [Extrude] command interface. Then, select the Front Plane to enter the sketching interface, draw the blade, and click Confirm (as shown in Figure 5.71). Since the blades on both sides are staggered, a new plane needs to be created before drawing the opposite side. Click the [Extrude] command, select the Front Plane, and in the pop-up command box, choose Offset. Hold down Ctrl, select the created rotation axis, and in the pop-up command box, choose Through. Set an appropriate offset angle to create a new plane (as shown in Figure 5.72).

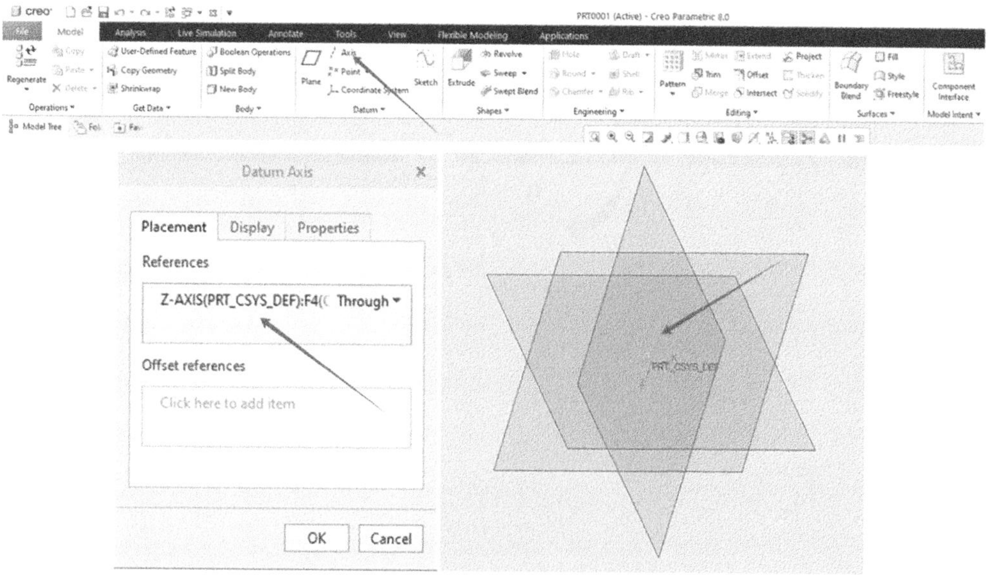

Figure 5.68 Rotation axis creation.

Then, click the [Extrude] button under the Shapes section in the Feature toolbar again, enter the [Extrude] command interface, select the newly created plane to enter the sketching interface, draw the opposite side blade, and click Confirm. After completing the upper and lower blades, select both blades and click the [Pattern] button in the toolbar. In the [Pattern] command, choose the type as [Axis], then click the rotation axis, set the quantity and angle in the first direction, preview the result, and click Confirm. The final impeller shape is shown in Figure 5.73.

5.4.2 Pump Casing and Pump Cover Modeling

Start Creo, and in the menu bar, click [File] → [New] to open the "New" dialog box, or directly click the button on the title bar. In the Type column, select [Solid], enter the desired file name in the Name field, uncheck [Use default template], and then click the [OK] button. A new file options window will then pop up, select the required template [mmns_part_solid_abs], and click the [OK] button.

Next, click the [Axis] button under the Datum section in the Feature toolbar. The [Datum Axis] dialog box will appear. Click the Z-axis, select the [Through] option, and then click the [OK] button to create a rotational axis. After creating the rotational axes for the X-axis and Y-axis, click the TOP plane, then click the [Revolve] button under the Shapes section in the Feature toolbar. This will switch to the sketching interface. In the sketching interface, draw the side view of the pump casing and pump cover. As before, only half of the side view needs to be drawn (as shown in Figure 5.74). After finishing the sketch, click Confirm to enter the [Revolve] command interface, set the rotation angle to 360°, click the rotation axis, preview the result, and then click Confirm. This will generate the basic shape of the pump casing and pump cover (as shown in Figure 5.75).

The pump casing typically has inlet and outlet pipes, and the pump cover needs corresponding notches for the inlet and outlet pipes. Therefore, it is necessary to add inlet and outlet pipes to the pump casing and cut away the excess portion of the pump cover. First, create a new plane by clicking the [Plane] command, selecting the TOP plane, and in the pop-up command box, choose Offset. Set an appropriate offset value to create the new plane (as shown in Figure 5.76). Then,

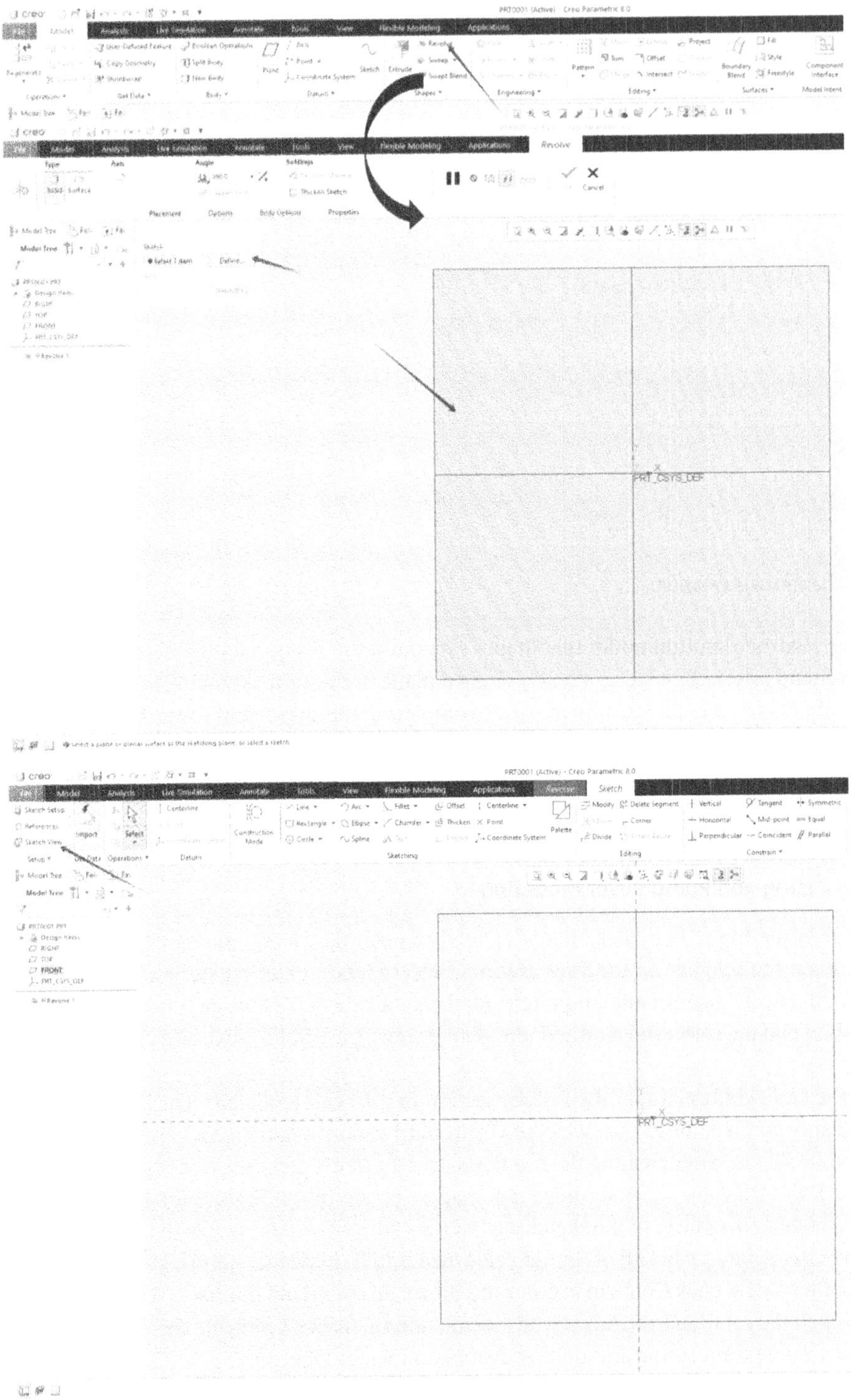

Figure 5.69 Revolve command.

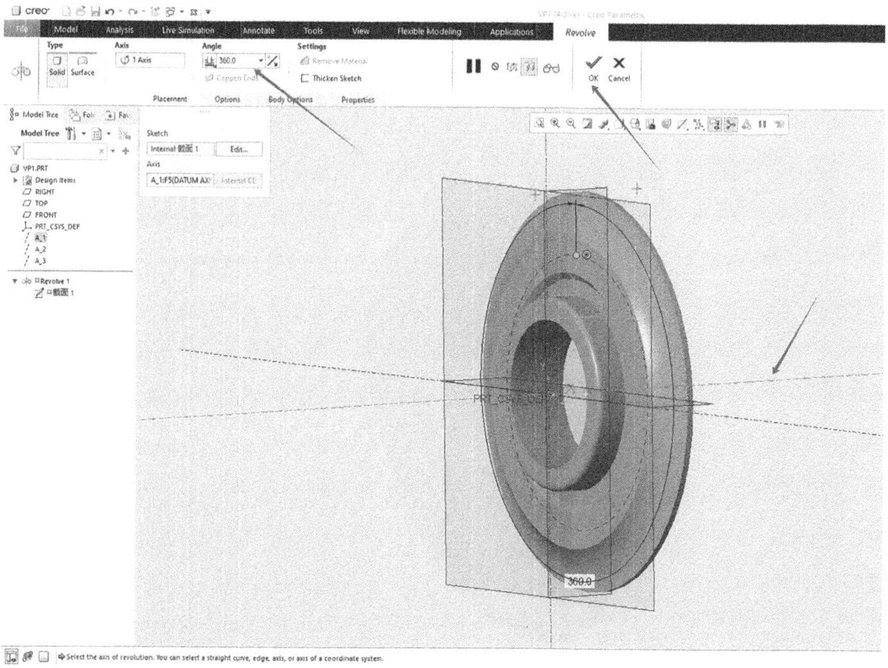

Figure 5.70 Creation of impeller basic shape.

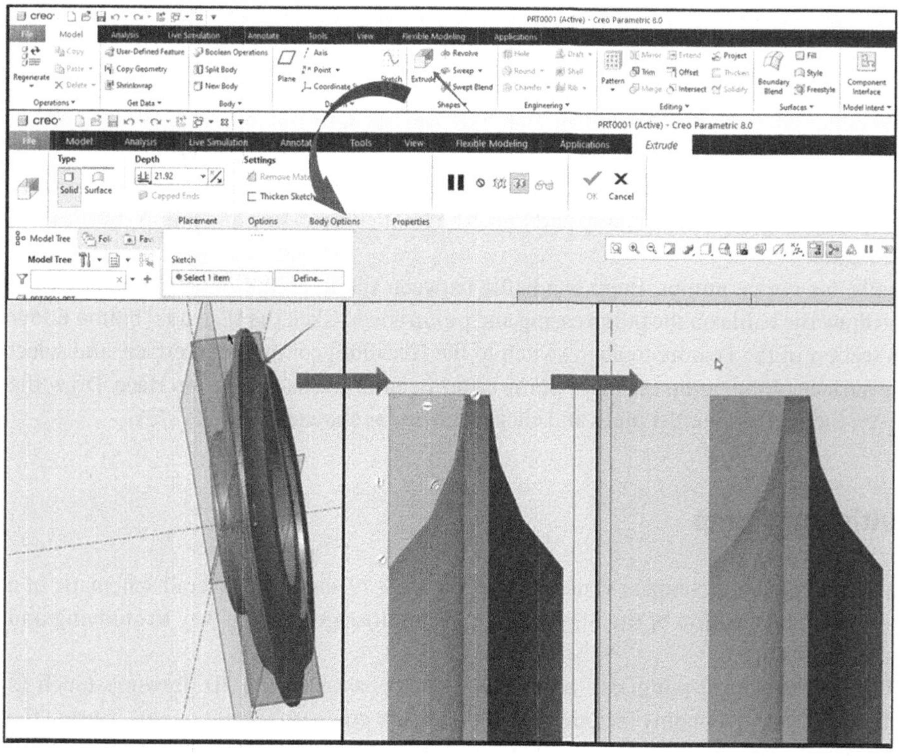

Figure 5.71 Blade sketching.

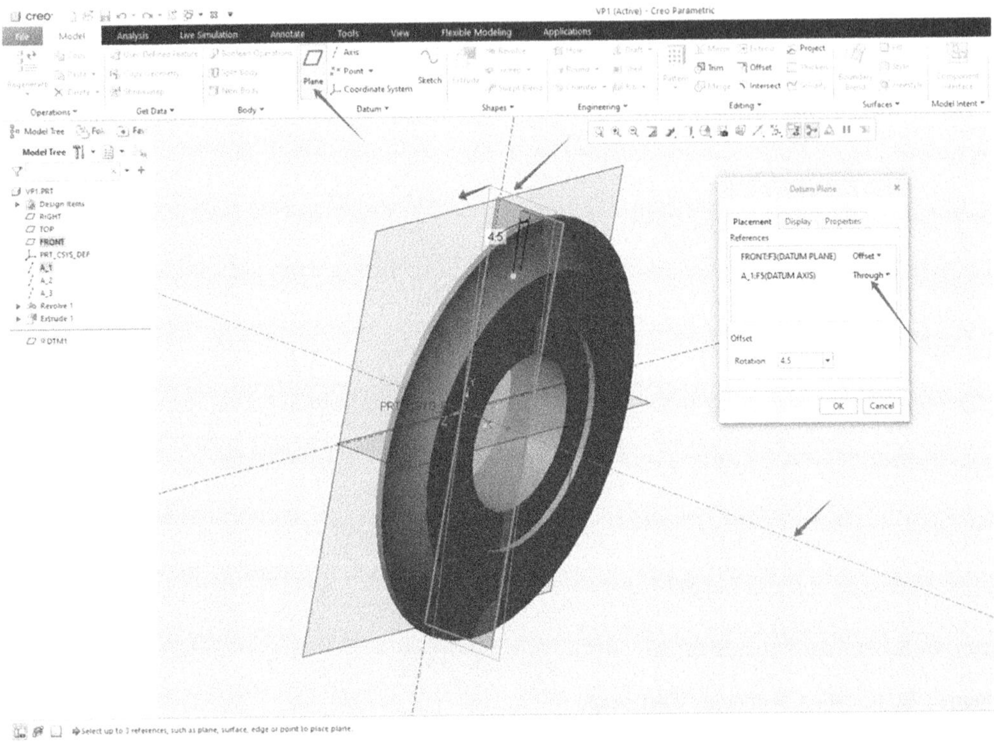

Figure 5.72 Plane creation and settings.

click the [Extrude] button under the Shapes section in the Feature toolbar to enter the [Extrude] command interface. Select the created plane to enter the sketching interface, draw the layout of the inlet and outlet pipes, and click Confirm. In the [Extrude] command interface, select "Extrude to next," and click Confirm. Next, use the [Extrude] command to create holes in the pump casing, and while using the [Extrude] command, choose to remove material. Similarly, use the [Extrude] command on the pump cover to cut away part of it. The final pump casing and pump cover model is shown in Figure 5.77.

Additionally, for vortex pumps, there is a baffle between the inlet and outlet. Therefore, it is necessary to draw the baffle on the pump casing and pump cover. Click the [Extrude] button under the Shapes section in the Feature toolbar, switch to the [Extrude] command interface, and select the concave surface of the pump casing or pump cover to enter the sketching interface. Draw the baffle between the inlet and outlet pipes and click Confirm (as shown in Figure 5.78).

5.5 Multistage Pump

The multistage pump has a complex structure, and Figure 5.79 shows the overall schematic of a multistage pump. The creation of the 3D model of the multistage pump is key to studying and optimizing its performance.

The multistage centrifugal pump can be modeled in 3D based on the 2D drawings (such as AutoCAD files). The model mainly includes the impeller and guide vane components. During the modeling process, the method for creating each stage's impeller and guide vane is the same. The following section describes the process using the first-stage impeller and guide vane as an example.

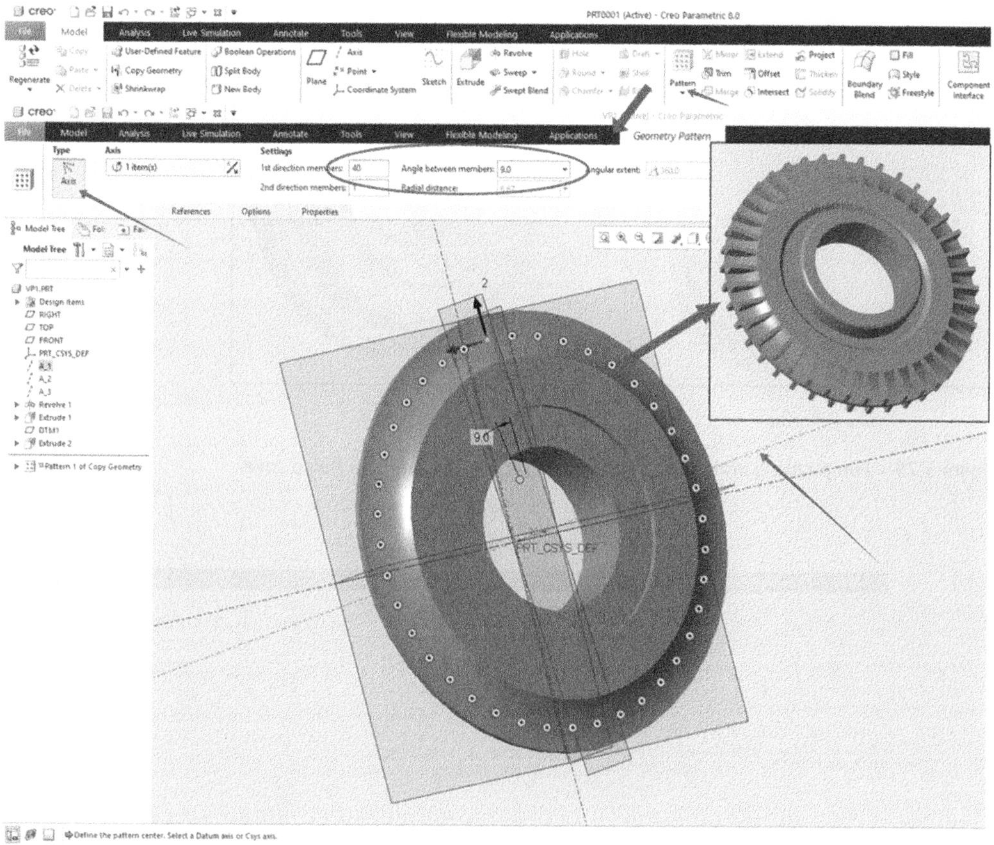

Figure 5.73 Final impeller shape.

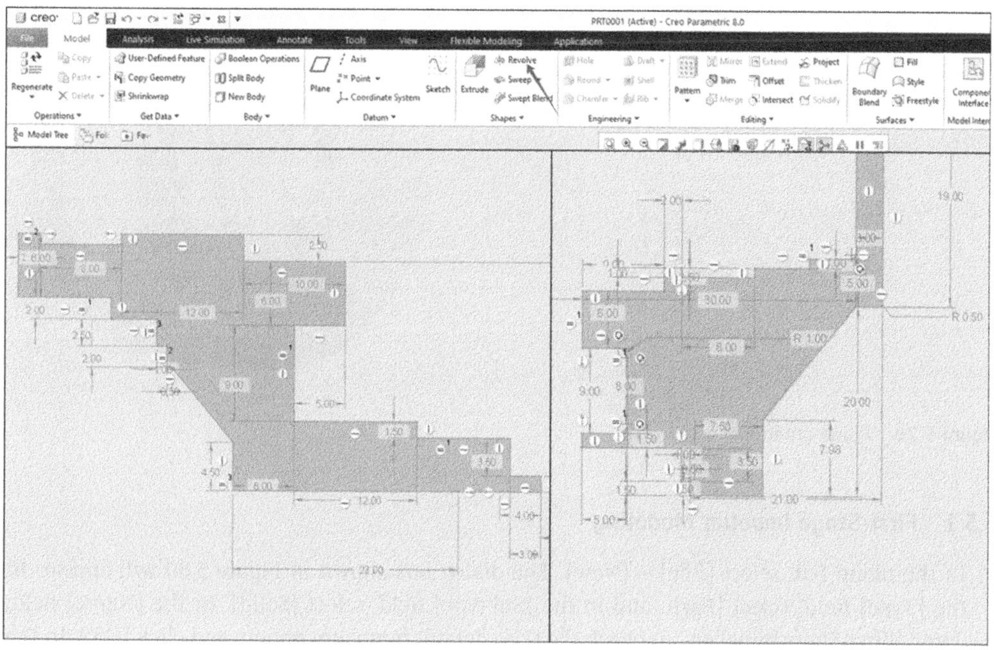

(a) (b)

Figure 5.74 Pump casing and pump cover revolve sketch: (a) Pump casing, (b) Pump cover.

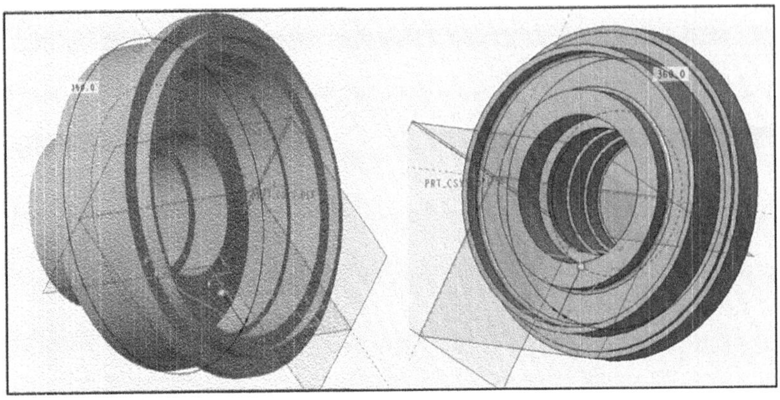

(a) (b)

Figure 5.75 Pump casing and pump cover shape: (a) Pump casing, (b) Pump cover.

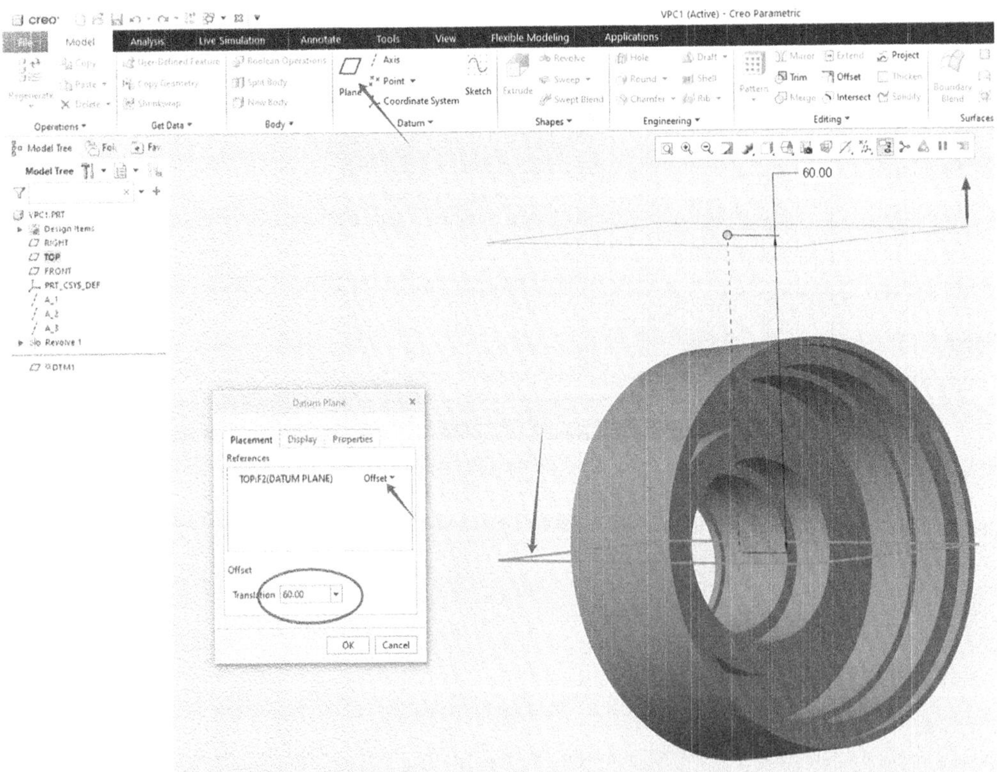

Figure 5.76 Plane creation.

5.5.1 First-Stage Impeller Modeling

1) In the menu bar, select [File] → [New]. The dialog box shown in Figure 5.80 will appear. In the [Type] field, select [Par]t, and in the [Subtype] field, select [Solid]. In the [Name] field, enter "First-Stage Impeller," uncheck the [Use default template] option, and click [OK]. In the

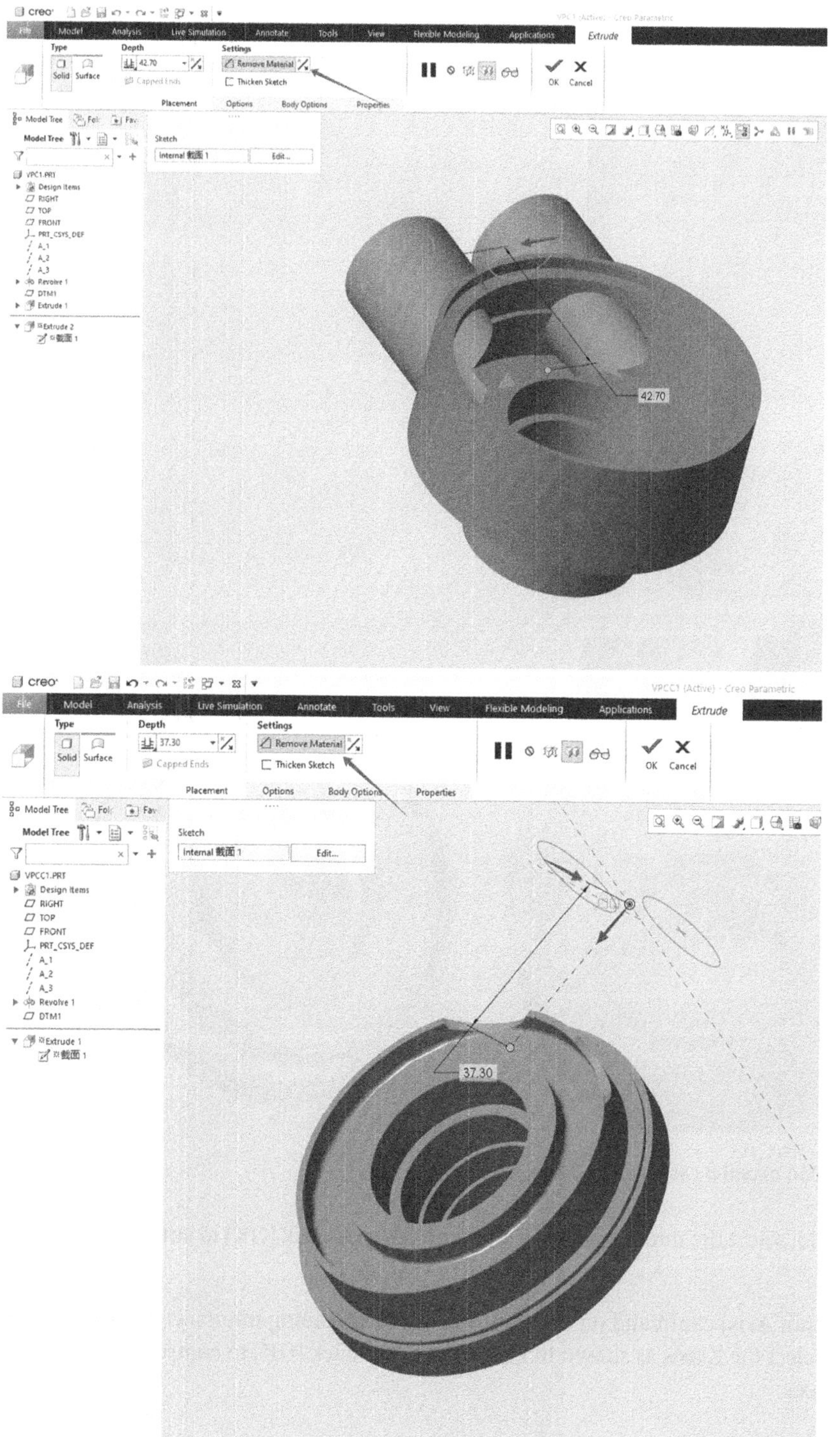

Figure 5.77 Positions of the target area and monitoring lines.

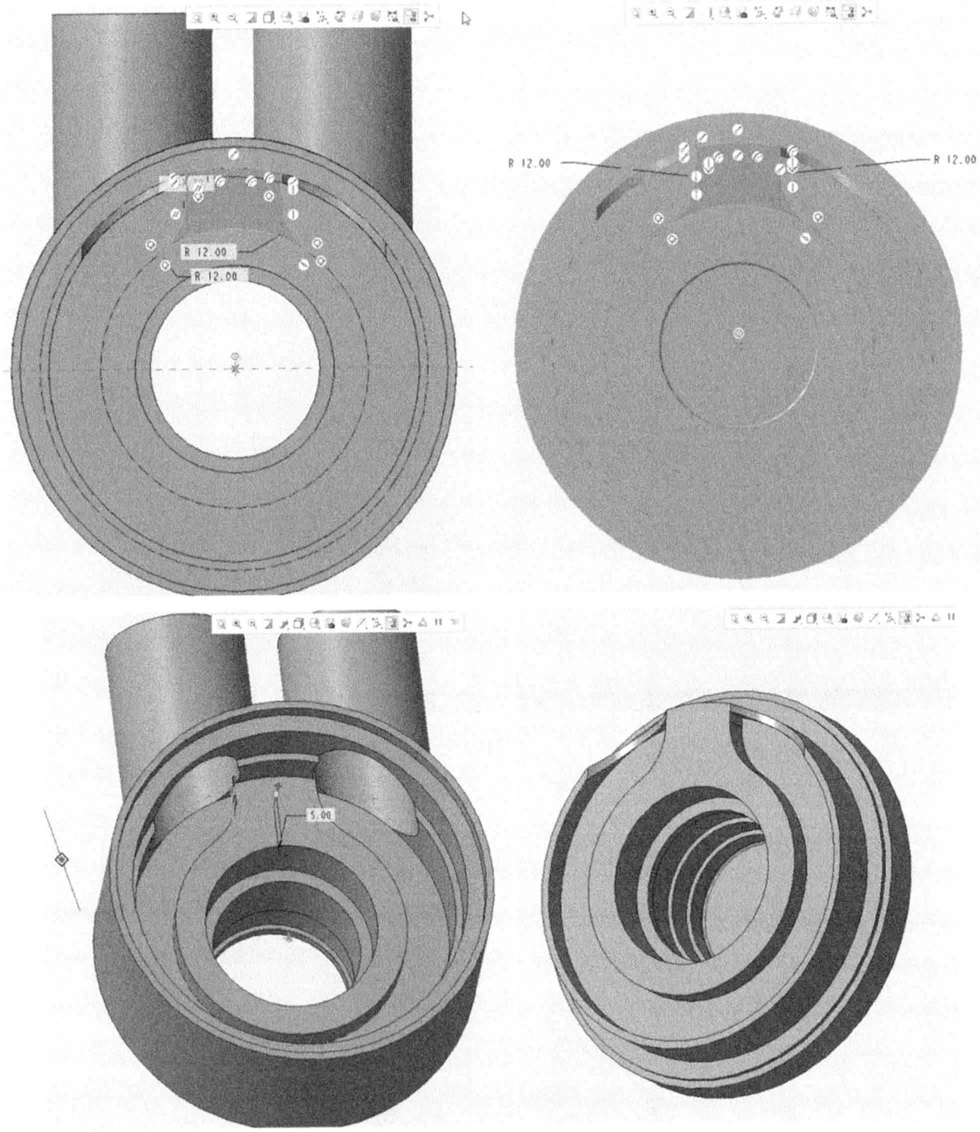

Figure 5.78 Creation of pump casing and pump cover baffle.

[Template] field, select the mmns-part-solid-abs template and click [OK] to enter the modeling interface.

2) Click the [Datum Axis] command on the right side of the modeling interface. After the dialog box pops up, select the Z-axis as shown in Figure 5.81, and click [OK] to complete the creation of the datum axis.

3) Drawing the Blade Working Surface and Back Surface

Use the point plotting method to draw the blade, based on the cylindrical coordinates provided in the blade template diagram, including the radial coordinate R, angular coordinate θ, and axial distance Z, as shown in Figure 5.82.

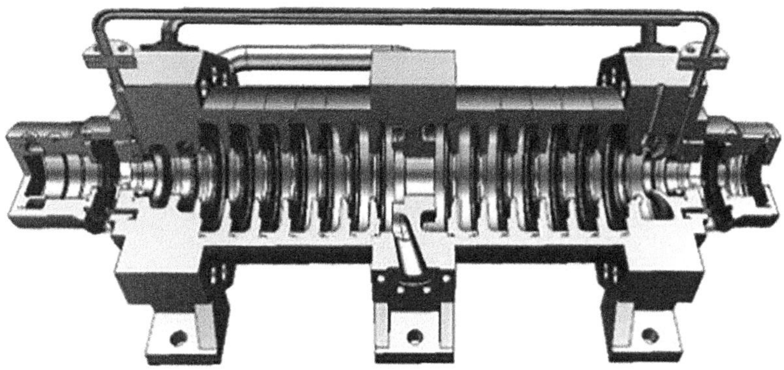

Figure 5.79 Schematic of the structure of a multistage pump.

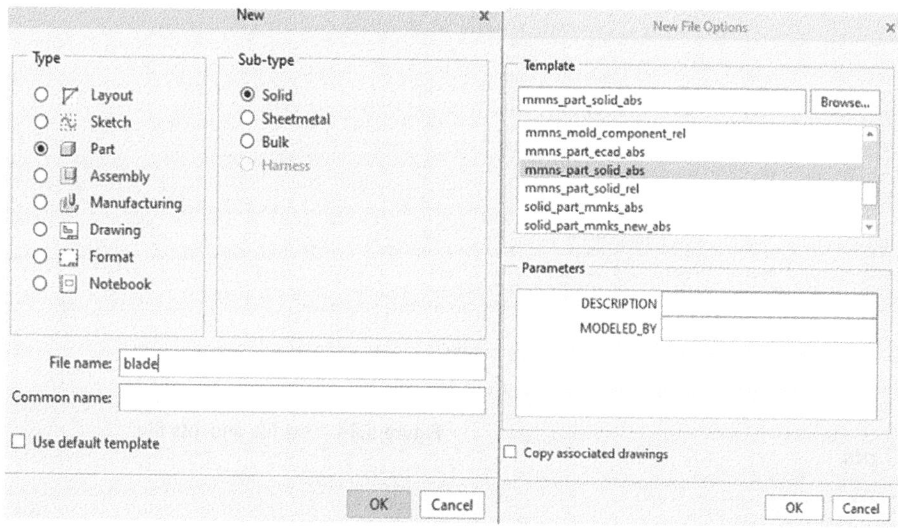

Figure 5.80 Establishing the first stage impeller .prt file.

Figure 5.81 Establishing the reference axis.

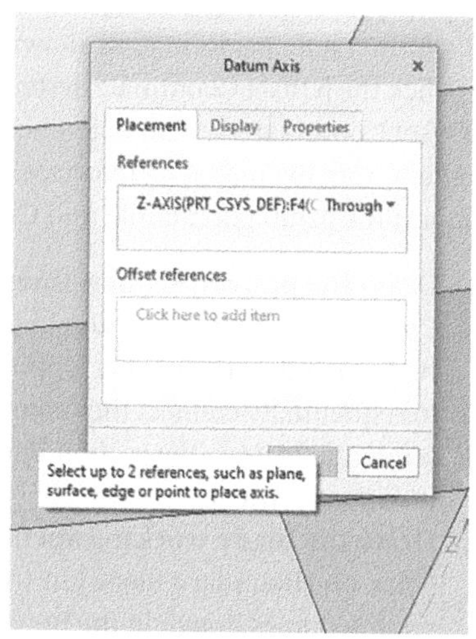

工作面 (径向?)

翼角(度)	?0?/角度	10	20	30	40	50	60	70	80	90	100	110	120	130	140	150
前缘?	55.3/4.5°	55.76	56.57	57.44	58.5	60.14	61.98	64.45	67.43	71.09	75.41	80.6	86.91	94.52	103.75	115
1	54/3.7°	55.27														
2	51.2/1.9°	52.51	54.28	56.32												
3	49.4/0.8°	50.59	52.23	54.03	56.22	58.7	61.62									
4	48.1/0.3°	49.3	50.86	52.56	54.64	56.94	59.78	62.87	66.35	70.33	74.88	80.32	86.81	94.51		
5			49.89	51.6	53.55	55.89	58.55	61.68	65.25	69.32	74.05	79.69	86.35	94.24	103.68	115
6						55.2	57.81	60.96	64.54	68.67	73.55	79.3	86.06	94.08	103.61	115
后缘?	47.6/0°	48.6	49.8	51.3	53	55.1	57.6	60.7	64.3	68.4	73.3	79.1	86	94	103.6	115

背面

翼角(度)	?0?/角度	10′	20′	30′	40′	50′	60′	70′	80′	90′	100′	110′	120′	130′	140′	150′
前缘?	54.8/4.5°	54.88	55.04	55.29	55.95	56.8	57.94	59.87	62.36	66.04	70.52	75.99	82.66	90.85	100.78	112.45
1′	51.2/2.4°	51.58	52.6	53.94	55.75											
2′	48.9/1.2°	49.38	50.22	51.43	53.02	55.06	57.63									
3′	47.2/0.5°	47.75	48.58	49.65	51.23	53.15	55.61	58.61	62.24							
4′	46/0.2°	46.54	47.37	48.48	49.95	51.88	54.24	57.1	60.67	64.92	69.79	75.46	82.38	90.76		
5′						50.98	53.25	56.16	59.72	63.96	68.89	74.7	81.64	89.91	99.81	111.91
6′								55.53	59.06	63.31	68.29	74.23	81.2	89.39	99.16	110.65
后缘?	46/0°	46.4	47	47.9	49.1	50.7	52.7	55.5	58.8	62.9	68	73.9	80.8	88.9	98.2	108.9

Figure 5.82 Data of the blade working surface and back surface.

55.3	4.5	0	55.3	4.5	0
55.76	10	0	55.76	10	0
56.57	20	0	56.57	20	0
57.44	30	0	57.44	30	0
58.5	40	0	58.5	40	0
60.14	50	0	60.14	50	0
61.98	60	0	61.98	60	0
64.45	70	0	64.45	70	0
67.43	80	0	67.43	80	0
71.09	90	0	71.09	90	0
75.41	100	0	75.41	100	0
80.6	110	0	80.6	110	0
86.91	120	0	86.91	120	0
94.52	130	0	94.52	130	0
103.75	140	0	103.75	140	0
115	150	0	115	150	0

```
55.3    4.5   0
55.76   10    0
56.57   20    0
57.44   30    0
58.5    40    0
60.14   50    0
61.98   60    0
64.45   70    0
67.43   80    0
71.09   90    0
75.41   100   0
80.6    110   0
86.91   120   0
94.52   130   0
103.75  140   0
115     150   0
```

Figure 5.83 Data file format of the curve coordinate points.

Figure 5.84 .txt file and .pts file.

According to Creo's requirements, import the data from the 2D hydraulic chart into Excel, and then copy the data into Notepad, editing it into the format shown in Figure 5.83. Convert the text file from .txt to .pts format, as shown in Figure 5.84.

Click the [Offset Coordinate System] button on the right side of the modeling interface to open the [Offset Coordinate System Reference Point] dialog. In the [Reference] dialog, select the coordinate system from the modeling interface, and in the [Type] dialog, choose Cylindrical. Click [Import] to import the gzm0.pts data file into Creo, as shown in Figure 5.85.

1) **Draw the pattern section lines:**
   Click on the [Datum] button in the modeling interface, which will open a dialog box. Click on [Curve] and in the [Curve Options] section, select [Curve through Points]. Use a spline curve to sequentially connect the imported points, as shown in Figure 5.86. Repeat the above steps to sequentially draw the section lines of the working surface from the pattern diagram, as shown in Figure 5.87a.
2) **Draw the blade working surface:**
   Click on [Boundary Blend] in the modeling interface, and sequentially select the wood mold sectional lines drawn in the previous step, as shown in Figure 5.87b.

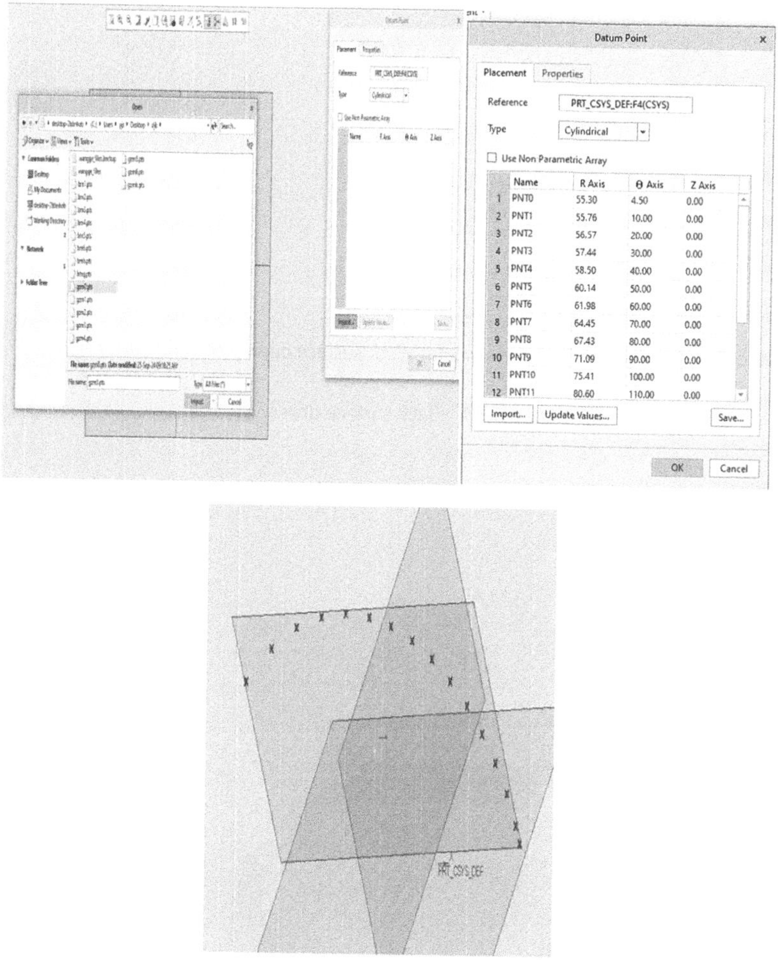

Figure 5.85 Data import into offset coordinate system.

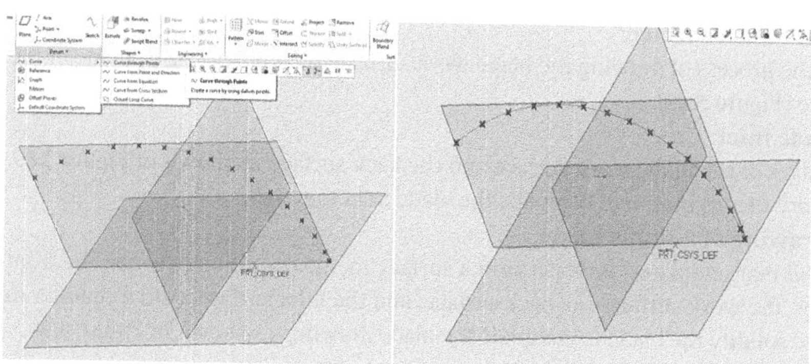

Figure 5.86 Work surface curve and work surface drawing.

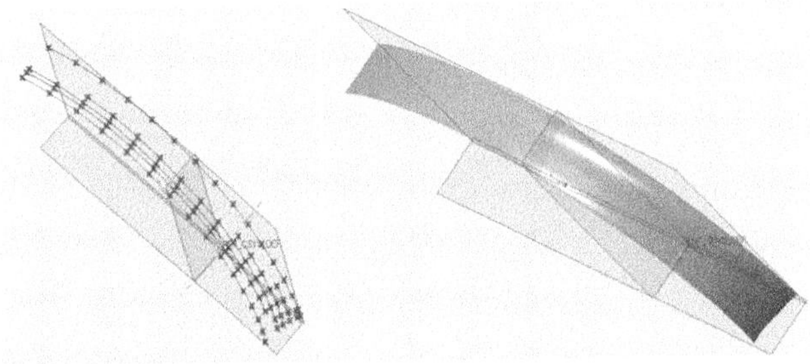

(a) (b)

Figure 5.87 Work surface curve and work surface drawing: (a) Work surface curve, (b) work surface.

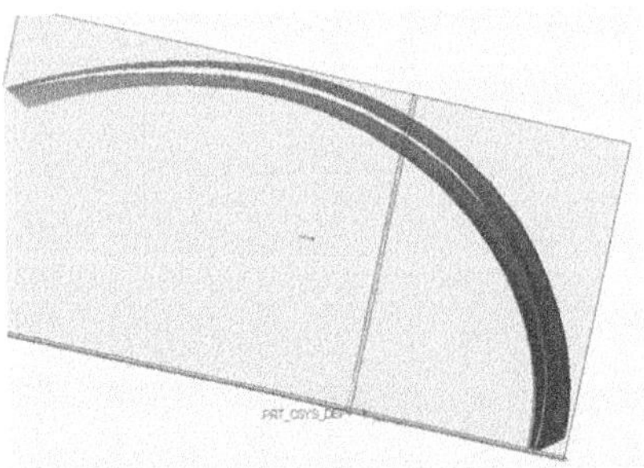

Figure 5.88 Back surface drawing.

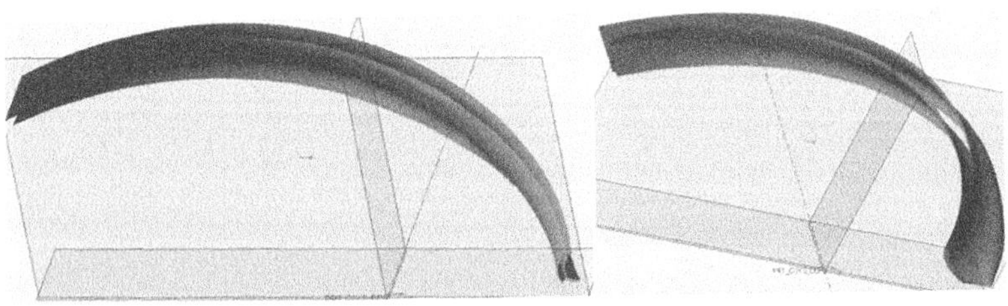

Figure 5.89 Inlet surface drawing.

3) **Drawing the blade back surface:**
   Similarly, repeat the process of drawing the blade work surface to complete the drawing of the blade back surface (Figure 5.88).
4) **Drawing the blade inlet:**
   Select the inlet edges of the blade work surface and the back surface as shown in Figure 5.89. Click the [Boundary Blend] button to complete the blade inlet drawing.
5) **Complete the drawing of a single blade:**
   Draw the front and rear endplates, then perform a surface merge operation between the front and rear endplates, the work surface, the back surface, and the inlet surface. Add a chamfer to the blade inlet and solidify the blade to complete the blade drawing, as shown in Figure 5.90.
6) **Array the blades:**
   Select all features, click the [Array] button, select the reference axis, set the total number of blades to 5, and set the array angle range to 360°. Complete the blade array, as shown in Figure 5.91.

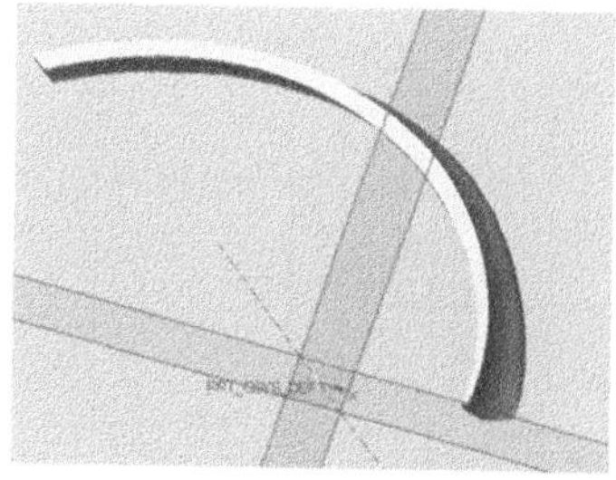

Figure 5.90 Single blade.

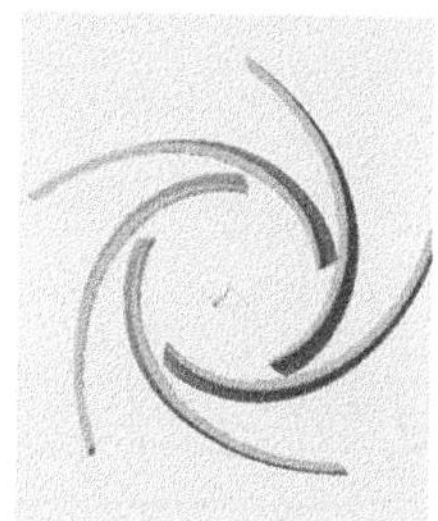

Figure 5.91 Blade array.

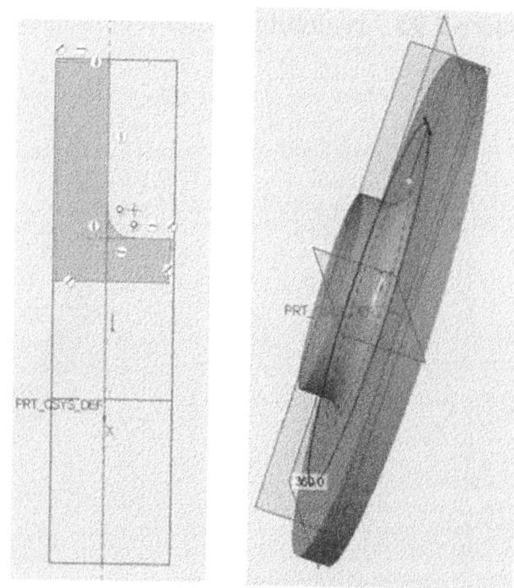

(a) (b)

Figure 5.92 Drawing of the impeller front and rear covers: (a) Cover plate profile sketch, (b) cover plate model.

7) **Drawing the front and rear cover plates of the impeller:**
   On the TOP reference plane, draw the outline sketch of the front and rear cover plates of the impeller, as shown in Figure 5.92a. Then, apply the thin-wall feature rotation to obtain the front and rear cover plates of the impeller, as shown in Figure 5.92b. The thickness of the cover plates can be adjusted during the rotation process to thicken the thin walls according to the actual requirements.

5.5.2 Radial Guide Vane Drawing

The radial guide vane mainly involves importing the cross-sections of the forward and reverse guide vanes into Creo or sketching the guide vane and determining the vane partition. The hydraulic diagram of the radial guide vane is shown in Figure 5.93.

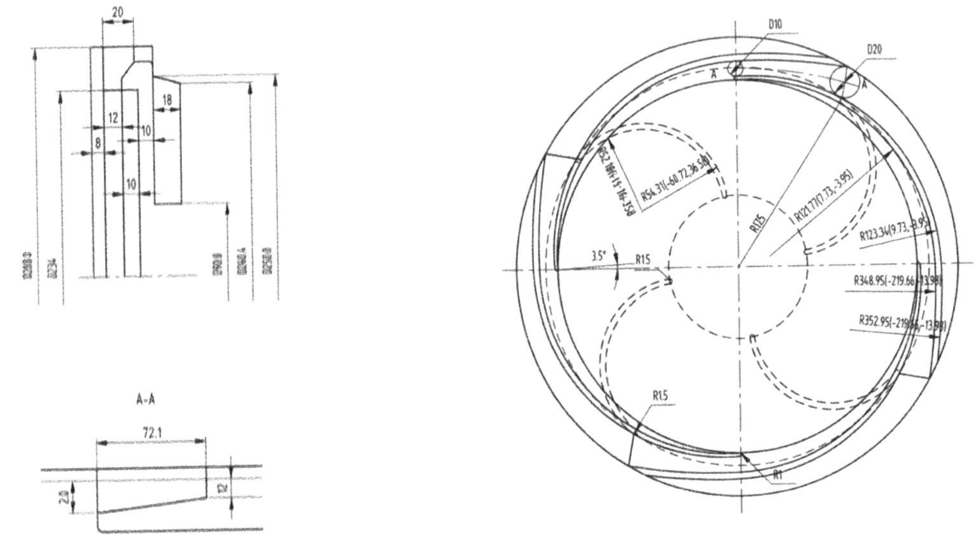

Figure 5.93 Hydraulic diagram of radial guide vane.

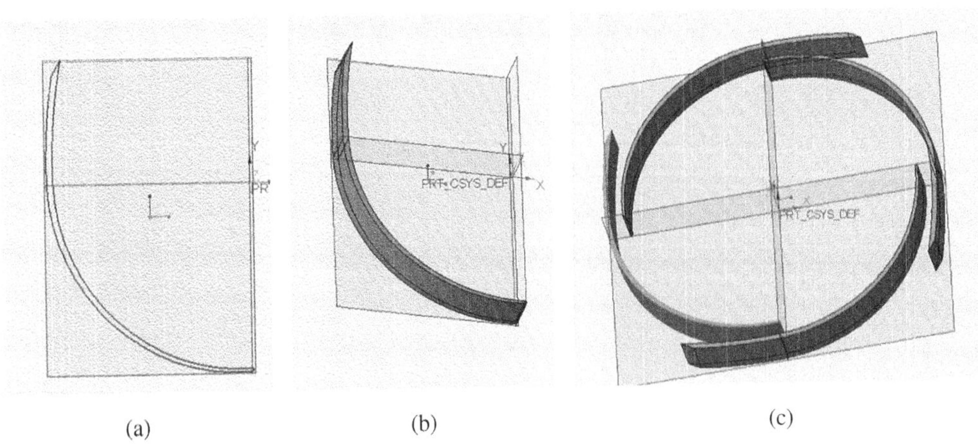

Figure 5.94 Forward guide vane modeling: (a) Inlet guide vane curve, (b) inlet guide vane, (c) guide vane model.

1) **Import the forward guide vane sketch:**
   Import the forward guide vane curve from the CAD file into the Creo file, as shown in Figure 5.94a.
2) **Stretch the forward guide vane:**
   Click the [Stretch] button, select the forward guide vane sketch, and the stretching result is shown in Figure 5.94b.
3) **Array the forward guide vane:**
   Click the [Array] button, select the array object (the stretched forward guide vane), choose [Axial Stretch], select the Z-axis as the reference axis, set the array count to 4, and set the angle to 90°. The array result is shown in Figure 5.94c.

Figure 5.95 Reverse guide vane modeling.

4) **Reverse guide vane modeling:**
   The reverse guide vane modeling steps are the same as for the forward guide vane. It is created based on the hydraulic diagram of the radial guide vane. The modeling result is shown in Figure 5.95.
5) **Guide vane partition modeling:**
   Draw the partition sketch on the right and upper reference planes, which are perpendicular to the front reference plane (XOY plane). The partition is represented as a rectangle in the projection on both planes. Guidelines are drawn on the front reference plane. Use the [Sweep] command to create the partition entity, as shown in Figure 5.96a. The partition is then arrayed with the forward and reverse guide vanes in the same way. The array result is shown in Figure 5.96b.
6) **Rear rib modeling:**
   Draw the projection of the rear rib on the right reference plane. Click the [Revolve] button to revolve the projection of the rear rib. Specify the axis of revolution as the Z-axis and the rotation angle as 360°. This results in the creation of the rear rib entity, as shown in Figure 5.97.
7) **Front rib modeling:**
   Create a new reference plane parallel to the front reference plane at the location of the forward guide vane. On this new reference plane, draw the projection cross-section sketch of the front rib. Click the [Extrude] button to obtain the front rib entity for the guide vane (Figure 5.98).

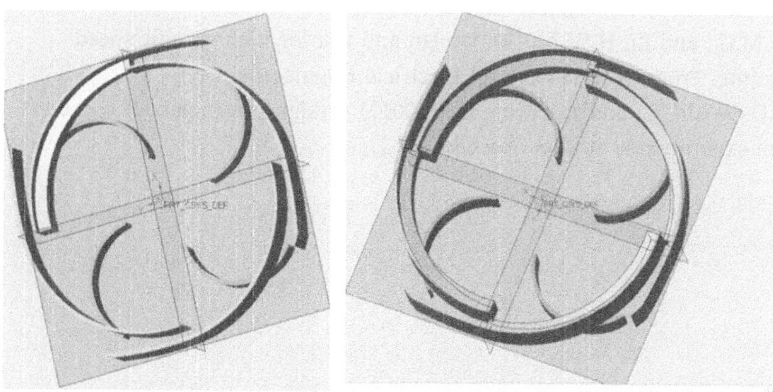

(a) (b)

Figure 5.96 Guide vane partition modeling: (a) Sweep of the partition, (b) array of the partition and guide vanes.

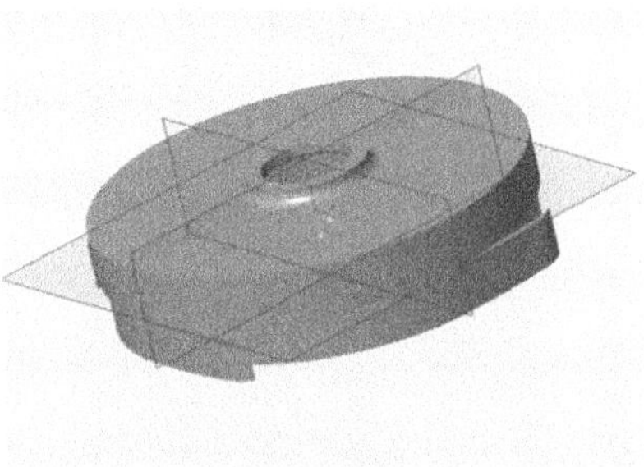

Figure 5.97 Rear rib modeling.

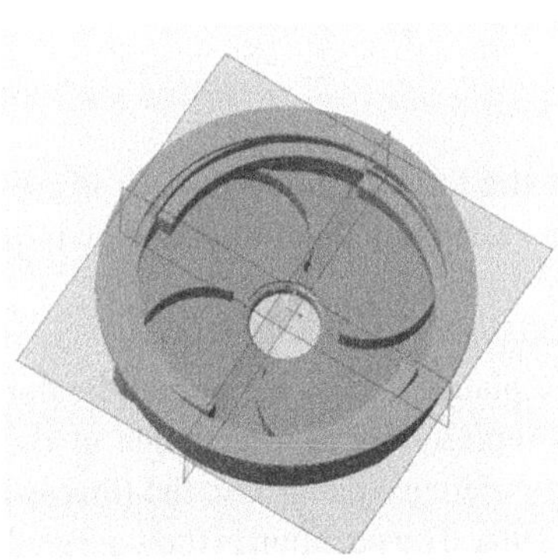

Figure 5.98 Front rib modeling.

References

1 Suhaimi, W.H.H.W., Sam, R., Johari, J., and Buniyamin, N. (2015). Design and analysis of a gate valve fluid flow using SolidWorks simulation software. 2nd International Conference on Electrical, Electronics and System Engineering (ICEESE), IEEE.

2 Wei, X.T., Du, Q., and Liu, J.C. (2010). The application of UG NX second development in centrifugal pump impeller vane 3D solid modeling. *Applied Mechanics and Materials* 20–23: 1286–1291.

3 Zhang, Q.H., Xu, Y., Shi, W.D., Lu, W.G., and Zhou, L. (2013). Research and development on the hydraulic design system of the guide vanes of multistage centrifugal pumps. 26th IAHR Symposium on Hydraulic Machinery and Systems, 15, 1–7.

4 Wu, X.F., Liu, H.L., Tan, M.G., and Li, H.W. (2012). Design and test for high specific speed mixed flow pump with volute. Progress in Power and Electrical Engineering.

5 Gerlach, A., Thamsen, P.U., Wulff, S., and Jacobsen, C.B. (2017). Design parameters of vortex pumps: a meta-analysis of experimental studies. *Energies* 10: 1.

6

Numerical Simulation and Analysis of a Centrifugal Pump

6.1 Computational Domain Modeling

6.1.1 Volute Profile Import

The following steps involve creating a 3D model of the volute and importing the .dwg file of the volute profile. The import process is the same as for the blade file, as shown in Figure 6.1. It is important to check whether the bottom of the profile aligns with the base circle profile; if not, adjustments should be made to ensure that they are aligned [1].

Next, click [Right Plane] → [Insert] → [DXF/DWG] to import the volute cross-sectional profile, as shown in Figure 6.2.

Next, follow the same procedure to sequentially insert the cross-sectional profiles, as shown in Figure 6.3.

6.1.2 Section Closure

After the volute profile is inserted, right-click the profile and click [Normal To] → [Edit Sketch] → [Line] to close the profile as shown in Figure 6.4.

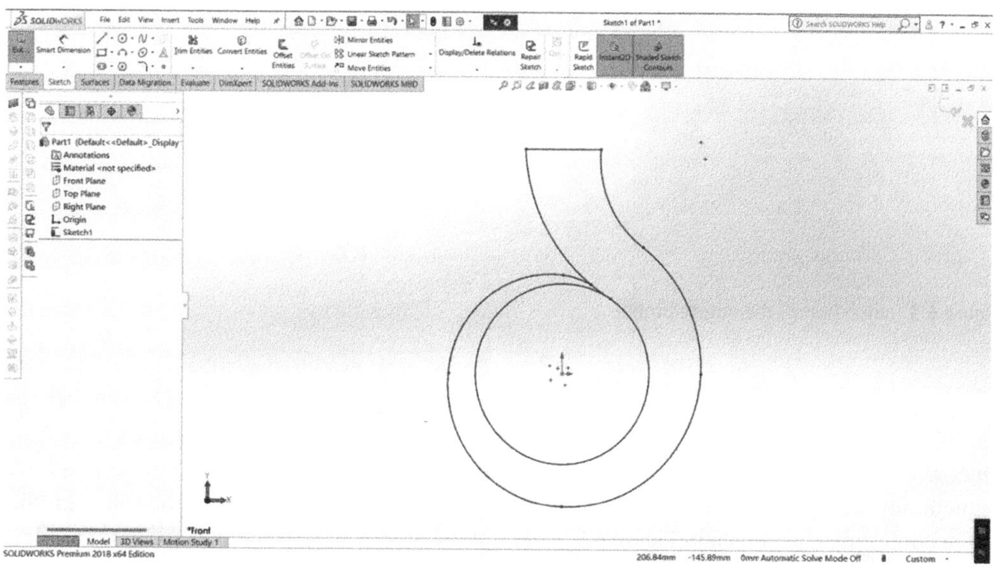

Figure 6.1 Profile of the volute.

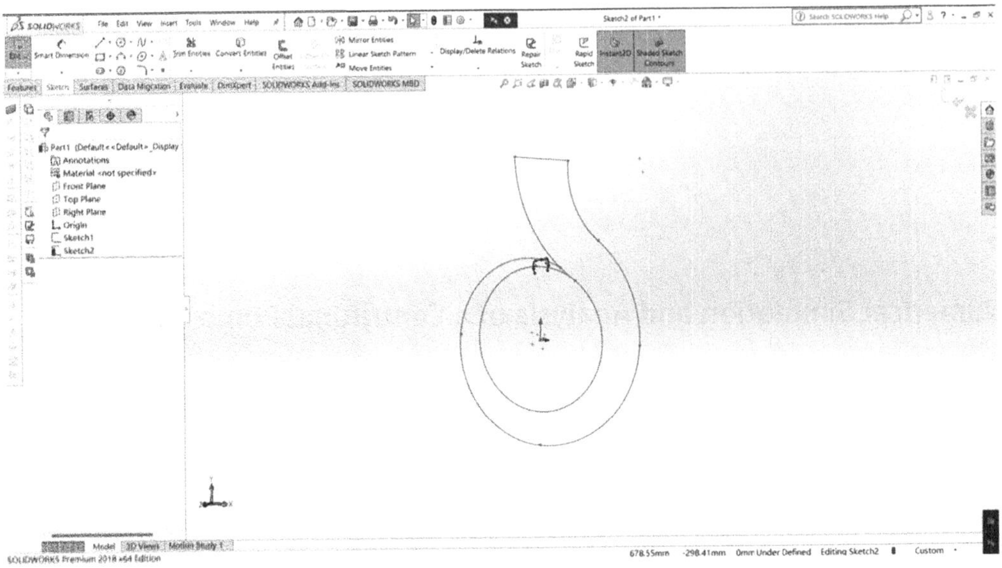

Figure 6.2 Importing the profile of the volute.

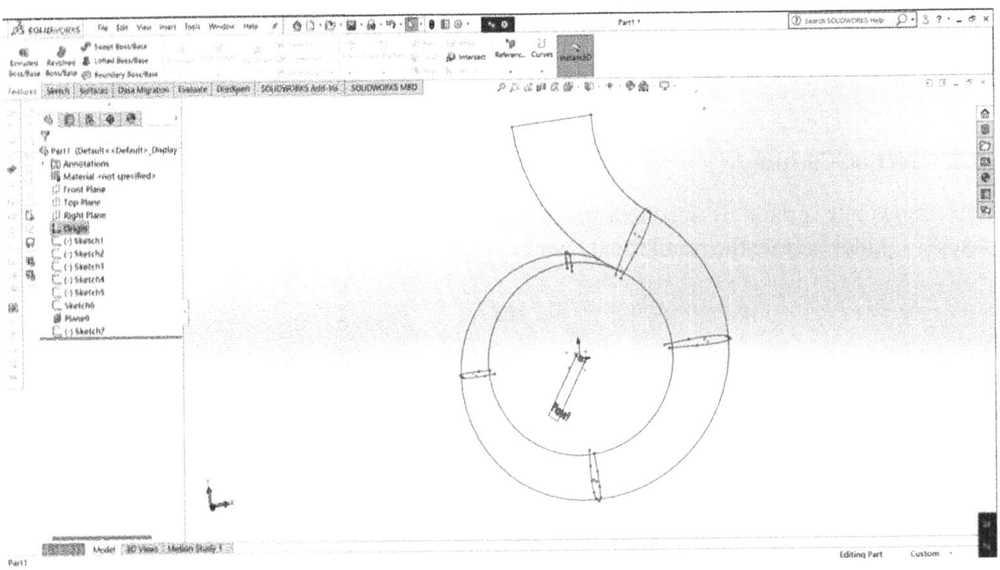

Figure 6.3 Insertion of the volute profile.

6.1.3 Sectional Lofting

Once the section is closed, click [Lofted Boss/Base] to create the loft. In the [Profiles] section, sequentially select the closed profiles, and in the [Guide Curves] section, choose the volute profile as the guide curve, as shown in Figure 6.5.

Then, follow the same steps to import the volute outlet section and create the loft, as shown in Figure 6.6.

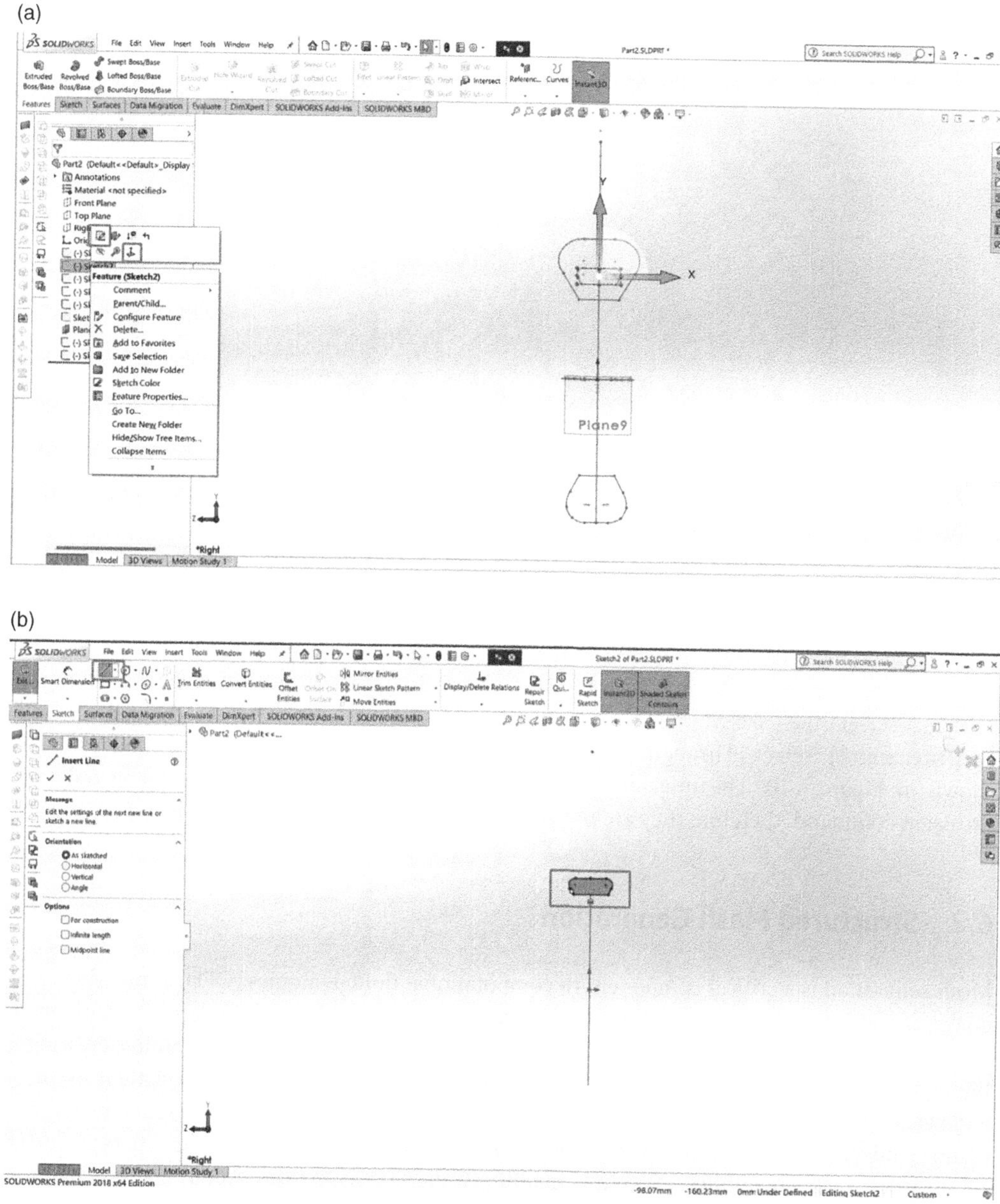

Figure 6.4 Closed section of the volute: (a) Normal section, (b) closed profile.

6.1.4 Tongue Molding

Click [Sketch] in the top menu bar, then select [3D Sketch] to create a 3D line. Sketch the base circle and the spiral curve between the two end faces, as shown in the two 3D sketches in Figure 6.7.

Next, click [Surfaces] → [Boundary Surface]. In [Direction 1], select the first imported volute section and the lower profile line of the right-side section, as shown in Figure 6.8. In [Guide Curves], choose the guide curve from Figure 6.8.

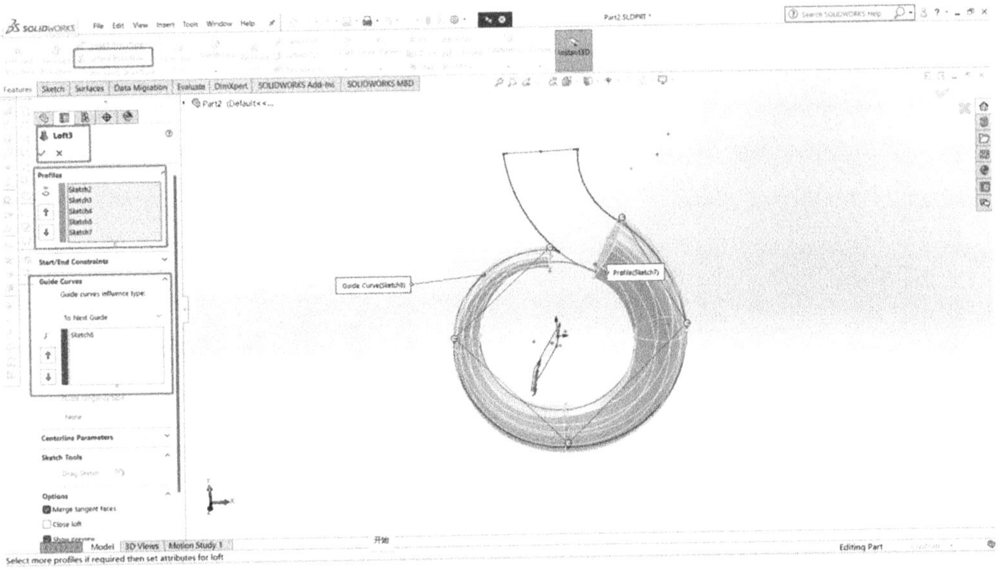

Figure 6.5 Layout of the volute section.

Then create a sketch at the junction of the flow passage on the left side of the volute and the outlet flow passage, as shown in Figure 6.9. Sketch according to the curvature of the septal tongue.

Then, select the sketch just created, click [Features], and choose [Extruded Cut]. In [Direction 1], select [Through All-Both] to perform the extrusion cut. The final volute is shown in Figure 6.10. The inlet and outlet flow passages can be completed using only the extrusion command, therefore they are not covered here.

6.2 Structured Mesh Generation

Mesh generation is a critical component in computational fluid dynamics (CFD) simulations and analysis, and its quality directly impacts the accuracy and reliability of numerical calculations. Due to the complex geometry and significant curvature variations of impeller pumps, creating a high-quality mesh that meets the computational requirements presents numerous challenges. Integrated Computer Engineering and Manufacturing (ICEM) CFD, one of the most popular mesh generation software, is widely recognized for its exceptional functionality. The software offers powerful CAD model repair capabilities, effectively handling imperfect geometries. Additionally, it features advanced mesh editing techniques, supports various solvers, and provides high-quality structured mesh generation for complex models, such as impeller pumps. These capabilities make ICEM CFD a widely used tool for mesh generation in impeller pump simulations, ensuring the accuracy of simulation results. This section provide a detailed introduction to different mesh generation methods for impeller pumps using ICEM CFD, including unstructured, hybrid, and structured mesh generation. Various application scenarios are explored describing advantages and disadvantages of each method to better understand these techniques in order to improve the overall effectiveness and accuracy of CFD analysis.

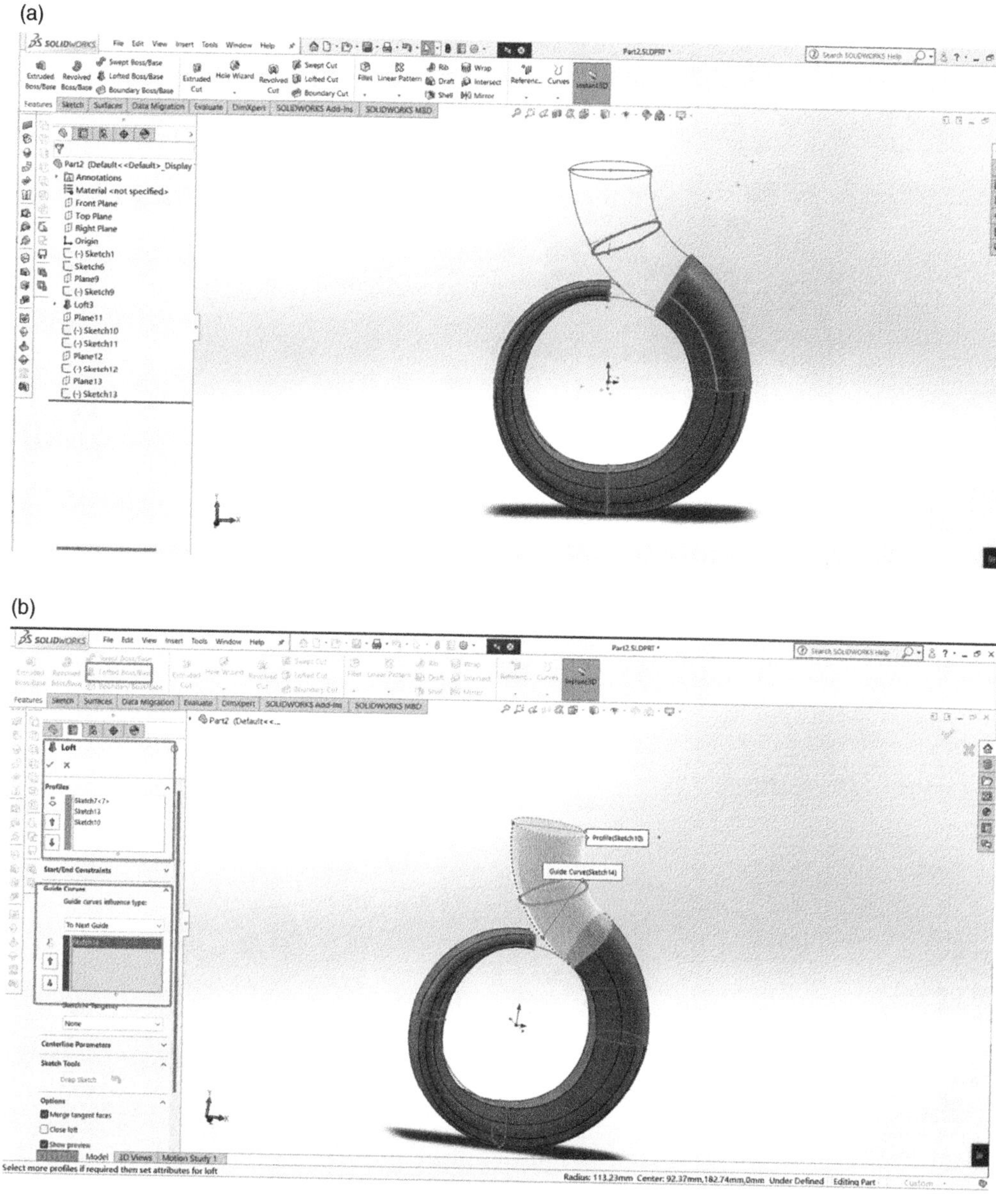

Figure 6.6 Outlet section of the volute: (a) Volute outlet section, (b) Create the loft feature.

6.2.1 Centrifugal Pump Impeller Structured Mesh Generation

6.2.1.1 Model Import

In this section, the methodology of conducting structured mesh generation for the impeller of the centrifugal pump is describe. Since the impeller is a rotating component with six identical flow passages, one can first generate the mesh for a single passage and then apply the rotational symmetry to obtain the mesh for the entire impeller domain. To begin, click [File] → [Import Model], select the impeller model for import, and set the unit for the impeller to [Millimeter] (Figure 6.11).

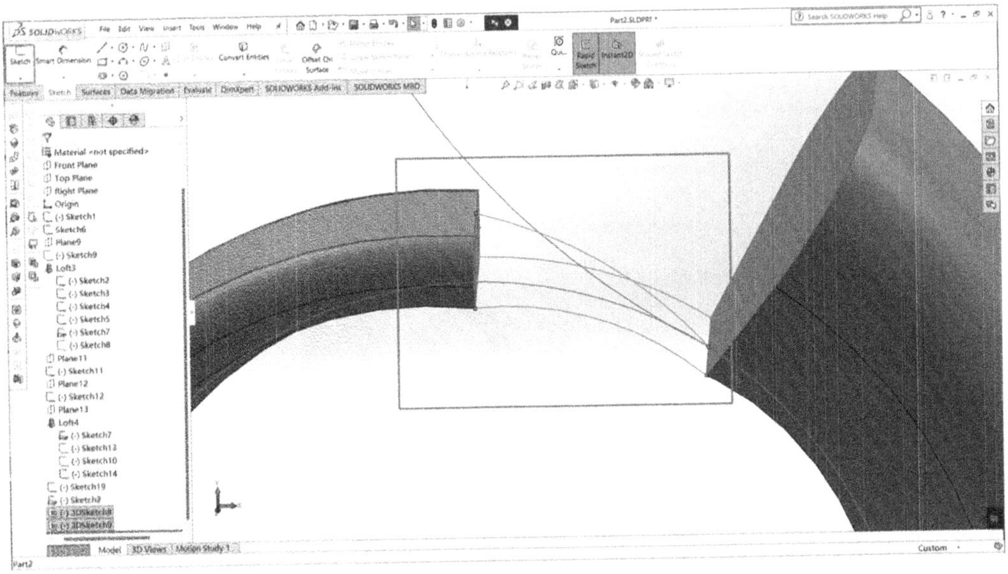

Figure 6.7 Drawing of guiding lines for tongue molding.

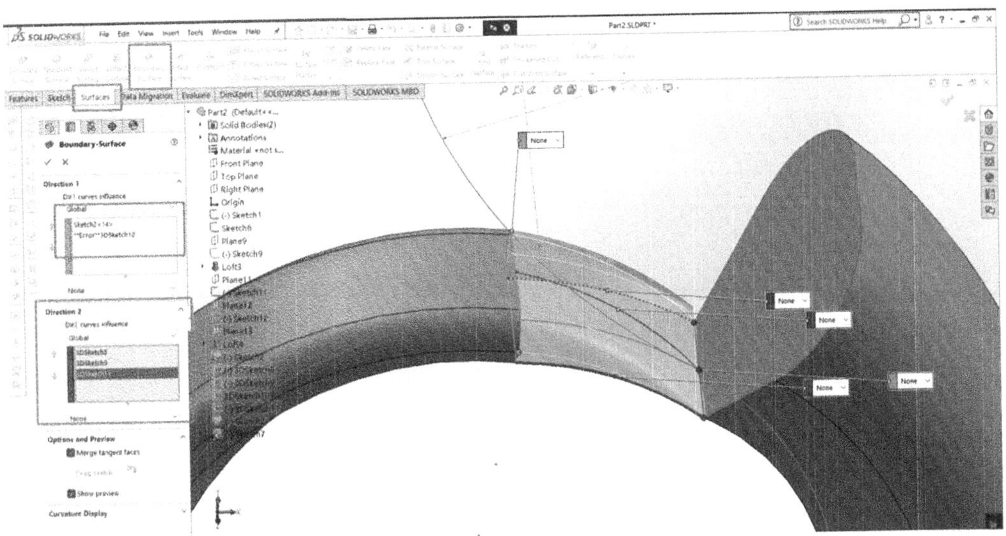

Figure 6.8 Creating a flow passage at the volute tongue.

As shown in the red-marked area in Figure 6.12, check the [Surfaces] option and then click  [Solid Simple Display] to solidify the model.

6.2.1.2 Geometric Topology

To create the geometric topology of the impeller flow passages, click  [Repair Geometry], then click  [Build Diagnostic Topology] in the appearing interface. Finally, click [OK] or [Apply] to automatically generate the geometric topology, as shown in Figure 6.13.

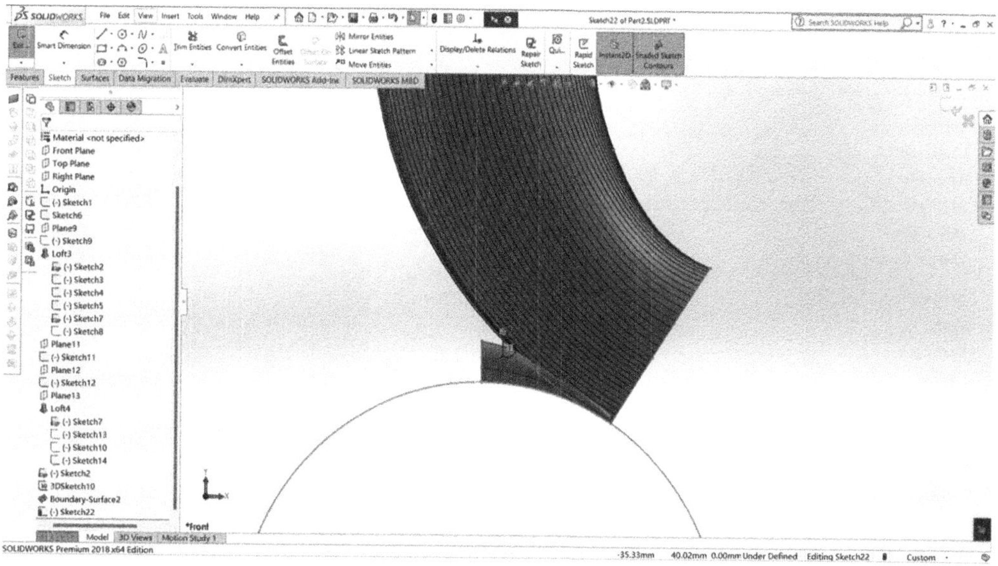

Figure 6.9 Sketch of the curvature of the volute tongue.

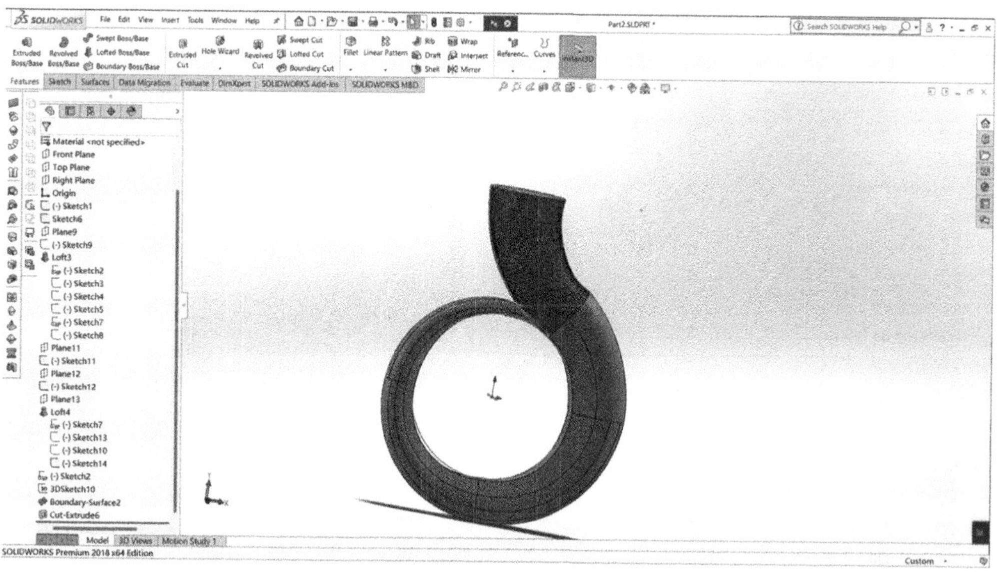

Figure 6.10 Three-dimensional diagram of the volute.

6.2.1.3 Create Body

To create the impeller single passage body, click  [Create Body]. In the dialog that appears, enter the name of the impeller single passage body as "Fluid-Impeller". Then, click [By Topology], and finally click the [Apply] button to create the impeller single passage body. The creation process and result are shown in Figure 6.14.

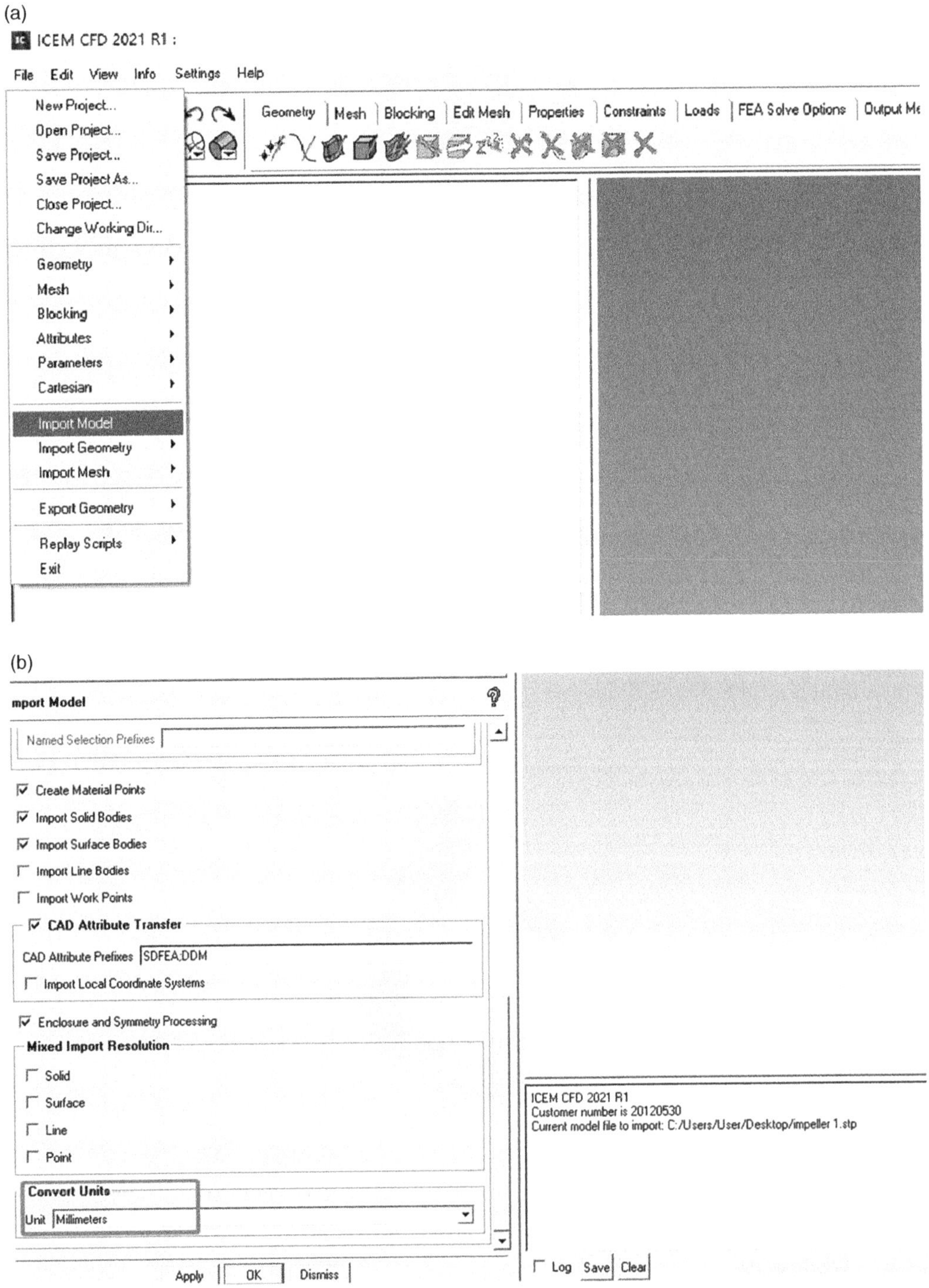

Figure 6.11 Importing a model: (a) Toolbar, (b) model unit settings.

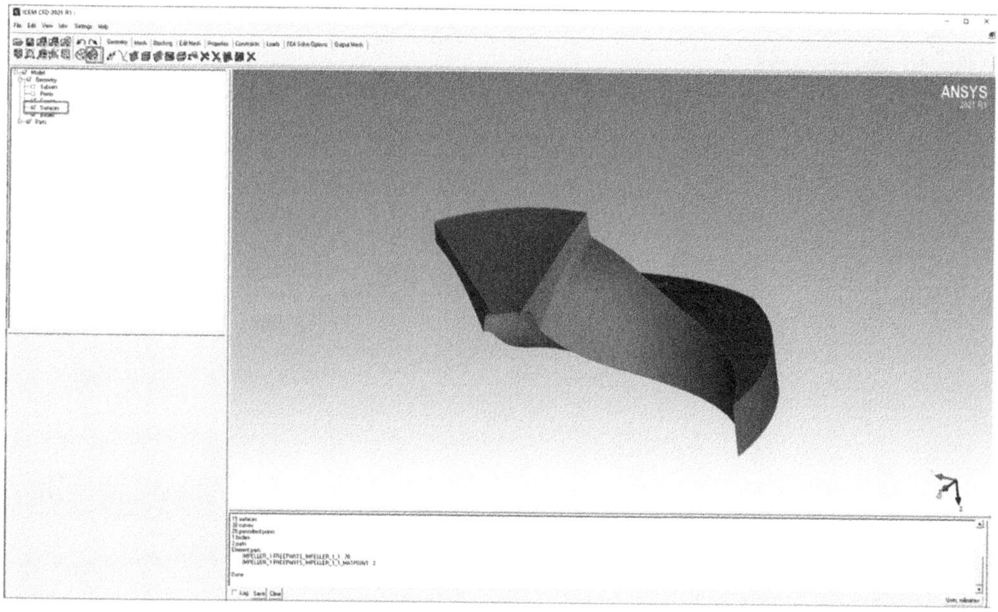

Figure 6.12 Model solidification.

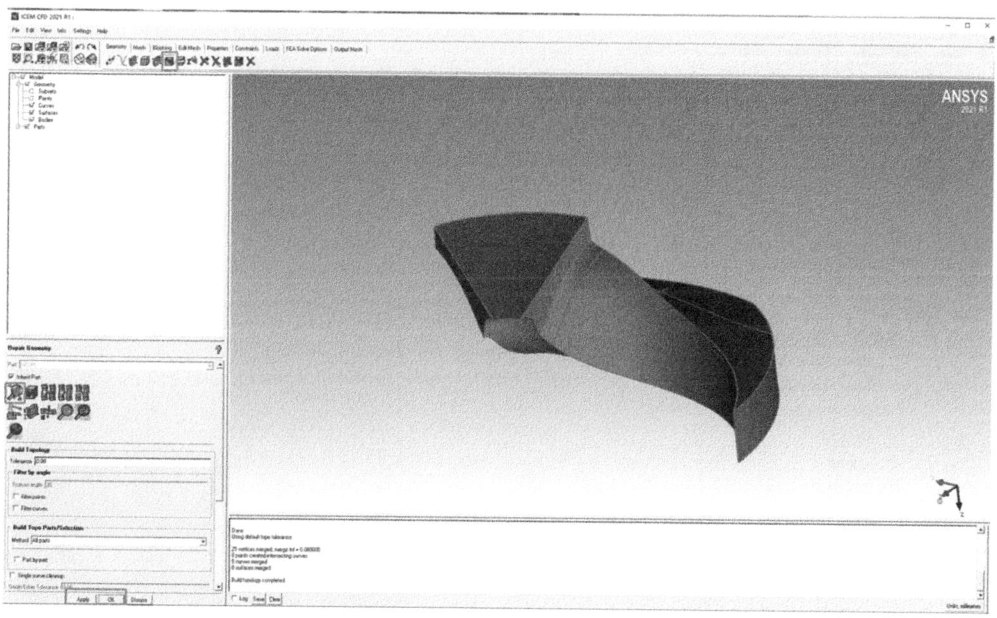

Figure 6.13 Geometry of the single channel.

6.2.1.4 Create Parts

Next, define the surfaces of the impeller flow passage. The impeller primarily includes the pressure surface of the blades, the suction surface of the blades, the interface between the impeller inlet and the inlet pipe, the interface between the impeller outlet and the volute, the periodic interfaces, the shroud, and the hub plate. Right-click [Parts], then select [Create Part], as shown in Figure 6.15.

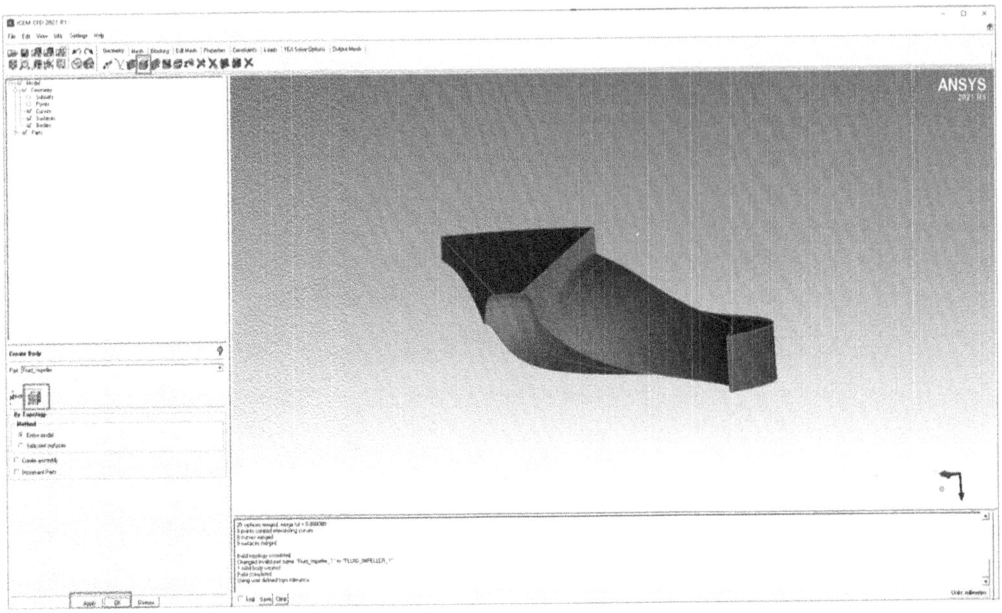

Figure 6.14 Creating a single flow channel.

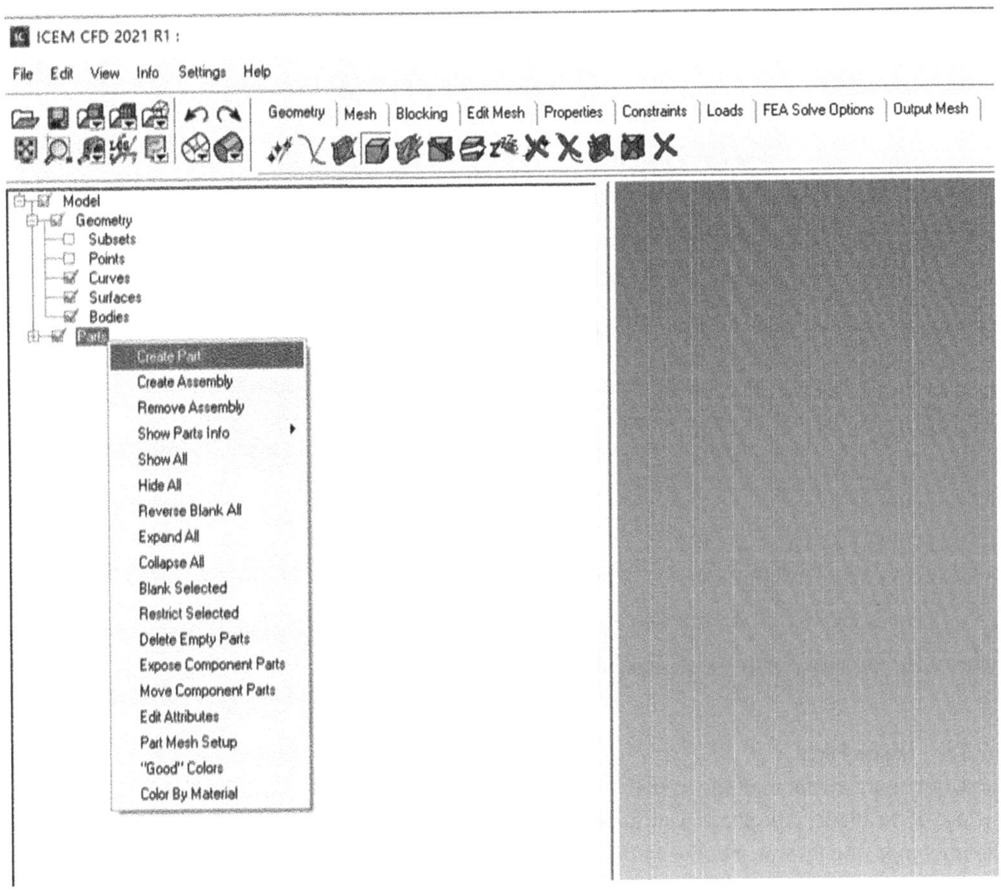

Figure 6.15 Creating a part.

In the dialog that appears, enter the names for different parts of the impeller single passage, for example, define the interface between the impeller inlet and the inlet pipe as "Inlet." Then, click [Create Part by Selection], select the inlet surface, and click the middle mouse button. The creation process and result are shown in Figure 6.16. Finally, sequentially define each surface of the impeller flow passage to create the corresponding parts.

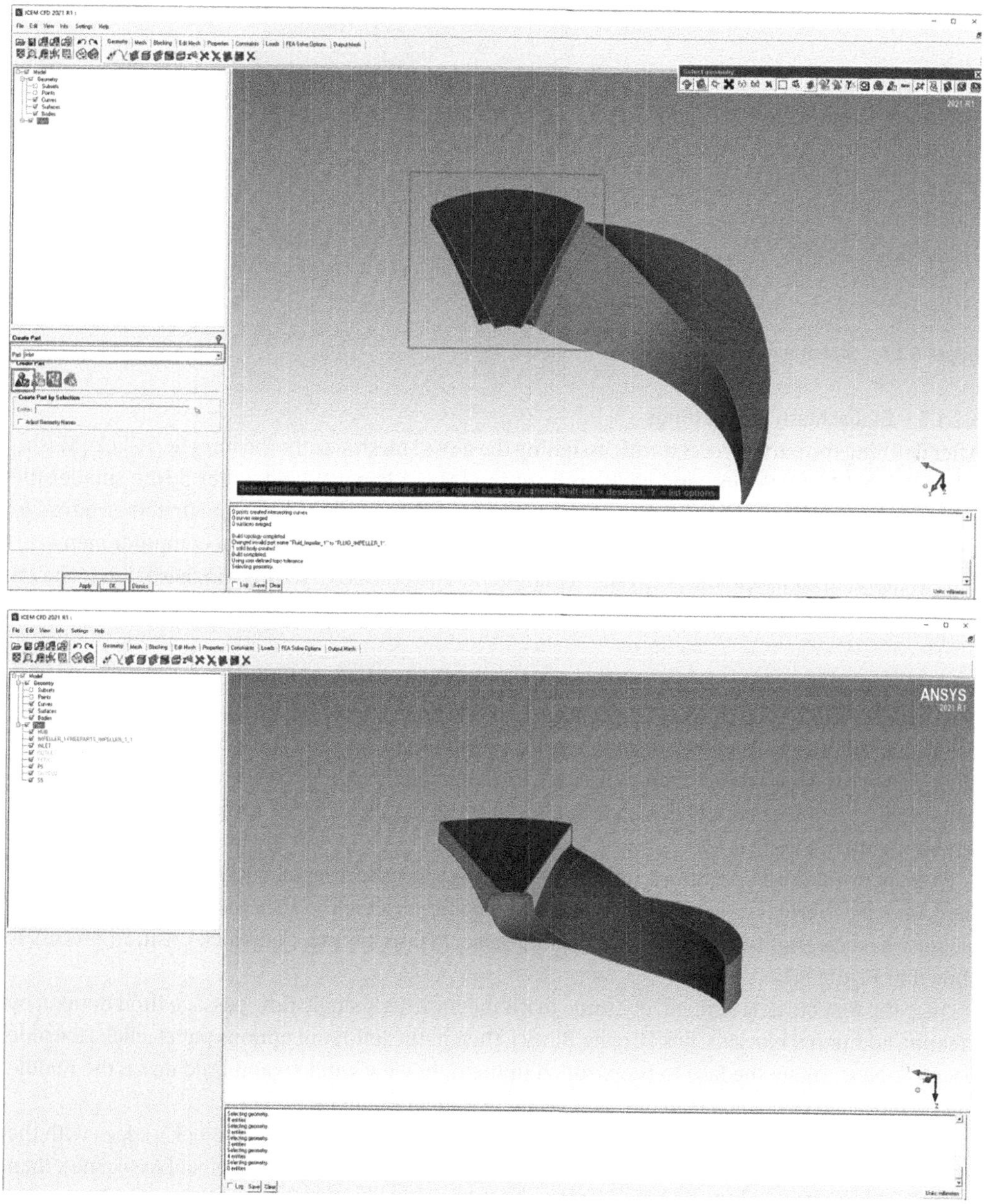

Figure 6.16 Process and result of creating parts.

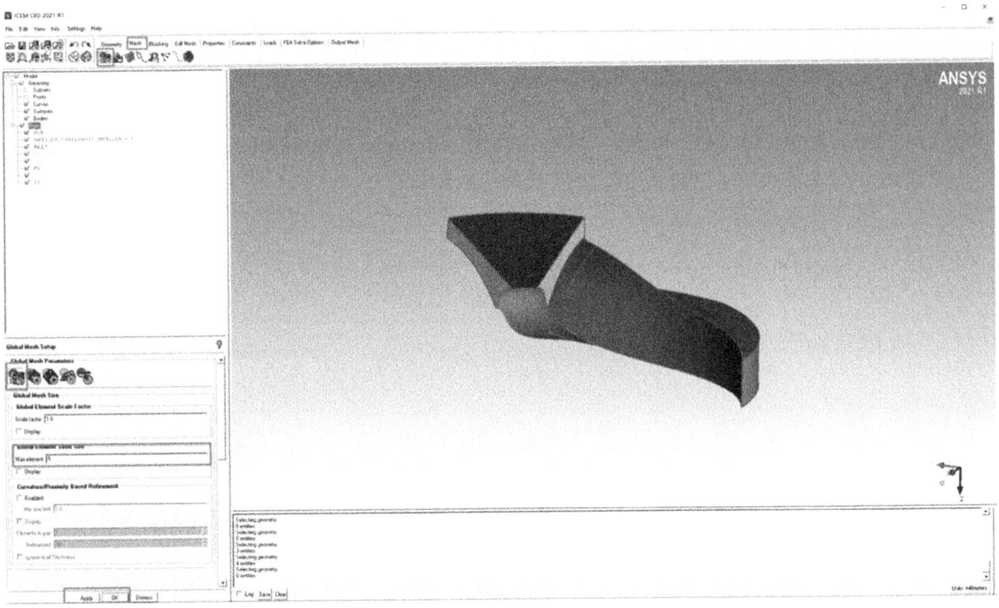

Figure 6.17 Global mesh dimensions.

6.2.1.5 Global Mesh Size Settings

After defining the parts, proceed with assigning the global mesh size for the impeller. Click [Mesh], select [Global Mesh Setup], and click on it. In the [Max Element] field, enter 5 (the smaller the value, the more the number of elements; it is recommended to set a larger value first, then adjust to a smaller value to avoid excessive mesh count, which could lead to insufficient computer memory). Click [Apply] to complete the mesh size assignment. The process of setting the mesh size is shown in Figure 6.17.

6.2.1.6 Create Blocks and Establish Mappings

Next, create the block by clicking [Blocking], then selecting [Create Block]. In the left-hand options panel, click [Initialize Blocks], then choose and click [Select Geometry] below. In the right window, click [Select All Appropriate Visible Objects]. Press the middle mouse button or click [Apply] at the bottom left. The program will generate the initial mesh topology for the impeller, as shown by the red boundary in Figure 6.18.

Then, perform the association by clicking [Blocking] and selecting [Associate], followed by clicking [Associate Vertex] to associate the points. To better observe the structure of the impeller single flow passage, right-click on [Surfaces] and select [Transparent]. The block creation process is shown in Figure 6.19.

Once the first block is defined, continue to fill the impeller's single flow passage fluid domain by creating additional blocks. Click [Create Block], then in the left-hand options panel, click [Extrude Face(s)]. Next, rotate the face to be extruded in the right view window and hold down the middle mouse button to drag and create the new block, as shown in Figure 6.20.

After creating the new block, perform edge association by associating the block's edges with the outer boundary of the impeller's single flow passage. Click [Blocking] and select [Associate], then click [Associate Edge to Curve] to establish the edge association. The association for the second block is shown in Figure 6.21. Then, repeat the process to fill the entire impeller single flow passage with blocks.

(a)

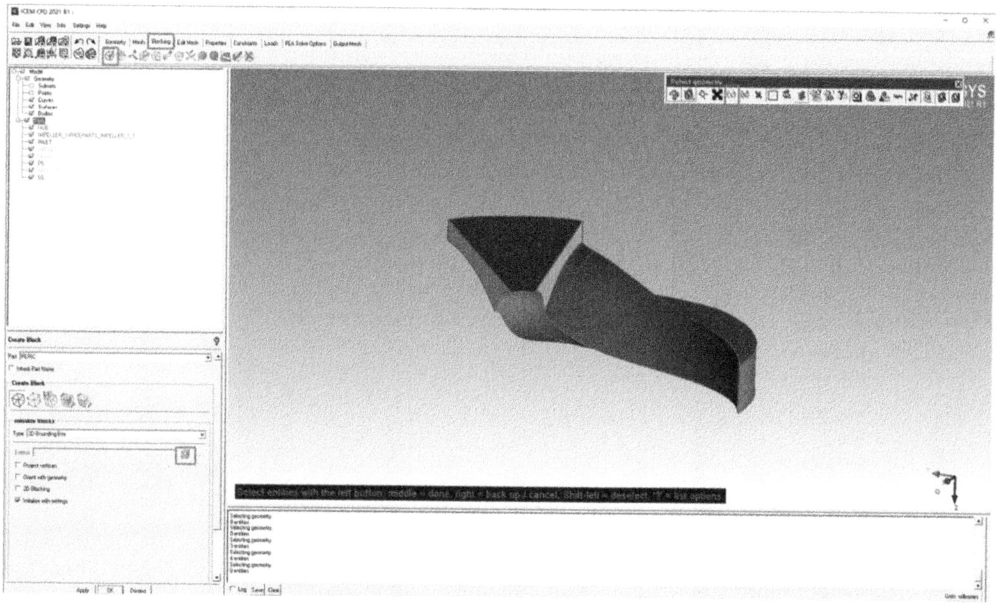

(b)

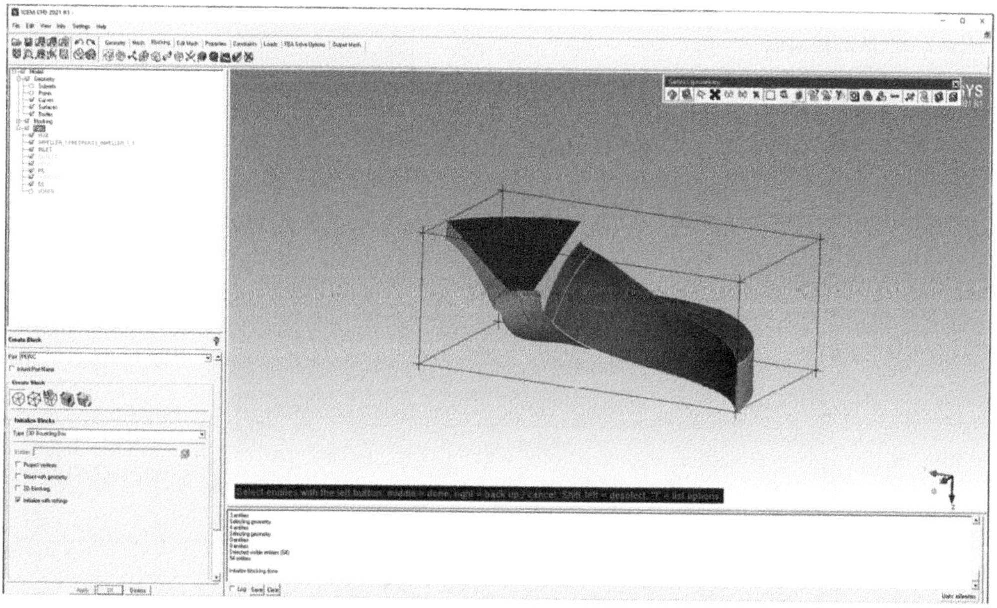

Figure 6.18 Block creation: (a) Panel operation sequence, (b) initial block.

6.2.1.7 Y-Block

Due to the sharp corner at the impeller outlet, a Y-Block division is required. Click [Blocking] and select [Edit Block], then click [Convert Block Type] in the bottom left options panel. In the [Set Type] menu, choose [Y-Block]. The final block structure is shown in Figure 6.22.

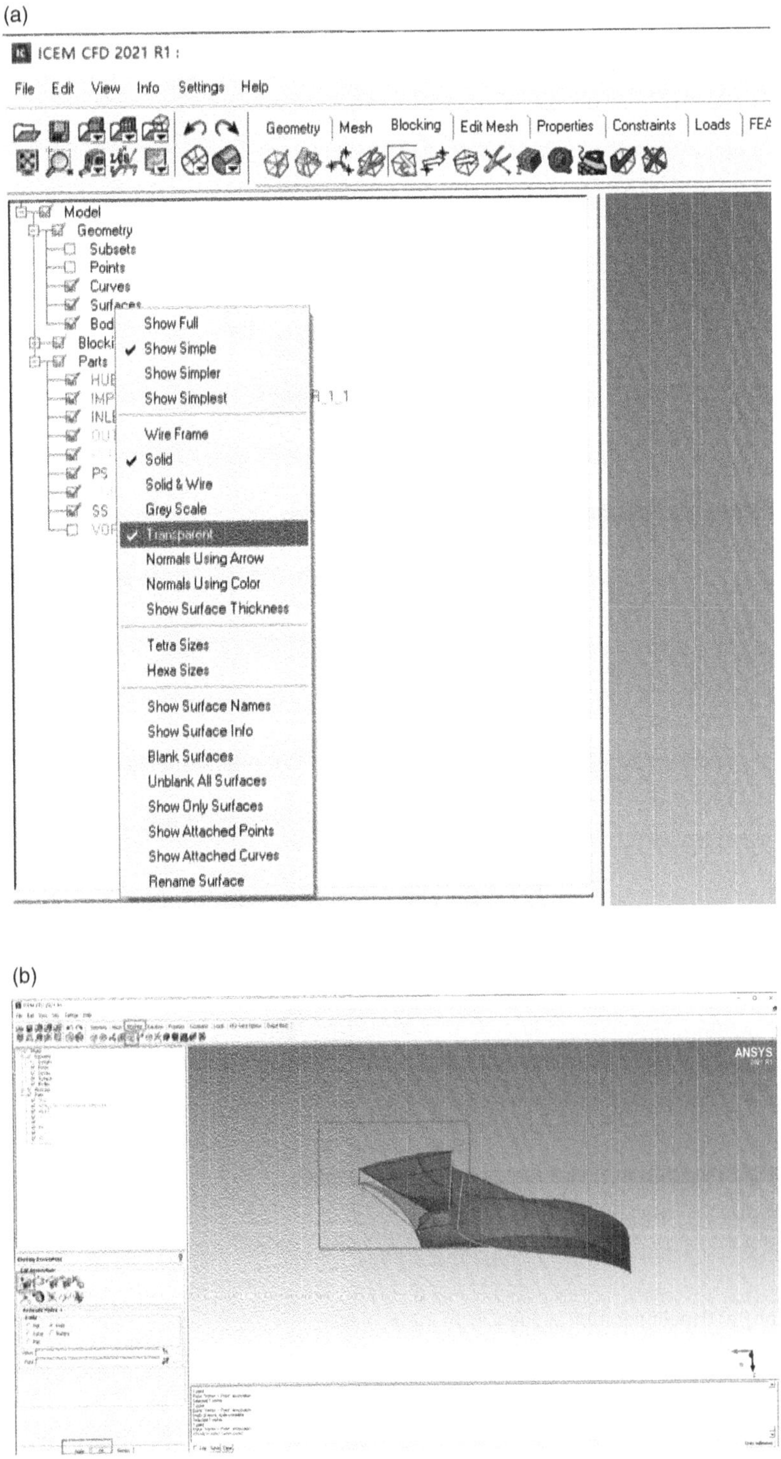

Figure 6.19 Block creation process 1: (a) Model transparency, (b) association settings.

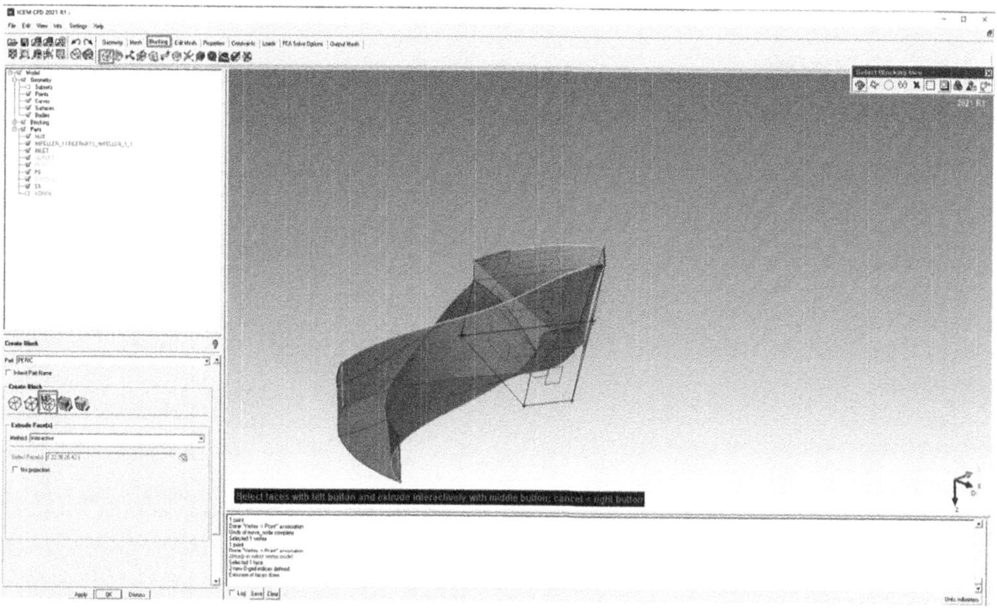

Figure 6.20 Block creation process 2.

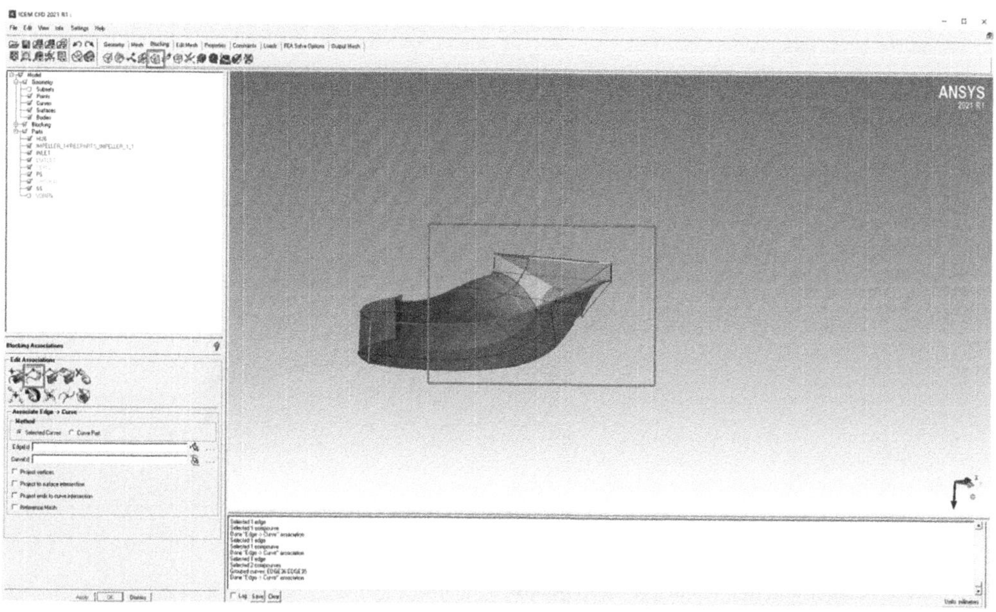

Figure 6.21 Block creation process 3.

6.2.1.8 Periodic Setting

Next, click [Blocking] and select [Pre-Mesh Params]. In the bottom left options panel, click [Edge Params]. In the [Edge] section below, click to select the edges on the periodic face, then, in the right window, click to choose the topological edges. Enter the number of nodes for the topological edges in the [Nodes] field, and finally, click [Apply] in the bottom left window, as shown in Figure 6.23. Then, sequentially define the number of mesh nodes for the edges.

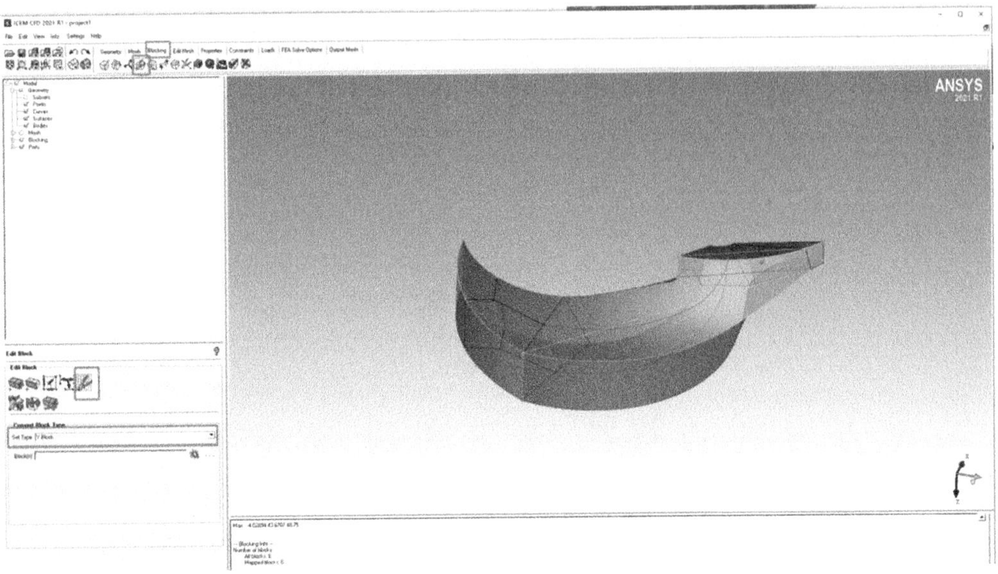

Figure 6.22 Block creation process 4.

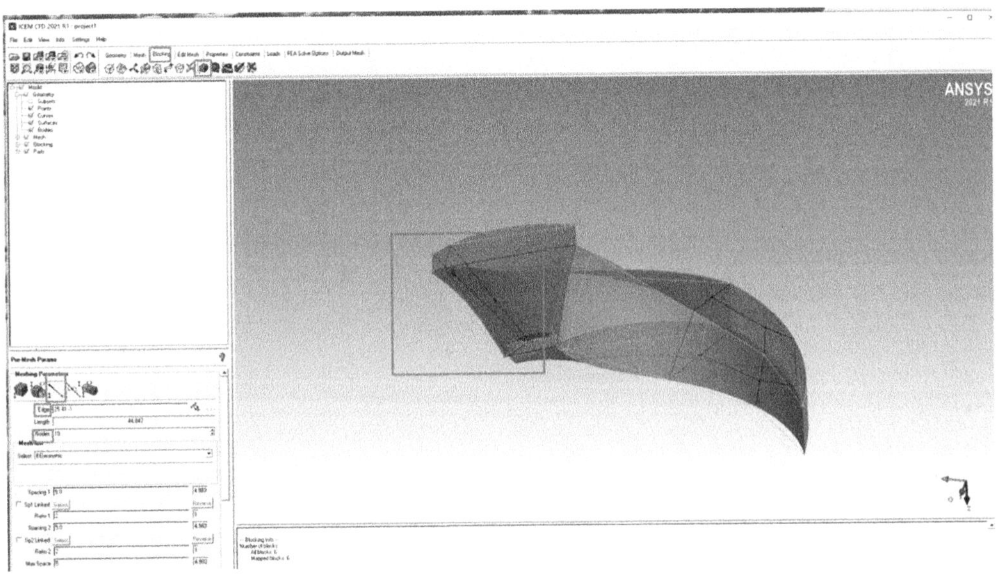

Figure 6.23 Confirming the number of mesh nodes.

Since only the mesh for the impeller single flow passage has been created, the impeller mesh for all passages needs to be obtained by periodic replication. Sequentially click [Mesh] → [Global Mesh Setup] → [Set Up Periodicity]. Set the periodicity type to "Rotational Periodic," as shown in Figure 6.24. The Axis setting defines the location of the rotational axis; for example, [0 0 1] indicates that the rotation center is along the Z-axis. The Angle setting defines the rotation angle for the single-pass flow path, which is related to the number of blades. For example, if there are 6 blades, enter 60.

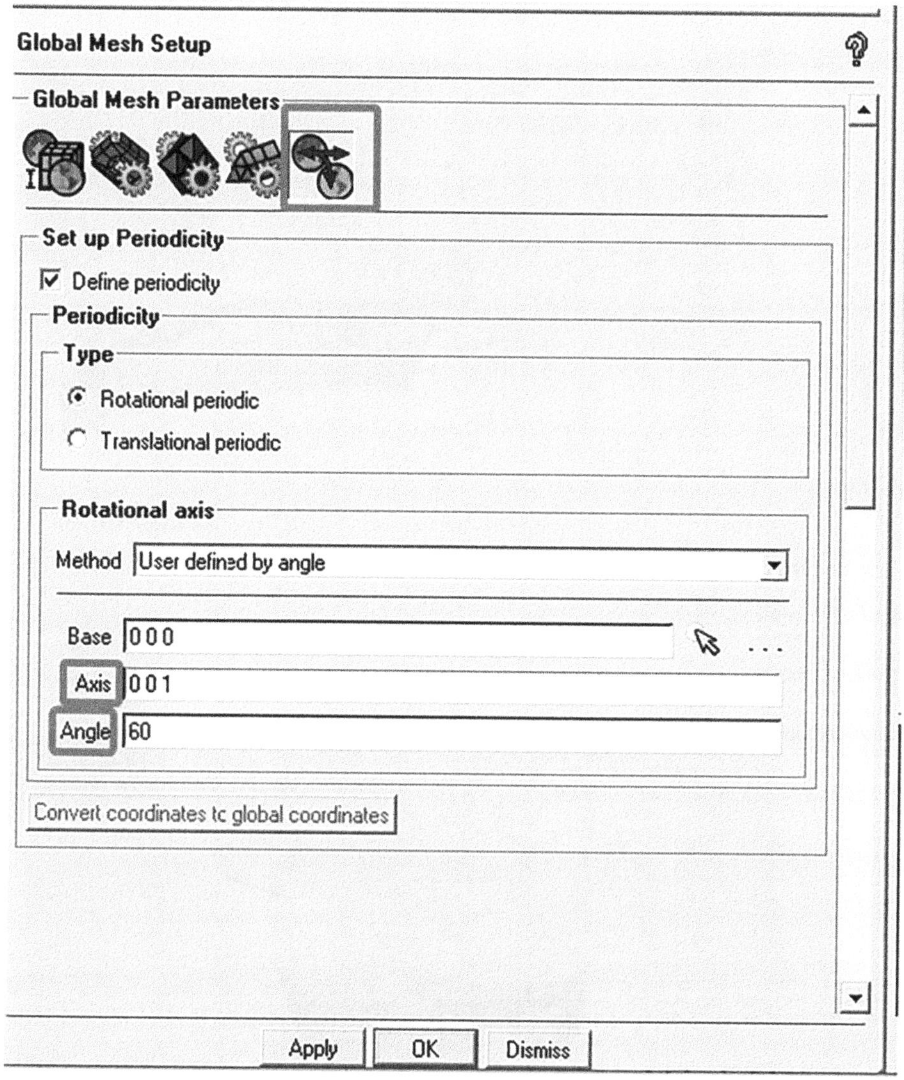

Figure 6.24 Setting the rotation period.

Next, set the periodic points by sequentially clicking [Blocking] → [Edit Block] → [Periodic Vertices]. Set the periodicity for the nodes on the interface, as shown in Figure 6.25.

6.2.1.9 Generate Mesh

Generate the structured mesh by selecting [Pre-Mesh] in the top left window. Click [Model] → [Blocking] → [Pre-Mesh] in the top left window, and choose [Convert to Unstructured Mesh]. Then, sequentially click [File] → [Mesh] → [Save Mesh] to save the current mesh file. Next, delete the mesh's periodic faces by clicking [Edit Mesh] → [Delete Elements] → [Select Items in a Part]. The "Select Part" dialog box will appear. In the dialog box, check the [PERIC] periodic face, click [Accept], and finally click [Apply] to complete the deletion of the impeller's periodic face mesh. The final mesh is shown in Figure 6.26.

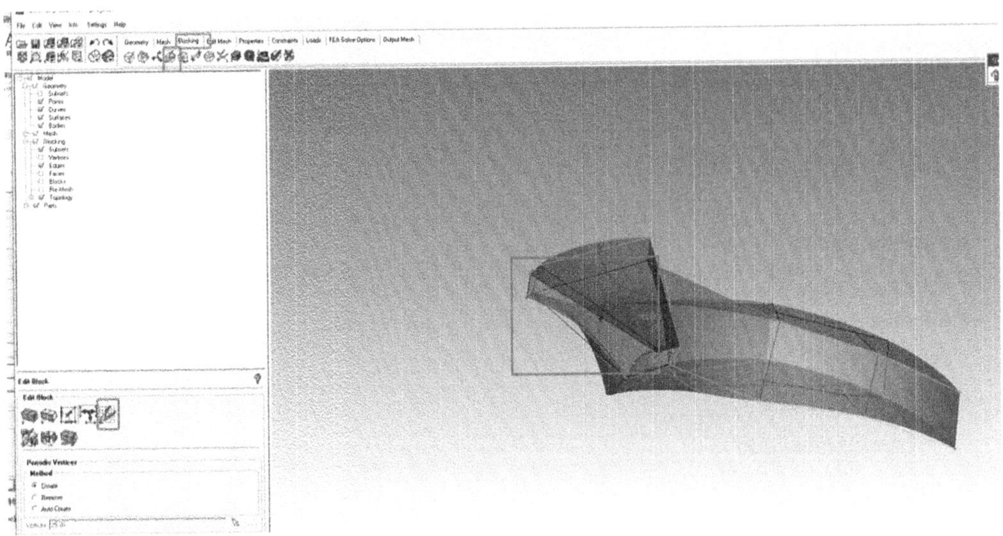

Figure 6.25 Setting the period point.

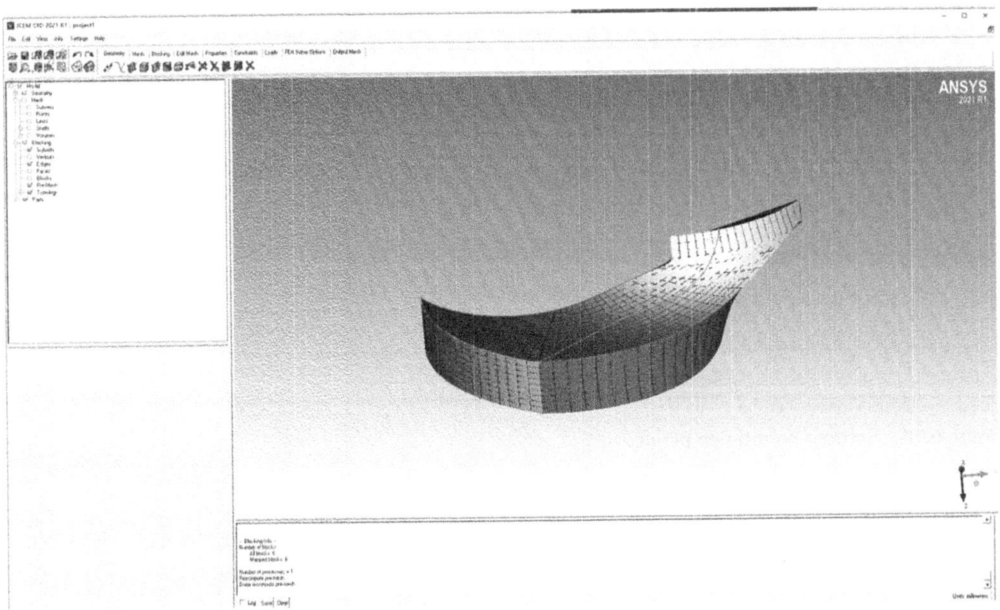

Figure 6.26 Single flow passage mesh of the impeller.

6.2.1.10 Checking Mesh Quality

Sequentially click [Mesh] → [Pre-Mesh Quality Histograms] → [Criterion] → [Quality] → [Apply] to check the mesh quality. If the mesh quality is not satisfactory, you can improve it by adjusting the mesh size, number of mesh nodes, mesh structure, and other related parameters (Figure 6.27).

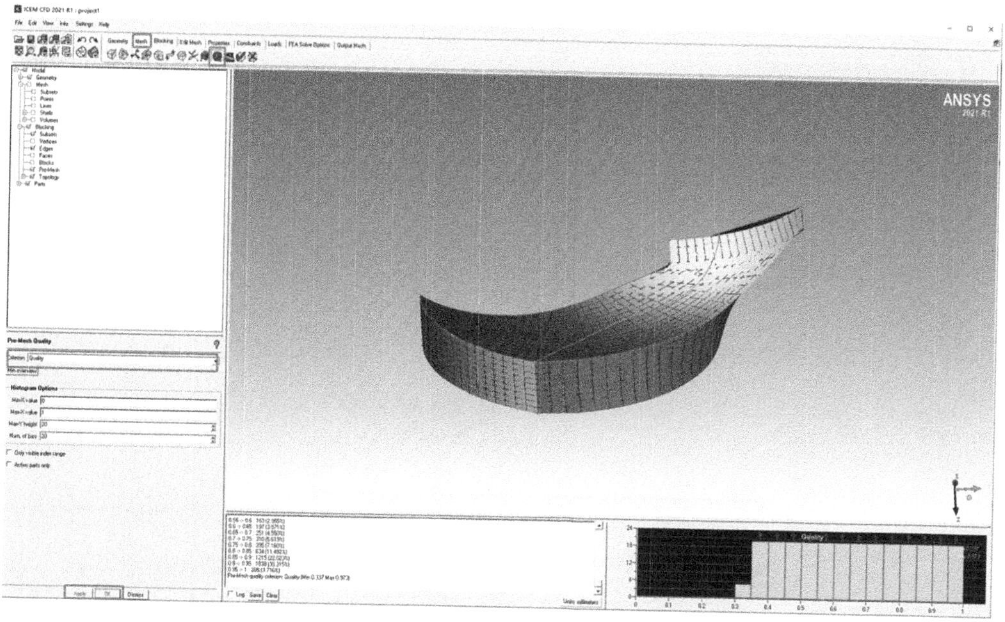

Figure 6.27 Mesh quality of the single flow passage.

6.2.2 Volute Meshing

6.2.2.1 Model Import and Simple Processing

The model import for the volute mesh division and the creation of Parts is the same as for the impeller, so it will not be repeated here. Sequentially click [Blocking] → [Create Block] → [Initialize Blocks] to create the blocks. After the blocks are created, they need to be associated. Since the volute block must be associated with the edges of the volute surface, any missing lines and points in the geometric model need to be created. To create these, click [Geometry] → [Create/Modify Curve] → [Surface Parameter], then click [Select Surface(s)] in the Surface(s) section to choose the surfaces where the lines need to be created. Finally, select the method for generating the lines in the [Method] section and specify the location for line generation in the [Parameter] section. For example, if the method is set to "U" and the parameter is set to 0.15, the result will be as shown in Figure 6.28.

6.2.2.2 Establish Mapping Relationship

After the lines are created, proceed with the block association. To better define the association locations, first perform vertex association. Sequentially click [Blocking] → [Associate] → [Associate Vertex], then select the corresponding points on the block and geometry for association. After the association, click [Apply]. The result is shown in Figure 6.29.

After the vertex association, proceed with the edge association by associating the block edges with the outer outline of the volute. Click [Blocking] → [Associate], then select [Associate Edge to Curve] to perform the edge association. If any association lines are missing, they can be created using the method shown in Figure 6.28. The result after edge association is shown in Figure 6.30.

6.2.2.3 Creation of a Lock Derivative Block

After the first block is created, proceed by creating additional blocks to fill the volute flow domain. The process for creating blocks is the same as in the impeller. Click [Create Block], then in the left-hand options panel, select [Extrude Face(s)]. Next, rotate the face to be extruded in the

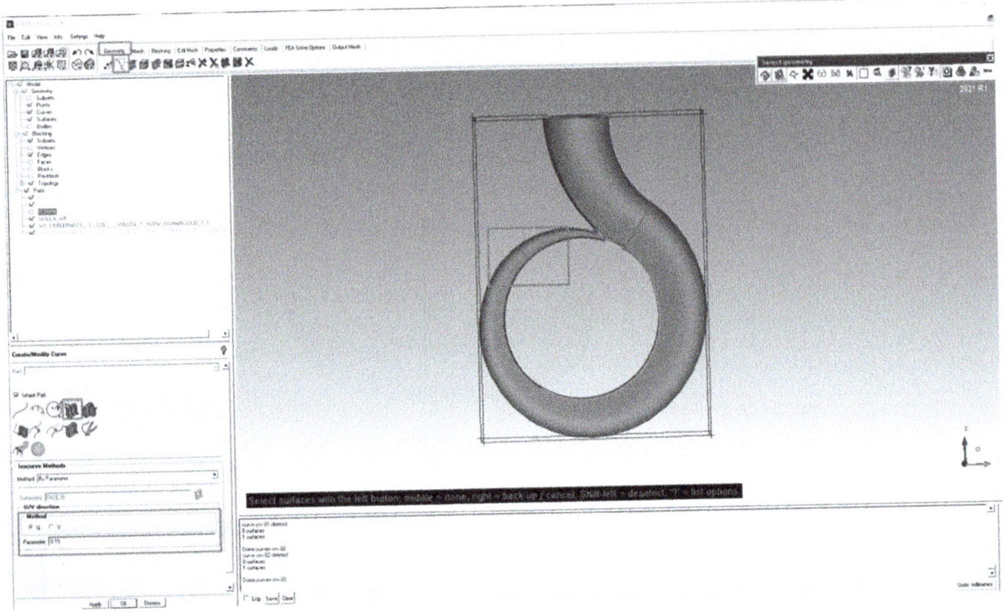

Figure 6.28 Creating lines on the surface.

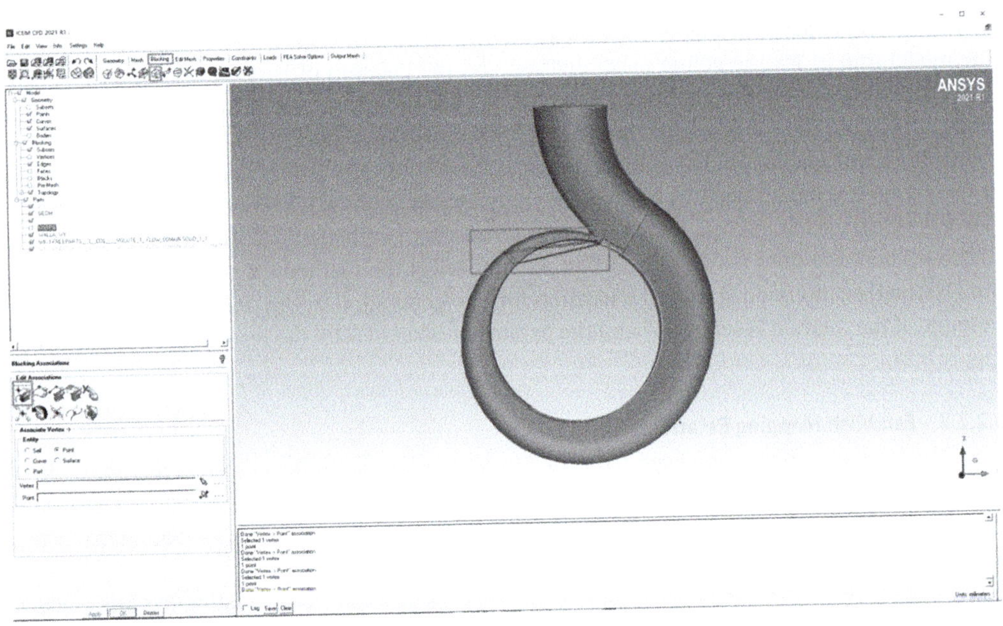

Figure 6.29 Point association.

right-side view window, and hold the middle mouse button to drag and create the new block, as shown in Figure 6.31.

Then, repeat the above steps until the volute flow passage is fully filled with blocks. The final spatial structure of the volute block domain is shown in Figure 6.32.

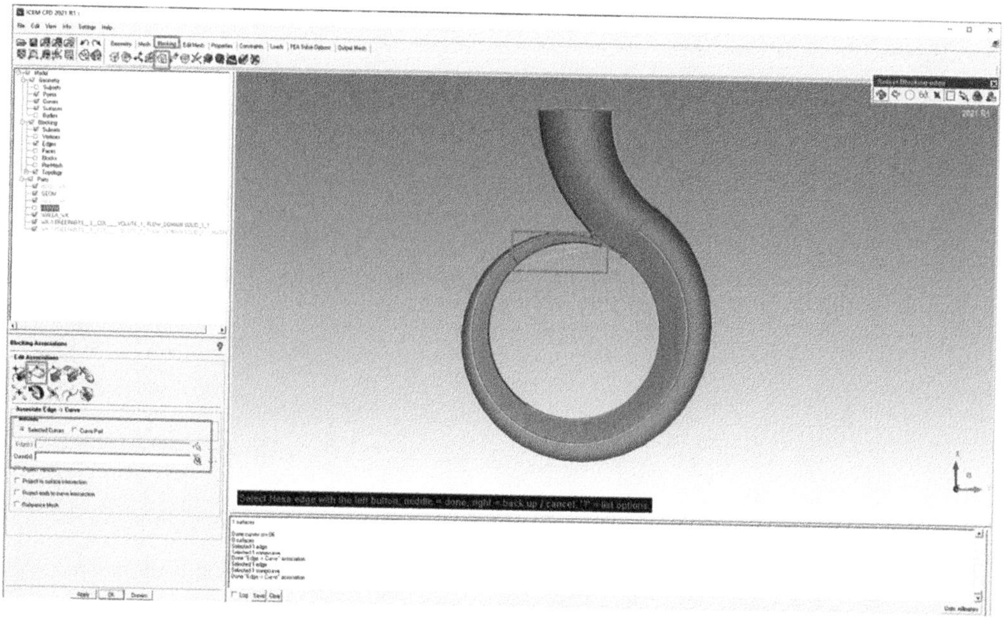

Figure 6.30 Line association.

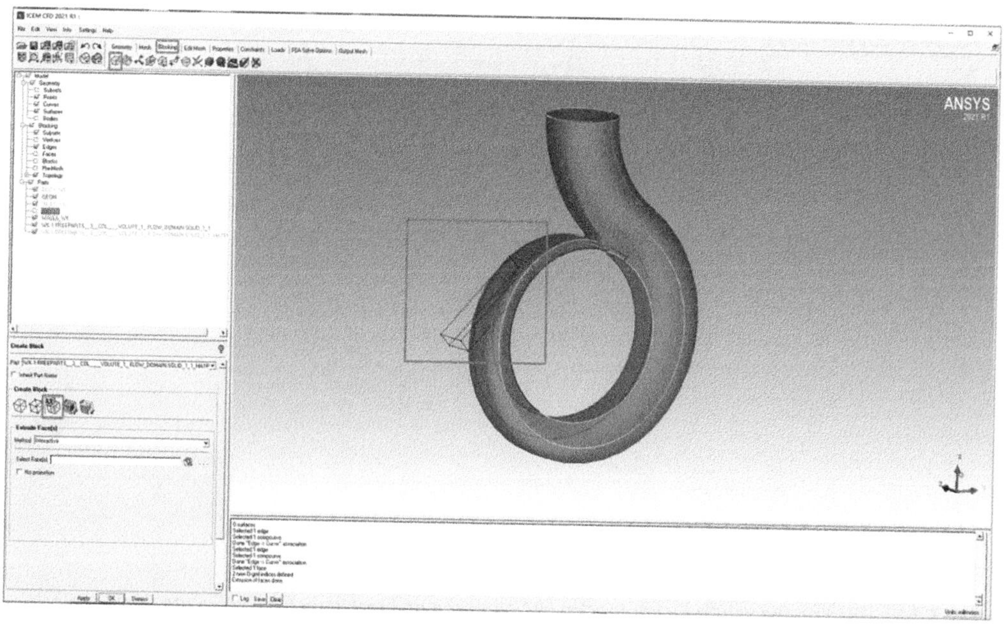

Figure 6.31 Creating a block derivative block.

6.2.2.4 Global Mesh Size Settings

Next, click [Mesh], then select [Global Mesh Setup]. In the [Max element] field, enter the mesh size, and click [Apply] to assign the mesh division size. The process of setting the mesh size is shown in Figure 6.33.

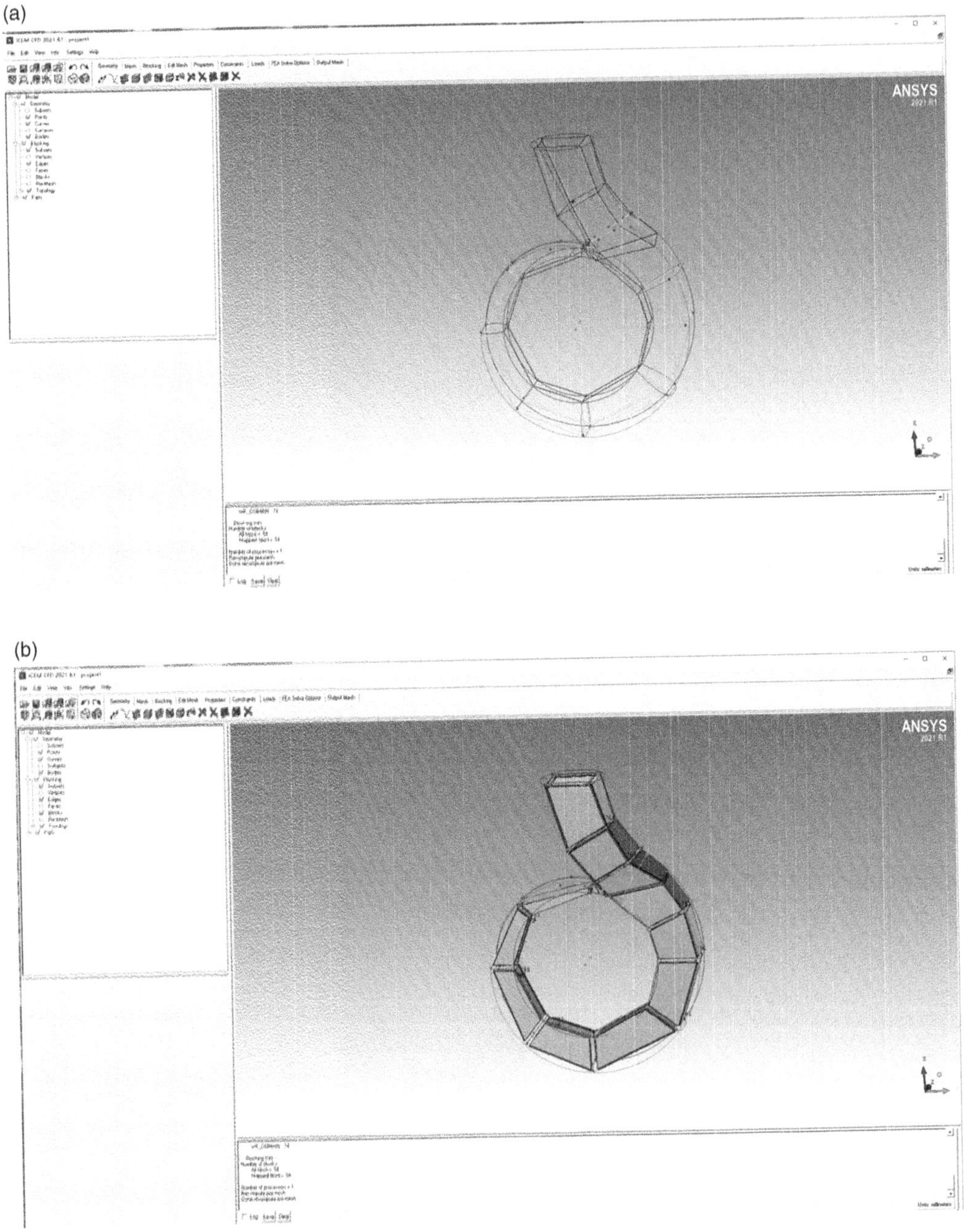

Figure 6.32 Dividing the volute runner blocks: (a) Volute region block structure, (b) volute domain block solid display.

6.2.2.5 Meshing

Next, follow the same steps as for the impeller passage mesh division. Click [Blocking] → [Pre-Mesh Params] → [Edge Params]. In the [Edge] section below, click on the desired edge, then in the right-side window, select the topological edge on the periodic face. Enter the number of nodes for the topological edge in the [Nodes] field, and finally click [Apply] in the bottom left window. Then,

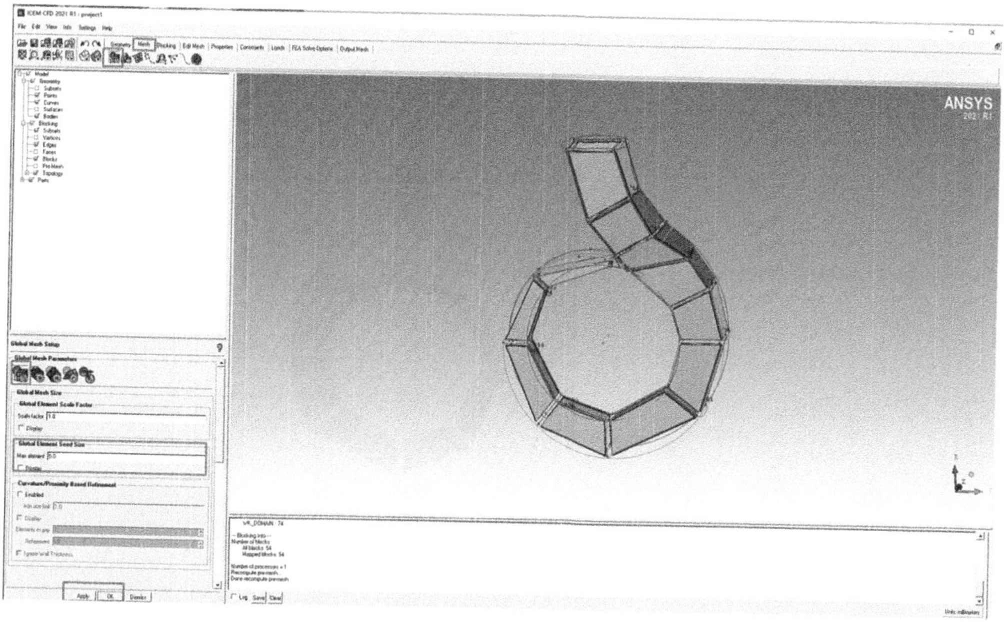

Figure 6.33 Determining the mesh size of the volute.

define the number of mesh nodes on the edges sequentially. Since the volute shape in a centrifugal pump is curved, creating an O-Block is necessary for better mesh quality. Click [Blocking] → [Split Block] → [Ogrid Block] → [Select block(s)], select all the blocks in the main volute flow passage, then click [Select Face(s)] to choose the two faces at the start and end of the selected blocks. Finally, click [OK], as shown in Figure 6.34.

6.2.2.6 Mesh Quality Check

Next, click [Model] → [Blocking] → [Pre-Mesh] in the bottom left window to view the mesh. Then, click [Mesh] → [Pre-Mesh Quality Histograms] → [Criterion] → [Quality] → [Apply] to check the mesh quality, as shown in Figure 6.35. A mesh quality of 0.3 or higher is generally sufficient for most calculations. If further improvement in mesh quality is required, this can be achieved by adjusting the number of nodes or modifying the block structure. Finally, select [Convert to Unstruct Mesh]. To save the current mesh file, click [File] → [Mesh] → [Save Mesh].

6.2.3 Inlet and Outlet Flow Passage Mesh Generation

The mesh division method for the inlet and outlet passages is the same. Below, consider the outlet passage as an example for mesh division. The process of importing the model and creating the Parts is the same as described earlier, so it is not repeated here. Click [Blocking] → [Create Block] → [Initialize Blocks], then select and click [Select geometry] below. In the right-side window, click [Select all appropriate visible objects], and either click the middle mouse button or click [Apply] in the bottom left corner. The result is shown in Figure 6.36.

Next, perform the edge association by clicking [Blocking] → [Associate] → [Associate Edge to Curve]. Select the edges on the block and the corresponding geometric edges of the outlet pipe, as shown in Figure 6.37.

If the outlet pipe is too long, it can be divided into multiple blocks. Click [Blocking] → [Split Block] → [Split Block], then click on the edge of the block. After that, click the middle mouse button to confirm the split. The block division is shown in Figure 6.38.

(a)

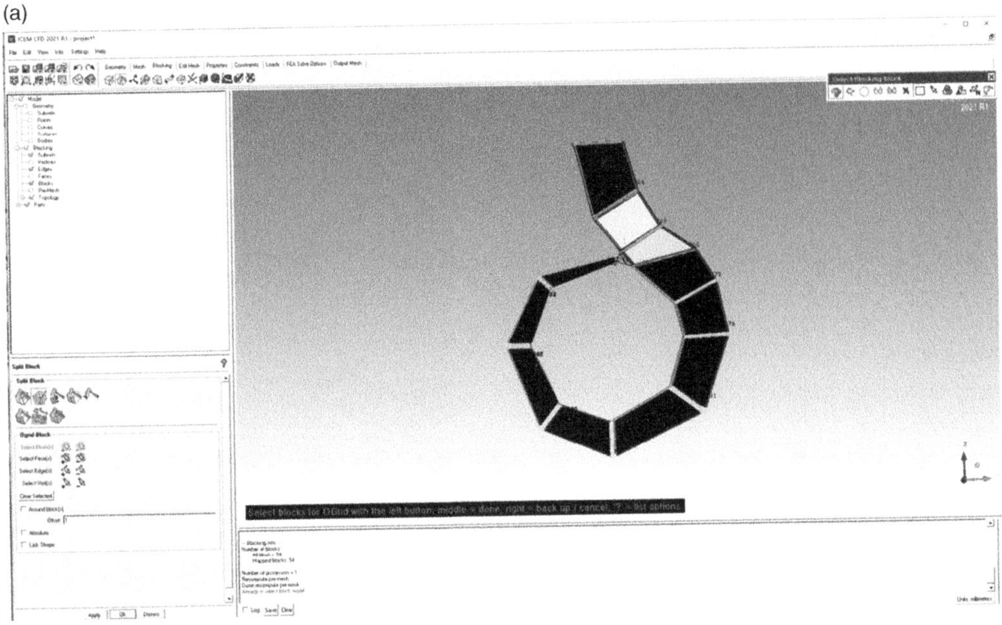

(b)

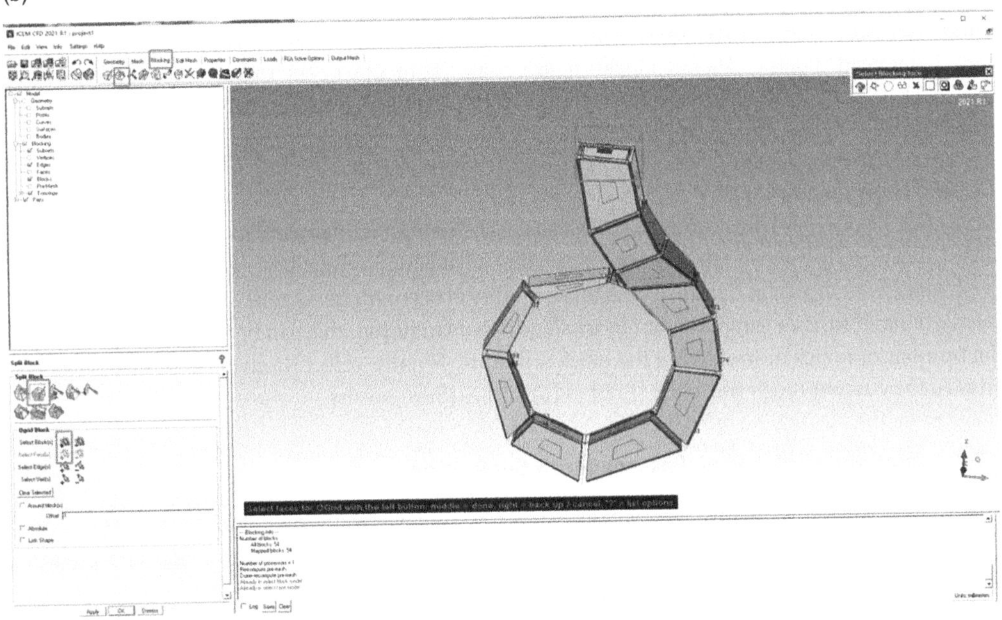

Figure 6.34 Meshing the volute: (a) O-grid block creation, (b) start and end faces of the block.

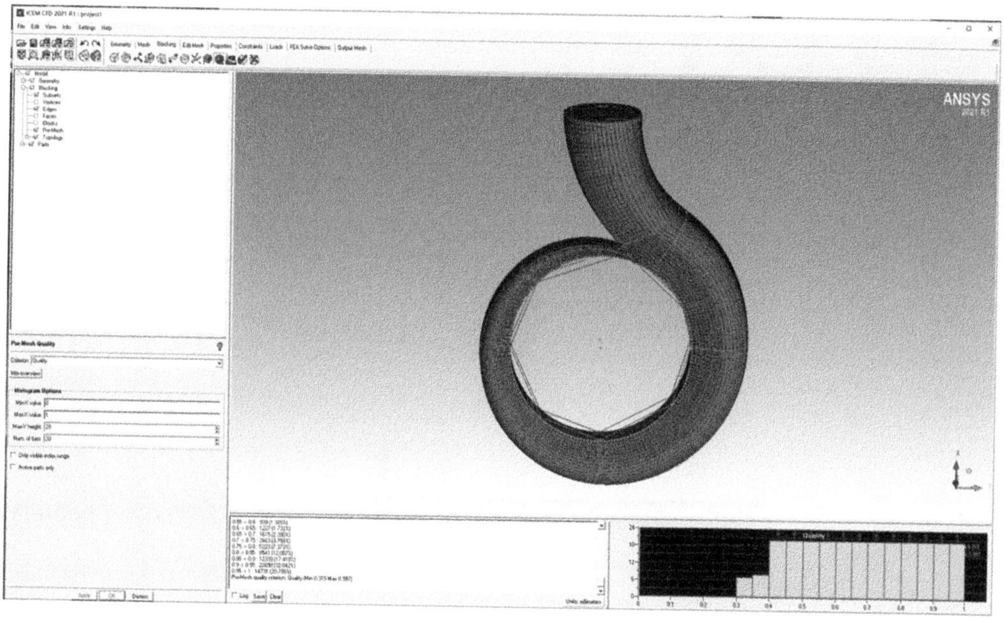

Figure 6.35 Volute mesh.

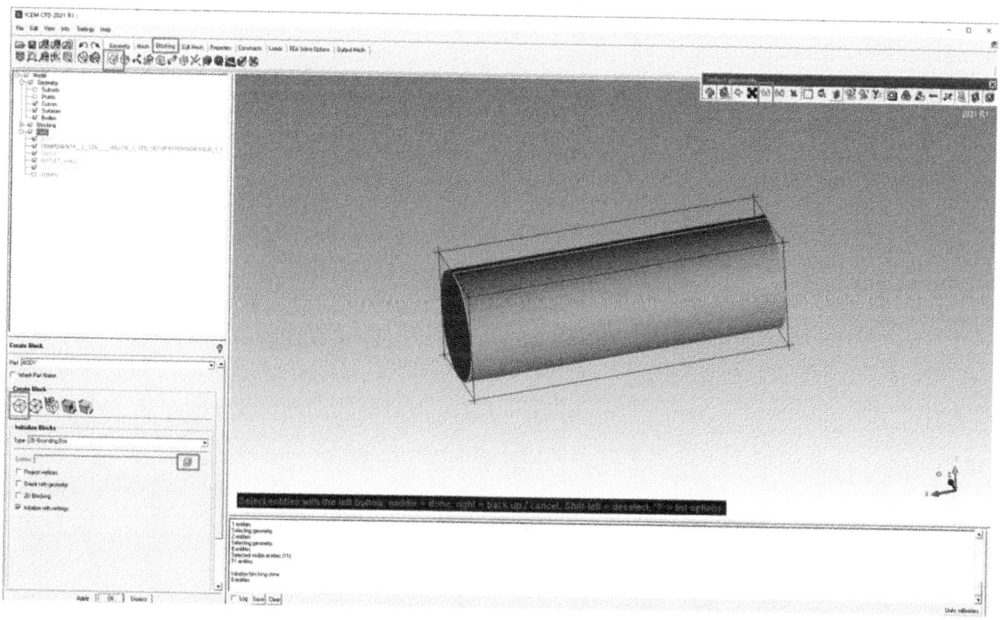

Figure 6.36 Creating an outlet pipe block.

Then, determine the mesh size and number of nodes for the outlet pipe based on the mesh size and node count established for the volute or volute's mesh. The process of determining the mesh size and node count will not be repeated here. After determining the mesh size and nodes, check the mesh quality, as shown in Figure 6.39.

As shown in the figure above, the mesh quality is relatively low. This can be improved by creating an O-Block. Click [Blocking] → [Split Block] → [Ogrid Block] → [Select block(s)], select all the

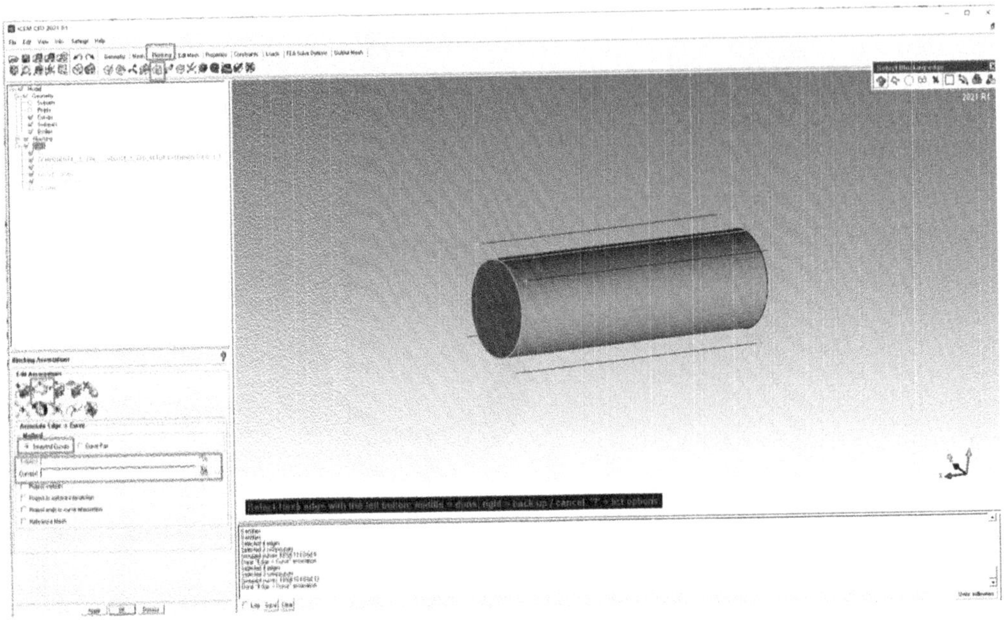

Figure 6.37 Cable association.

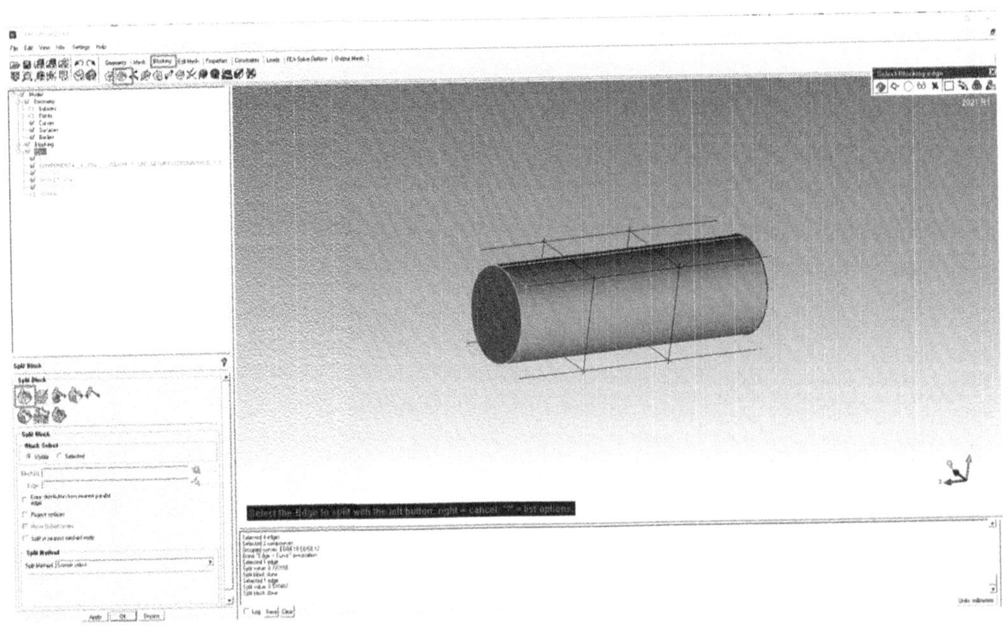

Figure 6.38 Block division.

blocks on the outlet pipe, then click [Select Face(s)] to choose the first and last faces of the selected blocks. Finally, click [OK], as shown in Figure 6.40.

After creating the O-Block, the number of nodes on the edges at the boundaries needs to be adjusted. Click [Blocking] → [Pre-Mesh Params] → [Edge Params], then click [Edge] below. In the right-hand window, click to select the edge just created. Enter the number of nodes on the

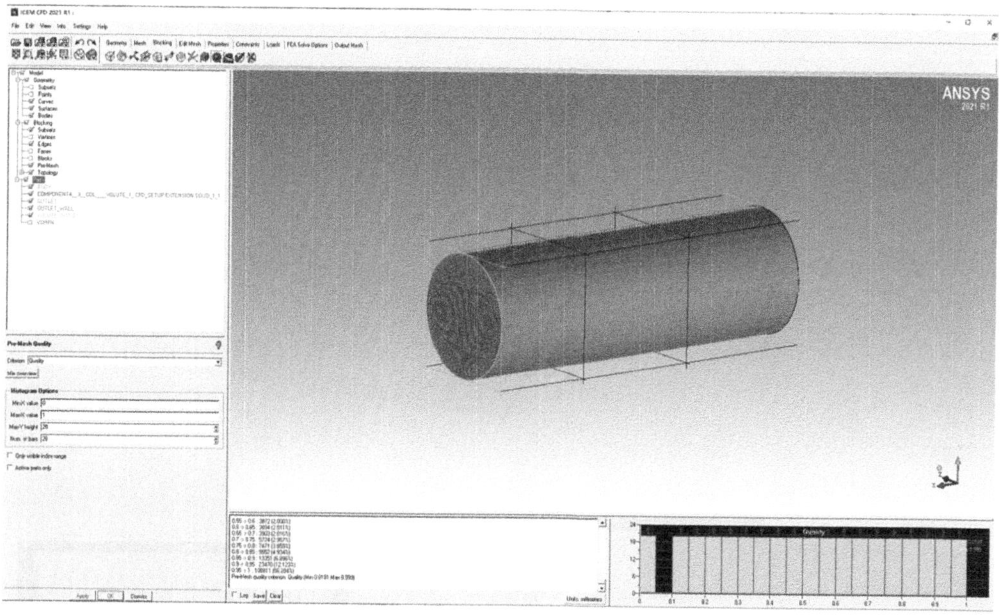

Figure 6.39 Mesh quality of the outlet pipe.

topological edge in the [Nodes] field, and finally click [Apply] in the bottom left window, as shown in Figure 6.41.

Finally, check the mesh quality, as shown in Figure 6.42. Then, proceed with mesh conversion and saving, using the same saving method as described earlier.

6.3 Steady Simulation Setup

6.3.1 Computational Domain Setting

6.3.1.1 Create a Case and Import the Mesh

Double-click the CFX software, then click [CFX-Pre] → [New Case] → [General] → [OK], as shown in Figure 6.43.

Right-click on [Mesh], select [Import Mesh] → [ICEM CFD] to import the mesh. The result is shown in Figure 6.44.

Since the impeller is divided into a single passage, after importing the mesh, it needs to be rotated to obtain the full passage impeller mesh. Right-click on the impeller mesh [Impeller.cfx5], then click [Transform Mesh]. In the [Transformation] section, select [Rotation]. Choose the rotation axis based on the position of the impeller mesh, then set the [Rotation Angle]. If the number of blades is 6, enter 60°. Check [Multiple Copies], and input 5 for the number of copies. Finally, click [Apply], as shown in Figure 6.45.

6.3.1.2 Fluid Domain Setting

Click on [Domain] to set up the fluid domain. In the [Name] field, enter the name of the fluid domain. For example, to define the impeller domain, enter "Impeller." Click the [Selection Dialog] button next to [Location] and select the mesh for the impeller domain, as shown in Figure 6.46.

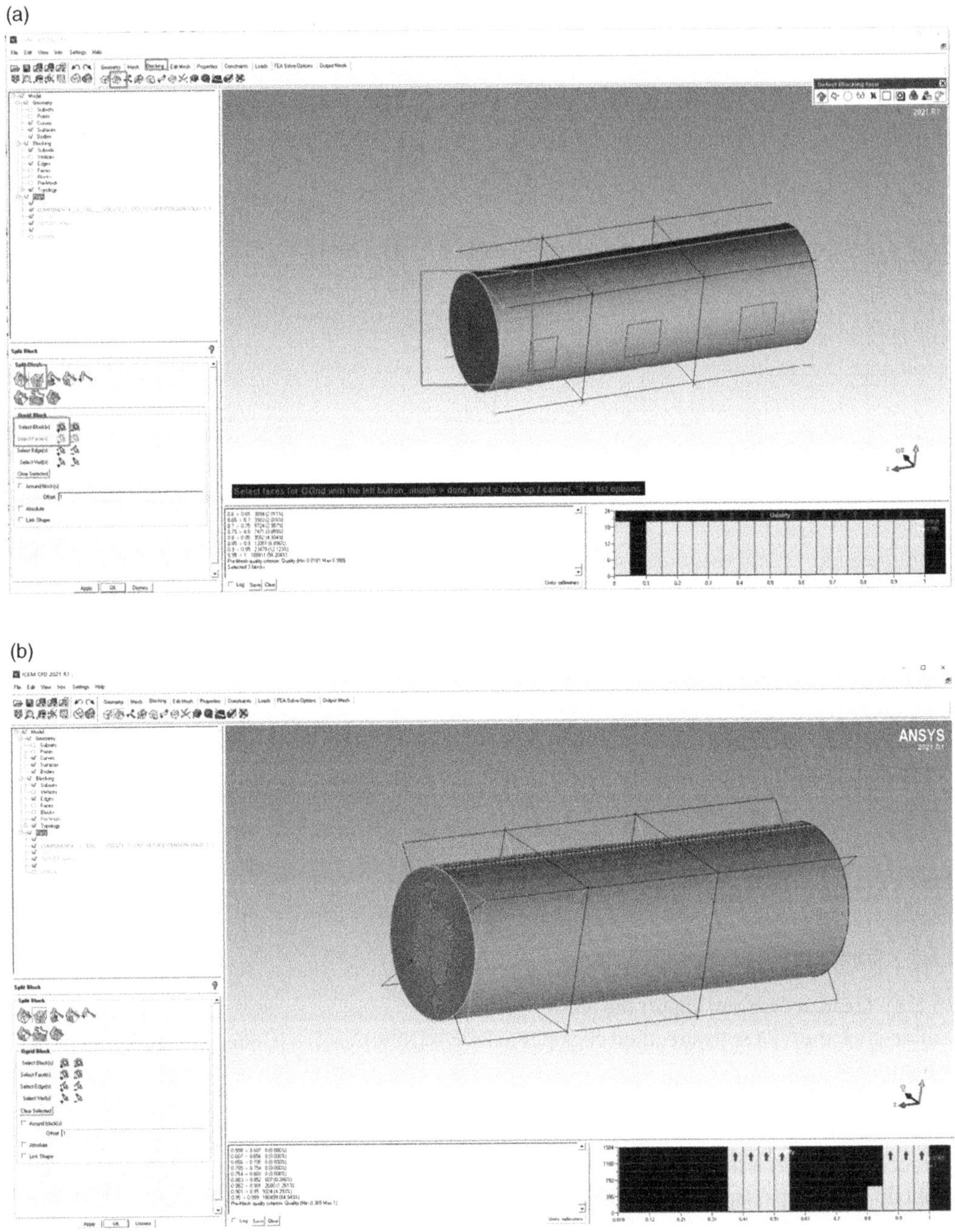

Figure 6.40 Creating an O-Block: (a) Start and end faces of the block, (b) optimized mesh quality.

Next, select the flow medium under [Material], such as [Water]. Enter the reference pressure in the [Reference Pressure] field. For [Domain Motion], choose the domain type. For a centrifugal pump, set the impeller domain to [Rotating] and the other domains to [Stationary]. After selecting [Rotating] for the impeller domain, enter the rotational speed in [Angular velocity], ensuring it is consistent with the impeller's rotation direction (following the right-hand rule). Select the rotational speed unit as [rev min^{-1}]. Finally, click [Apply], as shown in Figure 6.47.

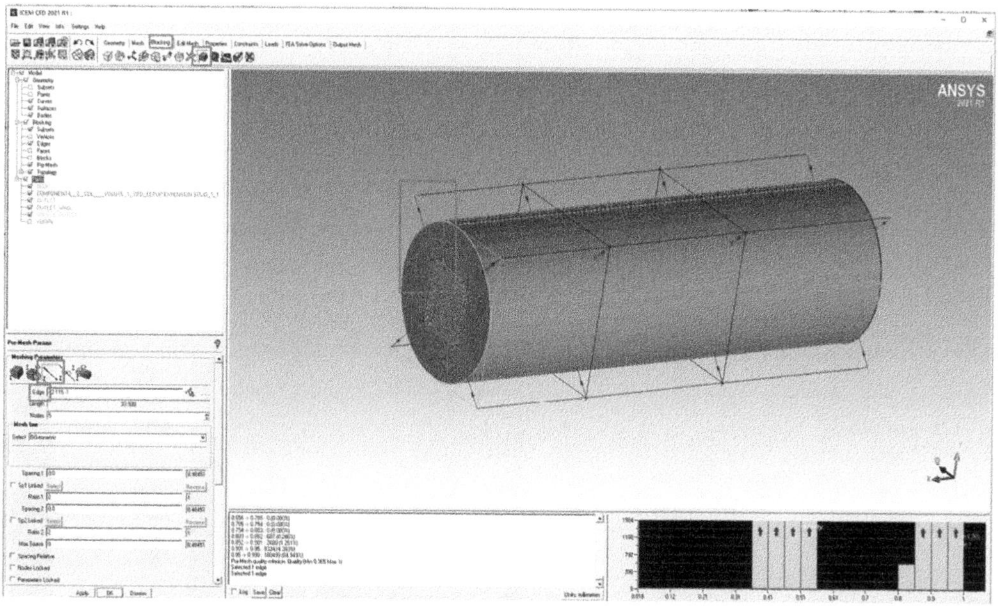

Figure 6.41 Confirming the number of nodes.

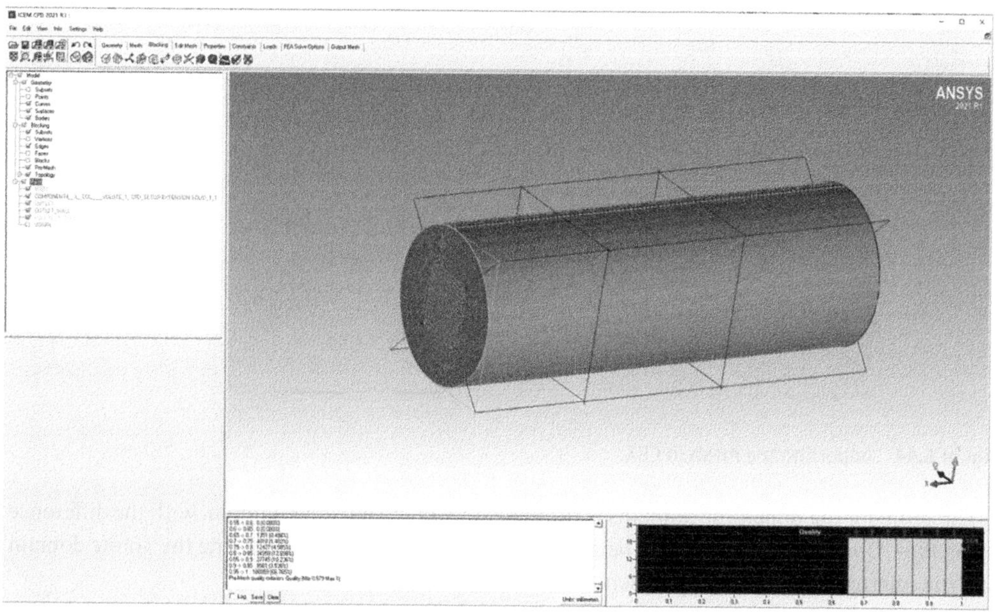

Figure 6.42 Final quality of the outlet pipe mesh.

Figure 6.43 Case creation in CFX.

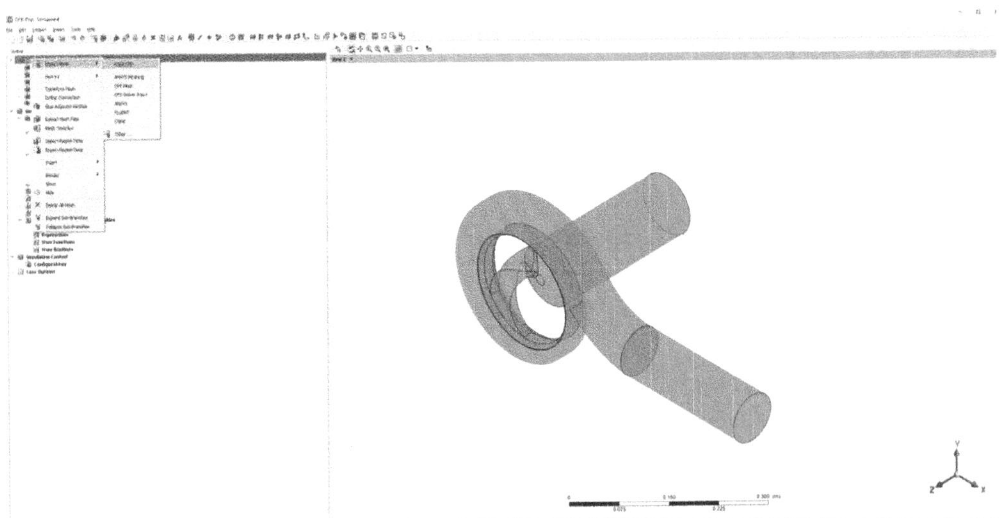

Figure 6.44 Importing the mesh in CFX.

The setup for the remaining domains is similar to that of the impeller domain, with the difference being that [Domain Motion] should be set to [Stationary]. Subsequently, define the volute domain and the inlet/outlet domains in sequence.

6.3.1.3 Turbulence Modeling

After setting up the basic computational domain, proceed to select the turbulence model. Click on the previously defined domain, such as the impeller domain. Click [Impeller], then [Fluid Models]. In the [Fluid Temperature] field, enter the fluid temperature. Under [Turbulence], select the turbulence model to be used for the calculation, as shown in Figure 6.48.

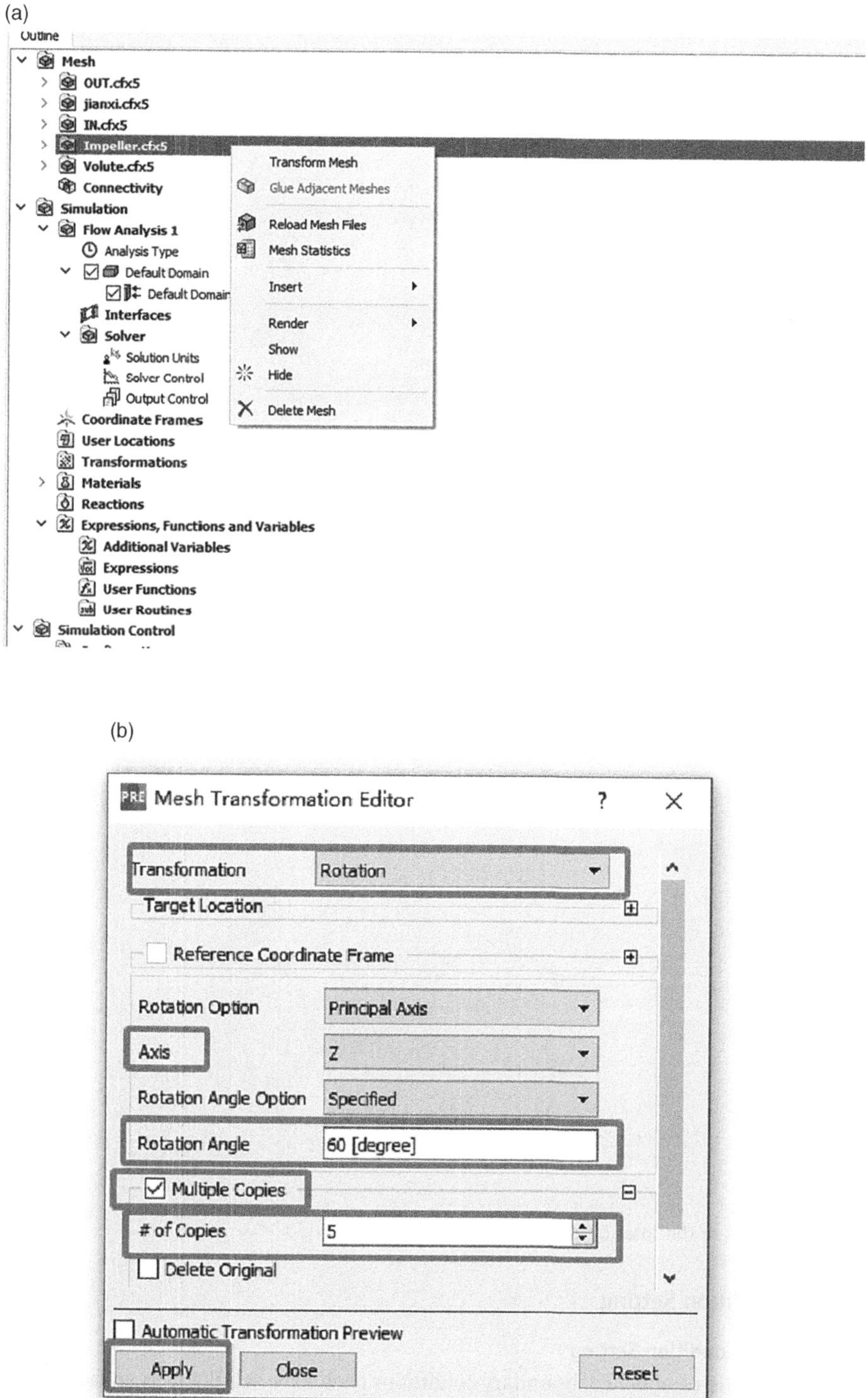

Figure 6.45 Mesh display of all impeller passages: (a) Transform mesh, (b) specific settings.

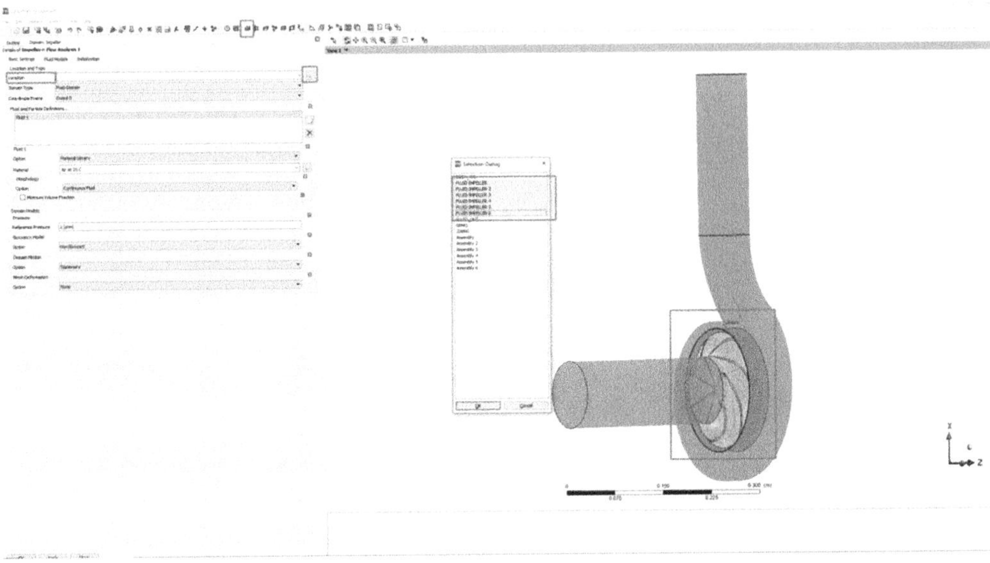

Figure 6.46 Selecting the fluid domain.

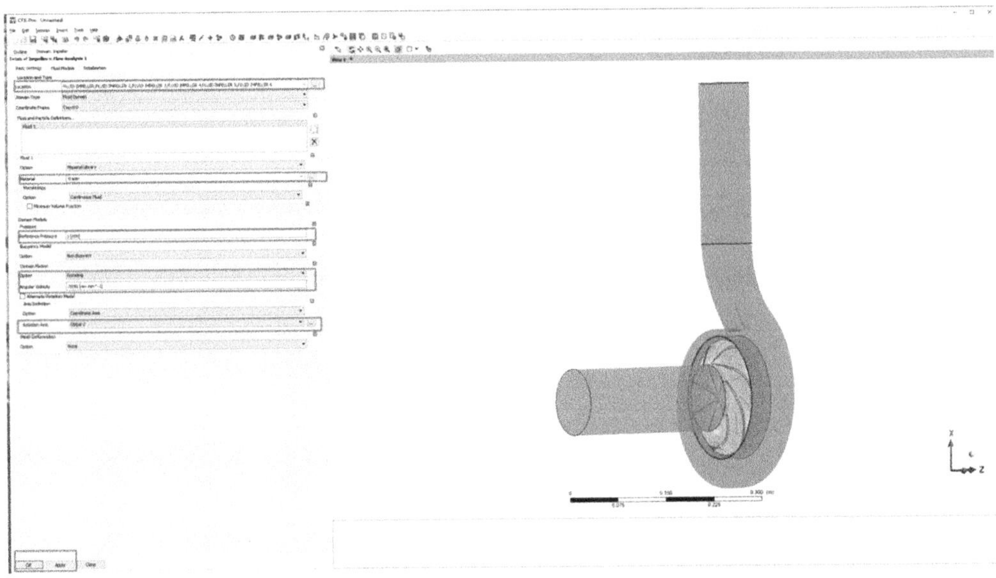

Figure 6.47 Basic settings of the rotation domain.

6.3.2 Boundary Condition Setting

6.3.2.1 Wall Boundary Condition Setting

Once the fluid domains are established, boundary conditions need to be set. Below is an example of setting the boundary conditions for the impeller domain. Click the downward triangle symbol next to [Boundary], select [Impeller], then enter the name of the wall, such as [Blade], in the [Name] field. In [Boundary Type], choose [Wall], and in [Location], select the blade's wall surface. For [Frame Type], select [Rotating], then click [Apply], as shown in Figure 6.49. Next, define the shroud and hub of the impeller in the same manner. The boundary definitions for the remaining domain

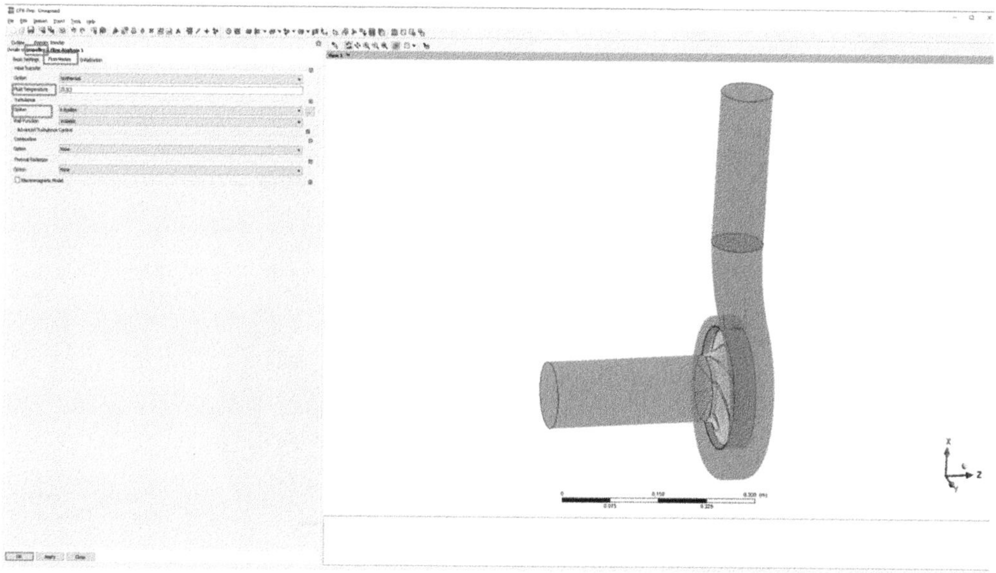

Figure 6.48 Selecting a turbulence model.

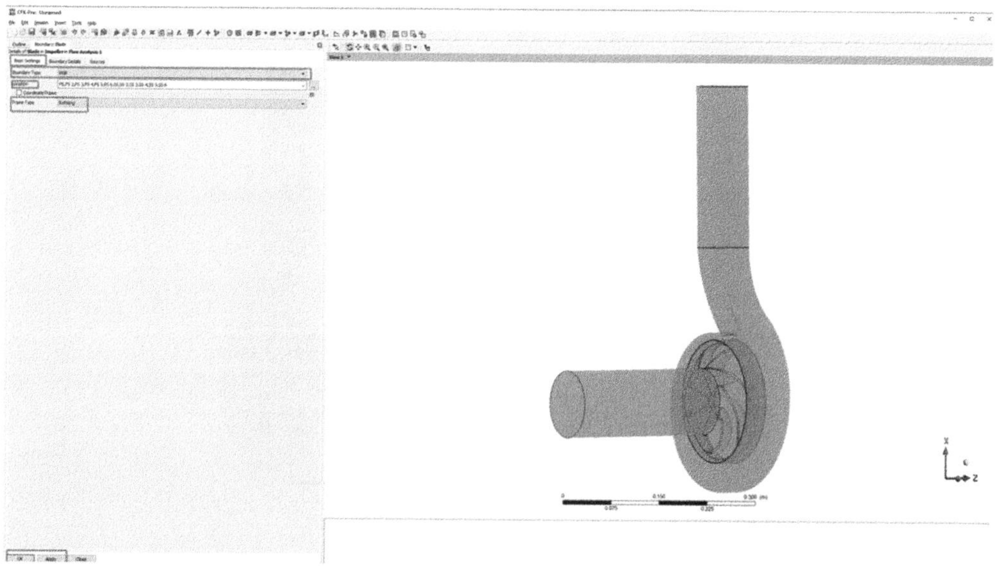

Figure 6.49 Setting the boundary conditions of an impeller.

walls are similar to those for the impeller wall, except that for [Frame Type], since the other domains are stationary, select [Stationary] instead.

6.3.2.2 Inlet and Outlet Boundary Conditions Setting

At the pump inlet, select [Inlet] for [Boundary Type], then click [Boundary Details]. Under [Mass and Momentum], choose the inlet boundary condition, such as [Normal Speed], and input the inlet velocity, as shown in Figure 6.50.

(a)

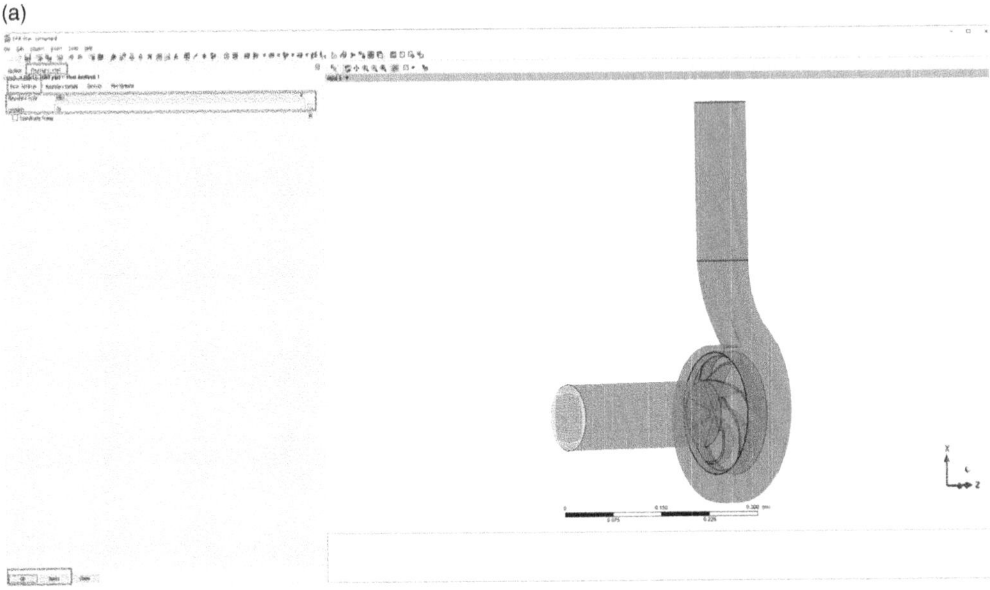

(b)

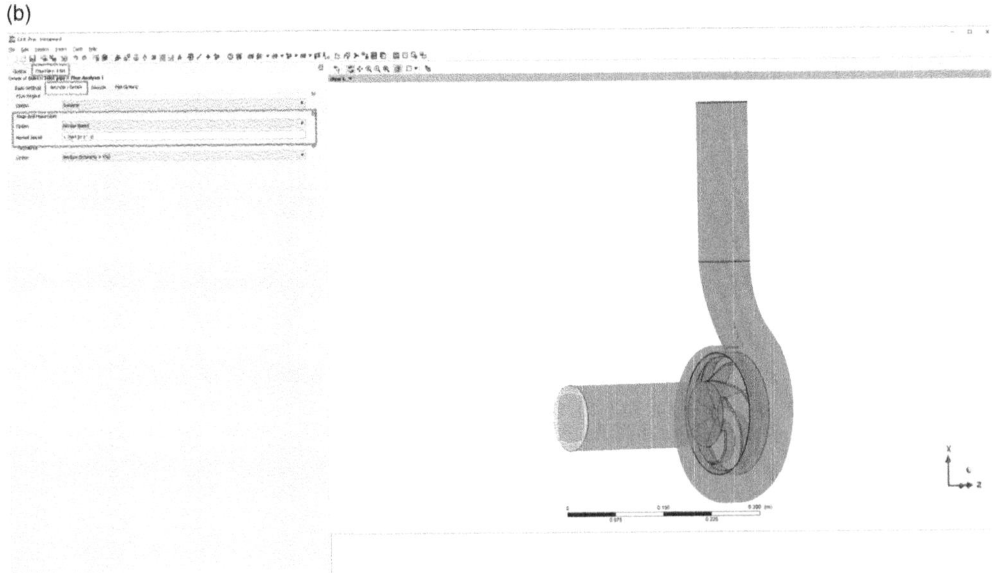

Figure 6.50 Definition of inlet boundary conditions: (a) Boundary type, (b) boundary condition.

At the pump outlet, select [Opening] for [Boundary Type], then click [Boundary Details]. Under [Mass and Momentum], choose the outlet boundary condition, such as [Opening Pres. and Dirn], and input the outlet pressure. Finally, click [Apply], as shown in Figure 6.51.

6.3.3 Interface Setting

6.3.3.1 Dynamic–Static Interface Settings

After defining the wall boundary conditions, the interface definitions need to be set. The interface between the rotating and stationary domains in a centrifugal pump includes the interface

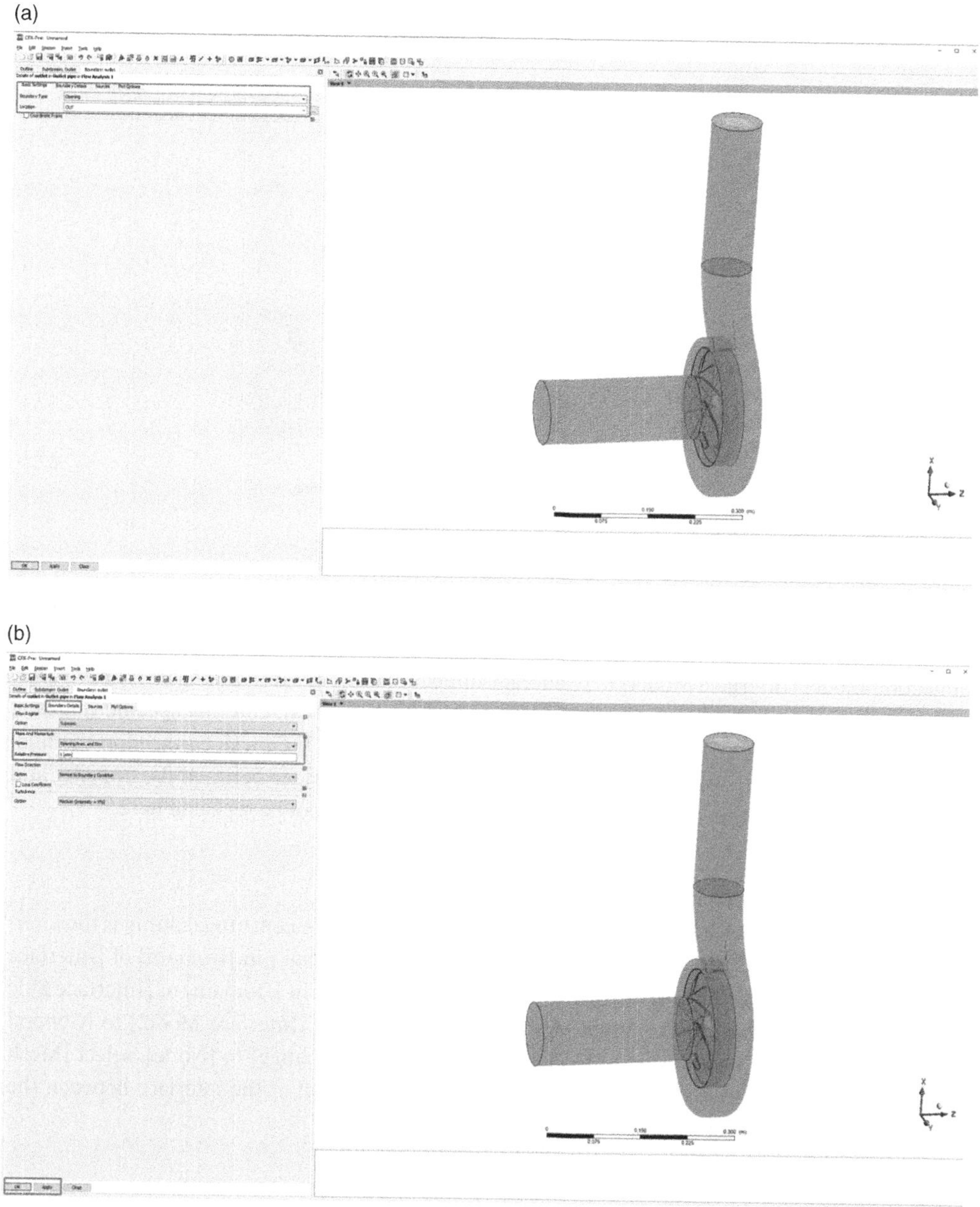

Figure 6.51 Setting the outlet boundary conditions: (a) Boundary type, (b) boundary condition.

between the inlet pipeline and the impeller, as well as the interface between the impeller and the volute [2]. Taking the interface between the inlet pipeline and the impeller as an example, click on  [Domain Interface] in the toolbar. In the pop-up dialog, enter [Inlet_Impeller] and click [OK]. In the first tab, [Basic Settings], select the inlet domain for [Interface Side 1] under [Domain]. In the pop-up dialog, choose the interface between the inlet pipeline and the impeller, which corresponds to the outlet of the inlet pipe domain. For [Interface Side 2] under [Domain], select the impeller domain, and in the [Region List], choose the inlet of the impeller domain. Set the interface

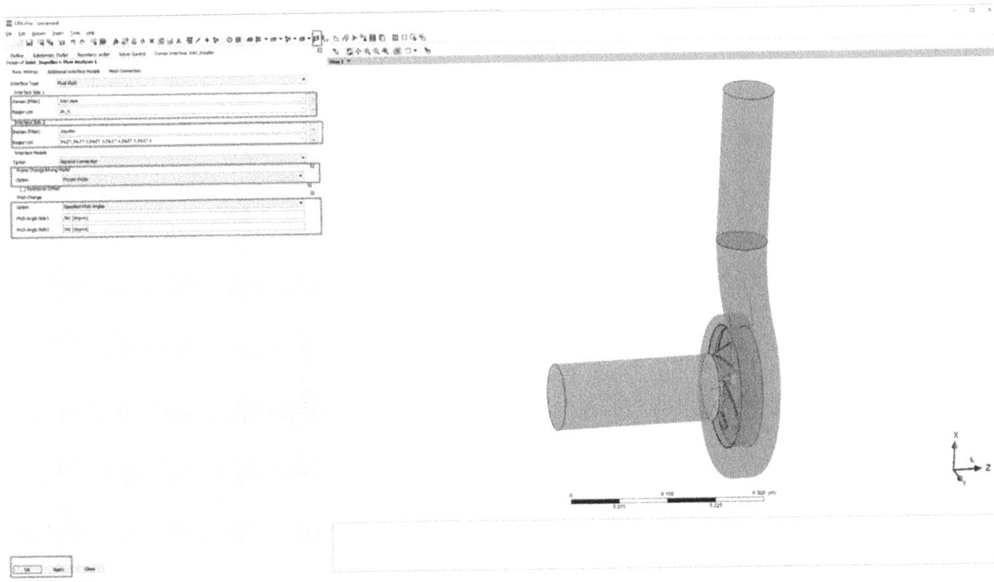

Figure 6.52 Setting the interface between the dynamic and static computing domains.

connection model [Interface Model] to [General Connection], the [Frame Change/Mixing Model] to [Frozen Rotor], and the [Pitch Change] to [Specified Pitch Angles], entering specific values of 360 for [Pitch Angle Side 1] and 360 for [Pitch Angle Side 2]. For the mesh connection method, select [GGI], then click [Apply], as shown in Figure 6.52. The interface between the impeller and the volute is defined in the same way as described above, and will not be repeated here.

6.3.3.2 Static–Static Interface Settings

The interface between the static and static calculation domains of the centrifugal pump is the interface between the volute and the outlet pipe. Select the volute Domain in [Domain] of [Interface Side 1], and select the outlet surface of the volute in [Region List]. In [Domain] of [Interface Side 2], select the inlet surface of the outlet pipe in [Region List]. Set [Interface Model] to [General Connection], [Frame Change/Mixing Model] to [None], [Pitch Change] to [None], select [Mesh Connection] as [GGI], and click [Apply] to complete the definition of the interface between the static and static computing domains, as shown in Figure 6.53.

6.3.4 Solution Setting

After the above Settings are completed, it is necessary to solve the Settings. Double-click [Solver Control] and set it in [Basic Settings]. For stream entry [Advection Scheme], select [High Resolution]. For the numerical item [Turbulence Numerics], select [First Order], and then set the total number of solution steps. In [Convergence Control], enter 2000 to [Max. Iterations] (ensure that the calculation results are stable). The time term [Timescale Control] of the solution parameter is selected [Physical Timescale], and the specific value is input, which is generally set as the reciprocal of the impeller speed. The [Convergence Criteria] is selected as the average value [RMS], and the specific value 1.E-4 is input. Finally, click [Apply] to complete the definition of solver parameters, as shown in Figure 6.54.

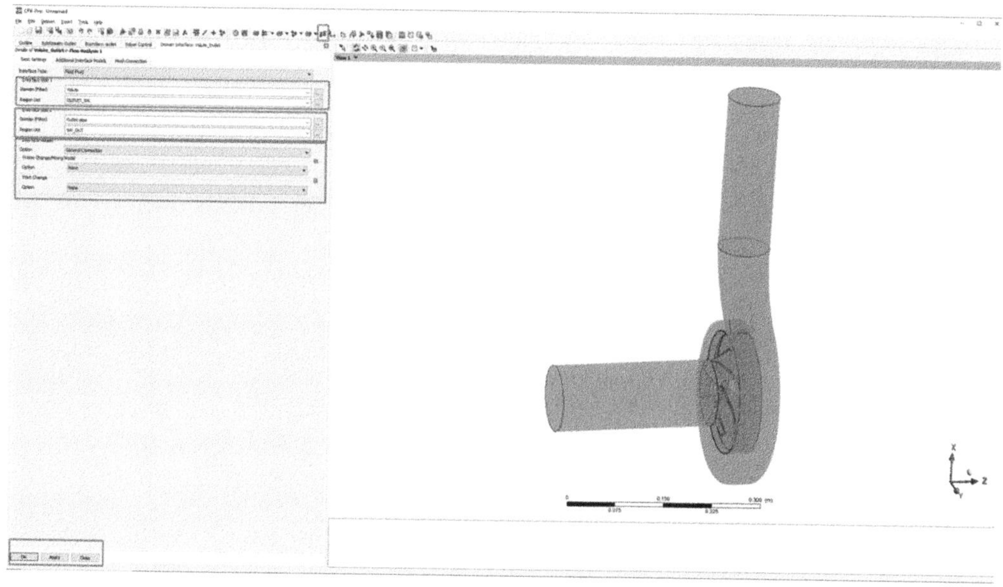

Figure 6.53 Definition of the interface between the static and static computing domains.

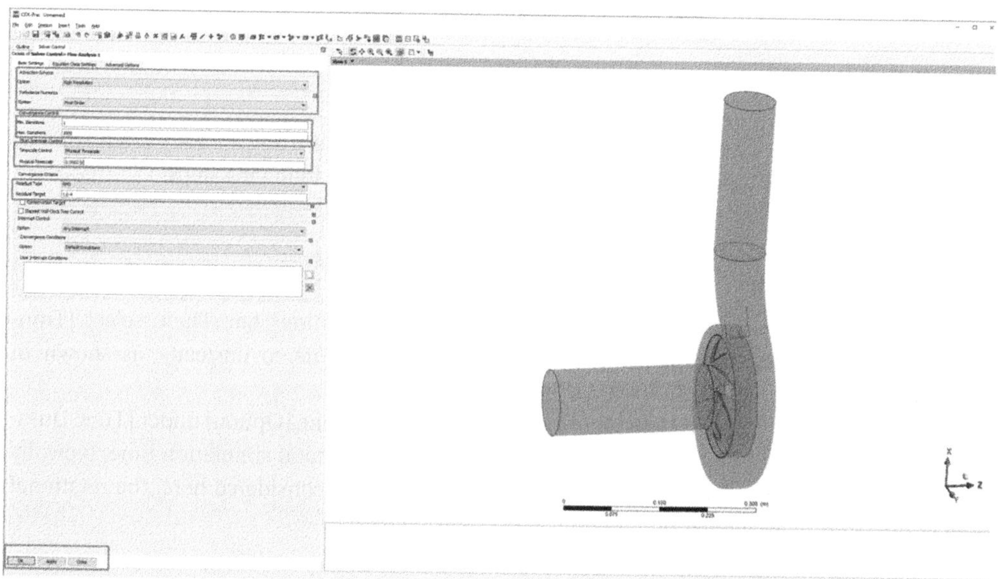

Figure 6.54 Solution settings.

6.3.5 Calculation Setting

Finally, save the current file, click the button on the menu bar, and generate a .def file for calculation. The first button generates a .def file and opens it directly; the last button selects the directory to save the .def, which needs to be opened manually. In the Define Run window, you can enable the Double Precision option when you want to calculate the precision. Run Mode: If the Intel Core is used locally, select Intel MPI Local Parallel. Enter the number of cores here. Select the folder needed to be saved in the Working Directory and click Start Run to perform the calculation (Figure 6.55).

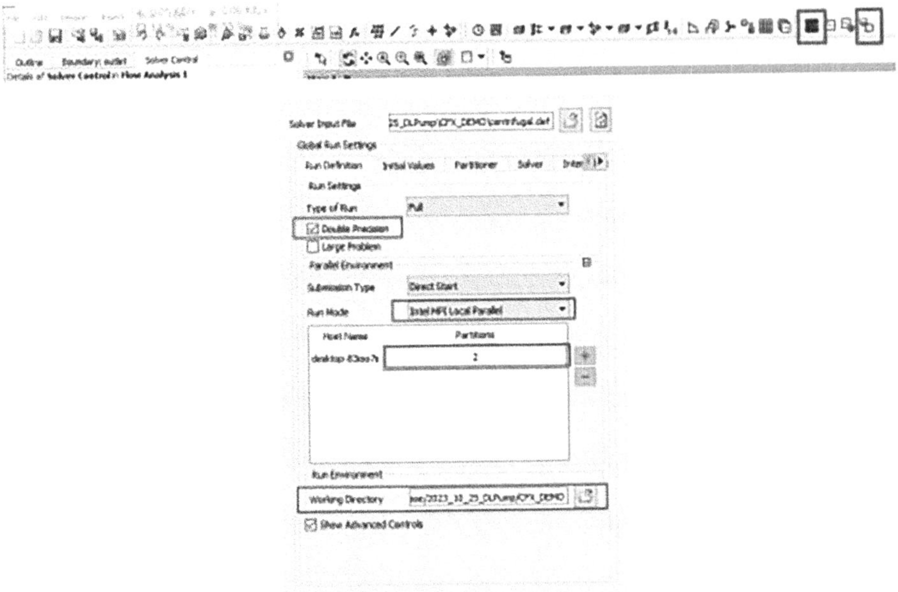

Figure 6.55 Case calculation and its saving in the directory.

6.4 Transient Simulation Setup

Transient simulation is usually obtained by running the flow solver using a steady calculation as initial data. The specific modification procedure is divided into the following steps.

6.4.1 Transient Setting

Click on [Analysis Type] under [Flow Analysis] in the [Simulation] tab. Then, select [Transient] under [Option] to switch the centrifugal pump's flow state to unsteady, as shown in Figure 6.56.

After selecting [Transient], the following dialog box will appear. The [Option] under [Time Duration] is set to [Total Time] by default. Here, one should enter the total simulation time, typically based on the rotational speed. For the centrifugal pump example considered here, the rotational speed is 2950 rpm. The calculation for [Total Time] is as follows:

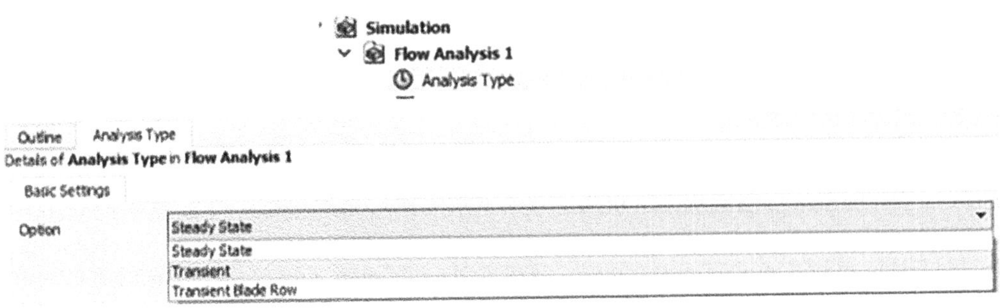

Figure 6.56 Flow state switching of the centrifugal pump.

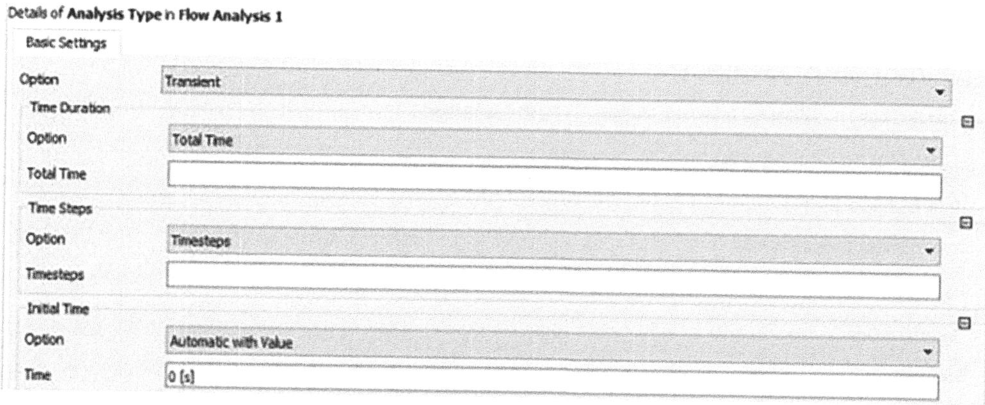

Figure 6.57 Setting the unsteady time for the centrifugal pump.

$2950/60 \times 360°$—How many degrees in a second;
$1/(2950/60 \times 360°)$—The time it takes to turn one degree;
$1/(2950/60 \times 360°) \times 360$—The time it takes to make one turn;

[Total Time] is typically an integer multiple of the time required for one full rotation. To ensure calculation accuracy, it is common to consider multiple rotations and use the results from the last few rotations for analysis. The [Option] under [Time Steps] is set to [Timesteps] by default, and it is usually an integer multiple of the time required for a one-degree rotation. The specific value can be determined based on the actual situation. Then, proceed with the solver settings as outlined in Section 6.3.4 (Figure 6.57).

6.4.2 Interface Model Modification

The interface settings for unsteady data differ significantly from those for steady-state calculations, primarily in the configuration of the rotor–stator interface model. When performing the operation, double-click the rotor–stator interface in the pump. For the interface model, although the [Interface Model] is still set to [General Connection], the [Frame Change/Mixing Model] needs to be changed to [Transient Rotor Stator] to meet the requirements of unsteady calculations. The remaining settings should remain consistent with those used for steady-state calculations. After completing the above configuration, click the [Apply] button to define the rotor–stator interface for unsteady calculations. The detailed steps are shown in Figure 6.58.

6.4.3 Calculation Setting

The same settings apply for unsteady calculations, but an initial flow field is typically required when performing unsteady simulations. The initial flow field in this case is the result of the steady-state calculation. Therefore, the foundation for unsteady calculations is the completion of the steady-state simulation. The specific steps are shown in Figure 6.59. First, check the [Initial Values Specification] box, which will display the [Initial Values 1 Settings] section. Locate the previous steady-state calculation result (.res file), and then click [Start Run] to begin the calculation.

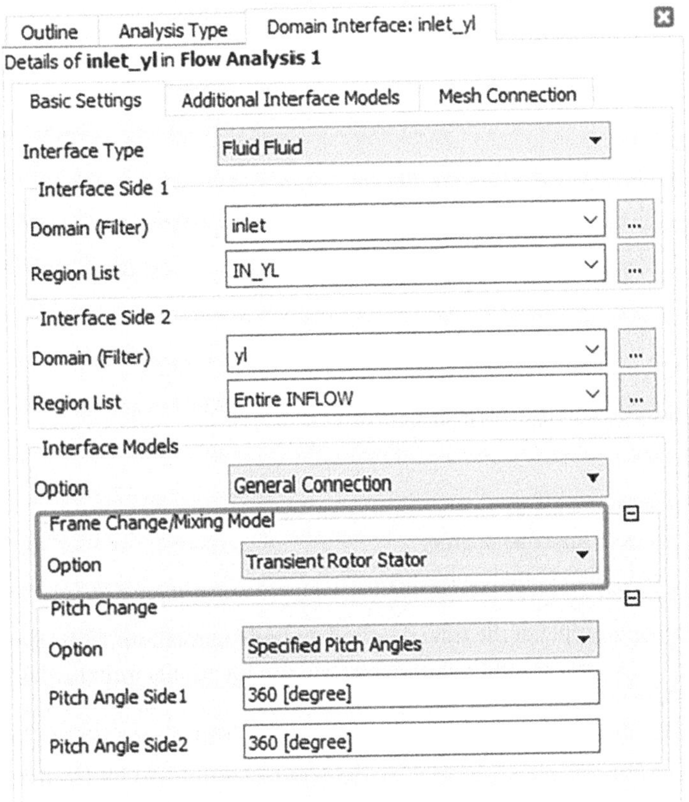

Figure 6.58 Setting the interface between the dynamic and static computing domains for unsteady computing.

6.5 Calculation Results and Postprocessing

Click on the CFD-Post software for postprocessing. Then, select [File] → [Load Results] and choose the .res result file obtained from the CFX simulation. Next, click [Insert] and select [Location]. The pop-up tree menu includes various postprocessing functions such as point, point cloud, line, surface, volume, isosurface, isoclip, vortex core region, and surface group, as shown in Figure 6.60 [3, 4].

Here is an introduction to some commonly used functions. Click [Plane] and name it, for example, [Plane 1], then click [OK]. In the [Domains] section, select the domain for the surface one wants to create, such as the impeller domain and the volute domain. Next, in the [Method] section, choose the location of the surface, such as [XY Plane], and enter 0.04 m in the Z field. Finally, click [Apply], as shown in Figure 6.61.

The creation of a surface is typically done to study the variation of relevant variables within the flow domain, such as pressure changes. Click on [Color] under [Plane 1], and in the [Mode] section, select [Pressure]. Then, in the [Range] section, choose [User Specified] to customize the pressure range. Finally, click [Apply], as shown in Figure 6.62. From the figure, it can be seen that the pressure is lower at the impeller section of the centrifugal pump, and pressure gradually increases along the flow direction, with the highest pressure located at the volute outlet.

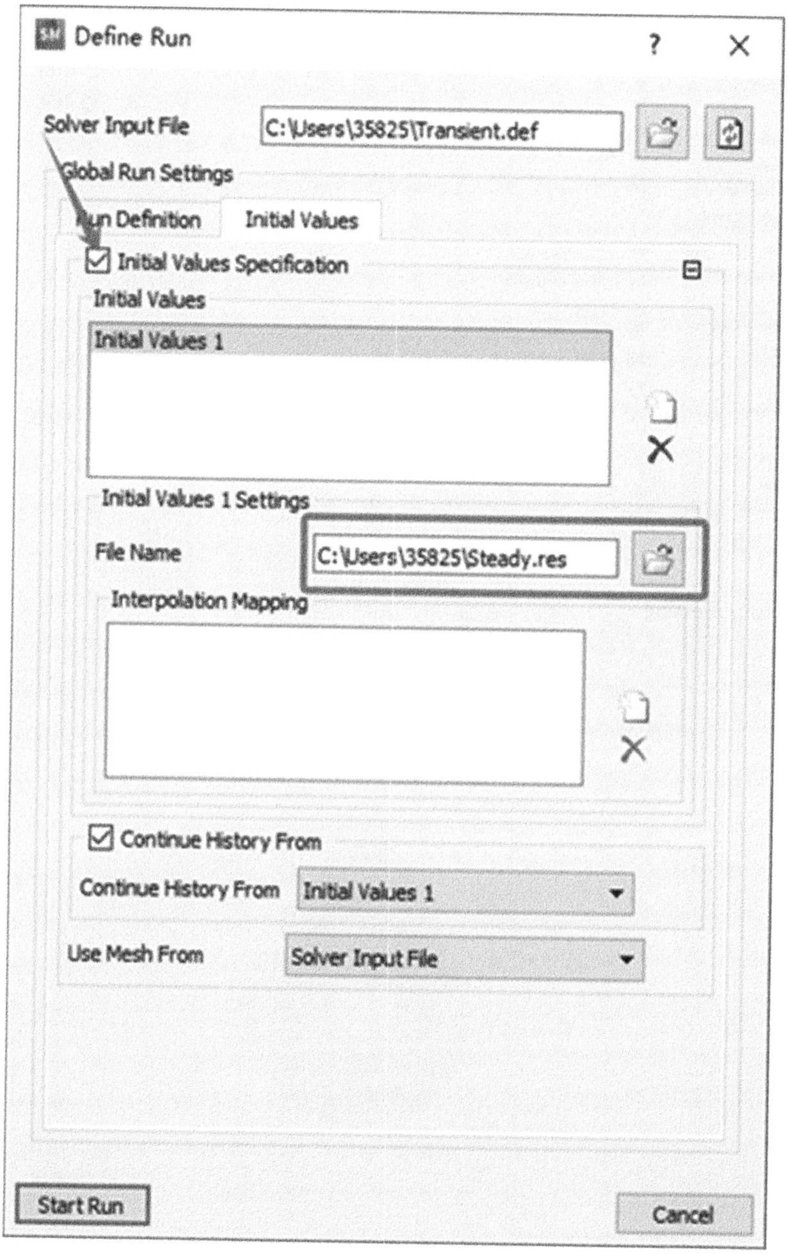

Figure 6.59 Unsteady calculation tool bar.

To create a line on the surface, click [Line] under [Location]. In the [Method] section, select [Two Points]. Click [Point 1] to create the first point, then click [Point 2] to create the second point. Finally, click [Apply] to generate the line, as shown in Figure 6.63.

Click [Insert] and select [Chart] from the options. In the [Name] field, enter a name, such as "Chart 1," and click [OK]. Next, click [Data Series], select the created [Line 1] in the [Location] section, choose [Chart Count] under [X Axis] for the [Variable], and select the desired variable, such as pressure, under the [Y Axis] section. Then, click [Apply] to generate the data chart, and

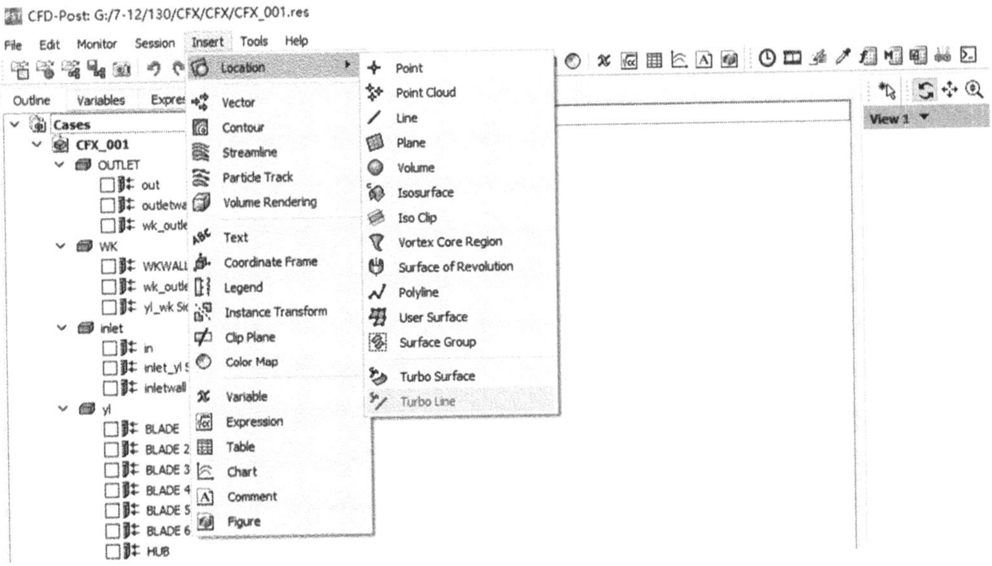

Figure 6.60 Toolbar of postprocessing.

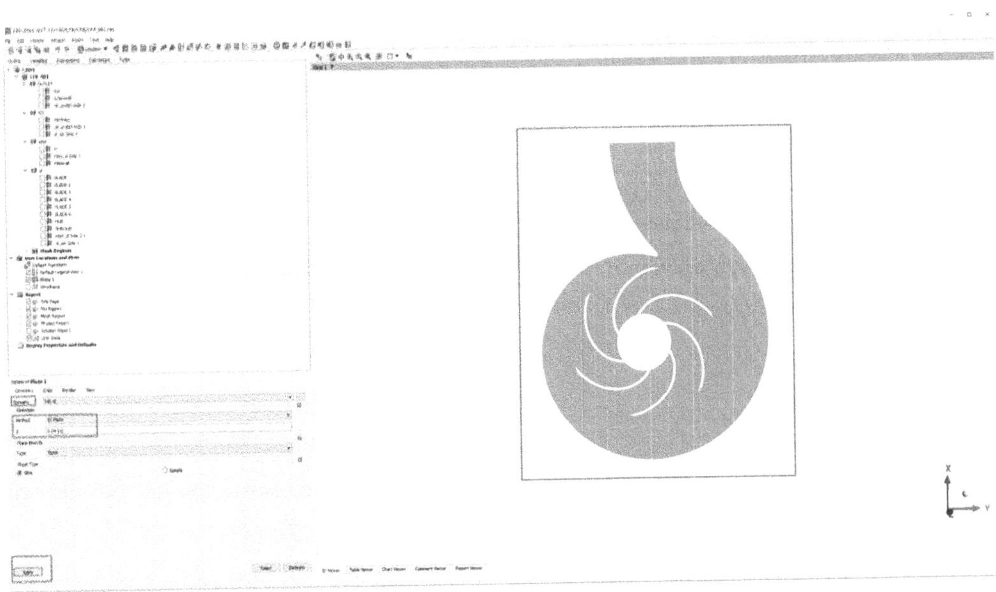

Figure 6.61 Creating a face.

finally, click [Export] to export the data, as shown in Figure 6.64. Clearly, the pressure along the entire measurement line shows a decreasing trend. This corresponds to Figure 6.62, where pressure gradually decreases from the volute compartment to the impeller domain.

Click [Surfaces Group] under [Location] to create a surface group. In the [Domains] section, select the domain for the surface you want to create, such as the impeller domain. Then, in the [Locations] section, select the surface within the impeller domain, such as the blades. Next, click [Color], choose [Pressure] in the [Mode] section, and select [User Specified] in the [Range] section

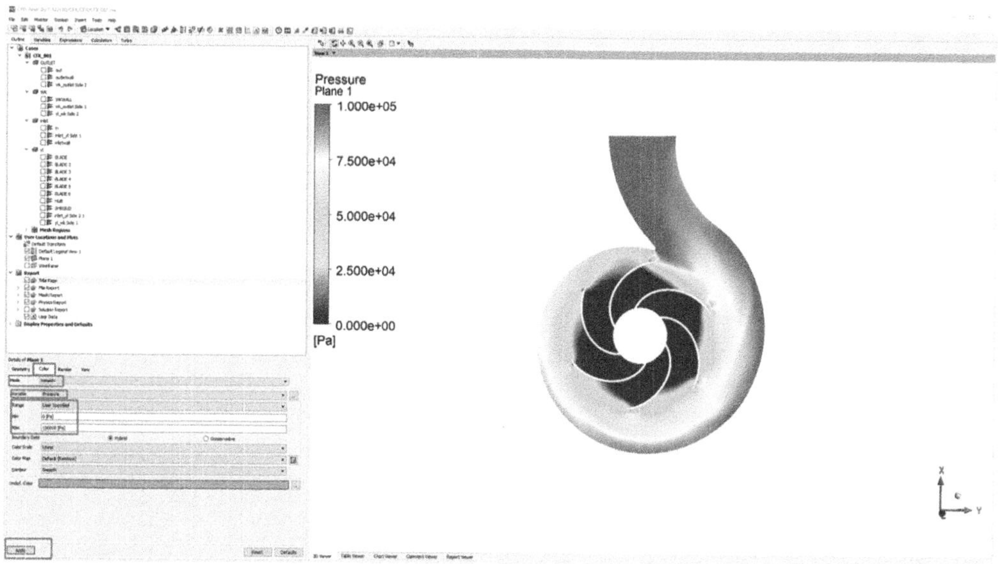

Figure 6.62 Pressure distribution on the surface.

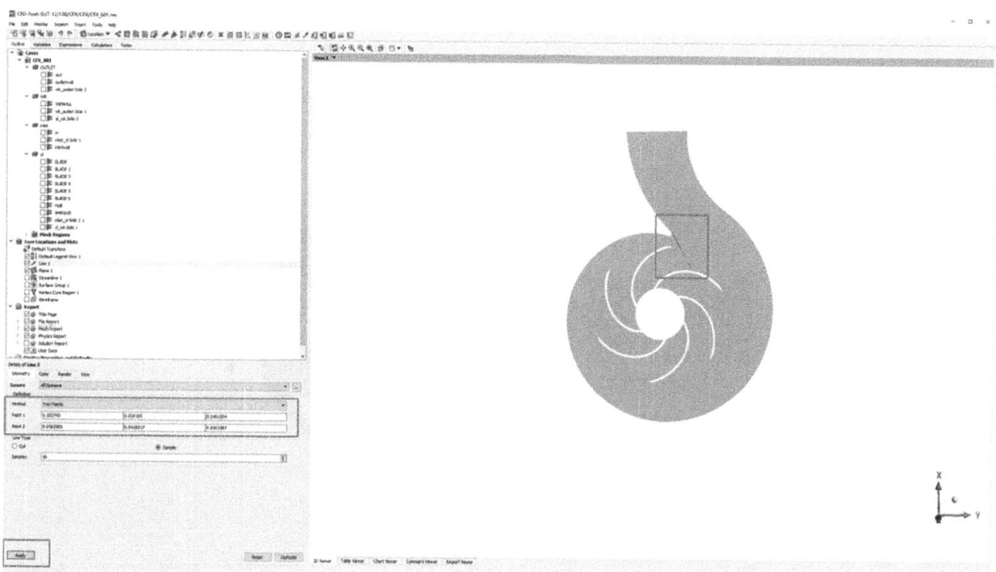

Figure 6.63 Creating a line.

to customize the pressure range. Finally, click [Apply], as shown in Figure 6.65. From the figure, it can be seen that the pressure gradually increases from the impeller inlet to the impeller outlet, which is related to the work done by the rotating impeller.

Click [Vortex Core Region] under [Location] to create the vortex. In the [Domains] section, select the domain where the vortex will be created, such as the impeller domain. In the [Method] section, choose the method for calculating the vortex, such as [Q-Criterion]. In the [Level] section, determine the vortex strength. Finally, click [Apply], as shown in Figure 6.66.

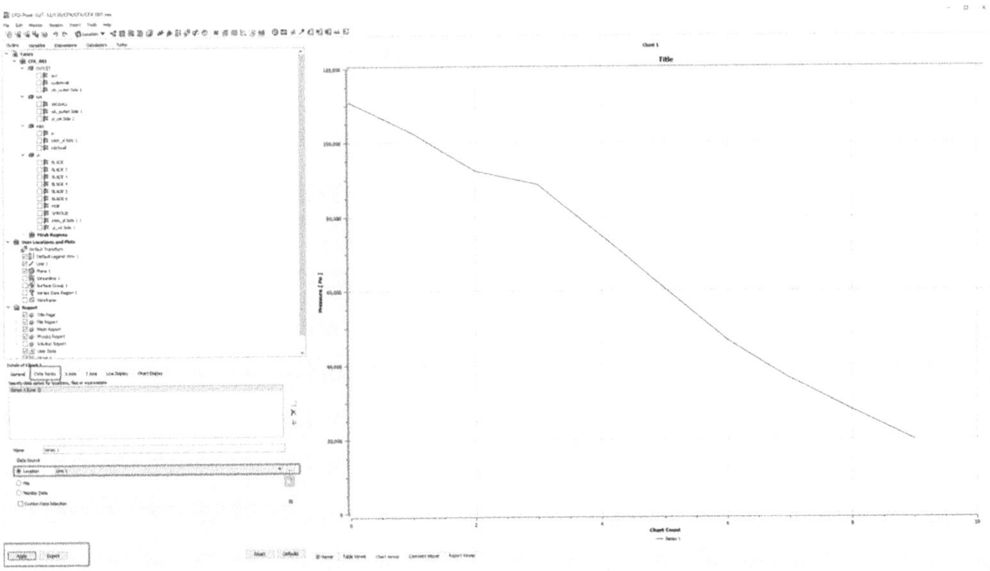

Figure 6.64 Data diagram.

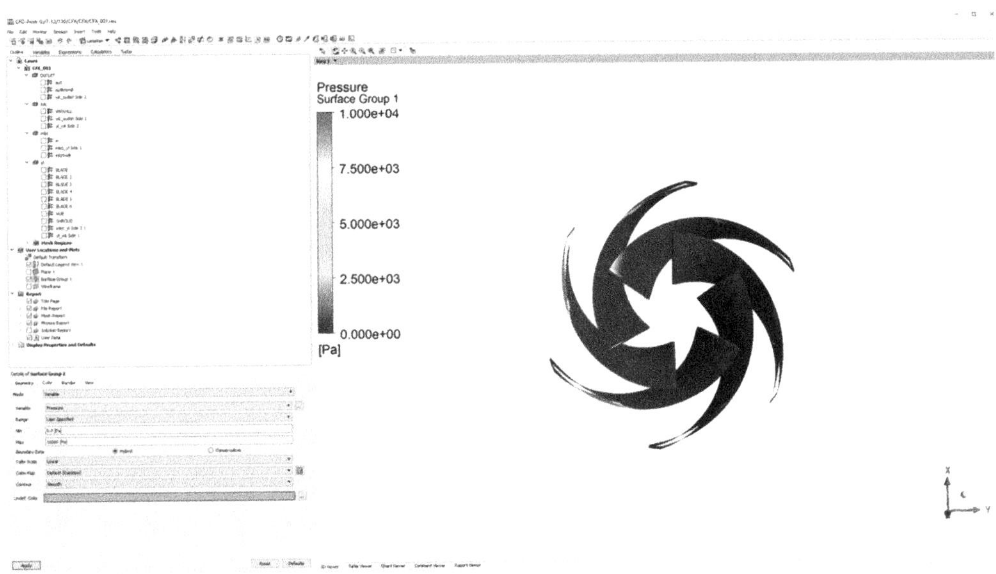

Figure 6.65 Pressure distribution on the impeller surface.

Click [Insert] and select [Streamline] to create streamlines. In the [Type] section, choose [3D Streamline]. In the [Domains] section, select the domain where the streamlines will be created, such as the impeller domain. For [Start From], choose the surface where the streamlines will begin, such as the impeller inlet surface. In the [Sampling] section, select [Vertex], and for [Reduction], choose [Max Number of Points]. Enter the number of streamlines in the [Max Points] field. Finally, click [Apply], as shown in Figure 6.67. From this figure, it can be seen that, along the flow direction, the fluid velocity gradually increases.

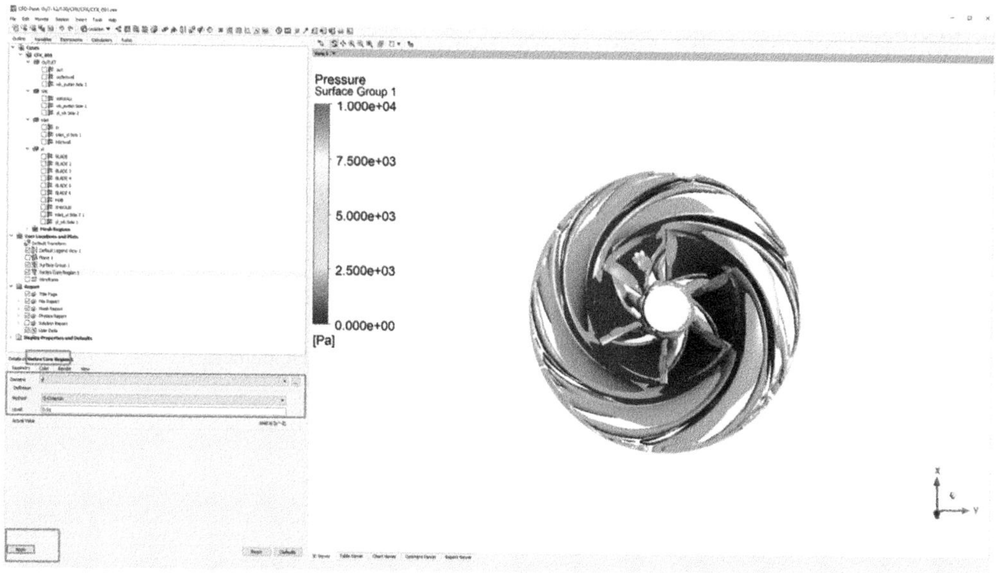

Figure 6.66 Creating the vortex.

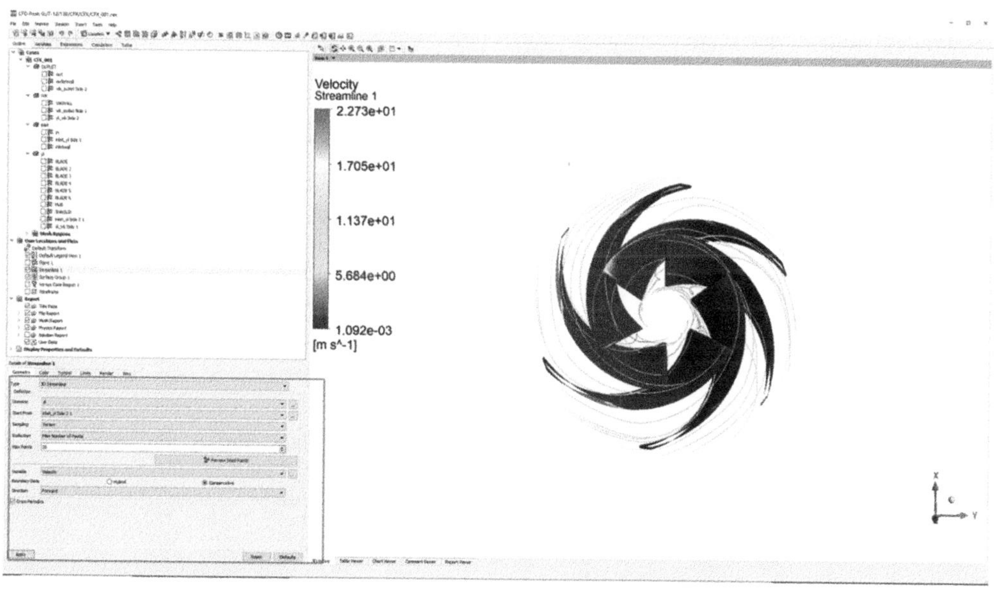

Figure 6.67 Creating the streamline.

References

1 Li, W., Pu, W., Ji, L., Zhang, X., and He, X. (2024). Hydraulic structure optimization of centrifugal pump based on orthogonal test. *Journal of Physics: Conference Series* 2707: 012056.

2 Li, W., Pu, W., Ji, L., Liu, M., Yang, Q., He, X., and Agarwal, R. (2024). Vortex dynamics characteristics in the tip region based on Wray-Agarwal model. *Physics of Fluids* 36: 013337.

3 Ji, L., Pu, W., Li, W., Shi, W., Yang, Y., Xiao, C., Tain, F., Zhou, J., and Agarwal, R. (2025). PIV experimental study on dynamic and static interference flow field of multi-operating centrifugal pump under the influence of impeller wake. *Experimental Thermal and Fluid Science* 161: 111355.

4 Johann, F.G. (2008). *Centrifugal Pumps*. New York: Springer Berlin Heidelberg.

7

Numerical Simulation and Analysis of an Axial Flow Pump

7.1 Computational Domain Modeling

Before performing a numerical simulation of the flow field in an axial flow pump, first the hydraulic domain modeling of an axial flow pump is conducted based on an established pump model. Next, the preprocessing software Space Claim is used to construct the computational domain of the axial flow pump. The computational domain primarily consists of the extended inlet/outlet flow channels and the pump body domain (including the impeller with inlet cone section as well as the guide vanes with outlet cone section).

7.1.1 Pump Volute Domain

The pump volute domain is obtained by extracting the internal volume from the established pump model. Use the Open SpaceClaim software and click [File] in the navigation panel. From the dropdown menu, select [Open] and choose the preexisting axial-flow pump model in the file explorer. After successful import, click [Prepare] in the Quick Access Toolbar. In the ribbon menu, select [Volume Extract] to enter the internal volume extraction interface. Click [Select Faces], then choose the cross-sectional faces at both ends of the pump model. Once selected, click [Select Seed Face] and pick any internal surface of the pump. Finally, click [Complete] to finalize the modeling of the pump volute domain as shown in Figure 7.1.

7.1.2 Inlet and Outlet Extension Domains

To ensure accuracy in the numerical simulations, it is a common practice to extend the pump's inlet and outlet sections. The extension length is typically 10 times the pipe diameter. For the inlet extension domain, apply the Extrude command to the inlet cross-section of the impeller and the inlet taper pipe computational domain. Similarly, construct the outlet extension domain using the same method. The modeled inlet and outlet extension domains are shown in Figure 7.2.

Finally, the complete model of the computational domain of the axial-flow pump is obtained as shown in Figure 7.3.

7.2 Structured Mesh Generation

A structured grid inside the pump model is typically generated using the ICEM CFD software, which features an intuitive user interface with well-arranged tools and functions in the main window. The core sections include the menu bar, toolbar, graphics display area, command console,

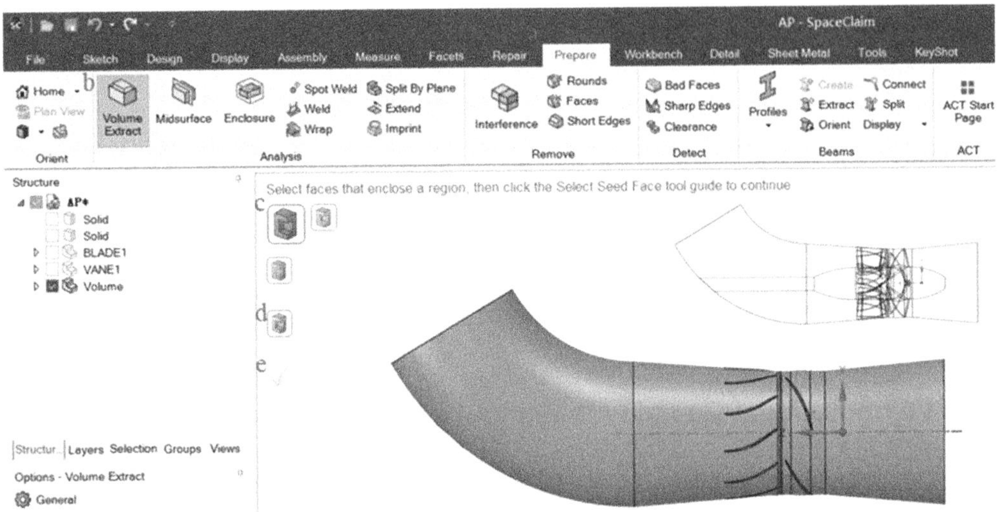

Figure 7.1 Pump volute domain model.

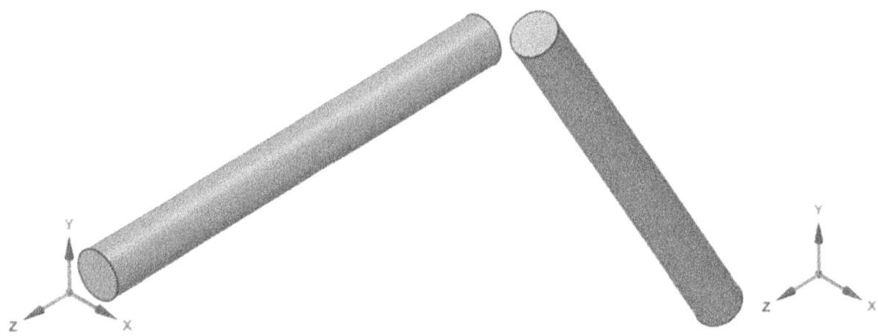

Figure 7.2 Models of the inlet and outlet extended flow domains.

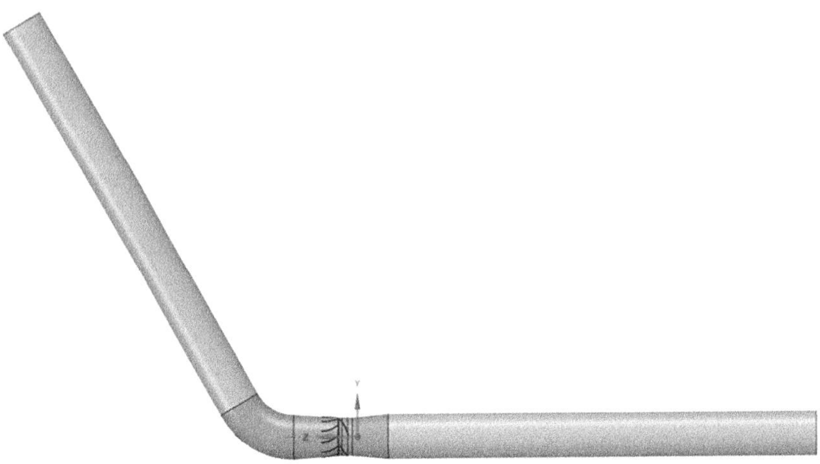

Figure 7.3 Complete computational domain of an axial-flow pump model.

and project tree. Through this interface, users can easily perform operations such as model import, mesh generation, boundary condition setup, and the file export.

Due to the complexity of the axial-flow pump computational domain, generating a structured grid for the entire model proves challenging. Therefore, a preprocessing software is used to divide the model into multiple sub-regions, which include the inlet extension region, inlet cone region, impeller region, guide vane region, outlet W-bend region, and the outlet extension region. A structured mesh is then generated for each sub-region sequentially, finally producing a complete structured mesh for the model axial-flow pump [1, 2].

7.2.1 Mesh Generation for the Impeller Flow Domain

Taking the impeller region as an example, this section provides details of the structured mesh generation process. Given the axisymmetric geometric features of the impeller, the preprocessing software splits the model into a single-passage geometric model based on the number of blades to simplify the mesh generation process as shown in Figure 7.4. A mesh is then generated only for this single passage, and the mesh for the full impeller region is obtained by utilizing the ICEM CFD's rotational copy function.

Launch ICEM CFD, click [File] → [Change Working Dir…] on the menu bar, and customize the working directory in the pop-up dialog box as shown in Figure 7.5. Next, click [File], select [Import

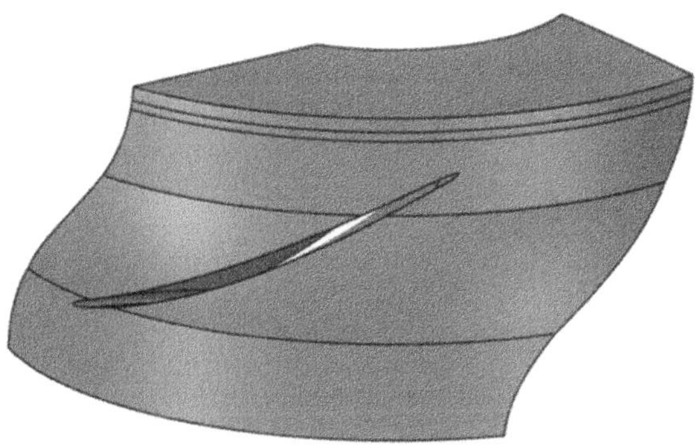

Figure 7.4 Single-passage geometric model of the impeller region.

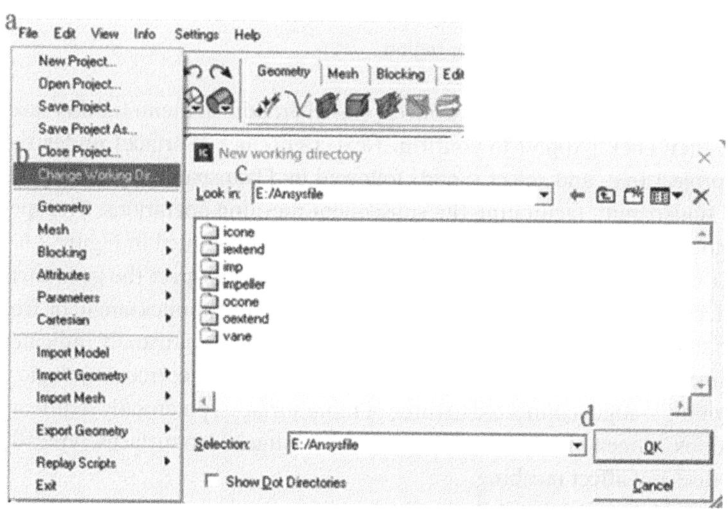

Figure 7.5 Customization of the working directory setup.

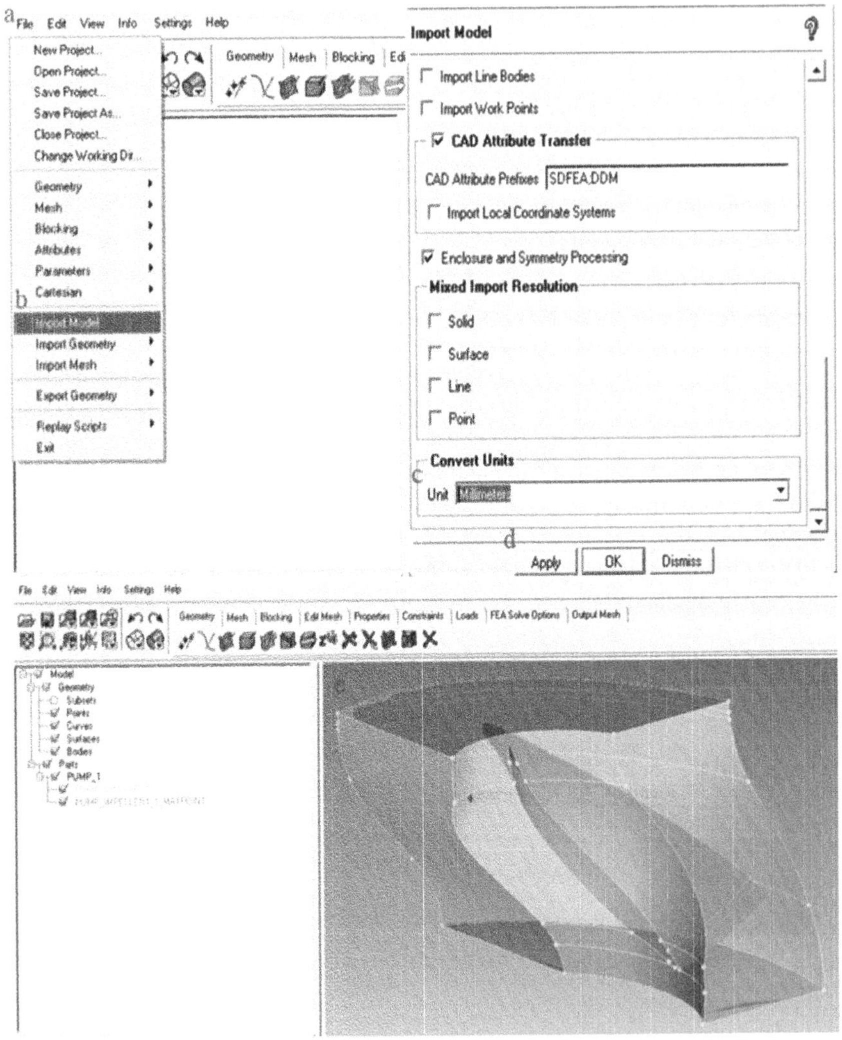

Figure 7.6 Importing the geometric model of the impeller region.

Model], choose the geometric model's unit from the [Convert Units] dropdown menu (in this case, millimeters [Millimeters]), then click [Apply] to confirm. Next, right-click [Surface] under the [Geometry] option in the project tree, and select [Solid] followed by [Transparent] to make the imported geometric model transparent, facilitating the subsequent meshing operations. The specific process for importing the geometric model of the impeller region is illustrated in Figure 7.6.

After importing the CAD model, the Build Topology function is used to inspect the geometric integrity through color-coded indicators in the graphics display: e.g. Yellow implies single or free edges (adjacent to only one surface); in 2D, it indicates boundary edges. In 3D, it usually indicates gaps requiring repair. Red implies double-sided edges shared by two surfaces—correct topological connections. Blue implies multiple edges connected to three or more surfaces, potentially requiring adjustment for the "T-junctions." Green implies standalone curves unattached to surfaces, typically an auxiliary geometry that does not affect meshing.

Before meshing, Parts must be defined to classify boundary conditions and interfaces (Figure 7.7): Right-click [Parts] in the project tree → Select [Create Part]. Enter the Part name. Click [Entities]

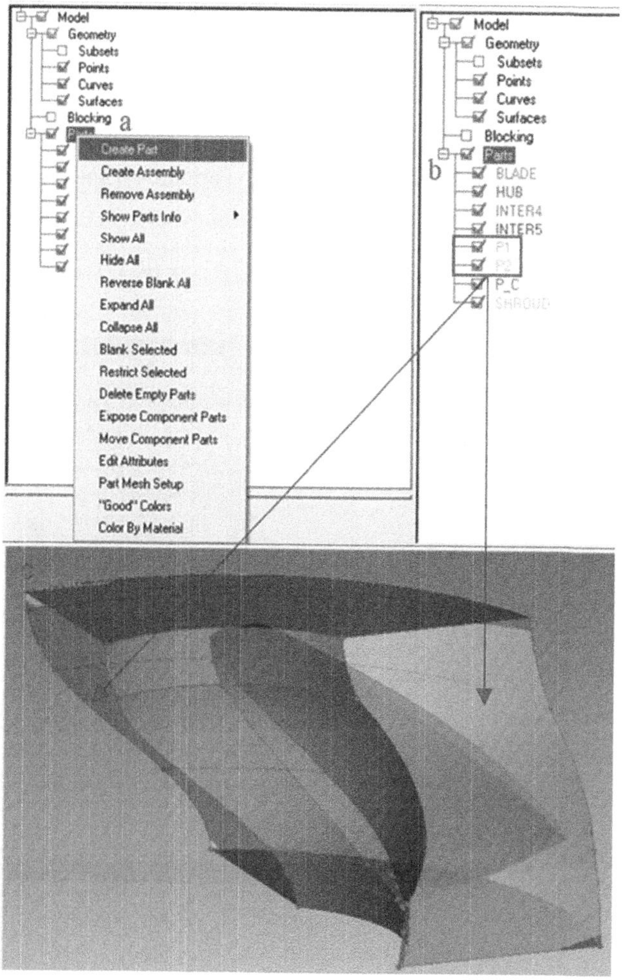

Figure 7.7 Creating a part for boundary identification.

under Create Part by Selection, pick the corresponding geometric entities in the display area, and confirm with [Apply]. Periodically symmetric surfaces must be explicitly named for subsequent rotational mesh replication.

In ICEM CFD, the creation of hexahedral structured meshes is typically achieved through Block-based topological decomposition. The software utilizes the fundamental topological principle that only tetrahedral or hexahedral elements possess geometric mapping capabilities. By manipulating virtual Blocks, topological structures are constructed, and their data is mapped onto the actual physical geometry during computational mesh generation, thereby realizing body-fitted structured meshing. Vertices: Corners of 2D blocks or vertices of 3D blocks. Edges: Lines connecting two vertices. Faces: Regions bounded by four edges (2D blocks) or six faces (3D blocks).

First, click the [Blocking] tab, then click the [Create Block] button, which will pop up the block creation dialog box. In the input field under [Part], enter the name of the block to be created. Under [Initialize Blocks], select [3D Bounding Box] from the [Type] dropdown menu to create a 3D block. Next, click the button on the right side of [Entities], then use the left mouse button to drag and select the imported impeller region geometry. After selection, confirm by clicking the middle mouse button—the software will automatically generate an initial block that fully encloses the impeller region. Finally, click the [Apply] button in the dialog box to complete the block creation. The specific process is illustrated in Figure 7.8.

Next, define the mesh model periodicity. Click the [Global Mesh Setup] button under the [Mesh] tab, then in the pop-up dialog box's [Global Mesh Parameters], select [Set up Periodicity] and check

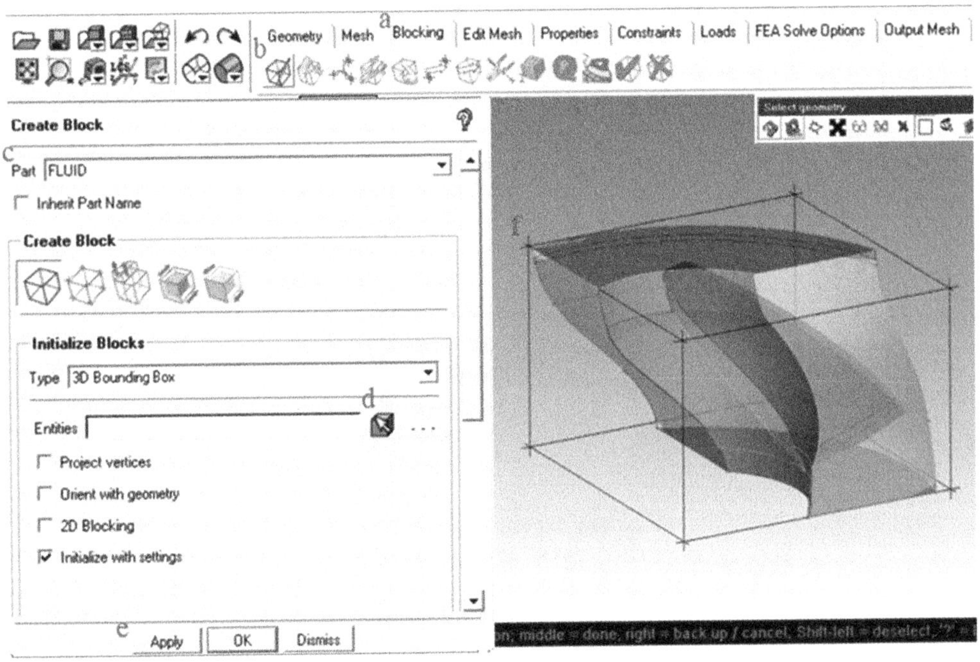

Figure 7.8 Generating a 3D bounding block.

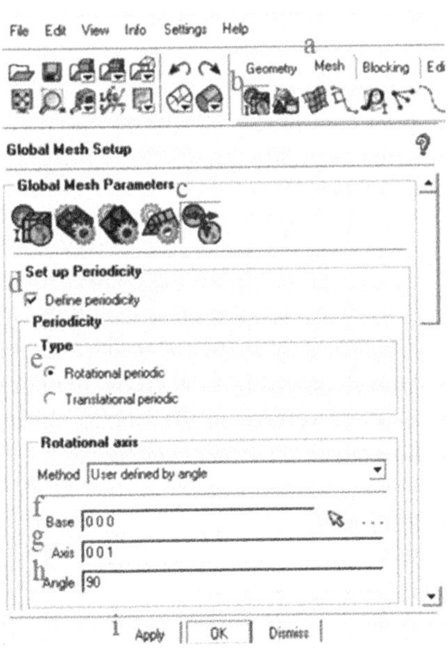

Figure 7.9 Configuring rotational periodicity.

[Define periodicity]. Under the [Type] section, choose [Rotational periodic], and in the [Method] dropdown menu, select [User defined by angle]. Click the arrow on the right side of [Base] to select the rotation center of the geometric model. In the [Axis] field, define the rotation axis (e.g., enter 0 0 1 for the Z-axis). In the [Angle] field, input the periodic angle (i.e., the corresponding sector angle of a single flow passage in the impeller region), as illustrated in Figure 7.9. Next, click the [Edit Block] button under the [Blocking] tab. In the pop-up dialog box, select [Periodic Vertices], then choose [Create] under the [Method] section. Click the arrow beside [Vertices], and sequentially select the vertex pairs that need to be matched for periodicity on the displayed Block in the graphics area. Once completed, click [Apply] to confirm and exit. Check the [Vertices] option under the [Blocking] branch in the project tree, then right-click [Vertices] and select [Periodic] in the prompt window.

This allows viewing the established periodic vertex pairs, as shown in Figure 7.10.

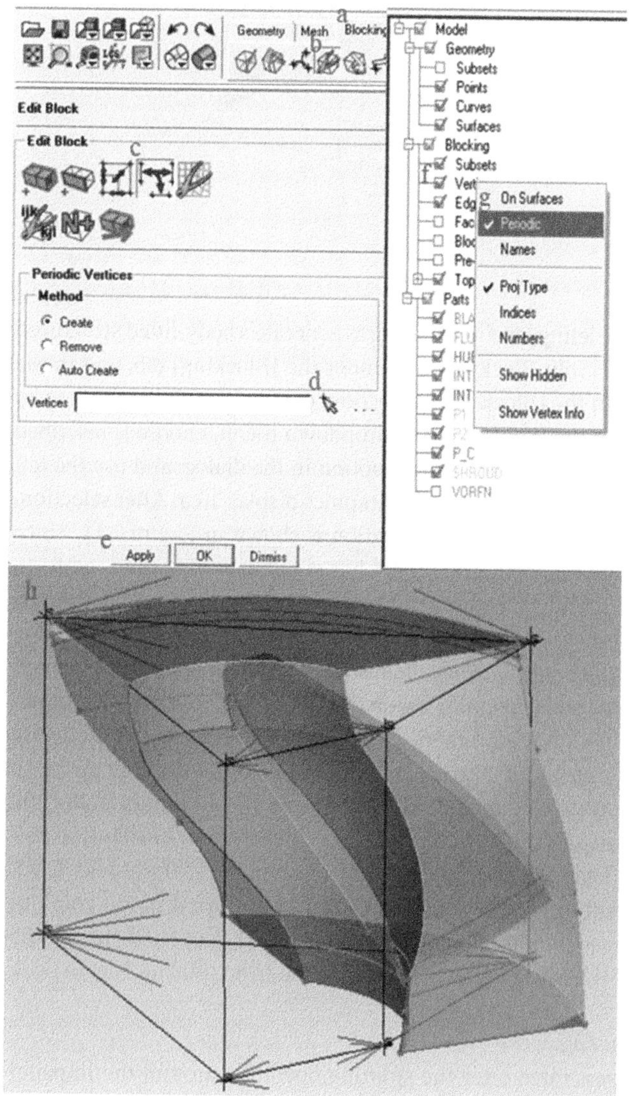

Figure 7.10 Periodic vertex pairing and validation.

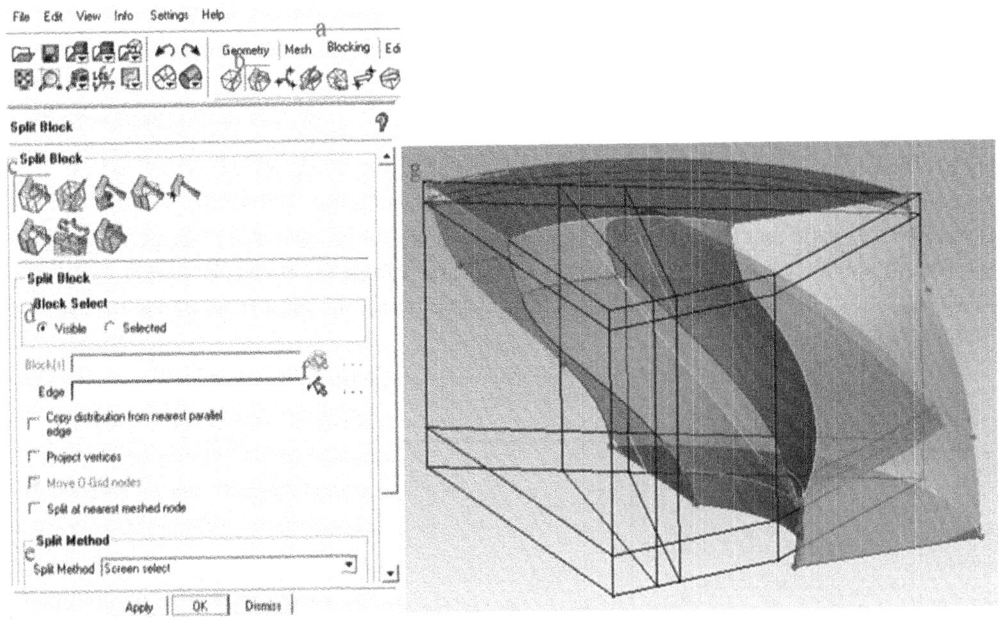

Figure 7.11 Splitting the block.

Process the initial Block by splitting, deleting, and associating to generate a body-fitted structured mesh for the impeller region. Click the [Split Block] button under the [Blocking] tab, and in the pop-up dialog box, select [Split Block]. In the [Block Select] section, ensure [Visible] is checked by default to only split visible Blocks. From the [Split Method] dropdown menu, choose [Prescribed point]. Then, click the button on the right side of the [Edge] option in the dialog, and use the left mouse button to select the edge of the Block to be split in the graphics display area. After selection, click the middle mouse button to confirm. The specific operation is shown in Figure 7.11. Since the impeller region contains blades and the geometric model has significant distortion, perform multiple splitting operations to ensure the Block better conforms to the geometric model, thereby obtaining higher-quality structured meshes.

In ICEM CFD, the structured mesh generated is hexahedral, which has relatively poor mapping capabilities for circular (arc shaped) and triangular geometries. This may result in unsatisfactory mesh quality for subsequent simulation requirements. To address these issues, the software introduces "Ogrid" and "Y-Block". Therefore, an "Ogrid" is employed near the impeller airfoil to accurately capture the geometric features of the model. Click the [Split Block] button under the [Blocking] tab. In the pop-up dialog box, select [Ogrid Block], then check [About block(s)] to perform an external "Ogrid" split. Set the layer height of the "Ogrid" mesh in the [Offset] field. Click [Select Block(s)], use the left mouse button to select the Block around the airfoil in the graphics display area, and click the middle mouse button to confirm. Then, click [Select Face(s)], use the left mouse button to select the faces of the Block at the airfoil that need to be split, and again click the middle mouse button to confirm.

The operation is illustrated in Figure 7.12.

Next, delete the redundant Blocks generated after the splitting operation around the impeller airfoil. Click the [Delete Block] option under the [Blocking] tab. In the pop-up dialog box, click the button next to [Entities], then use the left mouse button to select the Blocks to be deleted in the

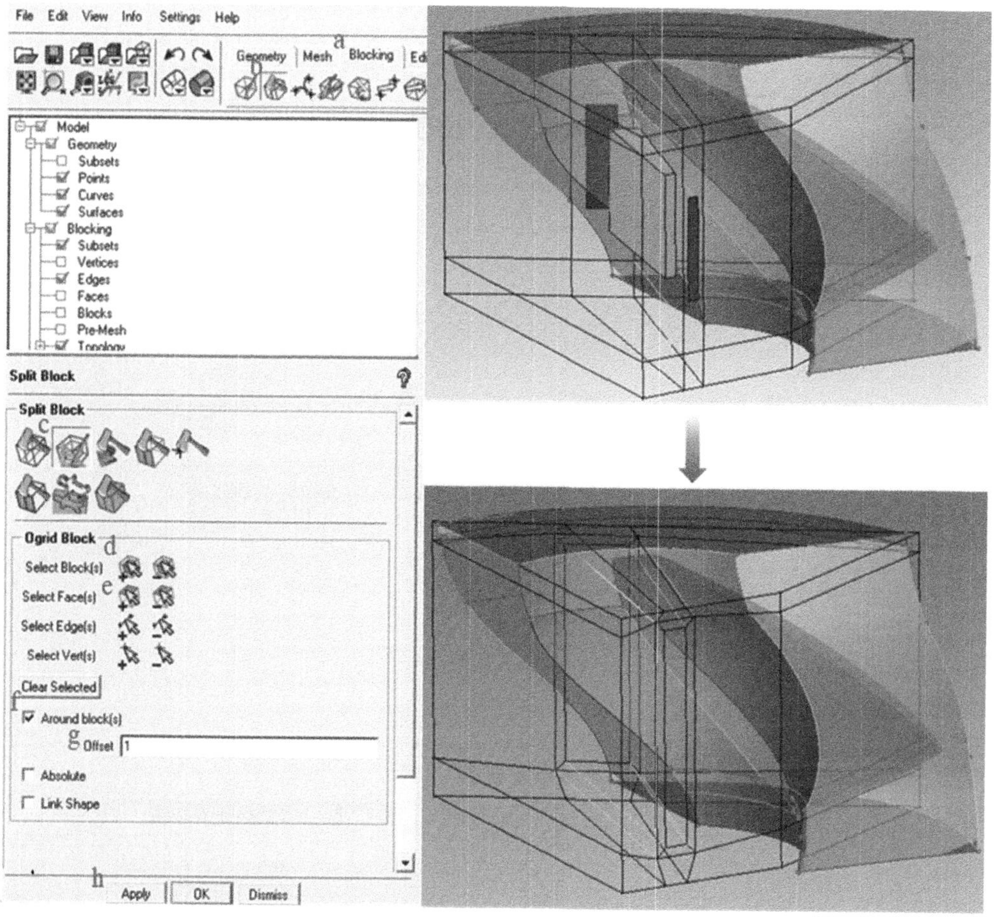

Figure 7.12 External "Ogrid" splitting.

graphics display area. After confirming the selection, click the middle mouse button to delete the selected Blocks. The process is illustrated in Figure 7.13.

Association is a frequent task in Block operations, primarily aimed at mapping data from the virtual Block to the actual physical geometry. Association is also one of the most critical tasks in ICEM CFD block meshing—poor mesh quality or meshing failures are often related to incorrect associations. Points and edges on a Block appear in three different colors: Green: Not associated, Black: Associated, Blue: Internal geometric elements generated after Block splitting.

Click the [Associate] button under the [Blocking] tab. In the pop-up dialog box, under the [Edit Associations] section, select [Associate Vertex]. Then, in the [Entity] section, choose [Point]. Click the arrow next to [Vertex], then use the left mouse button to select a vertex on the Block. The software will automatically switch to [Point] selection mode—use the left mouse button to pick a corresponding vertex on the geometric model. The software will then associate the two points. The procedure is shown in Figure 7.14.

Next, proceed with the edge association operation of the mesh model. Under the [Edit Associations] section, select [Associate Edge to Curve] and check [Project vertices] to map the associated edges on the Block to the geometric model. Click the arrow next to [Edge(s)], select the desired edges on the Block in the graphics display area, and confirm by clicking the middle mouse button. The software will automatically switch to [Curve(s)] selection mode—select the corresponding

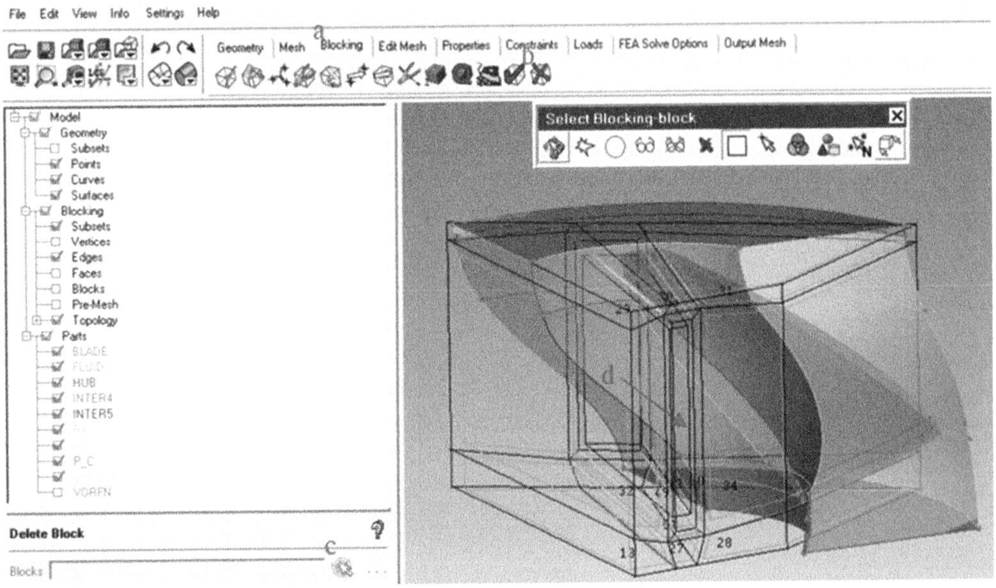

Figure 7.13 Block deletion.

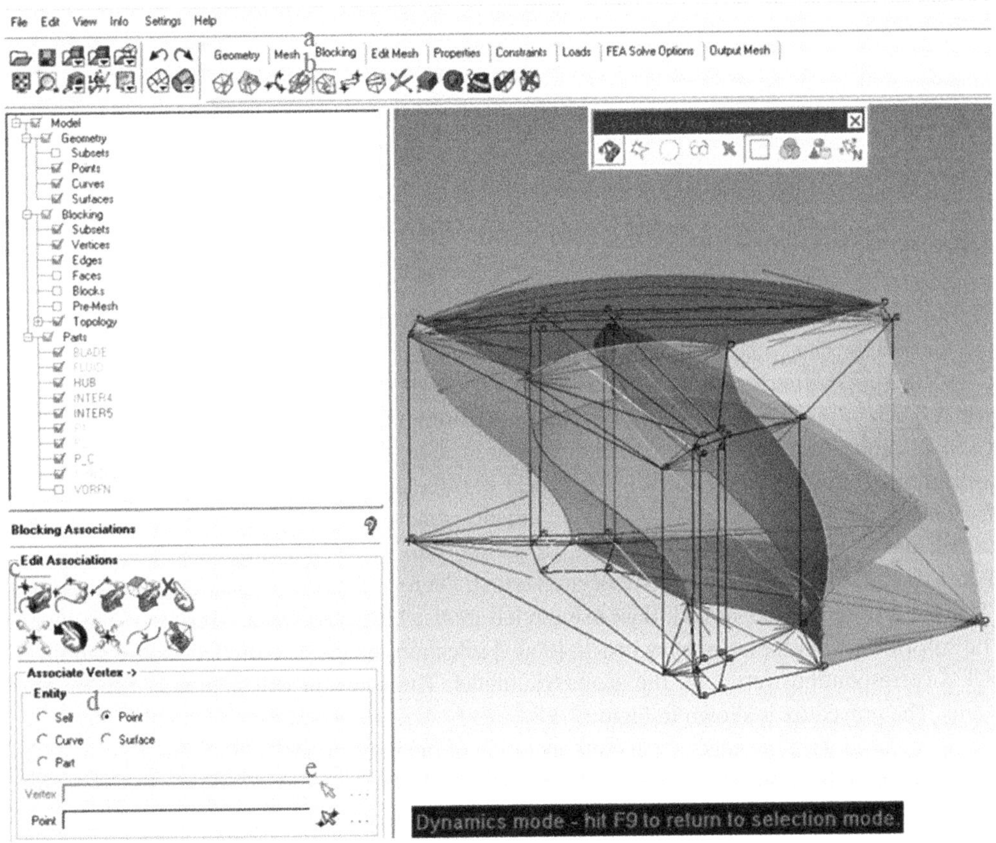

Figure 7.14 Vertex association in the mesh model.

edges on the geometric model, then click the middle mouse button again to confirm. The software will automatically associate them. Finally, click the [Apply] button at the bottom of the dialog to complete the association. The operation is shown in Figure 7.15.

After associating all points and edges on the Block, proceed with mesh sizing. First, click the [Mesh] tab, then [Global Mesh Setup]. In the pop-up dialog box, enter the desired maximum element size in the [Max element] field and click [Apply] to finish the global mesh size setting, as shown in Figure 7.16.

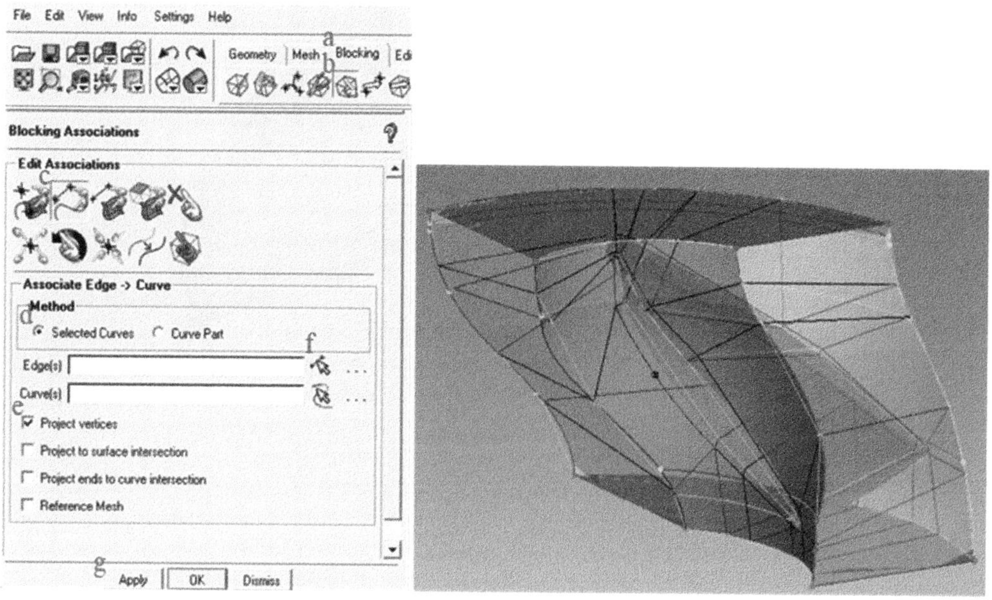

Figure 7.15 Edge association in the mesh model.

Figure 7.16 Mesh sizing setup.

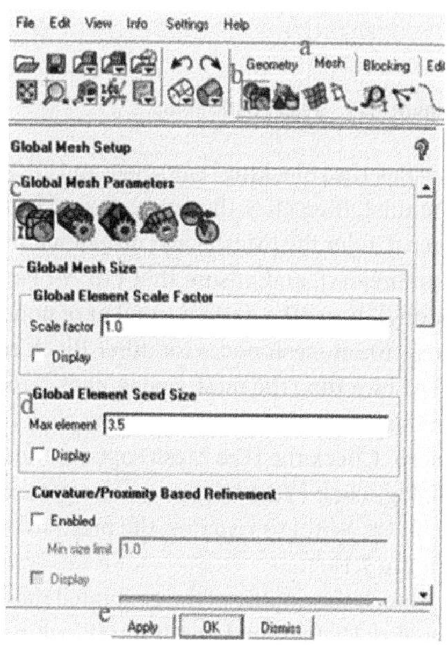

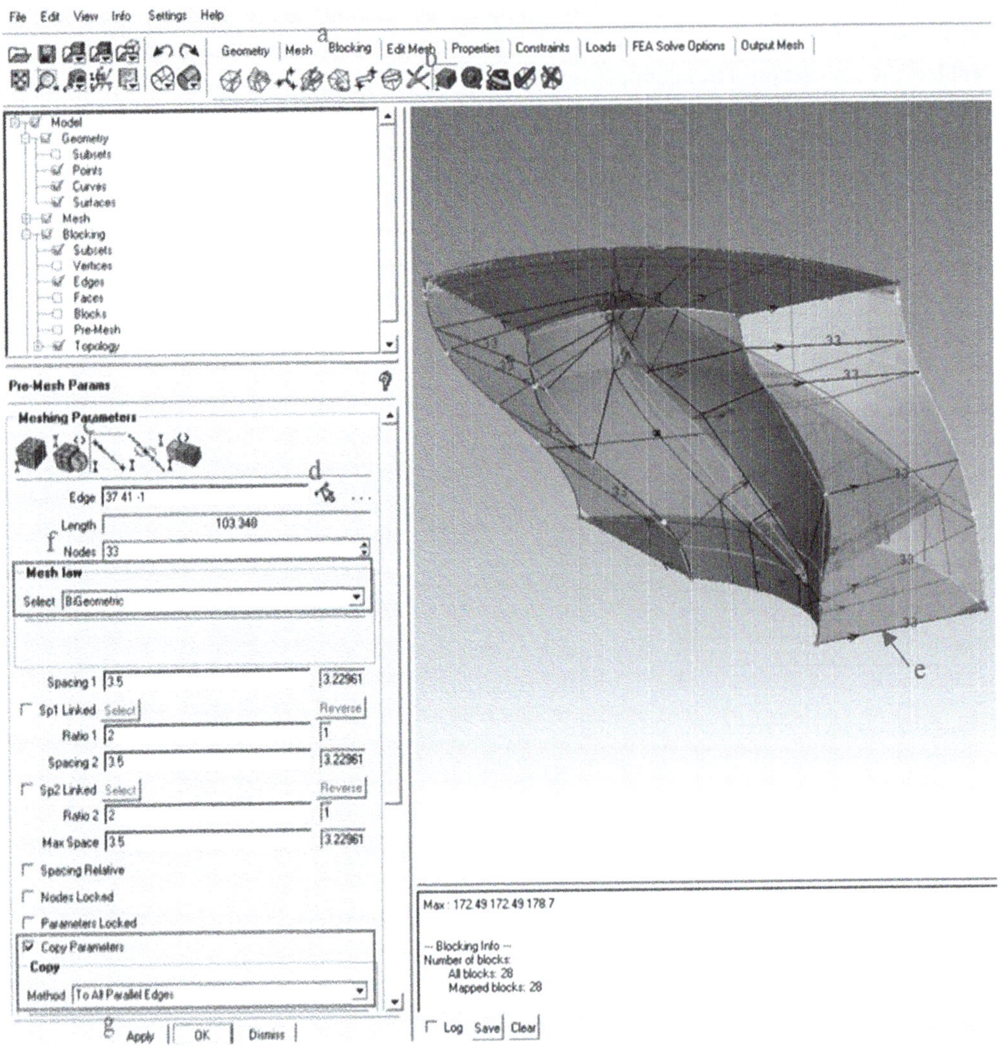

Figure 7.17 Mesh node setup.

Click the [Blocking] tab, select and click [Pre-Mesh Params]. In the pop-up dialog box, click [Edge Params], then click the arrow next to [Edge]. Select the edges on the Block in the graphics display area. Under the [Mesh Law] dropdown menu, keep the default [BiGeometric] option. Check [Copy Parameters], and ensure that [To All Parallel Edges] is selected in the [Method] submenu under [Copy]. Enter the desired number of nodes in the [Nodes] field and click [Apply]. Repeat the above steps to set mesh nodes for other Block edges. The operation is illustrated in Figure 7.17.

After setting the mesh nodes, click [Update Sizes] at the top of the dialog box, select [Update All] in the [Method] column, and then click [Apply] to confirm the mesh size settings and generate the mesh. Check the [Pre-Mesh] option under the [Blocking] branch in the project tree. In the pop-up dialog, click [Yes] to preview the generated structured mesh. Right-click [Pre-Mesh] and select [Solid & Wire] to visualize the mesh in both solid and wireframe modes. The details are shown in Figure 7.18.

Check the quality of the generated mesh and optimize any low-quality sections. Click [Pre-mesh Quality Histograms] under the [Blocking] tab. In the [Criterion] dropdown, select [Quality], then

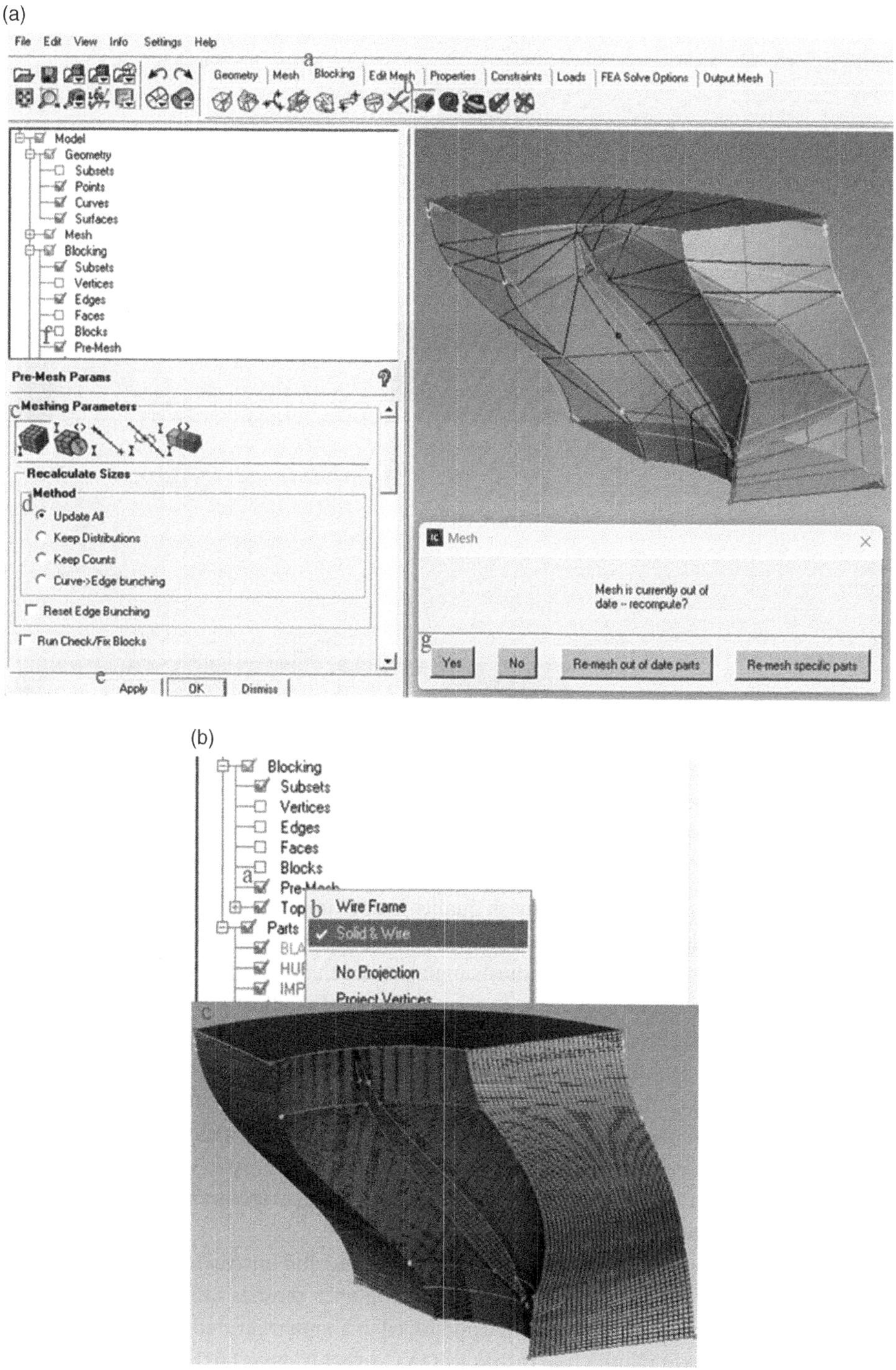

Figure 7.18 Mesh generation and preview: (a) Generated mesh, (b) previewed mesh.

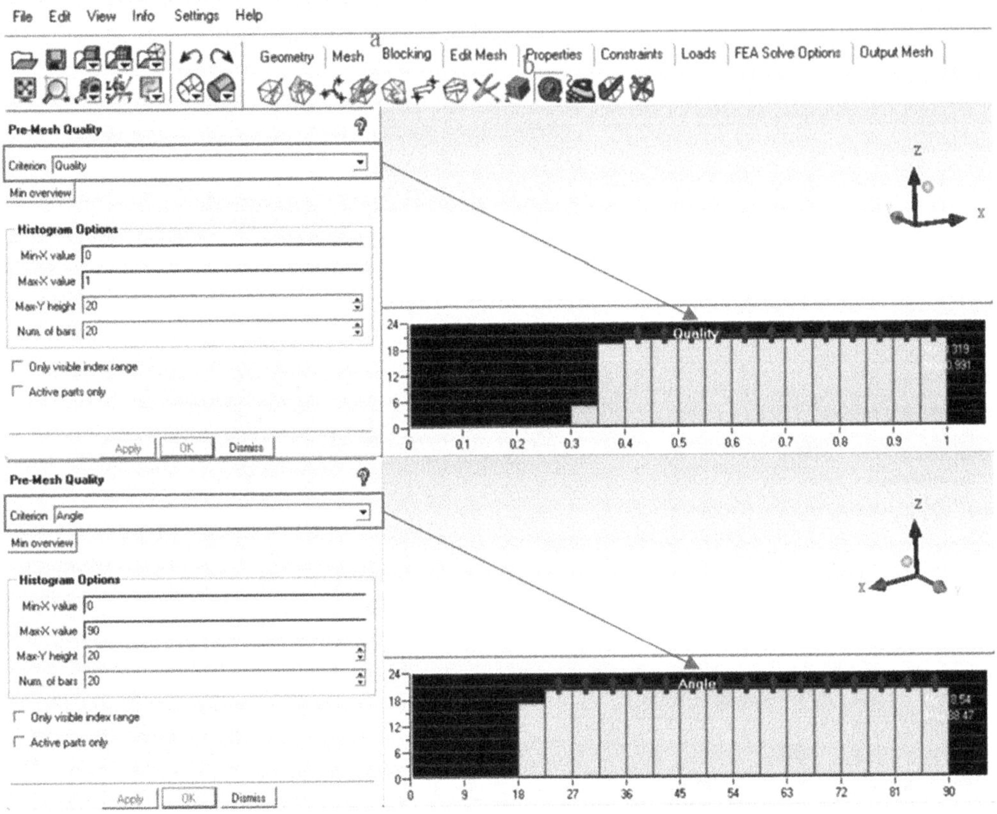

Figure 7.19 Check of mesh quality.

click [Apply] at the bottom of the dialog. A mesh quality histogram will appear below the graphics display area, as shown in Figure 7.19.

Generally, a mesh quality above 0.2 and a mesh angle greater than 18° meet simulation requirements. If the mesh quality is insufficient, it can be optimized by adjusting node positions and quantities, as well as applying mesh smoothing. Below are the basic steps for improving mesh quality using smoothing methods. Click [Pre-mesh Smooth] under the [Blocking] tab. In the pop-up dialog box, select [Quality] in the [Method] dropdown. Under [Smoothing iterations], enter the number of smoothing steps (e.g., 5). In [Up to quality], input the target mesh quality (e.g., 0.3). Then, select the smoothing criterion (e.g., [Angle]) in the [Criterion] dropdown. Click [Apply] to confirm, as shown in Figure 7.20. If the mesh quality still does not meet requirements after smoothing, repeat the steps for multiple iterations.

Since unstructured solvers are compatible with both structured and unstructured meshes, mainstream CFD software now predominantly employs unstructured meshes and solvers. Next, the generated structured mesh needs to be converted and saved in an unstructured format. Right-click [Pre-Mesh] under the [Blocking] branch in the project tree, select [Convert to Unstructured Mesh] from the pop-up dialog box, and the software will automatically convert the mesh into an unstructured form, as shown in Figure 7.21.

Since only the single-passage mesh for the impeller region has been generated, it is necessary to create the full-passage mesh for the impeller by rotational copying. Before performing the rotational copy operation, the mesh on the periodic surfaces must first be deleted. Click [Delete Elements]

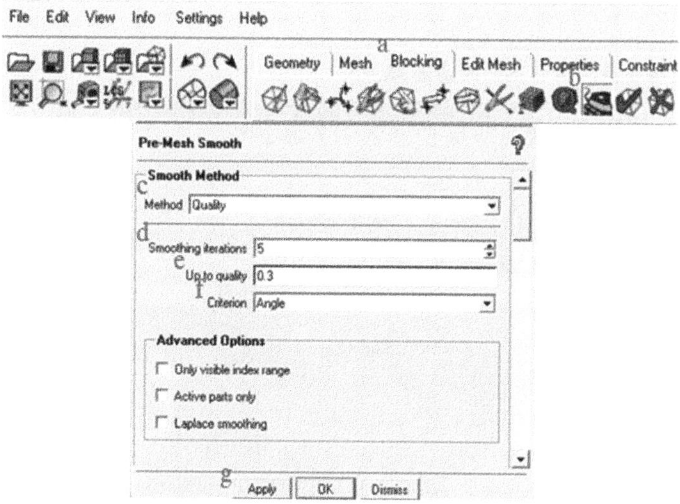

Figure 7.20 Mesh smoothing operation.

Figure 7.21 Structured mesh conversion to an unstructured mesh.

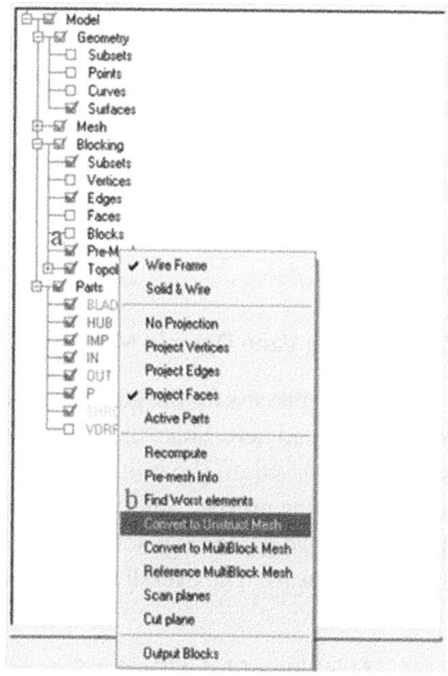

under the [Edit Mesh] tab, then click the arrow next to [Elements]. In the pop-up toolbar within the graphics display area, select [Select part]. In the subsequent dialog, check the previously created periodic surfaces and click [Accept] to confirm the deletion of the periodic surface mesh, as shown in Figure 7.22.

Ensure that [Shells] and [Volumes] under the [Mesh] branch in the project tree are checked. Click [Transform Mesh] under the [Edit Mesh] tab, then click the arrow next to [Select] and choose all mesh elements in the graphics display area. Next, click [Rotate Mesh] in the dialog, check the [Copy] option, and enter the number of copies (e.g., 3 in this case). Check the [Delete duplicate elements] option. Under [Rotation], set the axis to [Z], input the rotation angle (e.g., 90° in this case), and under [Center of Rotation], select [Origin] from the dropdown menu as the rotation

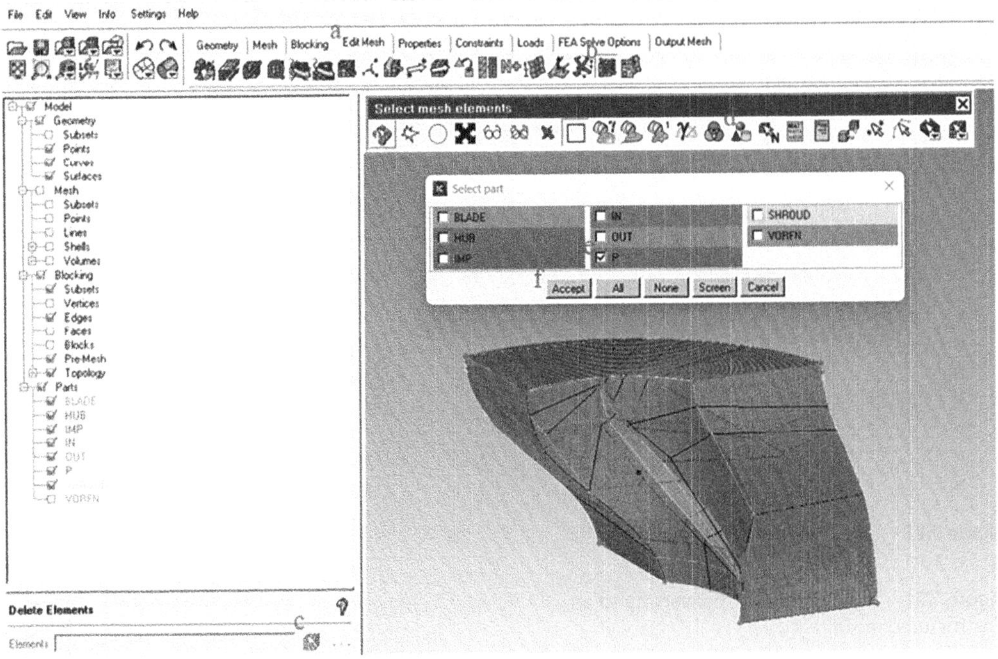

Figure 7.22 Deletion of periodic surface mesh.

center point. Finally, click [Apply] to complete the rotational copy operation for the single-passage impeller mesh, as shown in Figure 7.23. The final full-passage mesh model for the impeller region is shown in Figure 7.24.

7.2.2 Guide Vane Domain Meshing

The meshing process for the guide vane domain is identical to that of the impeller domain and will not be reiterated here. Since the guide vane region exhibits central symmetry, the geometric model was divided based on the number of impeller blades to obtain a single-passage model, simplifying the meshing process. Subsequently, only the single-passage geometric model was meshed, and the full-passage guide vane domain mesh was generated using the rotational copy function of the meshing software. The overall mesh model for the guide vane domain is shown in Figure 7.25.

7.2.3 Meshing of Inlet/Outlet Extension Domains

The geometric models of the inlet extension region, inlet cone region, outlet elbow region, and outlet extension region are relatively simple. The meshing method is similar to that used for the impeller region. The final mesh models for the inlet and outlet extension domains are shown in Figure 7.26.

7.3 Steady Flow Simulation Setup

Open the CFX simulation software. First, set up the working directory by clicking the [Browse] button next to [Working Directory], and either create a new folder or select an existing one in the

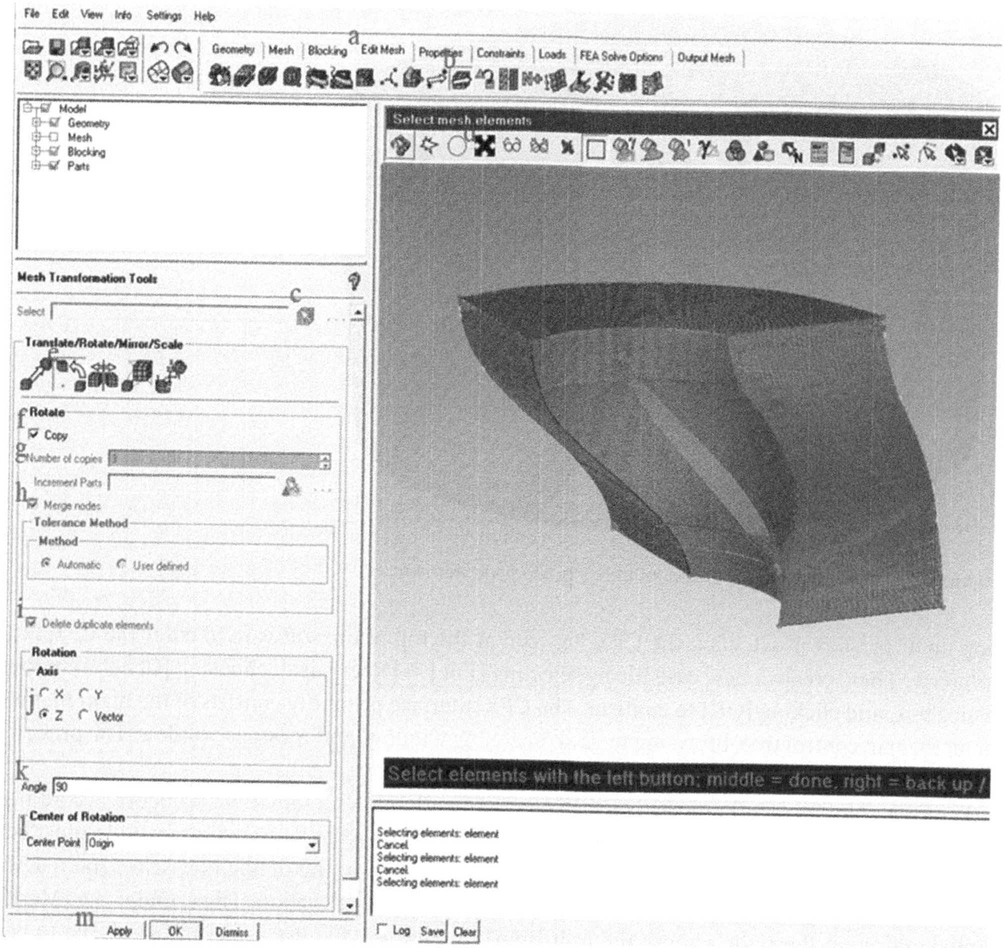

Figure 7.23 Periodic rotational copying of single-passage impeller mesh.

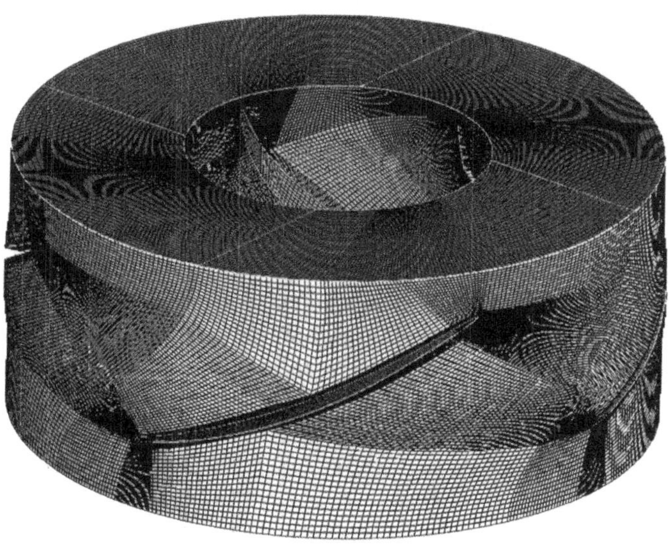

Figure 7.24 Full-passage mesh model of the impeller region.

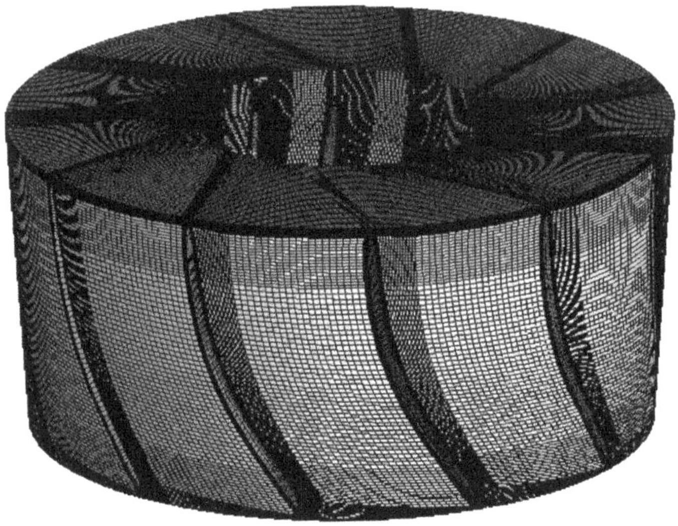

Figure 7.25 Full-passage mesh model of the guide vane domain.

pop-up dialog box. Next, click the CFX-Pre icon at the top of the software to enter the CFX-Pre interface. Then, create a new case file by selecting [File] → [New Case], choosing [General] in the dialog box, and clicking [OK] to confirm. The CFX interface primarily consists of the main menu, main toolbar, control tree, browsing toolbar, browsing window, and message window. The process is illustrated in Figure 7.27.

The CFX solution setup process is straightforward and follows a top-down sequence according to the control tree. Before configuring the solver settings, the axial flow pump mesh file must be imported. Click [File] → [Import] → [Mesh] in the menu bar. In the dialog box, select the ICEM CFD format (file extension: cfx or .cfx5) under the [Files of type] option. Then, under the [Mesh Units] option on the right, choose the unit [mm]. Browse the file list, select the mesh file to be imported, and click [Open] to confirm. Subsequently, the mesh model will be displayed in the browsing window, as shown in Figure 7.28.

Set the calculation type to steady-state. Expand the [Simulation] option in the control tree on the left, double-click [Analysis Type], and select [Steady State] from the dropdown menu under [Analysis Type Options]. Click [OK] to confirm, as shown in Figure 7.29.

7.3.1 Domain Configuration

Next, define the computational domains: Impeller region (IMPELLER), Guide vane region (VANE), Inlet extension region (INEXTEND), Inlet cone region (INCONE), Outlet elbow region (OUTELBOW), Outlet extension region (OUTEXTEND), Click [Insert] → [Domain], enter the domain name in the dialog box, and click [OK] to confirm. The detailed process is shown in Figure 7.30.

Next, configure the domain properties, primarily including basic attributes, motion properties, heat transfer model, and turbulence model. In terms of motion properties, the impeller region is a rotating domain, while the guide vane region, inlet extension region, inlet cone region, outlet elbow region, and outlet extension region are stationary domains.

For the impeller region, after defining its domain name, a domain dialog box will appear in the control tree on the left. Click the [Basic Setting] tab at the top of the dialog box, select the impeller region mesh from the [Location] dropdown menu, choose [Water] as the working medium from

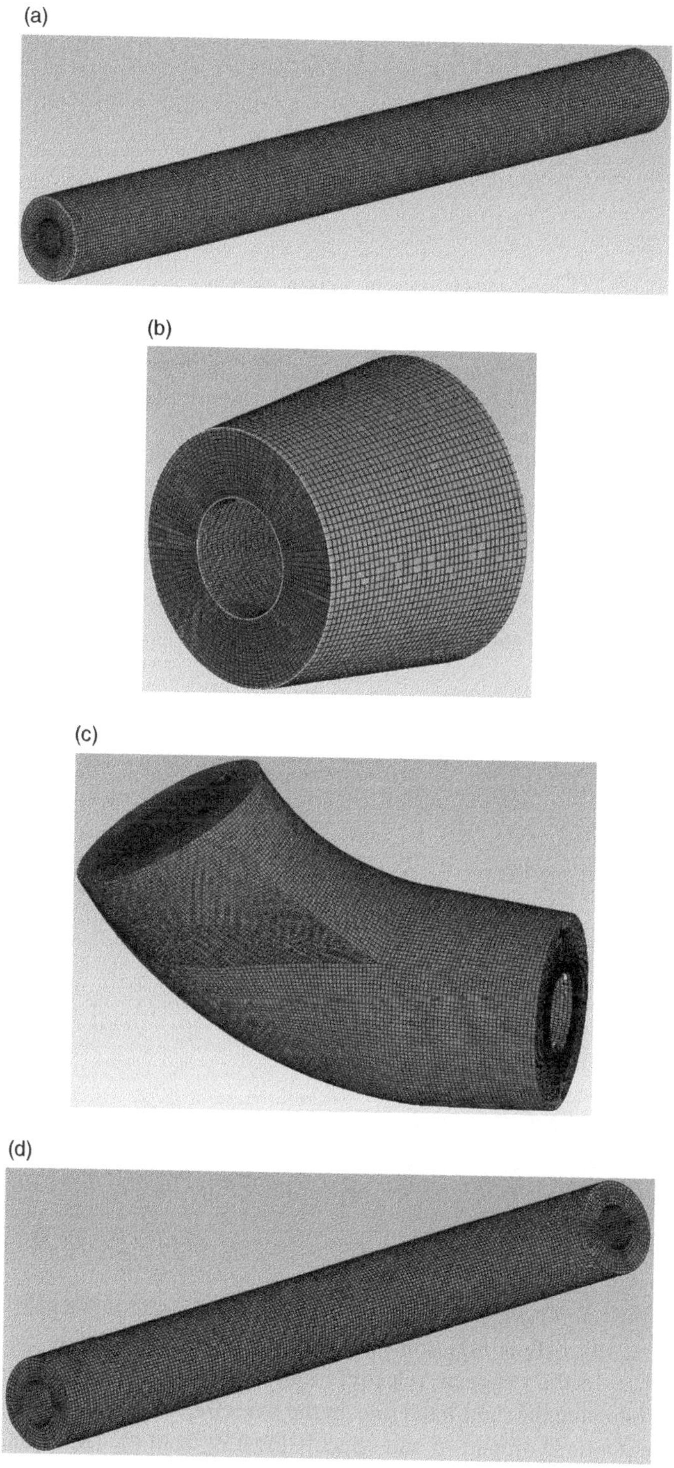

Figure 7.26 Mesh models of the inlet and outlet extension domains: (a) Mesh model of the inlet extension region, (b) Mesh model of the inlet cone region, (c) Mesh model of the outlet elbow region, (d) Mesh model of the outlet extension region.

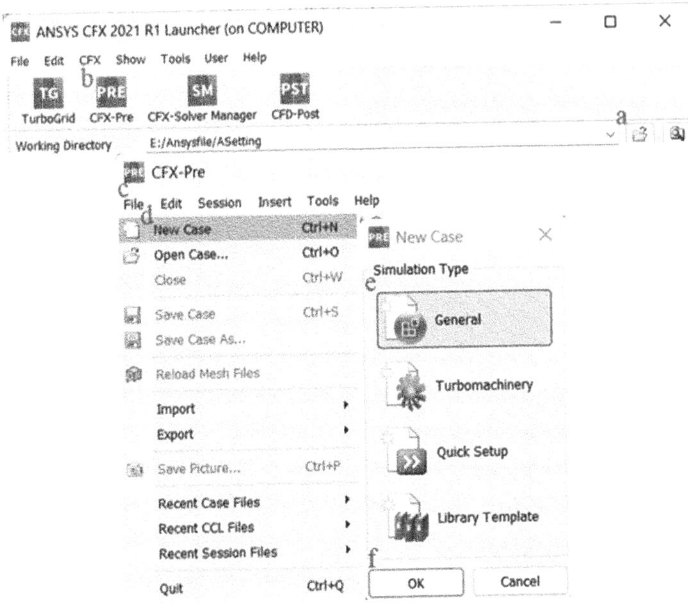

Figure 7.27 Setting the working directory and case file.

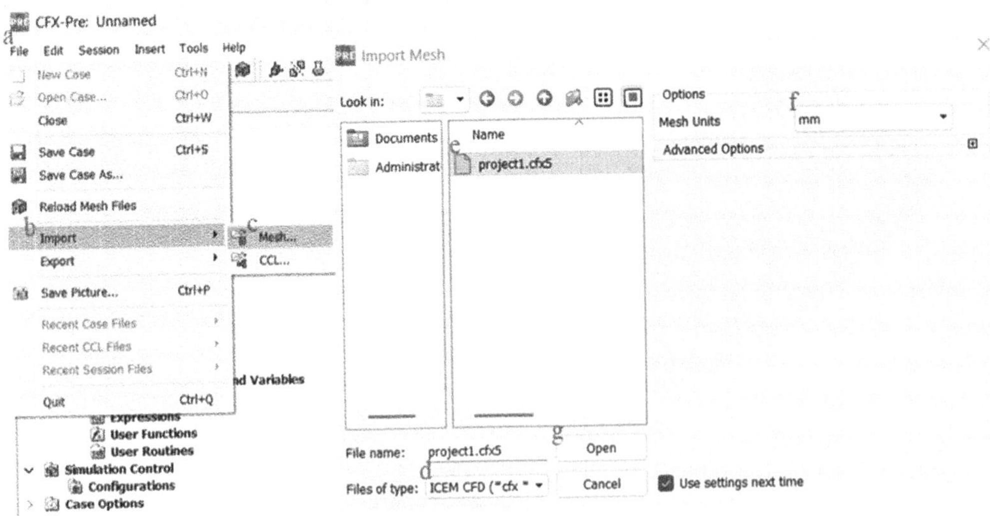

Figure 7.28 Importing the mesh file in CFX.

the [Material] dropdown, and set the reference pressure in the [Reference Pressure] input field. Under the [Domain Motion] option, select [Rotating] from the [Option] dropdown to define the impeller region as a rotating domain. In the [Angular Velocity] input field, enter the rotational speed of the impeller, with its sign following the right-hand rule. In the [Axis Deformation] option, choose [Coordinate Axis] from the [Option] dropdown and select [Global Z] from the [Rotation Axis] dropdown, designating the Z-axis as the impeller's rotation axis. The detailed settings are shown in Figure 7.31.

Additionally, since solving the Navier–Stokes equations requires a heat transfer model and turbulence model to close the governing equations, these must also be configured. Click the [Fluid

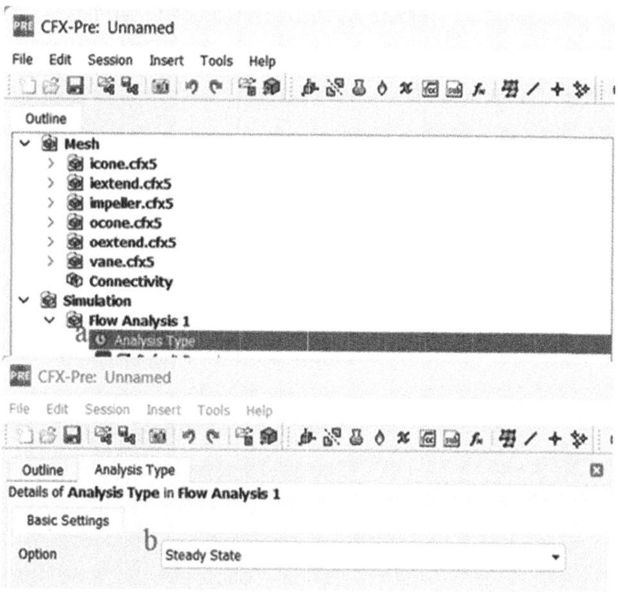

Figure 7.29 Setting the calculation type.

Figure 7.30 Naming of various domains.

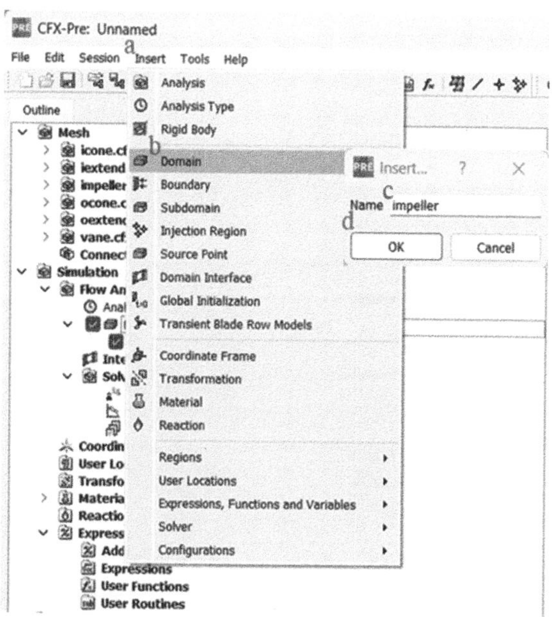

Models] tab at the top of the dialog box, set the heat transfer model and working medium temperature under the [Heat Transfer] option, and define the turbulence model and wall model under the [Turbulence] option. The specific settings for the impeller domain are shown in Figure 7.32.

The configuration of stationary domains is essentially the same as that of rotating domains, except for the motion properties. Under the [Domain Motion] option, select [Stationary] from the [Option] dropdown. Other settings are omitted here for brevity.

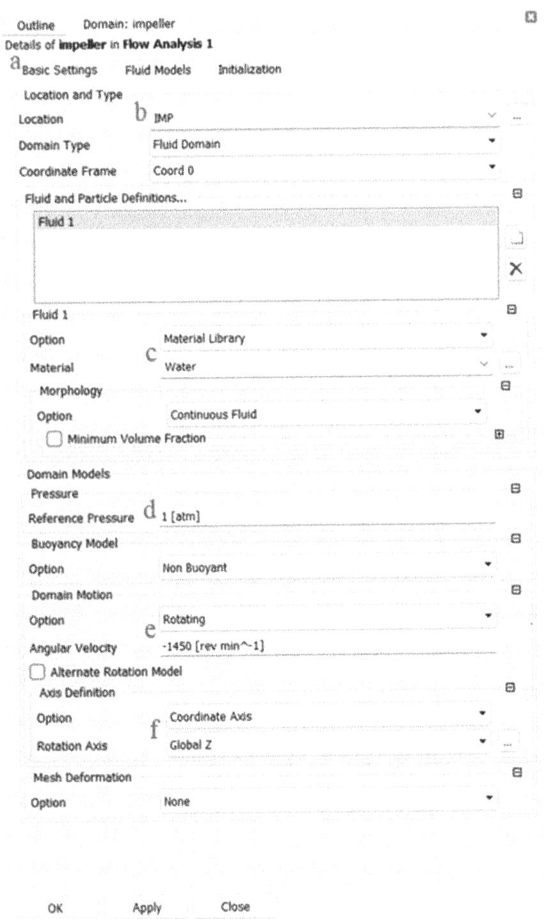

Figure 7.31 Domain property settings.

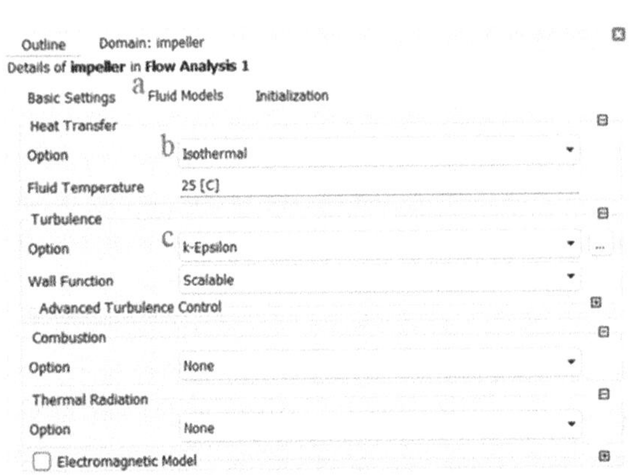

Figure 7.32 Domain heat transfer and turbulence model settings.

7.3.2 Boundary Condition Settings

In the CFD simulation setup, defining the inlet and outlet boundaries is a critical step. The inlet boundary is the end face of the inlet extension region, and the outlet boundary is the end face of the outlet extension region. First, right-click the inlet extension domain in the control tree on the left, select [Insert] → [Boundary], enter the inlet boundary name in the dialog box, and click [OK].

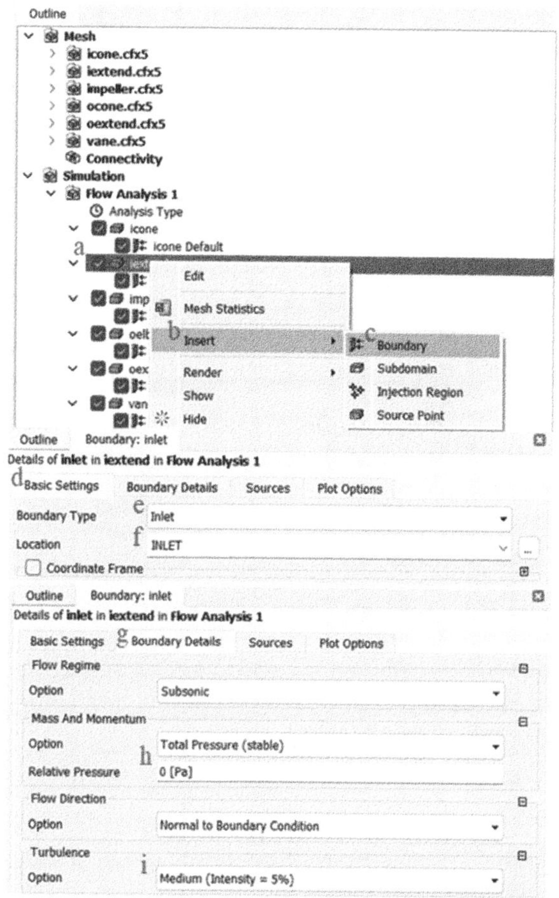

Figure 7.33 Inlet boundary condition settings.

This will open the boundary condition properties dialog. Under the [Basic Settings] tab, set the [Boundary Type] to [Inlet], click the selection button next to [Location], and choose the inlet face of the inlet extension domain from the dialog box. Then, click the [Boundary Details] tab, select [Total Pressure (Stable)] from the [Option] dropdown under the [Mass And Momentum] option to define a pressure inlet boundary. To specify the turbulence characteristics at the inlet, select [Medium (Intensity = 5%)] from the [Option] dropdown under the [Turbulence] option. After completing these settings, click [Apply] to confirm. The details are illustrated in Figure 7.33.

The outlet boundary setup follows a similar process. Right-click the outlet extension domain, select [Insert] → [Boundary], enter the outlet boundary name, and click [OK]. In the [Basic Settings] tab, set the [Boundary Type] to [Outlet], select the outlet face of the outlet extension domain under [Location], then switch to the [Boundary Details] tab. Under the [Mass And Momentum] option, choose [Mass Flow Rate] for a mass flow outlet boundary or [Normal Speed] for a velocity outlet boundary. Click [Apply] to finalize the outlet boundary settings. The process is detailed in Figure 7.34.

After defining the inlet and outlet boundaries, the wall conditions for each domain must be specified. Begin with the impeller rotating domain's wall conditions. Right-click the impeller domain (IMPELLER) in the control tree, select [Insert] → [Boundary], name the wall boundary, and click [OK]. In the [Basic Settings] tab, set the [Boundary Type] to [Wall]. Under [Location], select the impeller's corresponding wall faces and set the [Frame Type] to [Rotating]. In the

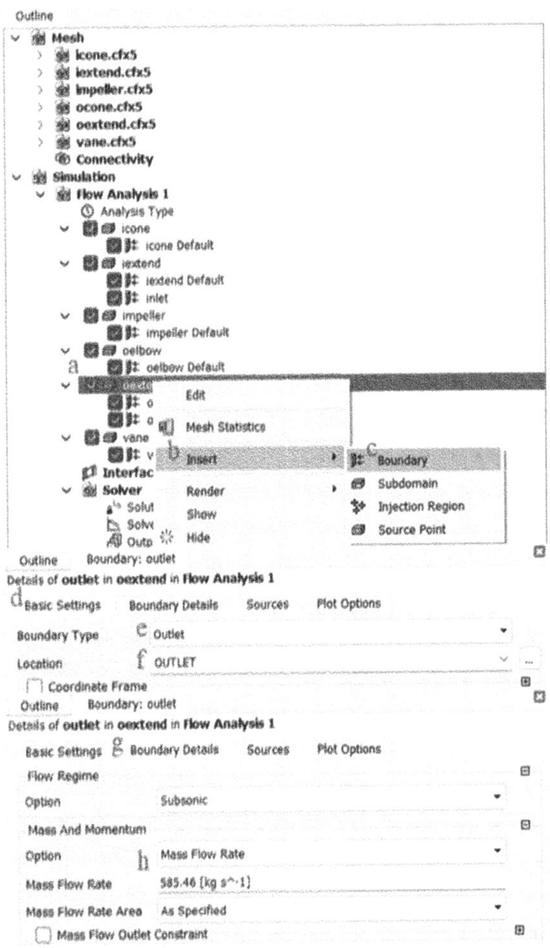

Figure 7.34 Outlet boundary condition settings.

[Boundary Details] tab, choose [No Slip Wall] under the [Mass And Momentum] option to define a no-slip wall. Enable [Wall Velocity], select [Rotating Wall] from the [Option] dropdown, and enter an angular velocity of 0. Under [Axis Definition], set the rotation axis to [Global Z]. For wall roughness, select [Smooth Wall] from the [Wall Roughness] dropdown, adjusting as needed based on actual conditions. Click [Apply] to confirm. The setup is shown in Figure 7.35. The remaining walls of the impeller domain follow the same procedure and are not detailed here.

Next, configure the wall boundary conditions for stationary domains. Taking the guide vane region (VANE) as an example, select and right-click the VANE domain in the control tree on the left, then click [Insert] → [Boundary], enter the wall boundary name in the dialog box, and click [OK]. The boundary condition properties dialog will appear. Under the [Basic Settings] tab, set the [Boundary Type] to [Wall]. Click the button next to [Location] to select the corresponding wall faces of the guide vane region. The [Frame Type] reference coordinate system defaults to the global stationary frame. Then, click the [Boundary Details] tab, select [No Slip Wall] under [Mass And Momentum], leave [Wall Velocity] unchecked, and choose [Smooth Wall] for the [Wall Roughness] option. Click [Apply] to complete the setup. The details are shown in Figure 7.36. The setup for other stationary domains follows the same method and is not repeated here.

In axial flow pump simulations, interfaces (Interface) include rotating–stationary and stationary–stationary types. Properly defining interface pairings ensures smooth data transfer

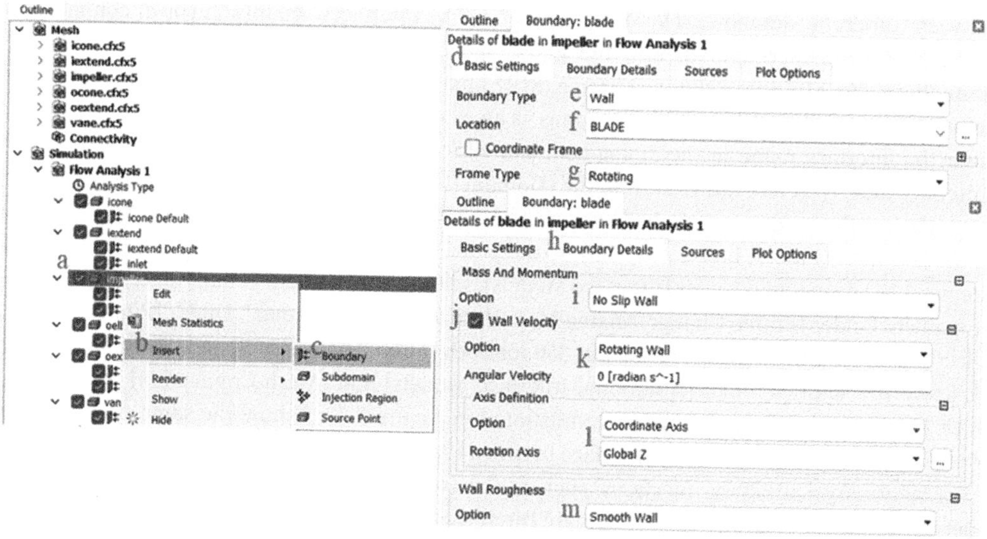

Figure 7.35 Impeller rotating domain wall boundary condition settings.

Figure 7.36 Guide vane stationary domain wall boundary condition settings.

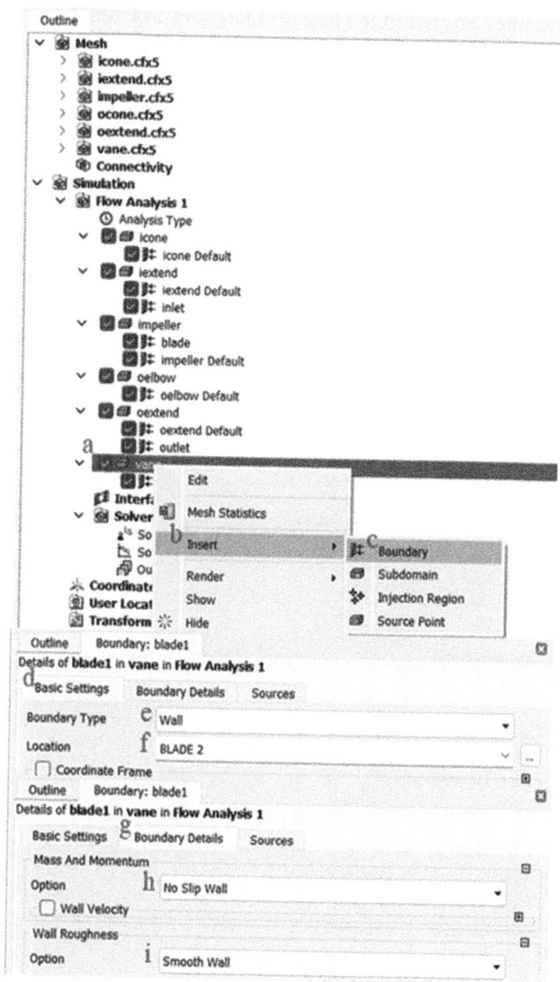

between different domains. The rotating–stationary interfaces mainly involve connections between the impeller domain and the guide vane or inlet cone domains. For steady-state simulations, the Multiple Reference Frame (MRF) method is typically used. Taking the interface between the impeller and guide vane regions as an example: Click [Insert] → [Domain Interface], enter the interface name in the dialog box, and click [OK]. Under the [Basic Settings] tab: For [Interface Side 1]: Select IMPELLER in the [Domain (Filter)] dropdown. Choose the impeller face connected to the guide vane in the [Region List] dropdown. For [Interface Side 2]: Select VANE in the [Domain (Filter)] dropdown. Choose the guide vane face connected to the impeller in the [Region List] dropdown. Under [Interface Models], select [General Connection] from the [Option] dropdown. Under [Frame Change/Mixing Model], choose [Frozen Rotor]. Under [Pitch Change], select [Specified Pitch Angles] and input 360 for both [Pitch Angle Side 1] and [Pitch Angle Side 2]. Leave the settings under [Additional Interface Models] and [Mesh Connection] unchanged. Click [Apply] to confirm. The setup is illustrated in Figure 7.37. Follow the same procedure to configure the rotating–stationary interface between the impeller and the inlet cone region.

For stationary–stationary interfaces, such as the connection between the inlet extension and inlet cone regions: Click [Insert] → [Domain Interface], name the interface, and click [OK]. Under [Basic Settings]: For [Interface Side 1]: Select INEXTEND in the [Domain (Filter)] dropdown. Choose the inlet extension face connected to the inlet cone in [Region List]. For [Interface Side 2]: Select INCONE in the [Domain (Filter)] dropdown. Choose the inlet cone face connected to the inlet extension in [Region List]. Under [Interface Models], select [General Connection]. Under [Frame Change/Mixing Model], choose [None]. Under [Pitch Change], select [None]. Keep the

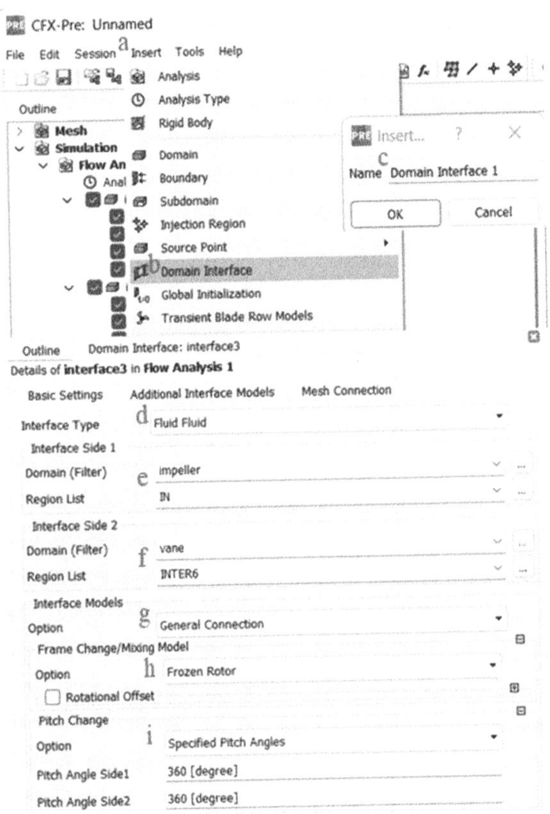

Figure 7.37 Rotating–stationary domain interface settings.

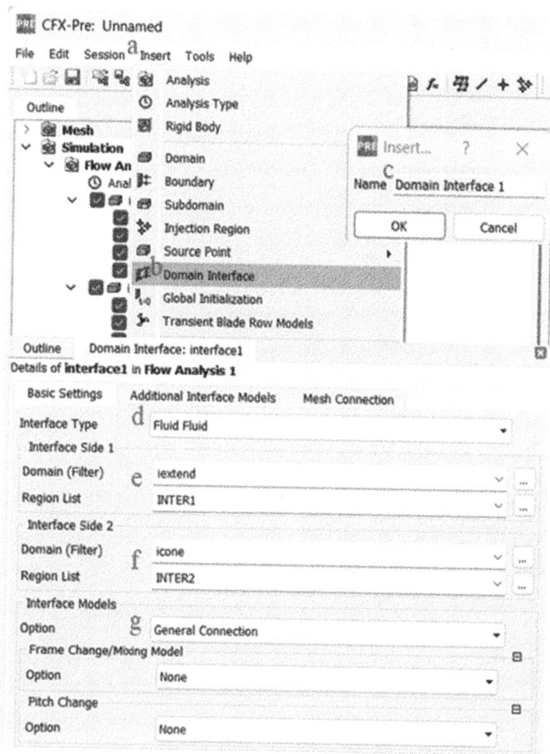

Figure 7.38 Stationary–stationary domain interface settings.

default settings for other options. Click [Apply] to finalize. The setup is shown in Figure 7.38. Other stationary–stationary interfaces follow the same approach.

7.3.3 Solver Settings

Expand the [Slover] node in the control tree on the left by left-clicking, then right-click [Slover Control], and select [Edit] from the pop-up dialog. This will open the solver parameter properties dialog. Click the [Basic Setting] tab: Under [Advection Setting], select [High Resolution] from the [Option] dropdown. Under [Turbulence Numerics], select [First Order]. In [Convergence Control], input the minimum and maximum iterations in [Min. Iterations] and [Max. Iterations], respectively. Under [Fluid Timescale Control], select [Physical Timescale] from the [Timescale Control] dropdown, and enter the time value (typically set as the reciprocal of the impeller speed). Under [Convergence Criteria], choose [RMS] from the [Residual Type] dropdown and set the [Residual Target] (usually 1e-4 suffices for engineering applications). Leave other settings unchanged. Click [Apply] to confirm. The specific settings are shown in Figure 7.39.

7.3.4 Calculation Setup

Export the Solver Input File: On the main toolbar, click [Write Solver Input File]. In the pop-up dialog, enter the solver file name and save (Figure 7.40). Save the Project File: Navigate to [File] → [Save Case As]. Save the CFX project file (with .cfx extension). The details are shown in Figure 7.41.

After saving, close the CFX-Pre interface, click the [CFX-Solver Manager] icon at the top of the CFX software to open the solver manager, and configure the simulation settings. Click the [Browse] button next to [Solver Input File] and locate the saved solver file (with the .def extension) in the

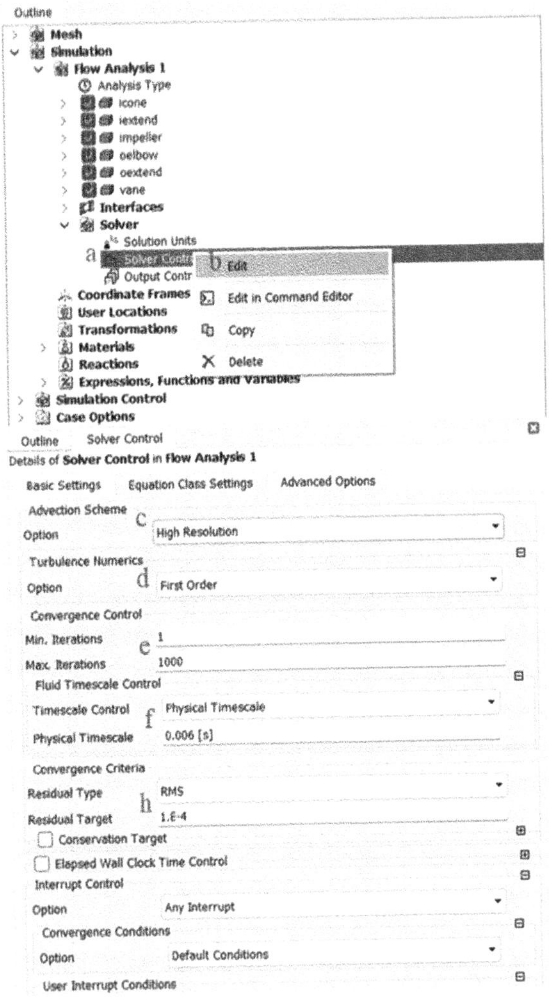

Figure 7.39 Solver control parameter settings.

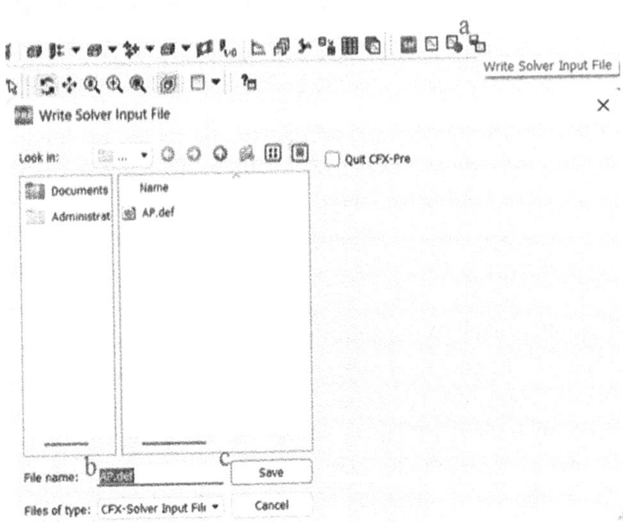

Figure 7.40 Exporting solver settings file.

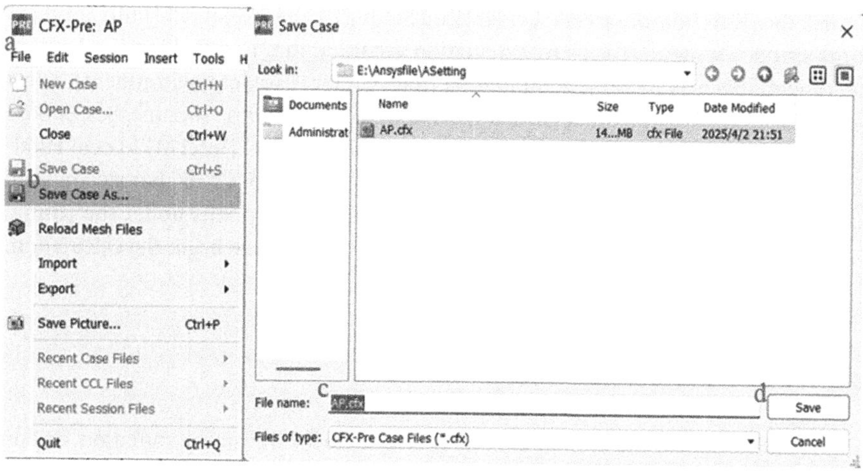

Figure 7.41 Project file save.

Figure 7.42 Steady and parallel computation settings.

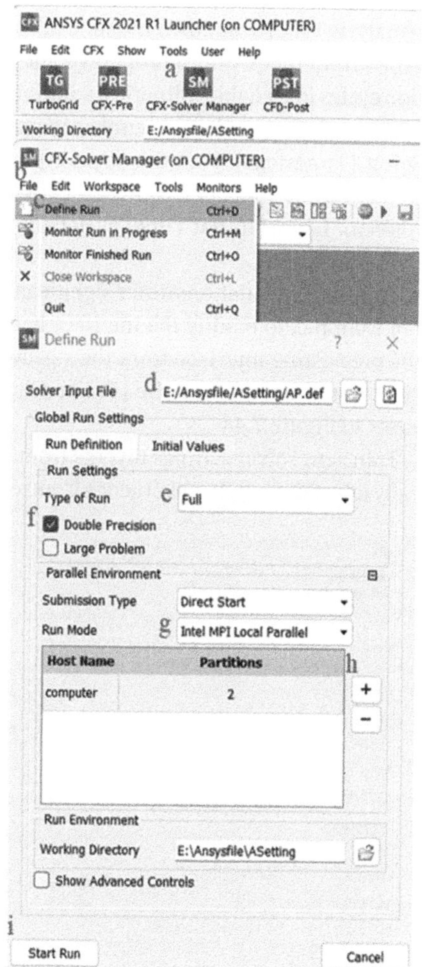

pop-up folder. Click the [Run Definition] tab: Under [Run Setting], keep the default [Full] option in the [Type of Run] dropdown menu. To ensure calculation accuracy, the [Double Precision] option can be checked, though this may increase computation time. Under [Parallel Environment], select the computation mode from the [Run Mode] dropdown menu: For cases with fewer meshes, choose [Serial] for sequential computation. To enable parallel computation, select [Intel MPI Local Parallel], then click the "+" button under [Partitions] to specify the number of CPUs. For steady-state computations, flow field initialization is typically unnecessary, so the options under the [Initial Values] tab can be left unconfigured. Finally, click the [Start Run] button to begin the calculation. The details are shown in Figure 7.42.

7.4 Transient Flow Simulation Setup

Steady simulation results serve as the foundation and prerequisite for transient simulation, ensuring a favorable initial flow field and faster convergence speed for transient simulations. The setup for transient calculations is largely similar to that of steady-state calculations and can be achieved by making minor modifications to the steady-state configuration. The specific modifications are as follows: Transient Settings: Locate and double-click [Analysis Type] in the left control tree to open the corresponding properties dialog. For transient calculations, select [Transient] from the [Analysis Type] dropdown menu. Under [Time Duration]: For total simulation time, choose [Total Time] from the [Option] dropdown and input the duration (typically covering 5–10 impeller rotation cycles for reliable numerical solutions). Alternatively, select [Maximum Number of Timesteps] and specify the maximum number of timesteps in the [Max. Timesteps] field. Under [Time Steps]: Select [Timesteps] from the [Option] dropdown and input the number of calculation steps per rotation cycle. This determines the angular increment (e.g., degrees per step) for the impeller's rotation. Leave [Initial Time] as default. Click [Apply] to confirm. Specific settings are illustrated in Figure 7.43.

Transient calculations differ significantly from steady-state in the setup of rotor-stator interfaces. For example, to modify the interface between the impeller and guide vane domains: Double-click the preexisting interface from the steady-state calculation. Under [Interface Model], select [Transient Rotor Stator] from the [Frame Change/Mixing Model] dropdown. Retain all other settings. Refer to Figure 7.44.

Transient solver settings diverge from steady simulation primarily in: [Transient Scheme]: Typically left at default. [Convergence Control]: Governs iterations per timestep (each timestep behaves

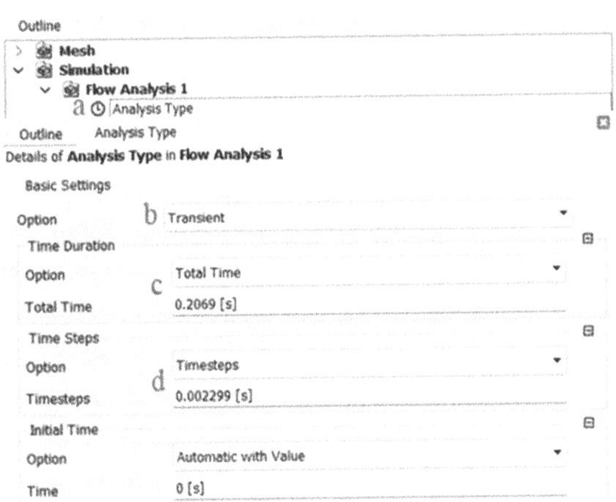

Figure 7.43 Transient calculation time parameter settings.

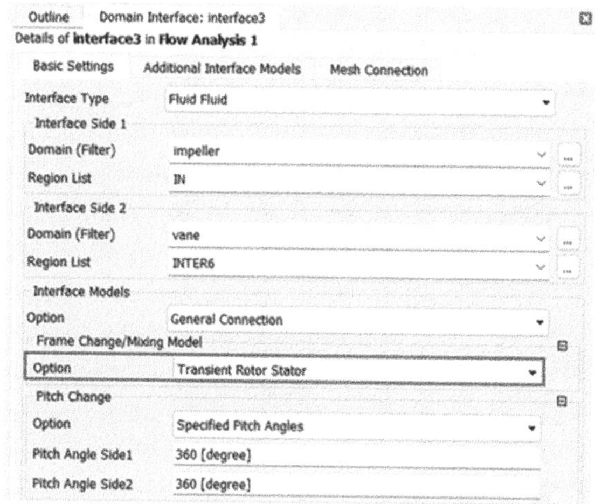

Figure 7.44 Rotor-stator interface settings for transient calculation.

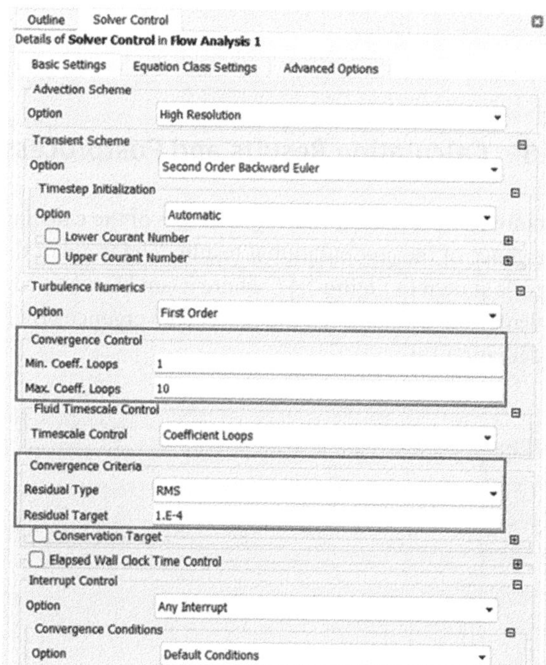

Figure 7.45 Transient solver parameter settings.

as a steady-state solve). [Min. Coeff. Loops]: Default=1 (minimum iterations per timestep). [Max. Coeff. Loops]: Set to 10 (once transient stability is achieved, convergence is rapid). Click [Apply]. See Figure 7.45.

During transient calculations, initialization is usually required. Here, the initial flow field is the result of the steady calculation. Open the solver manager and click [Define Run]. Under the [Run Definition] tab, enable [Initial Values Specification]. In the [Initial Values 1 Settings] section, browse and import the steady results file (*.res). Click [Start Run] to begin the calculation. Configuration details are shown in Figure 7.46.

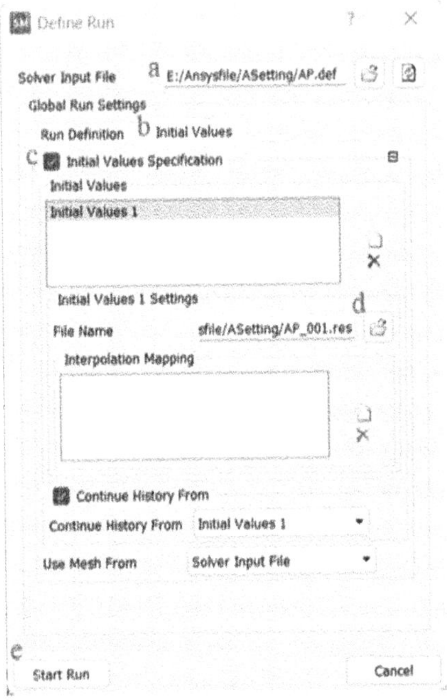

Figure 7.46 Transient calculation setup.

7.5 Calculation Results and Postprocessing

Before postprocessing and visualization of the calculation results, it is necessary to check the convergence of the computational results. The trend of the residual curve during the calculation process is shown in Figure 7.47, where it can be observed that the iteration residuals of the numerical calculation have reached the preset convergence criteria, indicating the normal completion of the calculation [3].

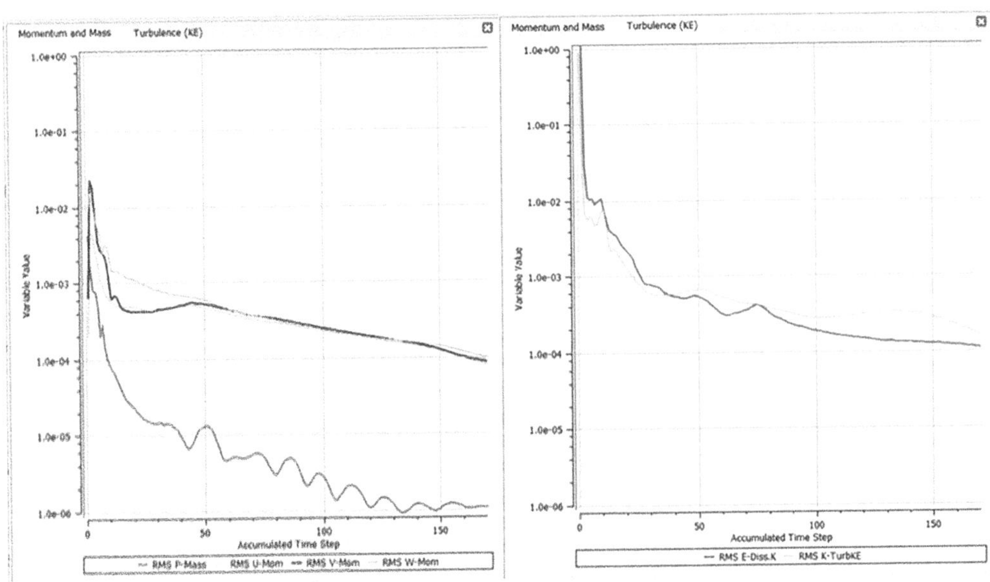

Figure 7.47 Numerical calculation iterative residual curve.

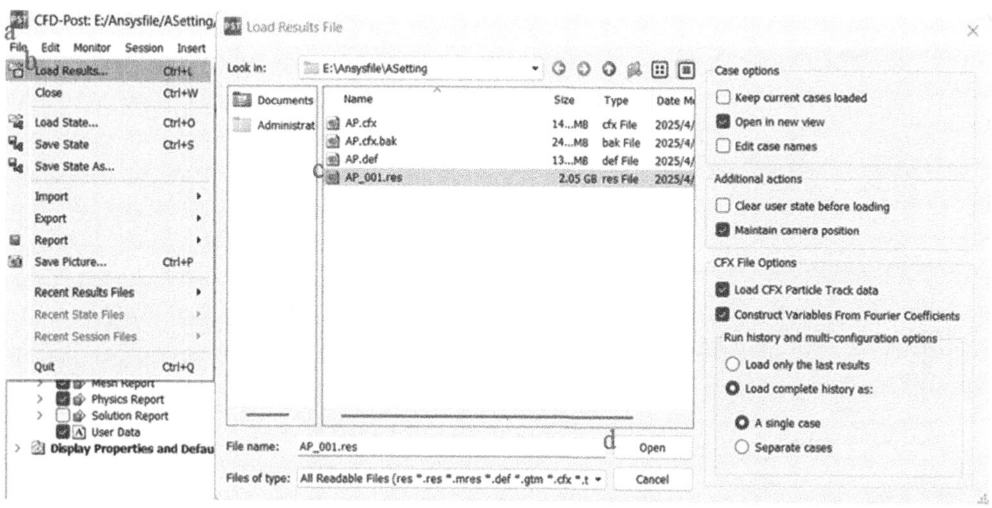

Figure 7.48 Import of numerical calculation results.

The CFX-Post software is used for postprocessing and visualization of numerical calculation results. Open the CFX-Post software, sequentially click File → Load Results, and in the pop-up folder browsing window, select the numerical calculation results of the axial flow pump (file with the .res extension). Click the Open button to import the results, as shown in Figure 7.48. The imported model is displayed in wireframe mode by default.

7.5.1 Section Plane Creation

Pressure contour plots and velocity vector plots are generally used to display the internal flow state of the axial flow pump, with the prerequisite being the creation of a section plane. Click the [Location] button to open a menu containing options for creating geometric elements such as points, lines, planes, volumes, and surfaces. Select [Plane], enter the name of the section plane in the pop-up dialog, and click [OK] to confirm. A property editing dialog will appear at the bottom of the left control tree. In the [Domains] option, specify the region to be cut by the section plane (default: [ALL Domains]). In the [Method] option, specify the orientation of the section plane (e.g., to create a YZ plane, select [XY Plane]). The X-coordinate value can be directly input or adjusted using the slider (default unit: meters [m]). Under [Plane Bounds], select [Rectangular] in the [Type] option and adjust the Y- and Z-axis coordinates in [Y Size] and [Z Size]. In the [Plane Type] option, select [Slice]. After completing the settings, click [Apply] to generate the section plane. For details, refer to Figure 7.49.

7.5.2 Contour Plot Creation

Based on the created section plane, a pressure contour plot can be generated. Click the [Contour] icon on the toolbar or select Insert → Contour from the menu. In the pop-up dialog, enter a name for the contour plot and click [OK]. A property editing dialog will appear at the bottom of the left control tree. In the [Domains] option, specify the region for the contour plot (select [All Domains]). In the [Locations] option, select the corresponding surface. In the [Variable] option, select [Pressure] to create a pressure contour plot. In the [Range] dropdown menu, choose [Local]. The [# of Contours] option controls the display precision of the pressure contour plot, i.e., the resolution

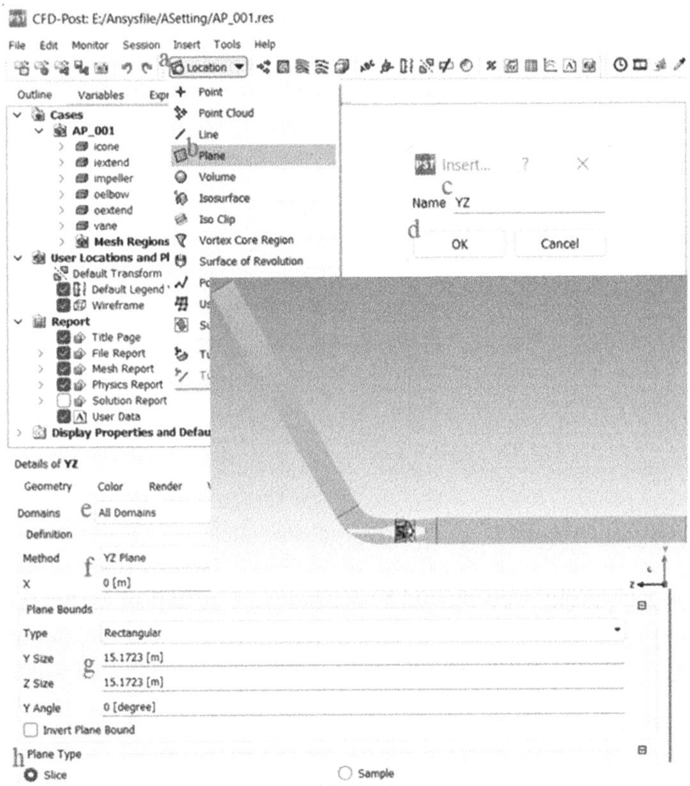

Figure 7.49 Creation of a section plane.

of the contour scale. By entering different values, the current variable range is divided into the corresponding number of intervals. Other settings can remain default. Click [Apply] to confirm, and the pressure contour plot can then be viewed in the graphics display area. For details, refer to Figure 7.50.

7.5.3 Creation of Streamlines

In addition to pressure contour plots, streamlines can be created on the section plane to illustrate the fluid motion inside the axial flow pump. Click the [Streamline] icon on the toolbar or select [Insert] → [Streamline] from the menu. In the pop-up dialog, input the name for the streamline plot and click [OK]. A property editing dialog will appear at the bottom of the left control tree. In the [Type] option, select [Surface Streamline]. In the [Surfaces] option, choose the corresponding surface. The [Start From] option can remain default or be adjusted as needed. In the [# of Points] option, enter a value—the higher the number, the more streamlines will be displayed. In the [Variable] option, select [Velocity] to create a velocity streamline plot. All other settings can remain default. Click [Apply] to confirm, and the velocity streamline plot will be displayed in the graphics area. For details, refer to Figure 7.51.

7.5.4 Creation of Velocity Vector Plot

To further analyze the flow inside the pump, a velocity vector plot can be created on the section plane. Click the [Vector] icon on the toolbar or select Insert → Vector from the menu. In the pop-up

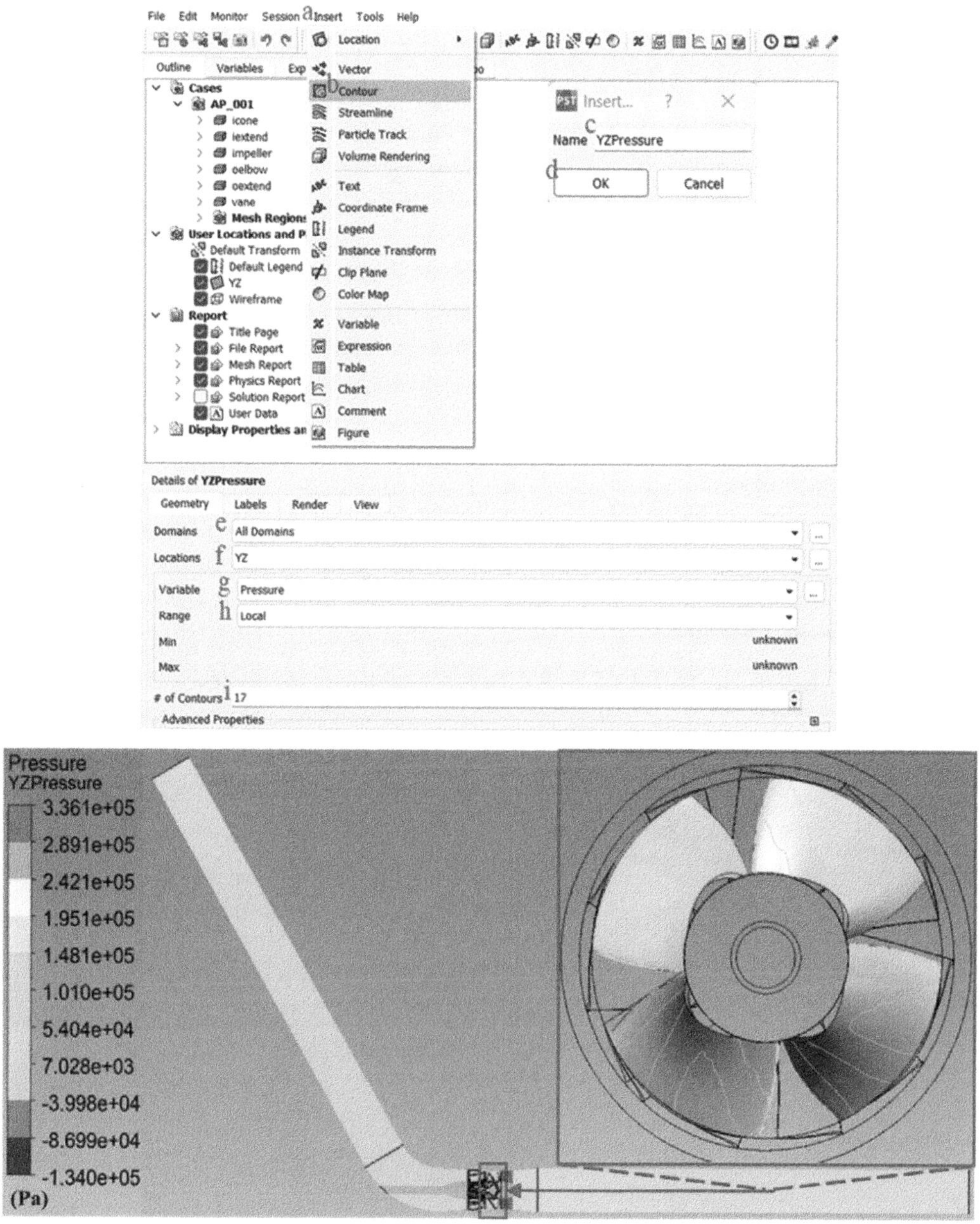

Figure 7.50 Creation of pressure contour plot.

dialog, input the name for the vector plot and click [OK]. A property editing dialog will appear at the bottom of the left control tree. In the [Domains] option, specify the region for the vector plot (select [All Domains]). In the [Locations] dropdown menu, choose the corresponding surface. In the [Variable] option, select [Velocity]. All other settings can remain default. Click [Apply] to confirm, and the velocity vector plot will be displayed in the graphics area. For details, refer to Figure 7.52.

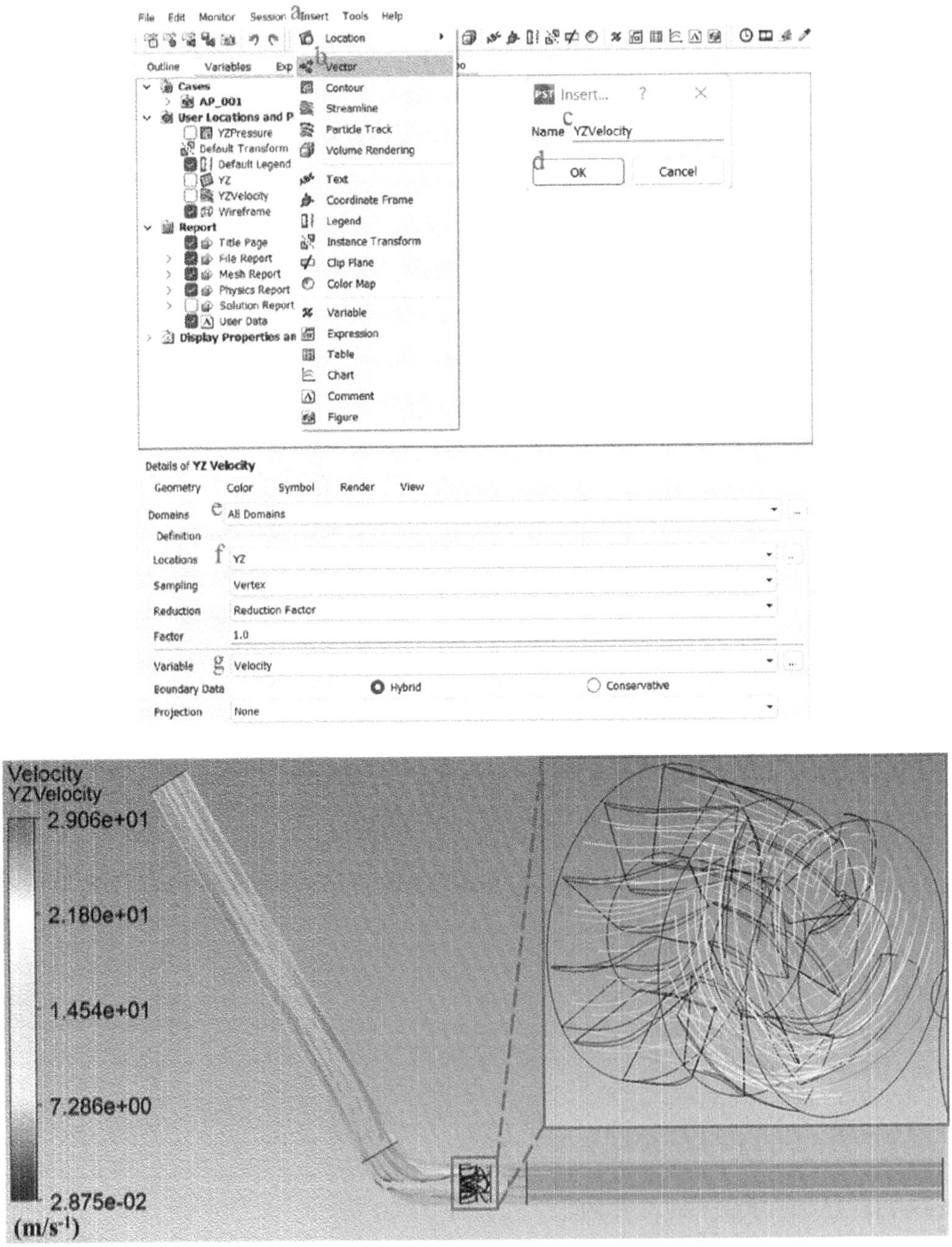

Figure 7.51 Creation of a velocity streamline plot.

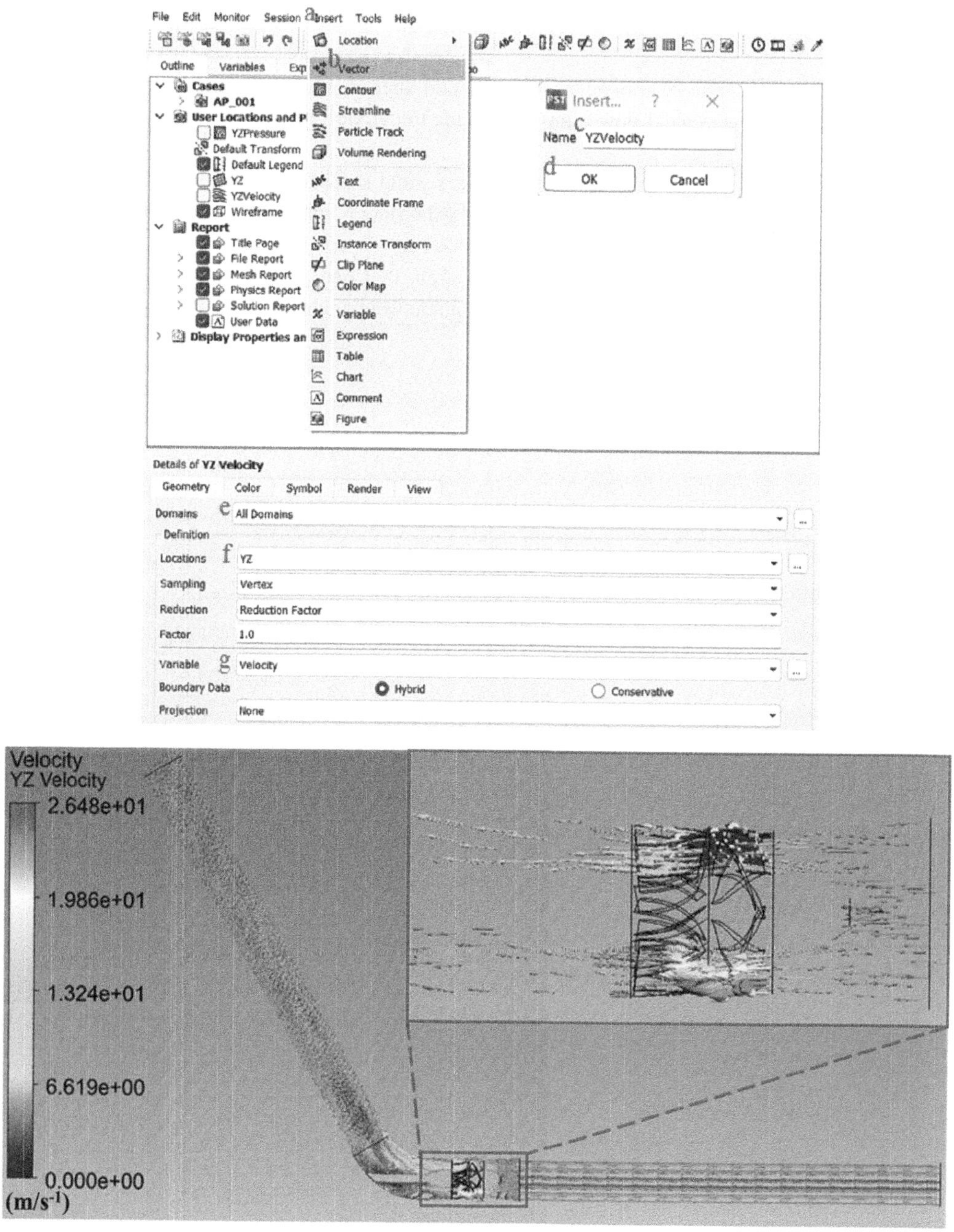

Figure 7.52 Creation of velocity vector plot.

References

1 Kan, K., Qingying, Z., Zhe, X., Zheng, Y., Gao, Q., and Shen, L. (2022). Energy loss mechanism due to tip leakage flow of axial flow pump as turbine under various operating conditions. *Energy* 255: 124532.

2 XiaoQi, J., ShengJi, S., ShuaiKang, Z., Hao, L., Zhe, L., and ZuChao, Z. (2024). Influence of tip clearance on internal energy loss characteristics of axial flow pumps under different operating conditions. *Physics of Fluids* 36: 015102.

3 Lijian, S., Yao, C., Li, W., Xu, T., Jiang, Y., Xing, J., Yan, B., Chen, Y., and Han Y. (2023). Numerical simulation and model test of the influence of guide vane angle on the performance of axial flow pump. *Physics of Fluids* 35: 015129.

8

Numerical Simulation and Analysis of a Mixed-Flow Pump

8.1 Computational Domain Modeling

8.1.1 Impeller Domain Modeling

1) **New fluid domain prt file**
   In the menu bar, click [File] → [New], open the "New" dialog box, or directly click the title bar of the button, jump out of the [New] window, in the type field select [Part]; enter a file name, for example, mix-fluid-impeller, remove the check box [use default template], click [OK] button. Subsequently, pop-up new file options window to select the required template, generally choose [mmns_part_solid], click [OK] button.
2) **Create a rotary axis**
   Click the [Axis] button on the feature toolbar to bring up the [Datum Axis] dialog box, select two planes in the work area RIGHT plane, TOP plane, select [Through], click [OK] button.
3) **Axial sketch**
   In the feature toolbar, select [Sketch] or click the icon to enter the sketch section. In accordance with the mixed-flow pump structure dimensions of the pump impeller wheel casing space rotation section boundary line, you can rotate to generate the shape of the wheel hub. After completion, click the OK button to complete the sketch.
4) **Rotate the axial surface drawing**
   In the feature toolbar, select [Rotate], select as a surface, and in [Placement], select Sketch and Rotate Axis A_1. Enter the Sketch interface, and then Project [Use Edge to Create Elements]. In the pop-up [Type] dialog box, select Single, and then select the perimeter lines drawn in the previous step one by one, and then click Close after completing the selection, and then click the Confirmation button to complete the drawing of the curves. Then select the option at the [Rotate as an entity], check the blank box, select the axis of rotation between the work area, and enter the rotation angle of 360°, as shown in Figure 8.1.
5) **Import the blade, cut the fluid domain**
   Select [Insert] [Import], select the built blade solid model.

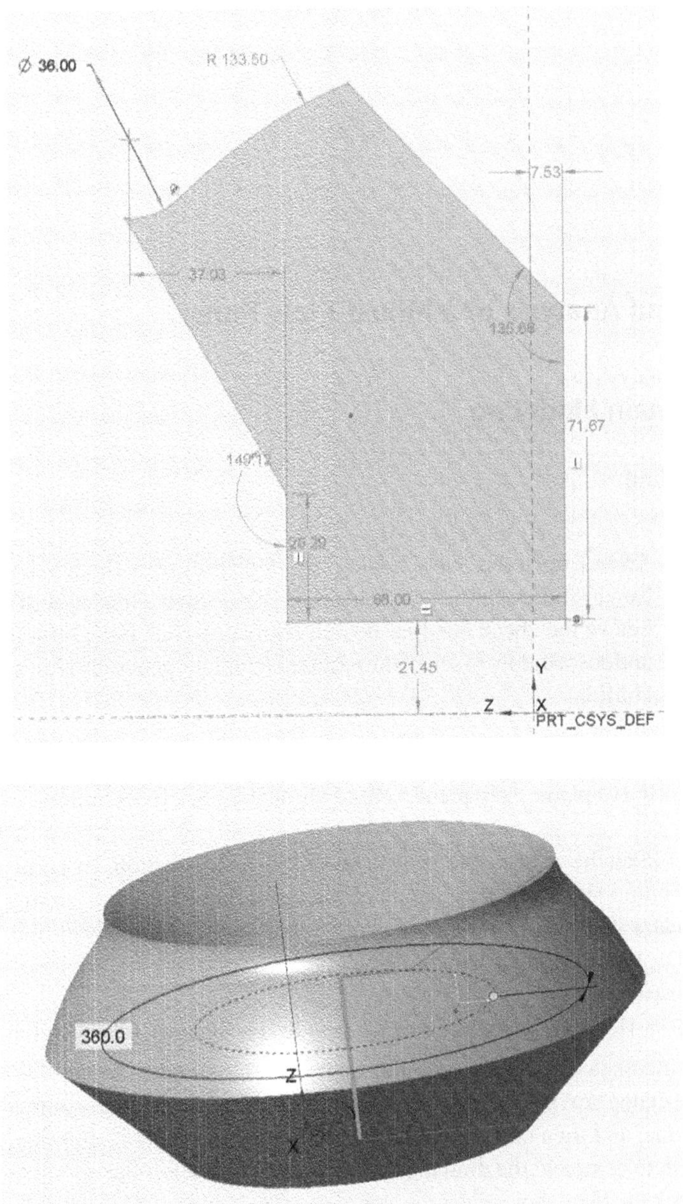

Figure 8.1 Impeller fluid domain modeling.

8.1.2 Modeling of Guide Vane Fluid Domains

The steps are similar to those of Section 8.1.1, with the similarities not being described further, and the differences being in the axial planes and the final modeling of the guide vane fluid domain. A brief description is given here.

The axial surface drawing can also be obtained by directly importing the CAD file. A sketch of the axial plane is shown in Figure 8.2. Import the vane shape and choose to excise the vane, leaving the fluid region. The final domain of fluid is shown in Figure 8.3.

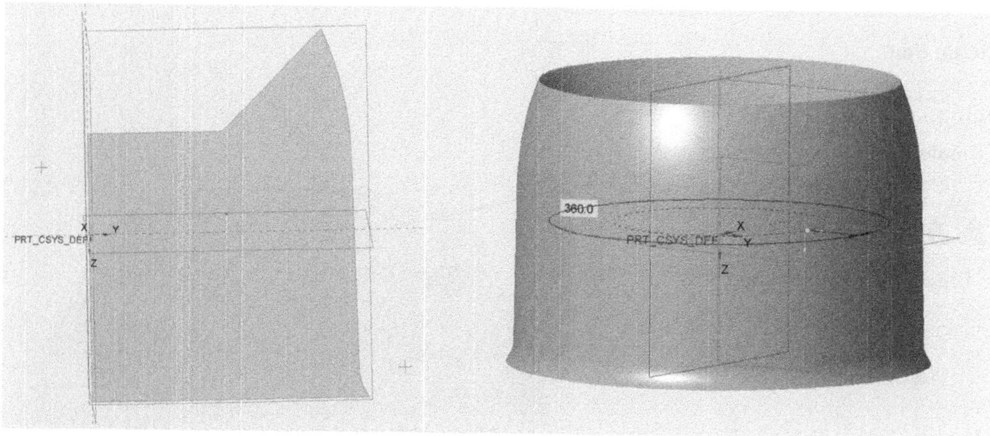

Figure 8.2 Axial planes and rotating domain.

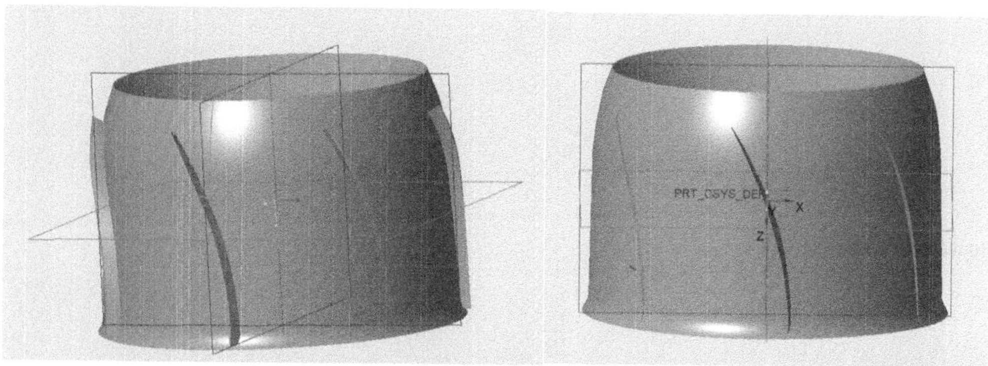

Figure 8.3 Guide vane.

8.2 Structured Meshing

8.2.1 Impeller Structured Meshing

1) Open ICEM CFD software. Set the working directory, [File] → [Change Working Dir], select the file storage path. In the menu bar, [File] → [Geometry] → [Open Geometry] or click the " " icon, open the mixing pump guide vane fluid domain file, and import impeller fluid domain parts.
2) Right-click the model tree Model → Geometry → surface, select Solid and Transparent, then the impeller fluid domain upper solid transparent.
3) Create Part. Right-click Model → Parts, select Create Parts, enter "IN" in the "Part" field, select "Create Part by Selection ", click " " to select the inlet side of the impeller fluid domain, as shown in Figure 8.4. Using the same approach, define the rest of the Part.
4) Create the topology. Select Geometry in the tab bar and click "Repair Geometry". As shown in Figure 8.5, click "Build Diagnostic Topology", keep the default settings, click [Apply] button to create the characterization of the geometric necessity of Point and Curve, the creation of the results as shown in Figure 8.5, Surface is displayed as Wire Frame.

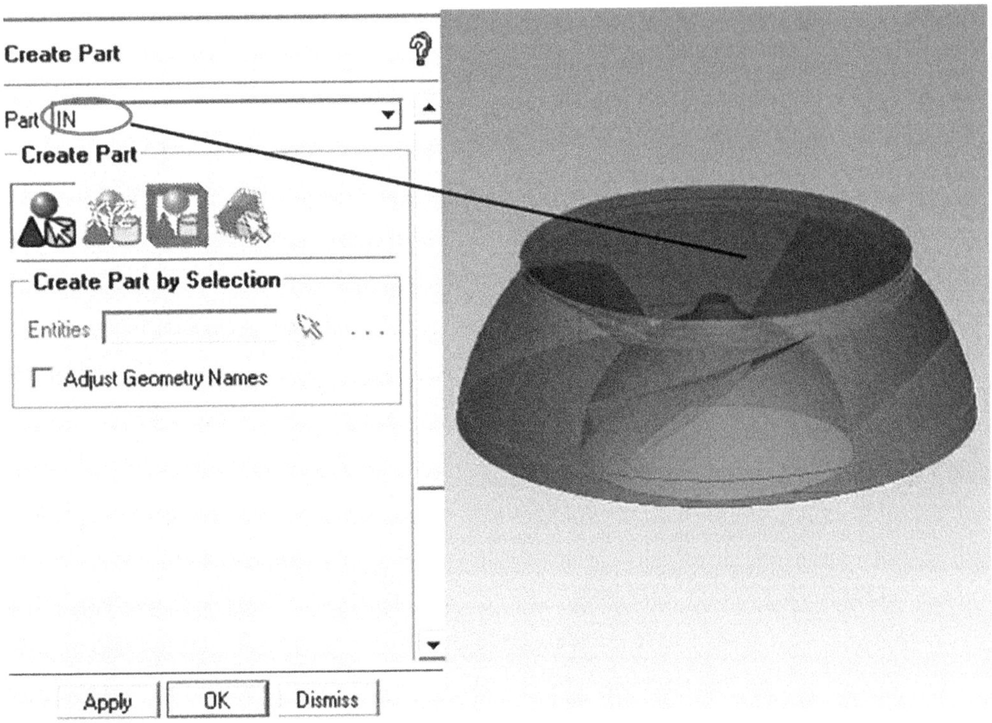

Figure 8.4 Creating a part.

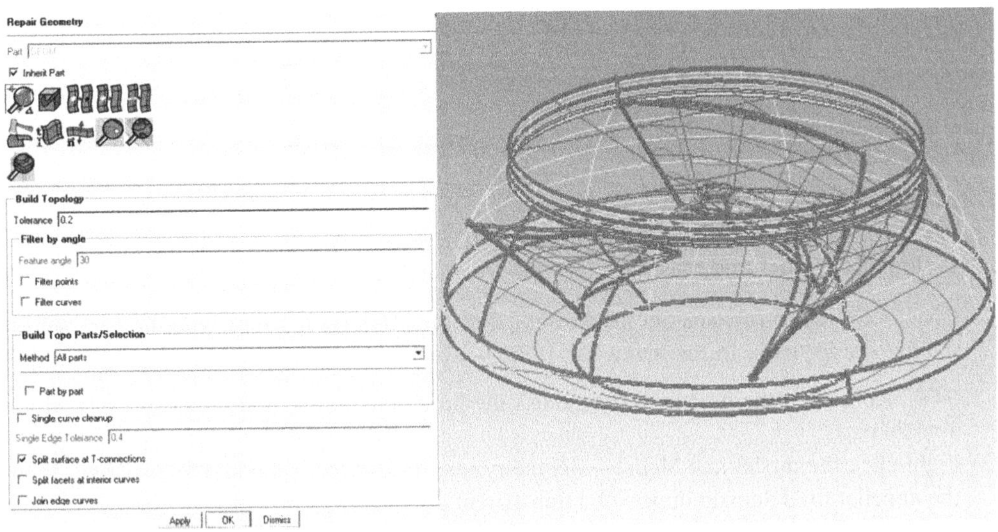

Figure 8.5 Creating a topology.

5) Define Body, select Geometry in the tab bar, and click “Create Body”. Enter “BODY” in the ‘Part’ field, click “ ”, check the “Entire model” checkbox, click [Apply], according to the topology of the entire geometric model to create the Body. The block is further segmented to split the approximate shape of the blade. The specific steps of division are not described in detail, only the more important parts are described.

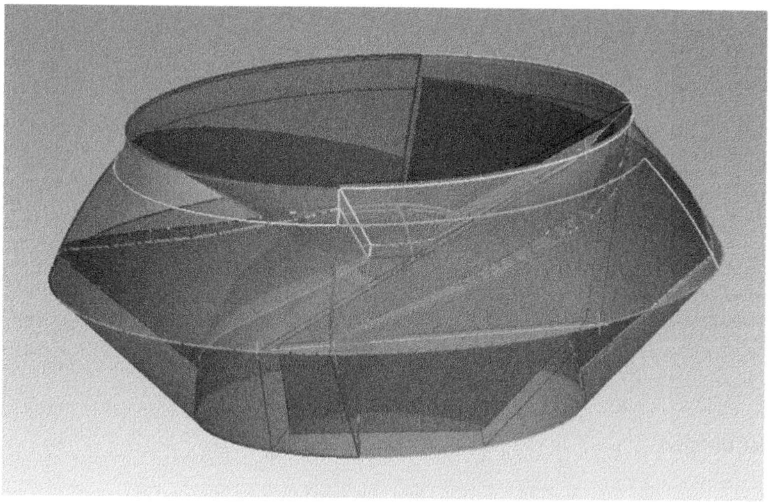

Figure 8.6 Initial block division.

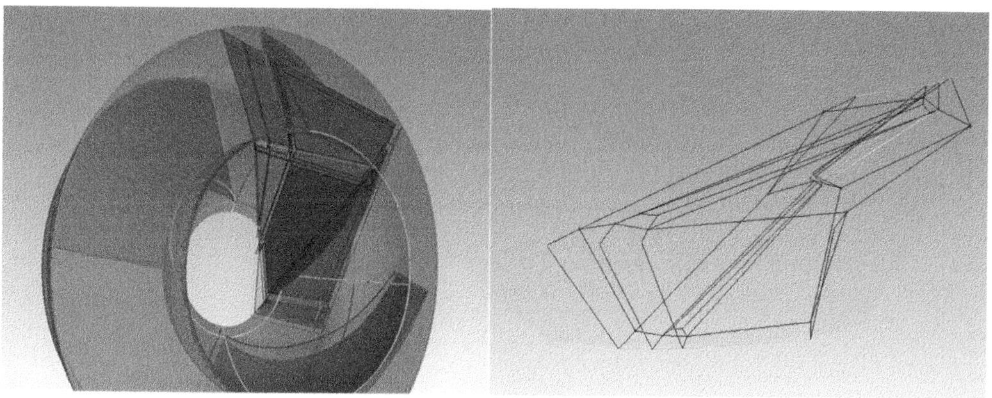

Figure 8.7 Cutting off the blocks of the blade section.

1. First, divide the block, dividing the approximate shape of the blade (Figure 8.6):
2. O-Block the middle blade section and delete the block corresponding to the blade entity (Figure 8.7)
3. And then points, lines associated with the blade are more twisted, so the block will be cut into small pieces, and perform the cycle setting. The final result is shown in Figure 8.8.

6) Set the global mesh size. Click the Mesh tab bar, click "Global Mesh Setup", select "Global Mesh Size", "Scale factor" column set "1", "Max element" column set "4", complete all the mesh Apply, complete all the settings of the grid size.
7) Generate mesh. Check the model tree Model → Blocking → Pre-mesh, the pop-up dialog box, click the "Yes" button to determine the generation of mesh as shown in Figure 8.9.
8) Check the mesh quality. Select the Blocking tab bar, click "Pre-Mesh Quality Histograms", "Criterion" column drop-down list, respectively, select "Determinant" and "Angle" in the drop-down list of the "Criterion" column, other defaults, click [Apply], and then get the mesh quality, as shown in Figure 8.10. The value of "Determinant $2 \times 2 \times 2$" of all meshes is greater than 0.4, and the value of "Angle" of all meshes is greater than 18°, so the mesh quality meets

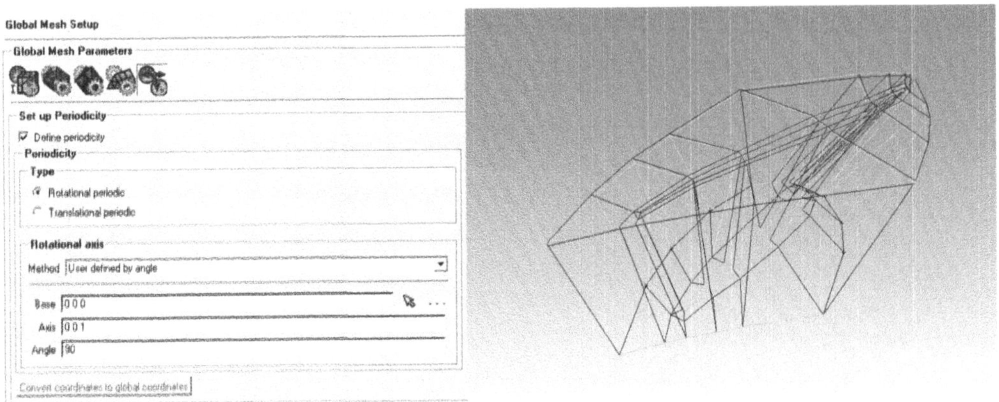

Figure 8.8 Results of block division.

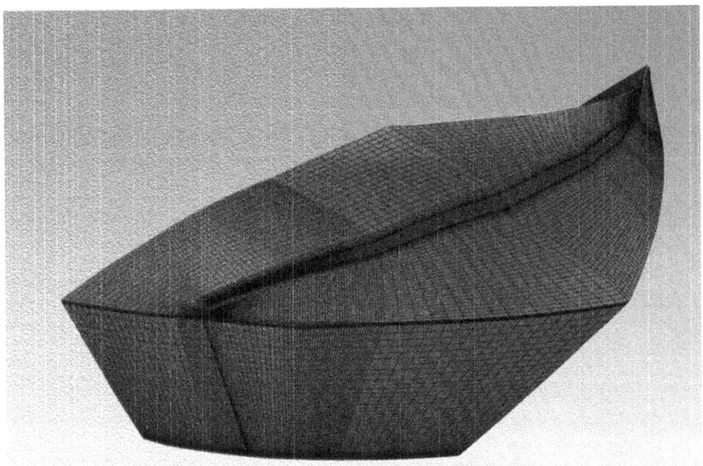

Figure 8.9 Generate mesh.

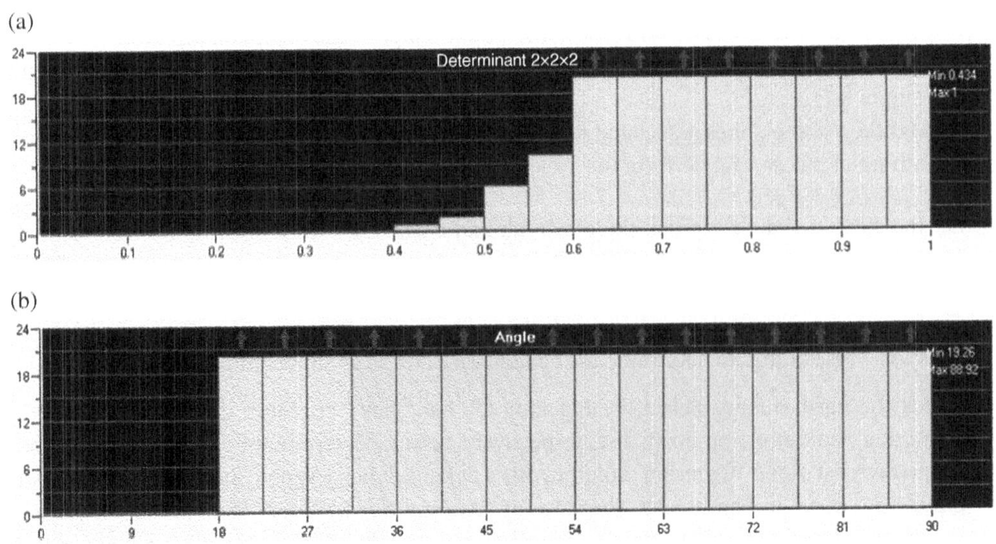

Figure 8.10 Mesh quality: (a) Mesh quality as determined by determinant $2 \times 2 \times 2$, (b) mesh quality using angle as a criterion.

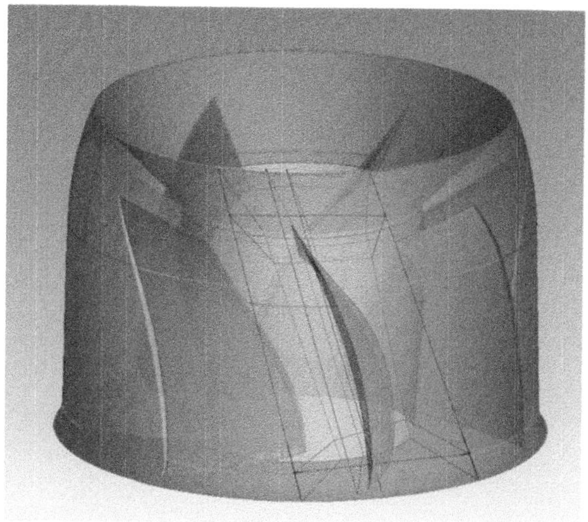

Figure 8.11 The preliminary division of blocks.

the calculation requirements. (If the mesh quality is not good, one can improve the mesh quality by adjusting the Block nodes.)

9) Save the mesh. Right-click the model tree Model → Blocking → Pre-mesh, select “Convert to Unstruct Mesh”. When the information window prompts “Current Coordinate system is global”, it indicates that the mesh conversion has been completed, generating a periodic mesh. According to the knowledge of the periodic mesh, the complete mesh is generated as shown in Figure 8.11 (no need to array Block). Click [File] → [Mesh] → [Save Mesh as] to save the current mesh as Impeller mesh.uns.
10) Select the solver. In the tab bar, click “Output”, click “Select solve”, open the “Select solve” dialog box. In the “Select solve” dialog box, “Output Solver” item, drop-down selection “ANSYS CFX”; “Common Structural Solver” item select “ANSYS”, [Apply] OK.
11) Export the mesh. In the toolbar, click “Output”, click “Write input”, open the “Save As” dialog box. “File name”: impeller.fbc, save, close the dialog box, the pop-up box. Select “No”. The “CFX5” dialog box pops up, in the “Output CFX5 file” column, change the file name to “impeller”, [done]. Finish exporting the mesh.

8.2.2 Guide Vane Structured Meshing

The steps of mesh delineation of the guide vane structure are the same as those of the impeller mesh delineation, and the main difference lies in the delineation of the blocks, so this section mainly describes the delineation of the guide vane blocks.

Preliminarily divide the blocks according to the shape of the guide vane blades, as shown in Figure 8.11. Establish mapping relationships by creating Edge-to-Curve associations. Select the Blocking tab, click “Associate,” then choose “Associate Edge to Curve.” Check the “Project vertices” option. Click the “ ” button under the “Edge(s)” column to select the Edge to be mapped, and click the “ ” button under the “Curve” column to select the corresponding Curve. This ensures that the Edge of the blade Block better conforms to the geometric model. Next, select the Blocking tab, click “Edit Edge,” and choose “Split Edge.” In the “Split Type” section, select “Spline” from the “Method” drop-down menu. Click the “ ” button to select points on the line that need adjustment, then hold the left mouse button to drag them to the appropriate positions. Use the same method for further adjustments. The final adjustment result is shown in Figure 8.12.

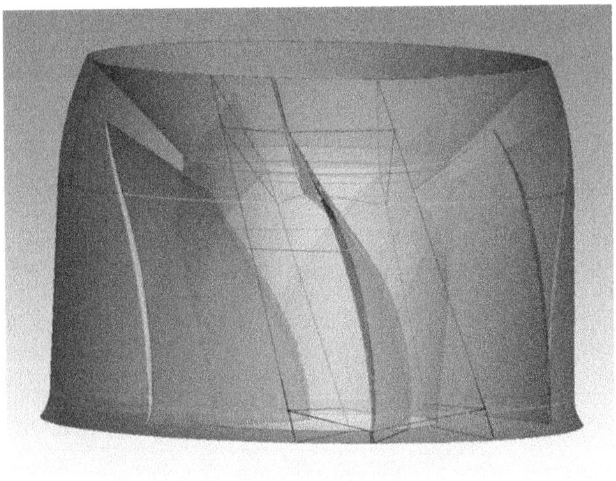

Figure 8.12 Creation result of the outer O-Block.

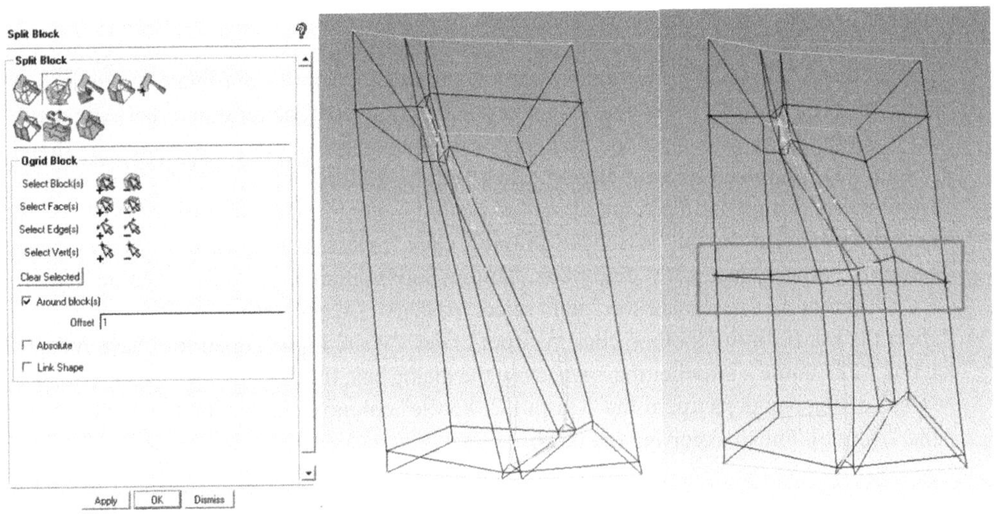

Figure 8.13 Adjusting the block shape.

Create the Outer O-Block. Click the Blocking tab, then select "Split Block" as shown in Figure 8.13. Choose "O-Block", click the " " button under the "Select Block(s)" column to pick the block corresponding to the guide vane blade shape. Next, click the " " button under the "Select Face(s)" column to select the hub contact surface and shroud surface of the guide vane blade. Check the "Around block(s)" option and click [Apply] to confirm. The resulting creation is shown in Figure 8.13. The subsequent steps are the same as those in Section 8.2.1, and the final guide vane mesh is obtained as shown in Figure 8.14.

8.2.3 Volute Structured Mesh Generation

This section continues to focus specifically on block division, with the overall procedure following the content described in Section 8.2.1. The volute structure in this book is relatively complex. It is

Figure 8.14 Guide vane mesh.

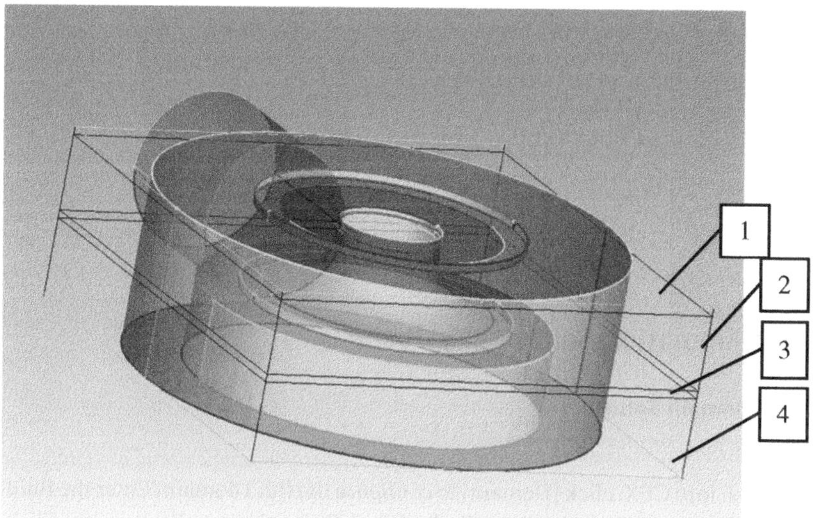

Figure 8.15 The preliminary division of blocks.

initially divided into four layers as shown in Figure 8.15, and then adjusted layer by layer. First, create an O-Block by selecting [Blocking] → [Split Block] → [O-Block]. Click the "Select Face(s)" field to choose the face(s) where the O-Block will be created, and enter "1" in the "Offset" field. Subsequently, delete the intermediate nonfluid domain blocks and verify the correctness of the vertex-to-curve and edge-to-curve mappings. To obtain higher-quality mesh, further divide the blocks into smaller subblocks, resulting in the configuration shown in Figure 8.16. Finally, process the outlet pipe section by creating new blocks. Select the outlet pipe portion and rename its part as FLUID2. Note that the outlet pipe blocks need to merge with the previously divided blocks through proper vertex and edge connections, with adjustments made according to the model geometry. The final block division is shown in Figure 8.17a. The next step is mesh generation, which can be performed following the method described in Section 8.2.1. The final mesh result is shown in Figure 8.17b.

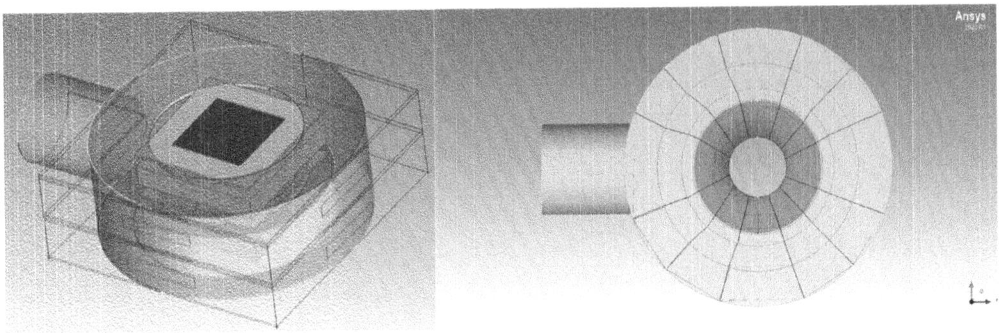

Figure 8.16 Creation result of the outer O-Block.

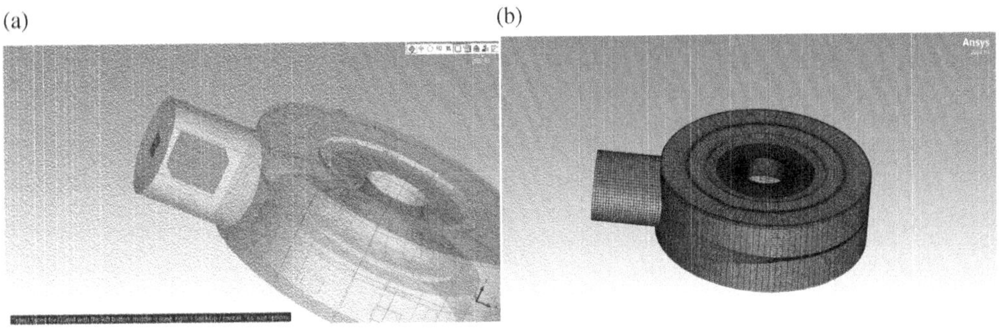

Figure 8.17 Final result of volute blocking: (a) outlet pipe section block division, (b) guide vane mesh.

8.3 Steady Flow Simulation Setup

8.3.1 Computational Domain Setup

1) Fluid Domain Setup

After importing the mesh into CFX, click [Domain] to configure the fluid domain. Enter the fluid domain name in [Name]—for example, input "Impeller" when defining the impeller domain. Click [Selection Dialog] on the right side of [Location] to select the mesh of the impeller flow domain. Then select the flow medium in [Material], such as [Water]. Input the reference pressure in [Reference Pressure]. In [Domain Motion], select the domain type:

- Impeller domain: [Rotating]
- Guide vane domain and volute domain: [Stationary]

As shown in Figure 8.18, eventually, the highlighted part is the impeller, which is set as the rotating domain, while the rest are all stationary domains.

2) Turbulence Model Setup

After completing the basic domain setup, proceed to select the turbulence model. Click on the previously established domain (e.g., the impeller domain). Select [Impeller], then navigate to [Fluid Models]. Enter the fluid temperature in [Fluid Temperature], and under [Turbulence], select the desired turbulence model for computation [1].

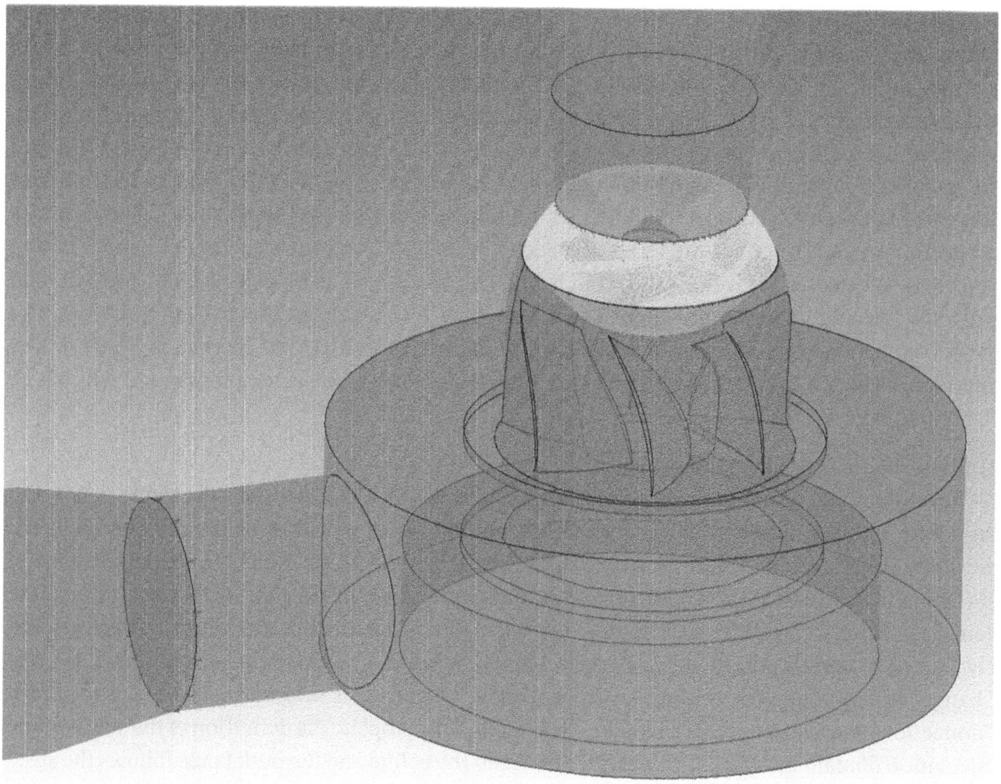

Figure 8.18 Fluid domain setup.

8.3.2 Boundary Conditions Setup

1) Wall Boundary Conditions Setup

After the fluid domain is established, boundary conditions need to be configured. Taking the impeller domain as an example for boundary condition setup: Click [Boundary], select [Impeller], then enter the wall name such as [Blade] in [Name], select [Wall] in [Boundary Type], choose the blade wall surface in [Location], select [Rotating] in [Frame Type], and finally click [Apply]. Then, define the front shroud and rear shroud of the impeller sequentially using the same method. The wall definitions for other domains are similar to those of the impeller, but differ in [Frame Type]. Since other domains are stationary, select [Stationary] in [Frame Type].

2) Inlet/Outlet Boundary Conditions Setup

At the pump inlet, select [Inlet] in [Boundary Type], then click [Boundary Details]. Under [Mass And Momentum], select the inlet boundary condition such as [Normal Speed] and enter the inlet velocity. At the pump outlet, select [Opening] in [Boundary Type], then click [Boundary Details]. Under [Mass And Momentum], select the outlet boundary condition such as [Opening Pres. and Dirn] and enter the outlet pressure, finally click [Apply].

8.3.3 Interface Settings

1) Dynamic–Static Interface Settings

After defining the wall boundary conditions, the interface definitions need to be made. The dynamic–static interface of the mixed-flow pump calculation domain includes the interface between the inlet pipe and the impeller, and the interface between the impeller and the volute. Taking the dynamic–static interface definition between the inlet pipe and the impeller as an example, click on [Domain Interface] in the toolbar, and enter [Inlet_Impeller] in the pop-up dialog box, then click [OK]. In the first tab [Basic Settings], select the inlet domain in [Interface Side 1] of [Domain], and in the pop-up dialog box, select the interface between the inlet pipe and the impeller, that is, the outlet of the inlet pipe domain; select the impeller domain in [Interface Side 2] of [Domain], and select the inlet of the impeller domain in [Region List]. The interface connection model [Interface Model] is selected as [General Connection], [Frame Change/Mixing Model] is selected as [Frozen Rotor], [Pitch Change] is selected as [Specified Pitch Angles], and enter the specific values, with Pitch Angle Side 1 as 360 and Pitch Angle Side 2 as 360. The mesh connection method [Mesh Connection] is selected as [GGI], and then click [Apply]. The interface definition between the impeller and the volute follows the same process and will not be elaborated here.

2) Static–Static Interface Settings

The static–static calculation domain interfaces are the interface between the guide vanes and the volute, and the interface between the volute and the outlet pipe. Taking the interface between the guide vanes and the volute as an example, select the guide vane domain in [Interface Side 1] of [Domain], and select the outlet surface of the guide vanes in [Region List]. Select the volute domain in [Interface Side 2] of [Domain], and select the inlet surface of the volute in [Region List]. The interface connection model [Interface Model] is selected as [General Connection], [Frame Change/Mixing Model] is selected as [None], [Pitch Change] is selected as [None], and [Mesh Connection] is selected as [GGI]. Finally, click [Apply] to complete the definition of the static–static calculation domain interface. The interface between the volute and the outlet pipe follows the steps above (Figure 8.19).

8.3.4 Solution Settings

After completing the above settings, solution settings need to be made. Double-click [Solver Control], and make settings in [Basic Settings]. For the convection term [Advection Scheme], select [High Resolution], and for the turbulence numerical term [Turbulence Numerics], select [First Order]. Then set the total number of solution steps. In [Convergence Control], enter 2000 for [Max. Iterations] (just to ensure the stability of the calculation results). For the time term of the solution parameters [Timescale Control], select [Physical Timescale] and enter a specific value, usually the reciprocal of the impeller speed. For the convergence criterion [Convergence Criteria], select the average value [RMS] and enter a specific value of 1.E-4. Finally, click [Apply] to complete the definition of the solver parameters.

8.3.5 Calculation Settings

Finally, save the current file, and click the button on the menu bar to generate the .def file for calculation. The first button generates the .def file and opens it directly, while the last button allows you to select a directory to save the .def file and open it manually. In the opened Define Run window, if you need to set the calculation accuracy, you can open the Double Precision option. If the Run Mode is for local use with an Intel core, select Intel MPI Local Parallel, and enter the number of calculation cores. In the Working Directory, select the folder where you want to save the file, and then click Start Run to start the calculation.

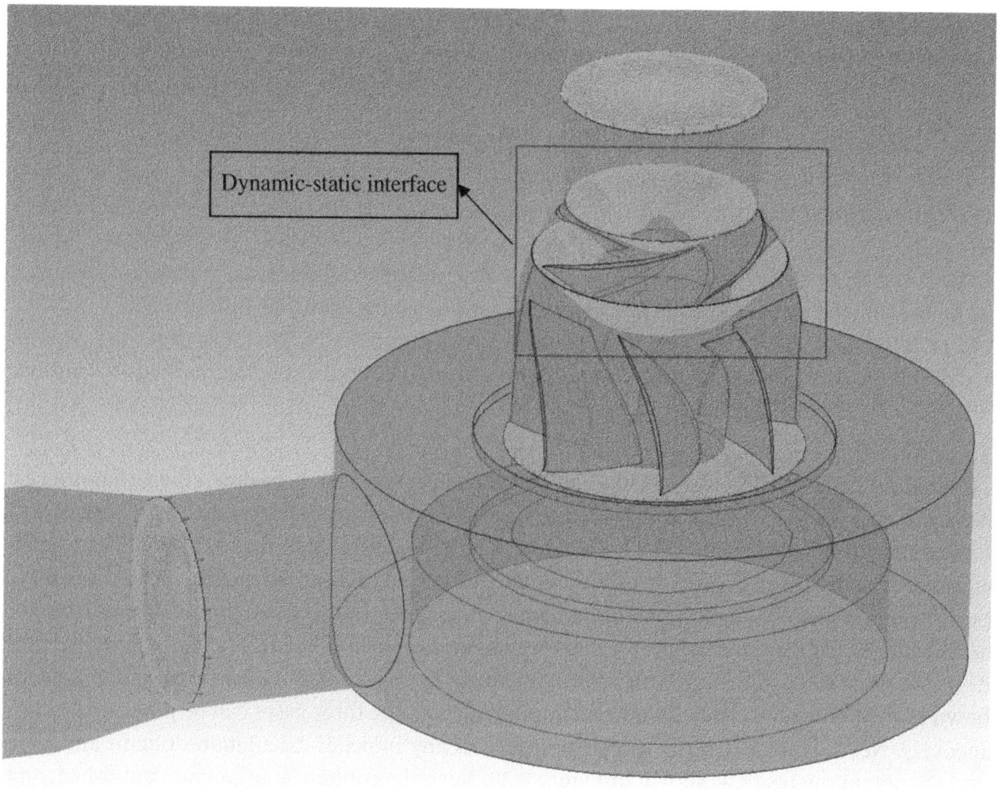

Figure 8.19 Interface settings.

8.4 Transient Flow Simulation Setup

The transient flow field calculation setup for mixed flow pumps is similar to that of centrifugal and axial flow pumps, with only minor differences in the model. Therefore, this section will only briefly describe the general steps, and for detailed information, please refer to the chapters on centrifugal and axial flow pumps.

First, change the value of [Analysis Type] to [Transient]. Next, define the total time for the transient calculation. Generally, to obtain a reliable solution, the transient calculation should cover 6–10 rotation cycles of the impeller. In the [Max. Timesteps] input box under [Maximum Number of Timesteps], enter the required maximum number of time steps. Then, define the number of calculation steps within each rotation cycle. This can be understood as determining how many degrees the impeller rotates before each calculation. Finally, click the [Apply] button to complete the transient setup for the unsteady calculation.

The interface settings for unsteady data differ significantly from those for steady-state calculations, mainly in the configuration of the connection model at the interface between the moving and stationary computational domains. Specifically, double-click the interface between the moving and stationary computational domains. For the connection model at the interface, although [Interface Model] is still selected as [General Connection], the coordinate transformation [Frame Change/Mixing Model] needs to be changed to [Transient Rotor Stator] to meet the requirements of unsteady calculations. The rest of the settings remain the same as for steady-state calculations.

After completing the above configuration, click the [Apply] button to define the data interface between the moving and stationary computational domains for unsteady calculations [2].

8.5 Calculation Results and Postprocessing

The result obtained after the calculation converges is just a file. To observe the flow field change information more intuitively, one needs to use a postprocessing software for visualization. Simply put, it is the process of converting the source file into pictures, animations, and data results that can be understood by people. This process is the postprocessing of the CFD simulation.

After detailing the postprocessing steps for centrifugal pumps and axial flow pumps, the steps for mixed flow pumps are similar. Therefore, this stage briefly introduces some methods to improve the efficiency and quality of postprocessing [3, 4].

1) Multifile Mode

Open the CFD-POST interface, click File, click Load Results File, hold down the Ctrl button on the keyboard, select the three result files that need to be processed simultaneously, and check Edit case names. In this way, the name of each result file will be displayed in the view display box, making it easier to distinguish. The character length is displayed and can be modified. After opening, it will be automatically divided into three display interfaces. In this case, the working surface of the impeller is taken as an example for demonstration, as shown in the figure. If one wants to process these three cases in the same way simultaneously, double-click the Case Comparison item, as shown in the figure, and check the relevant options. Then, the three cases can be processed simultaneously. Next, double-click the working surface of any impeller calculation domain and click Apply to obtain the result as shown in Figure 8.20. Next, the contour map can be established, and the scale can be modified, etc., all of which can be done simultaneously for the three cases. The detailed process of contour map processing is not elaborated here.

2) Transient Simulation Animation Processing

This section provides the animation of the pressure contour map as an example. First, the pressure contour map of the impeller blade is established in advance, and this process is not described

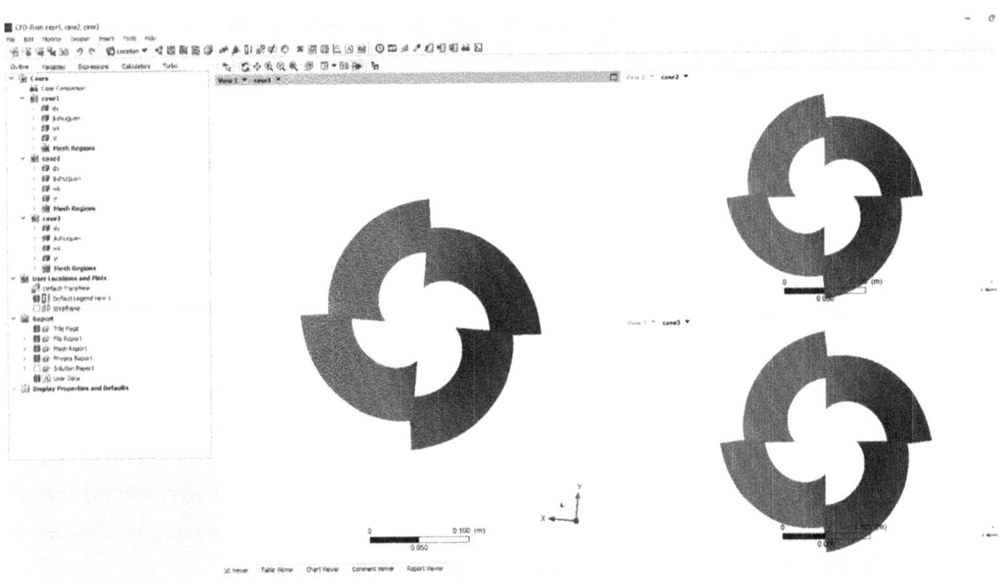

Figure 8.20 Multifile mode.

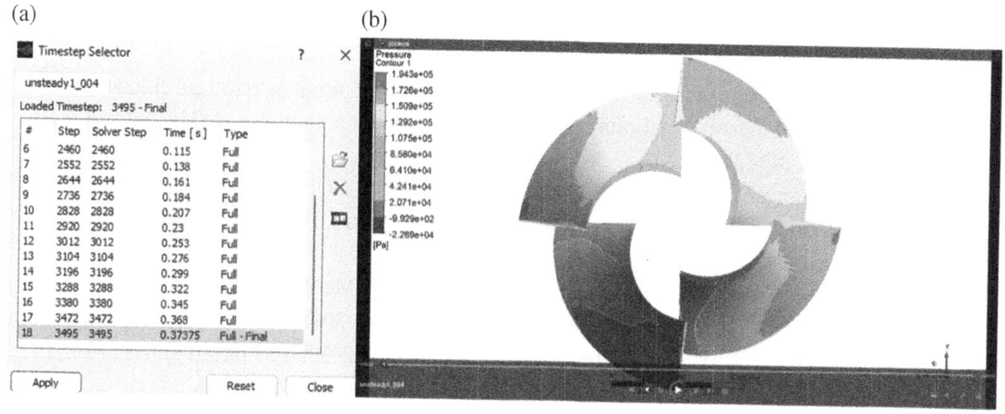

Figure 8.21 Transient simulation animation processing: (a) Calculation time, (b) simulation animation video.

here. Click on Tools, Timestep Selector to view the calculation time, as shown in Figure 8.21a. Next, click on Animation, set the animation save path, and click on Format to set the save format. For more specific settings, such as image quality, click on Options for further settings. Finally, click on the play button and obtain the simulation animation video, as shown in Figure 8.21b.

3) Vorticity Field Processing and Analysis

Common vortex identification methods include the vorticity method, Q criterion, λ_2 criterion (Lambda-2), Δ criterion (Delta), λ_{ci} criterion (Lambda-ci), and Ω criterion (Omega). The specific principles are not described here; this chapter only presents the final results.

Step 1: Click on Location, Vortex Core Region to establish vorticity analysis. Step 2: Select the computational domain; in this case, the impeller is taken as an example. Step 3: Choose the vorticity analysis model; here, the vorticity method (Vorticity) is selected. The desired vorticity contour map can be obtained by adjusting the Level. Step 4: Click on Color, set Mode, Variable, and Range. After setting, click Apply to obtain the result as shown in Figure 8.22. From the contour map, obvious clearance leakage vortices and leading edge leakage vortices can be observed.

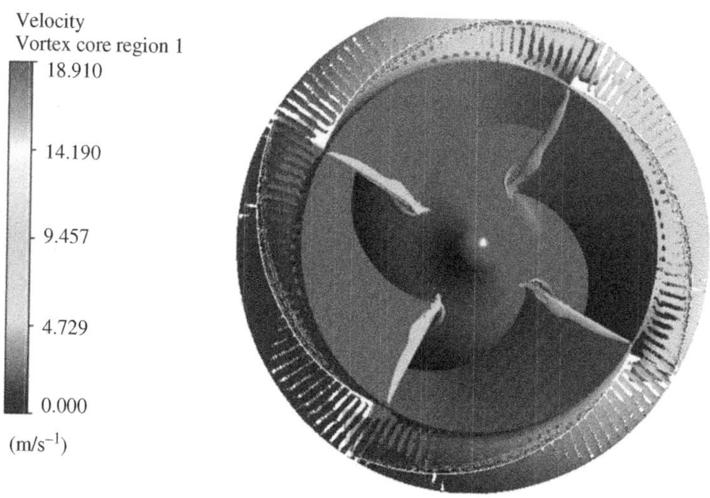

Figure 8.22 Vorticity contour map.

References

1 Ji, L., Li, W., Shi, W., and Agarwal, R. K. (2020). Application of Wray-Agarwal turbulence model in flow simulation of a centrifugal pump with semi-spiral suction chamber. *ASME Journal of Fluids Engineering* 143 (3): 031203.
2 Li, W., Yang, Q., Yang, Y., Ji, L., Shi, W., and Agarwal, R. (2024). Optimization of pump transient energy characteristics based on response surface optimization model and computational fluid dynamics. *Applied Energy* 362: 123038.
3 Li, W., Pu, W., Ji, L., Yang, Q., He, X., and Agarwal, R. (2024). Mechanism of the impact of sediment particles on energy loss in mixed-flow pumps. *Energy* 304: 132166.
4 Li, W., Long, Y., Ji, L., Li, H., Li, S., Chen, Y., and Yang, Q. (2024). Effect of circumferential spokes on the rotating stall flow field of mixed-flow pump. *Energy* 290: 130260.

9

Numerical Simulation and Analysis of a Vortex Pump

9.1 Computational Domain Modeling

Based on the impeller, pump body, and pump cover created using Creo in the previous chapters, the computational domain of the vortex pump was extracted and modeled, as shown in Figure 9.1. The main components of the computational domain include the inlet pipe, impeller flow passage, front volute flow passage, rear volute flow passage, and outlet pipe [1–3]. The inlet and outlet pipes, as well as the front and rear flow passages, can be modeled directly using the [Extrude] command. The extraction of the impeller flow passage can be performed by following the steps below:

For the impeller flow domain, the extraction is based on the previously drawn impeller. First, use the [Extrude] command and click [Remove Material] to cut away the excess hub portion, as shown in Figure 9.2.

Next, select the remaining part, click Copy, then click Paste to enter the SURFACE: Copy interface, and then click OK to generate a series of planes, as shown in Figure 9.3.

Click [Extrude], select a side plane to enter the Sketch interface, then draw the side view of the impeller flow passage. Finally, click Confirm, as shown in Figure 9.4.

Select the previously copied plane, click [Solidify], then select the previously copied plane again, and choose [Remove Material] to finalize the impeller flow passage drawing, as shown in Figure 9.5.

9.2 Structured Mesh Generation

9.2.1 Mesh Generation for Impeller Flow Domain

The generation of a block-structured grid is typically accomplished using ICEM software. After launching ICEM, the first step is to click [File] in the menu bar, then select [New Project] to create a new impeller mesh file and assign it a name. Next, click [File] again and choose [Import Model] to import the geometric data of the impeller. During the import process, the appropriate file format exported from the 3D modeling software (in this case, the STP format is used) should be selected. The specific workflow is illustrated in Figure 9.6.

Before proceeding with mesh generation, a crucial step is to create Parts and clearly define the boundary and interface surfaces. The specific steps are as follows: First, right-click on the [Parts] option, then left-click on [Create Part] to initiate the Part creation process. This step is illustrated in Figure 9.7. Next, in the [Part] input box at the bottom of the interface, enter a clear name for the Part to be created. After naming, click the [Create Part by Selection] button and carefully select the body representing the impeller flow passage in the right-hand window. This step is visually demonstrated in Figure 9.7. After completing the selection, click [Apply] to confirm. Subsequently, follow the same procedure to define and set all other boundary and interface surfaces one by one.

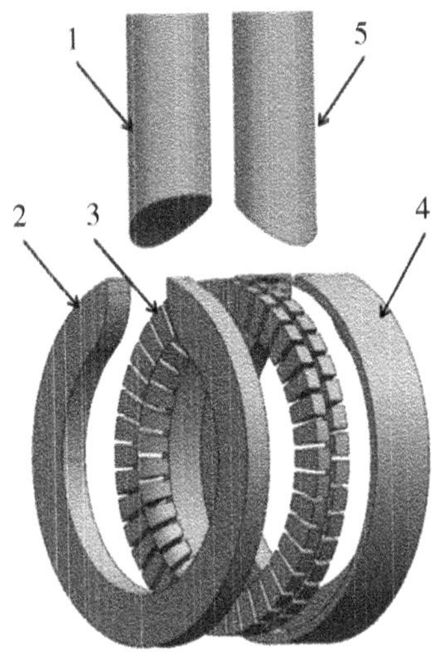

Figure 9.1 Computational domains of vortex pump: 1. Inlet. 2. Front flow passage. 3. Impeller. 4. Rear flow passage. 5. Outlet.

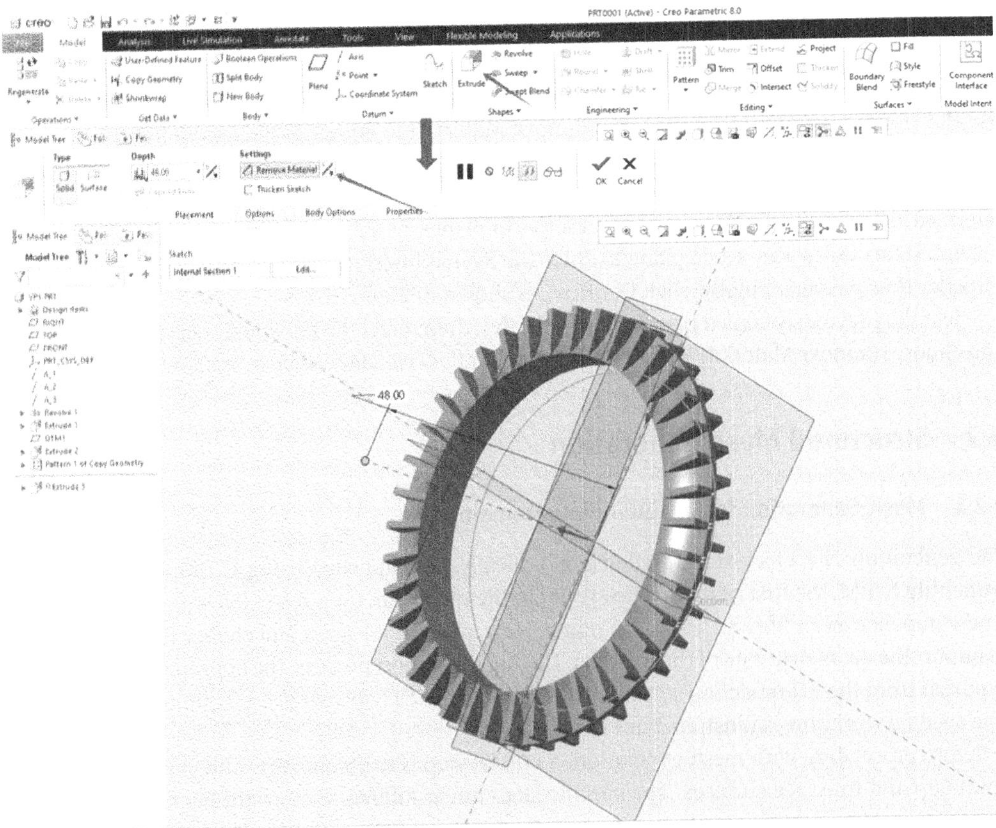

Figure 9.2 Hub removal.

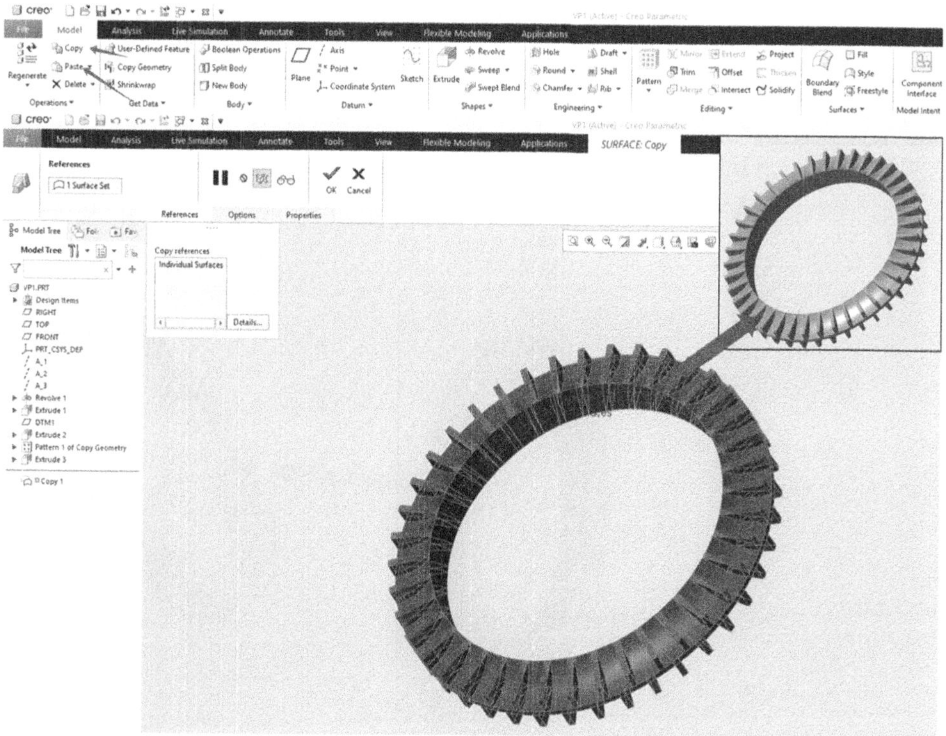

Figure 9.3 Surface copy.

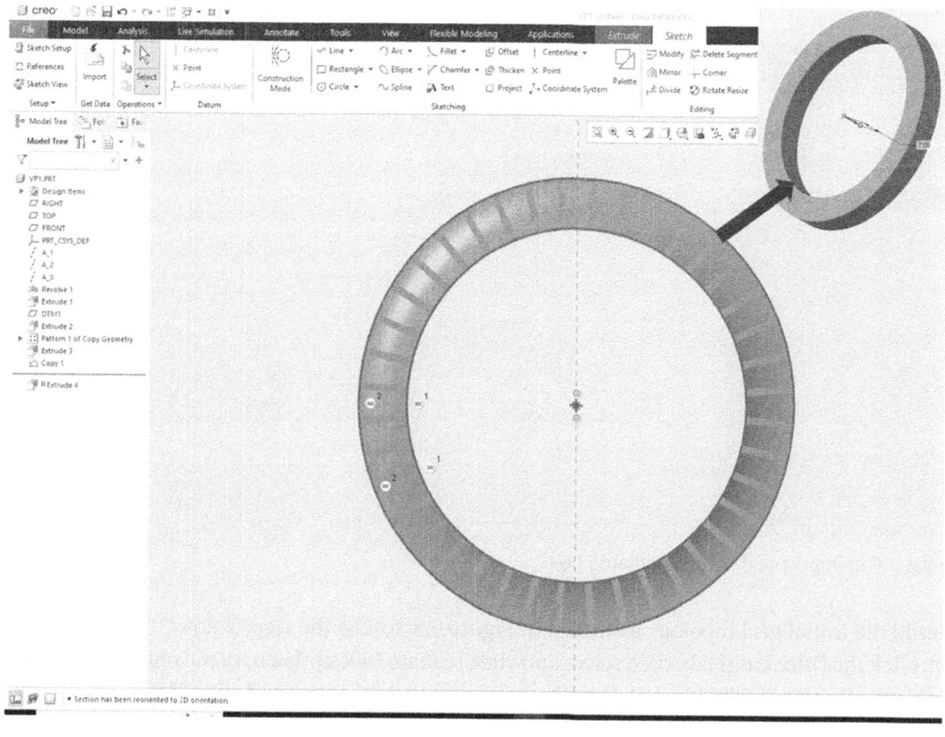

Figure 9.4 Annular flow passage drawing.

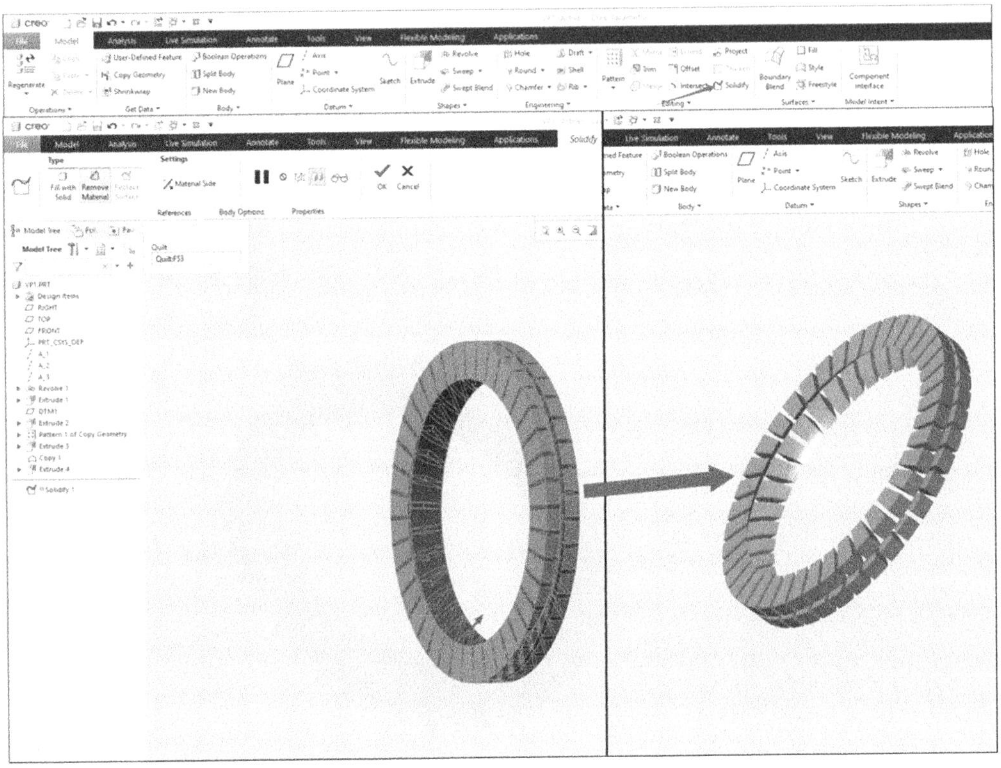

Figure 9.5 Impeller flow passage drawing.

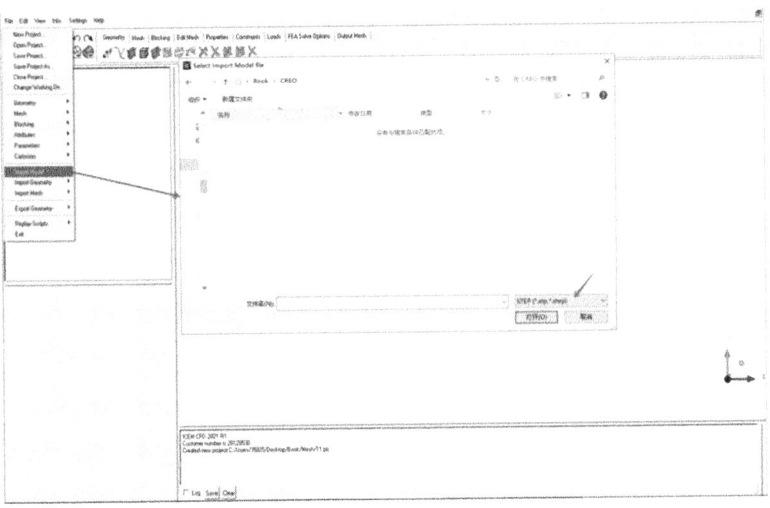

Figure 9.6 File import and selection dialog box.

To build the initial grid topology as shown in Figure 9.8, follow the steps below:

First, click the [Blocking] tab, then select and click [Create Block]. Then, in the option bar on the left, find and click [Initialize Blocks]. Next, in the options below, select and click [Select geometry]. At this time, in the window on the right, click with the left mouse button to select the watershed area of the impeller. After selection, confirm by clicking the middle mouse button or clicking the

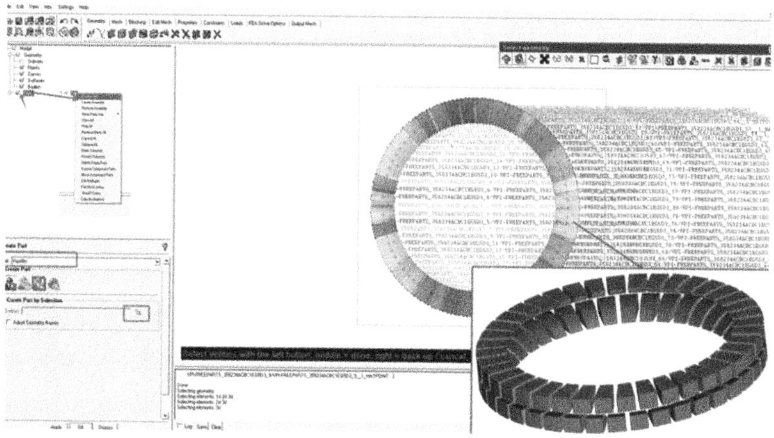

Figure 9.7 Part creation.

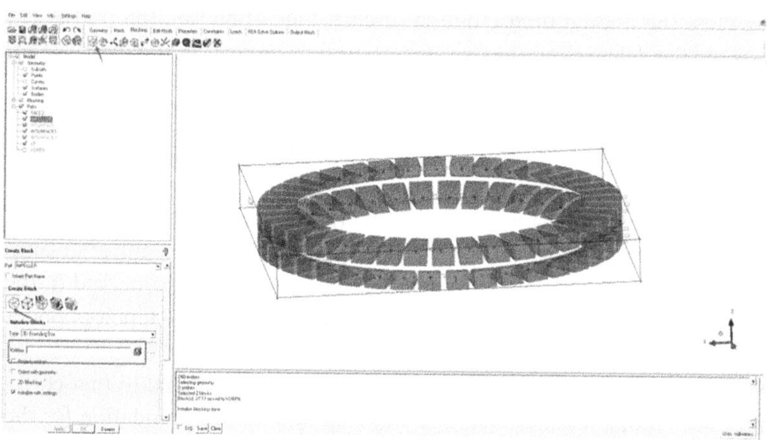

Figure 9.8 Block creation.

[Apply] button in the lower left corner. The program will then generate the initial mesh topology of the impeller, which is shown in the form of black wireframe in Figure 9.8.

After completing the construction of the topology of the impeller fluid domain, it is necessary to further adjust and cut the fine operations to generate the main body of a single impeller passage Block; the specific operation flow is shown in Figure 9.9. First, click the [Blocking] tab, and select and click [Split Block]. Next, click [Split Block] again in the option bar on the lower left side of the interface. At this point, in the [Block Select] option below, check [Visible] to ensure that only visible blocks are operated. Afterwards, click on the [Edge] option in order to select the cut edge. Next, in the right window, use the left mouse button to click exactly on the edge of the selected Block that needs to be cut. Finally, by clicking the middle mouse button or clicking the [Apply] button on the lower left side of the interface, the precise cutting operation of the initial topology can be completed. After successfully completing the block cutting operation, the next step is to clean up the redundant blocks, at this time, just click on the [Delete Block] option, and then select those blocks that are no longer needed in the interface, after confirming that the selection is correct, click the middle button of the mouse or click on the [Apply] button on the lower left side of the interface to perform the deletion operation, so that all the redundant blocks are accurately removed. After

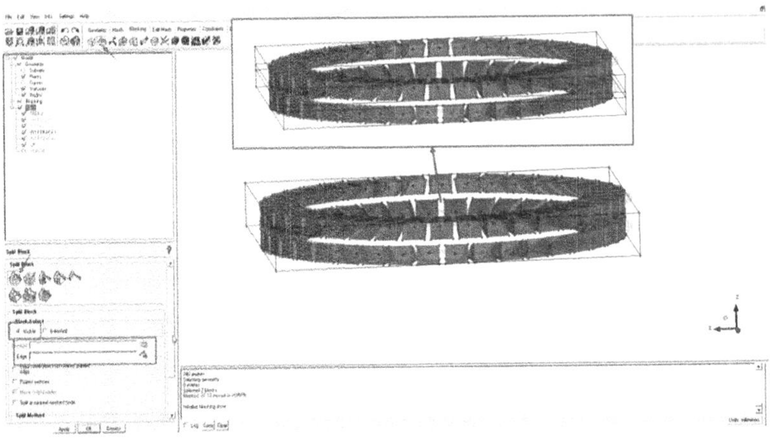

Figure 9.9 Block splitting.

confirming the selection, click the middle mouse button or click the [Apply] button on the left side of the interface to execute the delete operation, so as to accurately remove all the unnecessary blocks.

Since the shape of a single impeller passage is not hexagonal, it is necessary to adjust the topology accordingly by merging two topology nodes, and the detailed operation of this process is shown in Figure 9.10. First, click the [Blocking] tab, and select and click [Merge Vertices]. Next, in the left-hand column, click [Merge Vertices] again to start the merge operation. Then, in the right window, click to select the two topology nodes marked in the red box in the figure, and after completing the first selection, then continue to click to select the two topology nodes marked in the blue box. After the completion of this series of operations, the merging of the initial topology nodes is realized.

In the preliminary stage of constructing the impeller passage mesh, it is crucial to first correlate the initial topology with the impeller structure, and this step lays a solid foundation for the subsequent adjustment of the topology. The specific operation flow is shown in Figure 9.11:

First, click the [Blocking] tab, and then select and click [Associate] from it. In the option bar on the lower left side of the interface, find and click [Associate Vertex]. Immediately below, click

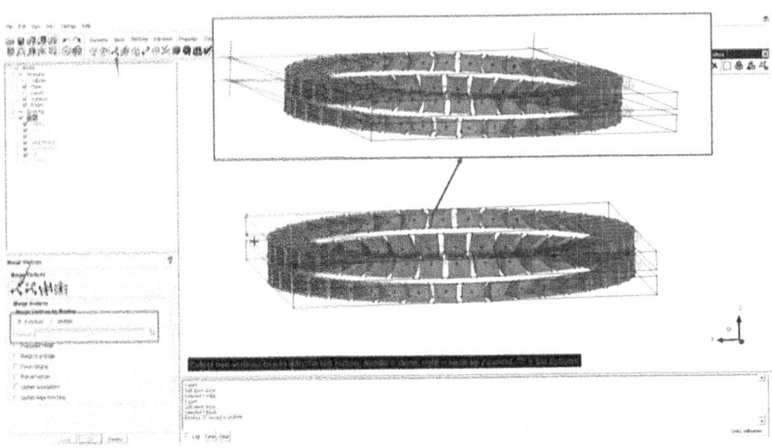

Figure 9.10 Block nodes merging.

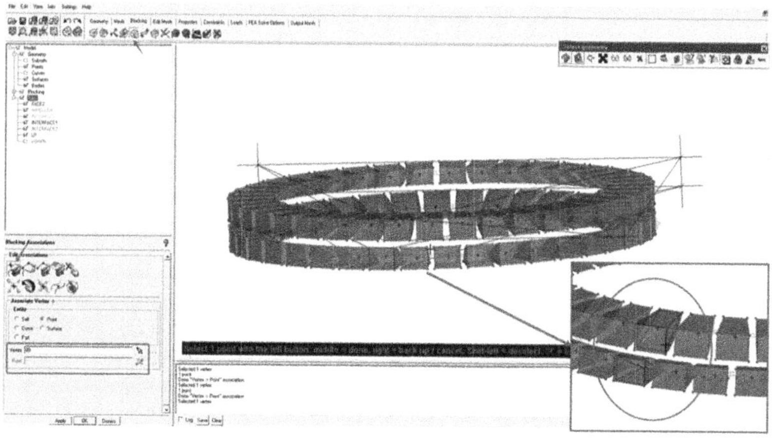

Figure 9.11 Impeller topology association.

on the [Vertex] area to activate the point selection function. Subsequently, in the window on the right, precisely click and select the topological point in the illustration. Next, turn the view back to the lower left window, perform the appropriate action, and then, in the right window, precisely click and select the point corresponding to the impeller structure. After completing this series of clicks, you can confirm and complete the point association operation between the initial topology and the impeller structure by clicking the middle mouse button or directly clicking the [Apply] button in the lower left window. Follow the above steps to associate the remaining topology nodes with their corresponding points on the impeller structure one by one, making sure that each node is accurately connected to its corresponding structure point.

Given that the shape of a single impeller flow passage is approximately triangular, it is necessary to construct a Y-shaped block topology to ensure the quality of the generated impeller flow passage mesh and avoid negative mesh volume. The detailed steps are shown in Figure 9.12:

First, click on the [Blocking] tab, then select and click [Edit Block]. In the options on the lower left side, find and click [Convert Block Type]. Next, in the [Type] section at the bottom, select [Y-Block] as the target type. Then, click in the [Blocks] area to activate the selection function for the blocks.

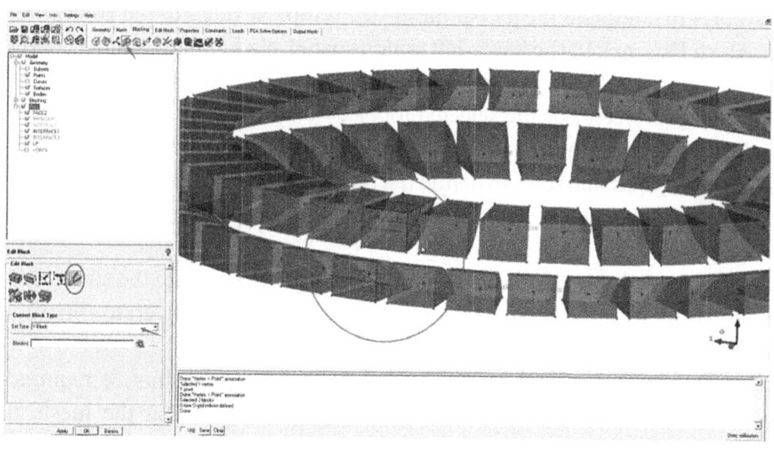

Figure 9.12 "Y"-type block creation process.

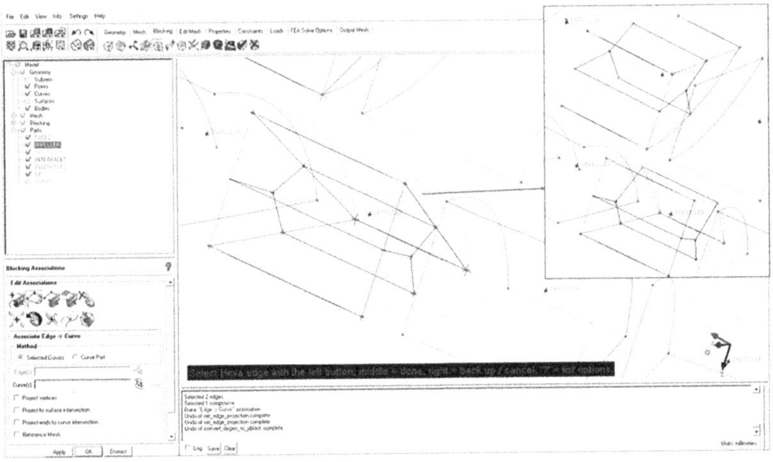

Figure 9.13 Topology line association.

At this point, in the right window, accurately click and select the block from the initial topology structure.

After completing the above steps, you can confirm and finalize the conversion of the initial topology structure into the [Y-Block] by clicking the middle mouse button or directly clicking the [Apply] button in the lower left window. This step is crucial for improving the quality and stability of the impeller flow passage mesh.

Next, perform the line association operation for the impeller flow domain topology, as shown in Figure 9.13. First, click on the [Blocking] tab and select the [Associate] function. Then, in the options on the lower left side, click [Associate Edge to Curve]. Next, click in the [Edges] section at the bottom, and in the right window, select the red topology line shown. Afterward, click in the [Curves] section, and in the right window, select the corresponding impeller flow passage edge. Finally, complete the operation by clicking the middle mouse button or clicking the [Apply] button in the lower left window.

After setting up the Block structure, proceed with mesh generation. The first step in this process is to set up the global mesh size, as shown in Figure 9.14. First, click on the [Mesh] tab, then select [Global Mesh Setup] and click on it. In the left options panel, find the [Max element] option, input the value 2, and click [Apply] to complete the global mesh size setup. Next, go to the [Blocking] tab, select and click [Pre-Mesh Params]. In the lower-left options panel, click [Edge Params]. Then, click the corresponding button under the [Edge] option, and in the right window, and select the topology edges on the periodic surfaces. For the selected topology edges, input the desired number of nodes under [Nodes]. After entering the values, click [Apply] in the lower-left window to apply the settings. Finally, follow the same steps to define the mesh nodes for the remaining edges, ensuring that each edge is correctly assigned its node count, thereby completing the entire mesh generation process.

Check [Pre-Mesh] in the upper-left window, as shown in Figure 9.15, to generate the structured mesh. Afterward, right-click [Pre-Mesh] in the upper-left window and select [Convert to Unstructured Mesh], as shown in Figure 9.15.

Since only one impeller flow passage mesh has been created, periodic replication is required to obtain a full impeller mesh. The final step involves rotating and replicating the mesh of a single impeller flow passage, as shown in Figure 9.16. Click on the [Edit Mesh] tab, select [Transform Mesh], and click on it. In the lower-left options panel, click the appropriate button.

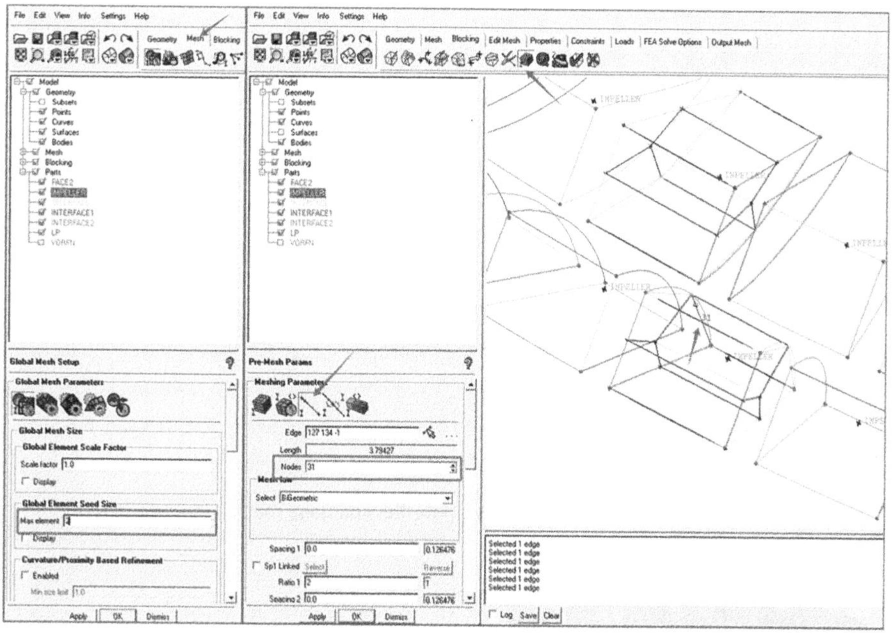

Figure 9.14 Mesh size setup.

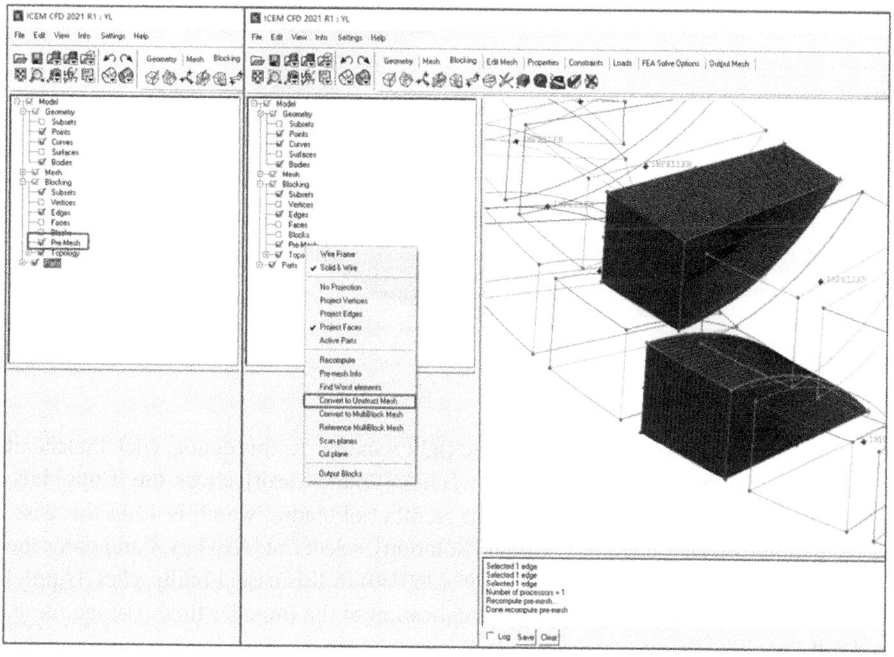

Figure 9.15 Structured mesh generation and conversion to unstructured mesh.

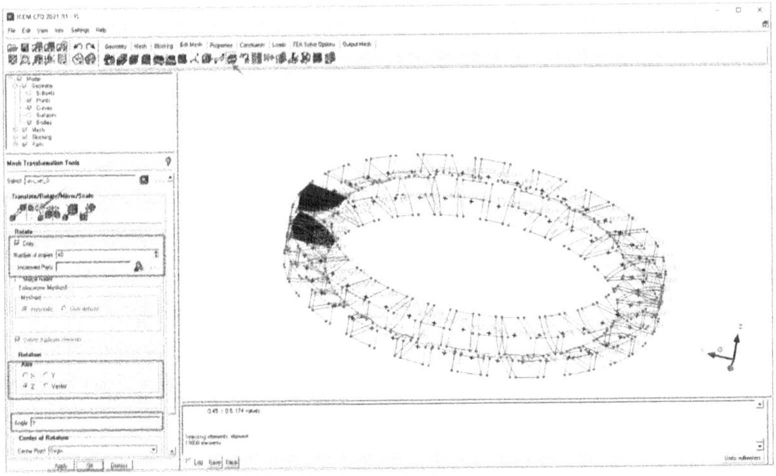

Figure 9.16 Periodic rotational replication of impeller's single-flow passage mesh.

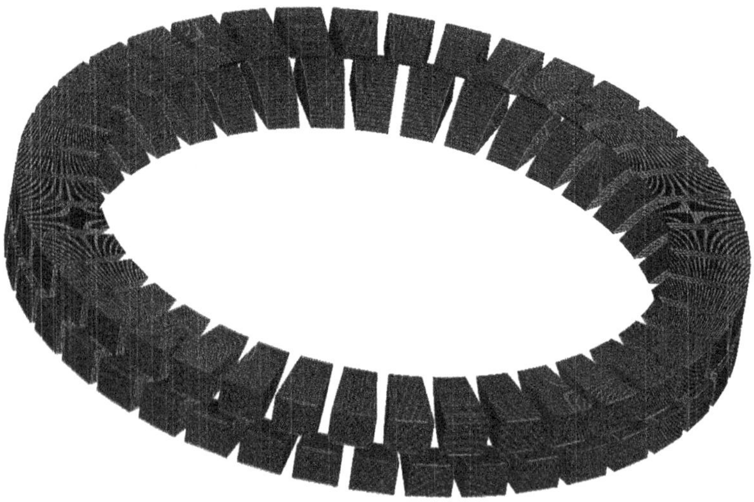

Figure 9.17 Full impeller mesh.

A Select mesh elements dialog box will appear in the right window. In this dialog, click [Select all appropriate objects]. Then, in the lower-left window, click [Rotate Mesh], check the [Copy] box, and in the [Number of copies] input field, enter the number of blades, which is 40 in this case. Check the [Delete duplicate elements] box. Under [Rotation], select the [Axis] as Z and enter the periodic angle in the [Angle] input field, which is 9° (360/40) in this case. Finally, click [Apply] in the lower-left window to complete the periodic replication of the impeller flow passage mesh. The replicated full-flow impeller mesh is shown in Figure 9.17.

9.2.2 Mesh Generation for Volute and Inlet/Outlet Flow Domains

When creating the mesh of volute structure, the process of geometry import and part creation is similar to the step division of impeller structured mesh, therefore it will not be repeated here. The

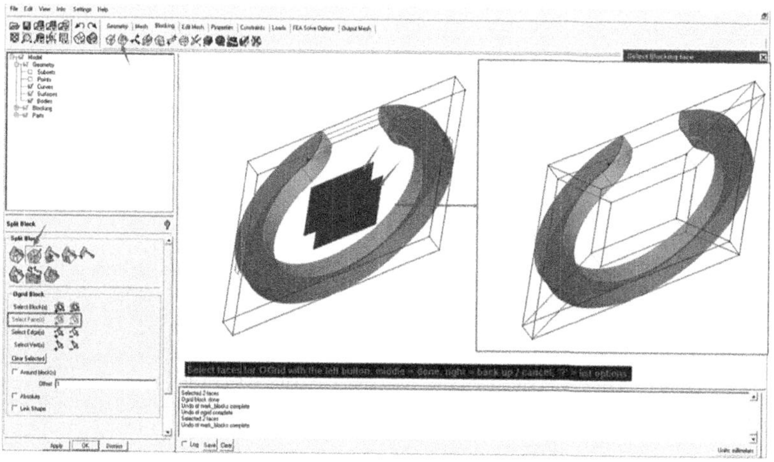

Figure 9.18 O-grid block splitting.

reader should refer to the previous section for details. This chapter focuses on the core steps for the mesh generation of the front and rear volute passages.

Firstly, the meshing of the flow passage in front of the volute is started. In order to build the Block, since the shape of the front flow passage of the volute is similar to that of the circular ring, an "O"-shaped cutting is required for the Block.

The operation steps are as follows: First, click the [Blocking] tab and select the [Split Block] function for clicking. Then, in the option bar at the lower left side of the interface, click [O-grid Block]. Then, click [Select Face (s)] in the lower operation area and select the corresponding face in the right window. When the selection is completed, confirm the operation by clicking the middle mouse button or clicking the [Apply] button in the lower left window, as shown in Figure 9.18.

Then, the middle and top blocks need to be deleted. Since the shape of the top of the front passage is approximately triangular, the vertices need to be merged at the top part and a "Y" block topology is constructed. When these operations are completed, enable the [Associate] function to associate the topological boundary of the Block with the geometric boundary. Block status after association is shown in Figure 9.19.

Finally, the global size of the mesh and the number of nodes on the topology side are set accurately and the meshing operation is performed accordingly. The grid results after division are shown in Figure 9.20.

When generating the mesh for the volute rear flow passage, it is also necessary to perform an O-grid cut on the Block. Next, the Block undergoes stretching and further cutting operations until the final Block topology structure and its associated topology results are formed, as shown in Figure 9.21. Then, the global mesh size and the number of nodes on the topological edges must be precisely defined, and mesh generation is performed according to these parameters. The resulting mesh is shown in Figure 9.22. As for the inlet and outlet pipes, since they are regular circular pipes, the processing is relatively simple. Only an O-grid cut is needed for the block, followed by the association operation and the setting of the global mesh size and the number of nodes on the topological edges. After completing these steps, the required mesh can be generated, as shown in Figure 9.23.

Once the grid has been successfully created, it is then exported. Since CFX is used for numerical simulation, the grid files generated during export must meet CFX requirements. The specific steps

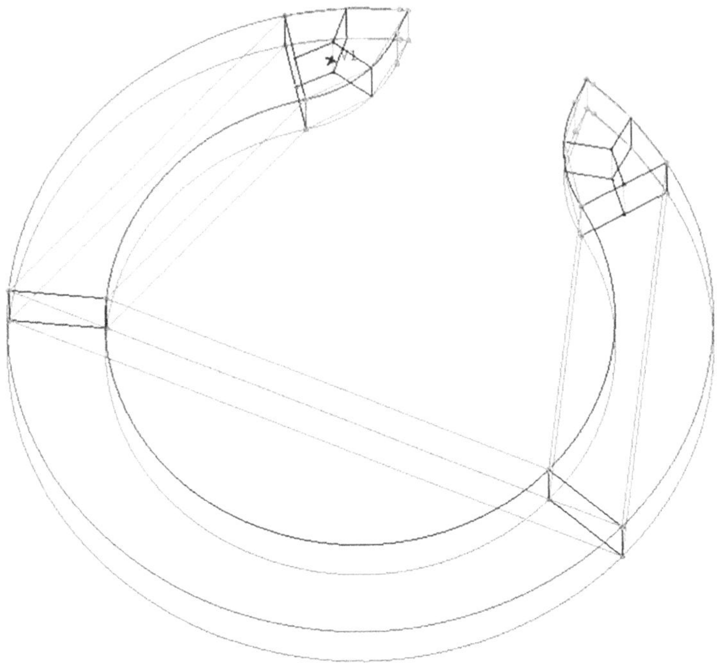

Figure 9.19 Volute front flow passage block association.

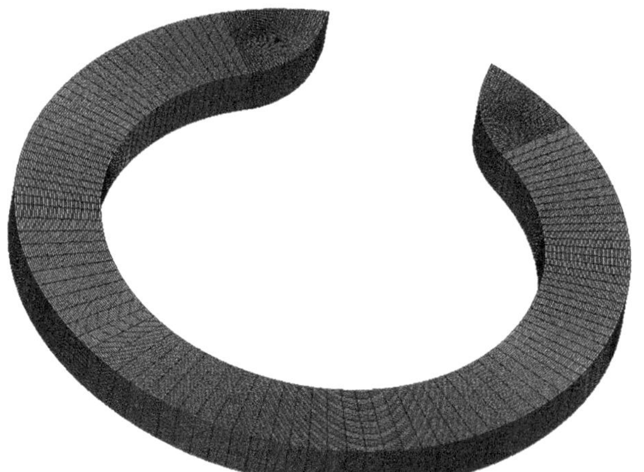

Figure 9.20 Volute front flow passage mesh.

are as follows: First, as shown in Figure 9.24, click the [Output Mesh] tab, then select [Select Solver]. An options bar will appear. In the [Output Solver] option, select ANSYS CFX. Then, click [Output] and [Write Input] to open a dialog box. In this dialog, you can modify the name of the calculation grid file to be exported. Once the naming is complete, simply click [Done] to successfully export the cfx5 grid file for CFX calculations.

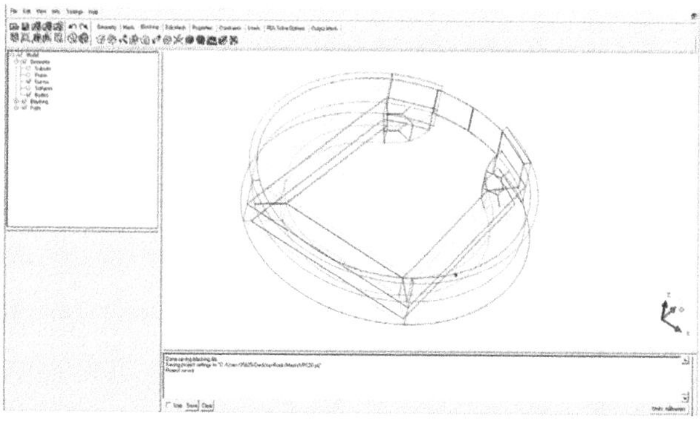

Figure 9.21 Volute front flow passage block association.

Figure 9.22 Volute rear flow passage mesh.

Figure 9.23 Volute inlet and outlet pipe mesh.

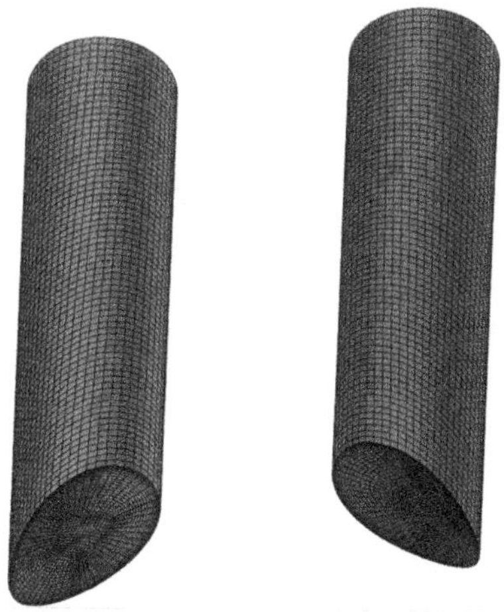

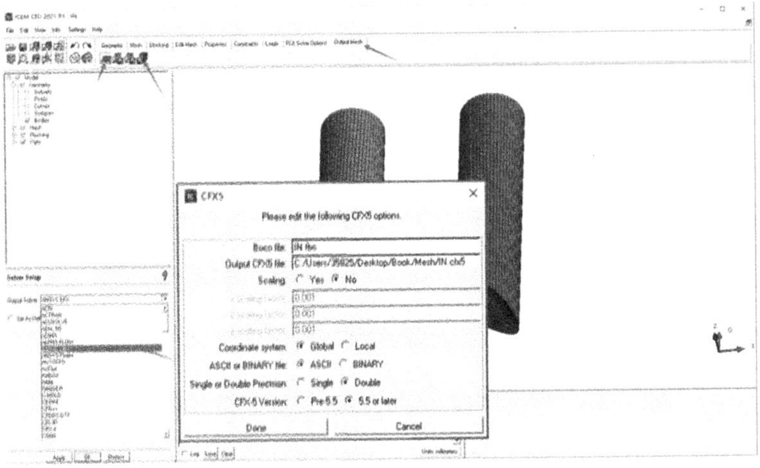

Figure 9.24 Mesh export setup.

9.3 Steady-State Flow Simulation Setup

9.3.1 Computational Domain Setup

Before performing a steady-state flow calculation, the mesh file needs to be imported into CFX. The specific steps are as follows: First, click [File] in the menu bar, then select [Import], and choose [Mesh]. A dialog box will appear. In the "Files of type" option of the dialog box, select the ICEM CFD format (file extensions. cfx or cfx5). Then, in the "Mesh Units" section on the right side of the dialog, select millimeters (mm) as the unit. Next, choose the mesh file to be imported from the file list and click the [Open] button. After the import is complete, the mesh file will be displayed in CFX, as shown in Figure 9.25.

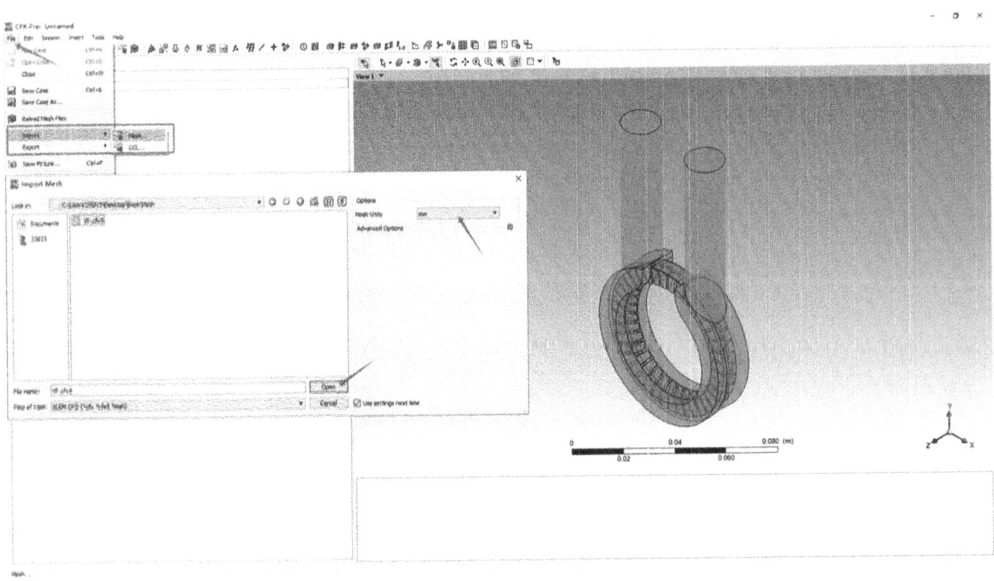

Figure 9.25 Mesh import setup.

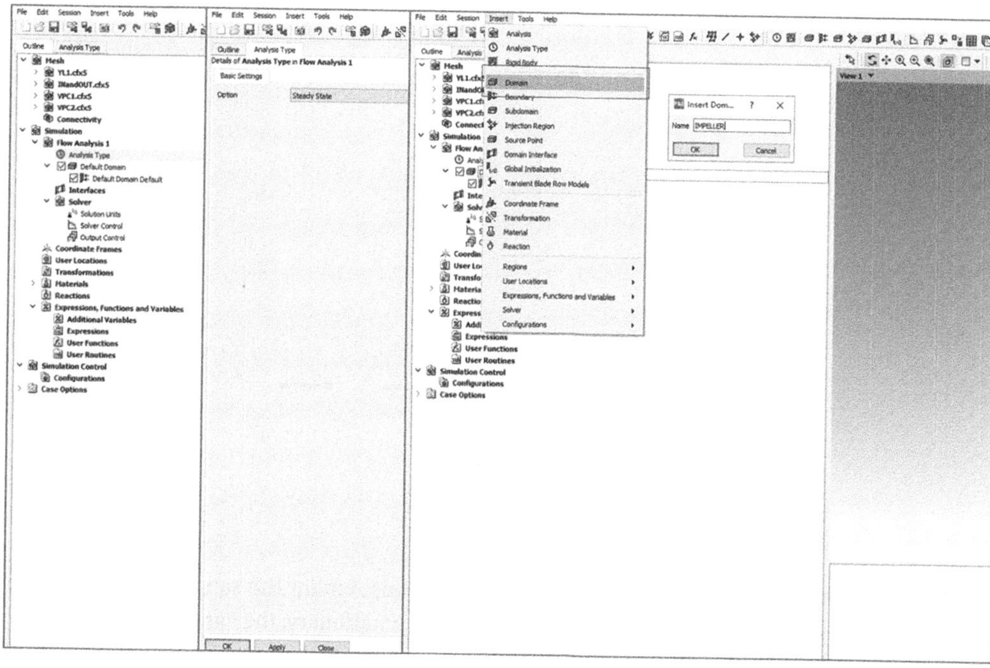

Figure 9.26 Computational domain definition.

When performing a steady-state calculation, set the [Analysis Type] to [Steady State] and click the [Apply] button to complete the calculation type setting. Next, define each calculation domain. Click [Insert] → [Domain], and a dialog box will pop up. In this dialog box, please enter the name of the calculation domain, the specific procedure is shown in Figure 9.26. The computational domains to be defined at this stage include the impeller (IMPELLER), front volute flow passage (VOLUTE1), rear volute flow passage (VOLUTE2), inlet pipe (IN), and outlet pipe (OUT).

In the definition of the computational domain properties, the core content covers the basic properties of the computational domains, the motion properties, the heat transfer model, and the turbulence model. For the four computational domains of the vortex pump, the impeller is categorized as the rotational domain, while the front and rear flow passages of the volute, the inlet pipe and the outlet pipe belong to the stationary domain. For the rotary domain of the impeller, after defining the domain name "IMPELLER" and clicking on [OK], the control tree on the left will bring up a tab for the calculation domain. Afterwards, click [Location] and in the pop-up selection dialog box, select the calculation grid of the impeller and click [OK]. In the Fluid Properties setting, "Water" is selected as the material. Meanwhile, set 1 atm as the reference pressure in the Reference Pressure input box. In the "Domain Motion" area of the Control Tree, click [Rotating] under [Option] to set the rotation property of the impeller. In the "Angular Velocity" input field, enter the rotating speed of the impeller, the positive and negative of which follow the "right hand rule," meaning the direction of the "thumb" is pointing to the axis of rotation. In the "Axis Deformation" field below, the position of the rotation axis is defined. In addition, since the CFD solution of the Navier-Stokes equations requires a heat transfer model and a turbulence model to close the equations, it is necessary to define these two models by selecting "Fluid Models" in the control tree [1, 4]. The process of defining the impeller calculation domain is shown in Figure 9.27.

As for the static computational domain of the vortex pump, it includes the front flow passage of the volute (VOLUTE1), the back flow passage of the volute (VOLUTE2), the inlet pipe (IN), and the outlet pipe (OUT). The definitions of the basic properties of the computational domain, the heat

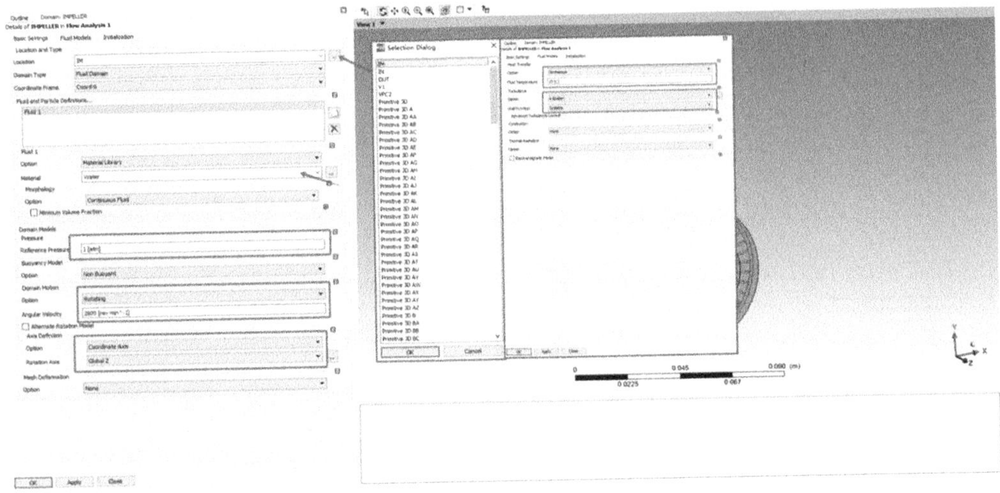

Figure 9.27 Computational domain definition.

transfer model, and the turbulence model for these sections remain the same. However, in the definition of the motion properties, since these sections are stationary, they are set to [Stationary] under "Domain Motion."

9.3.2 Boundary Conditions Settings

In the CFD calculation process, it is important to specify and define the "inlet boundary" and the "outlet boundary." The inlet boundary is usually set at the starting edge of the inlet pipe. The operation steps are as follows: First, select the inlet pipe, then right-click to bring up the dialog box, click [Insert] and [Boundary], in the new pop-up dialog box, enter "inlet" as the name of the boundary. Successfully create an imported boundary named "inlet," click [OK], then the left side of the control tree will show a boundary condition property editing tab. Under the [Basic Settings] tab, select [Inlet] for [Boundary Type]. Click on the option after [Location] and select the corresponding inlet surface of the inlet extension precisely from the pop-up dialog box. Next, go to the second tab [Boundary Details], and in the [Mass And Momentum] section, select [Total Pressure (stable)] as the boundary condition type, and set 1 atm. In order to define the turbulence characteristics of the inlet, select [Medium] for the [Turbulence] option. [Turbulence] option, [Medium] is selected. After completing the above settings, click [Apply] to finalize the definition of the inlet boundary conditions. The specific steps are shown in Figure 9.28.

The process of creating an outlet boundary is similar to creating an inlet boundary, but there are some key differences. First, in the same way as in the creation of the inlet boundary, right-click on the outlet pipe and select [Insert] → [Boundary], and then enter a name for the new boundary "outlet." After successfully creating an outlet boundary named "outlet," click [OK] to open the Boundary Condition Properties Edit tab. Under the [Basic Settings] tab, select [Outlet] for [Boundary Type]. Click on the option after [Location] and select the exact outlet surface of the corresponding outlet pipe from the pop-up dialog box. Next, go to the second tab [Boundary Details], in the [Mass And Momentum] section, there are two types of boundary conditions to choose from: select [Mass Flow Rate] as the boundary condition and enter a specific value. This value should be defined according to the pump flow rate to ensure the accuracy and realism of the simulation. Alternatively, [Normal Speed] can be selected as the boundary condition and a corresponding value for the outlet speed can be entered. This depends on the specific needs of the simulation and the

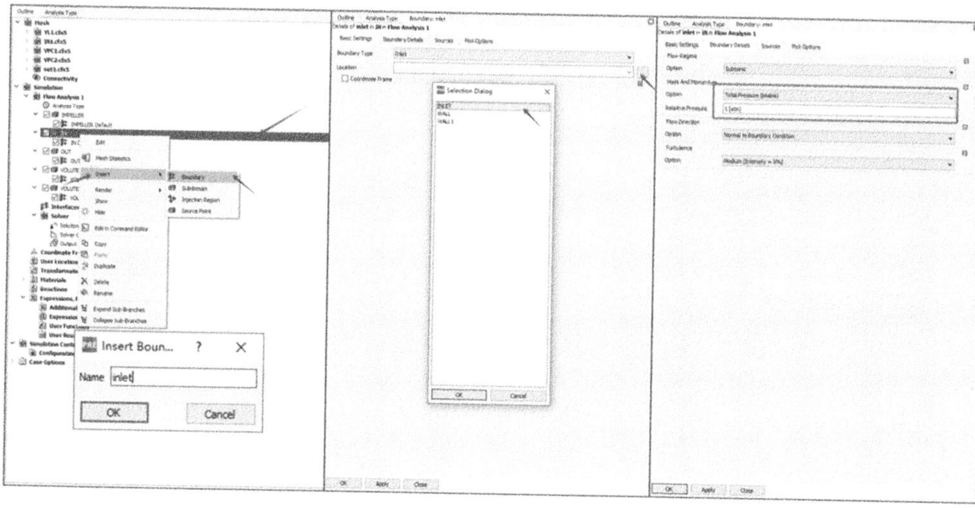

Figure 9.28 Inlet boundary condition setup.

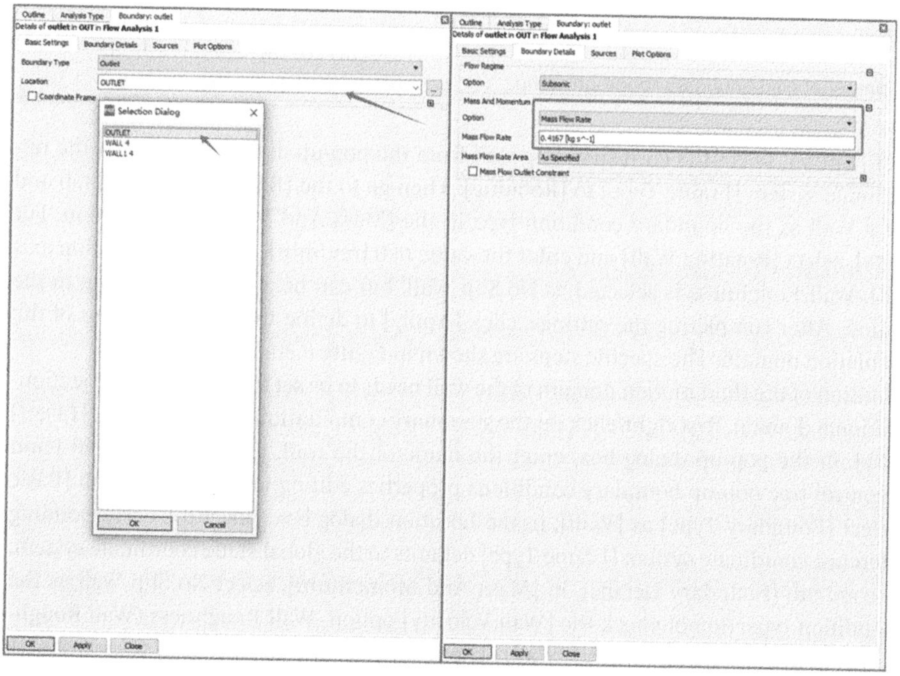

Figure 9.29 Outlet boundary condition setup.

operating conditions of the pump. Then click [Apply] to finalize the definition of the outlet boundary condition. The specific steps are shown in Figure 9.29.

After successfully setting the inlet and outlet boundary conditions, the next step is to define the wall conditions of each computational domain, and the first task is to define the wall conditions of the rotational domain. Similar to the insertion of inlet/outlet boundary conditions, insert [Boundary] in [IMPELLER] and assign a name to it. The left control tree then brings up the Boundary Condition Properties Edit tab. In the [Basic Settings] tab, select [Wall] for [Boundary Type]. Click

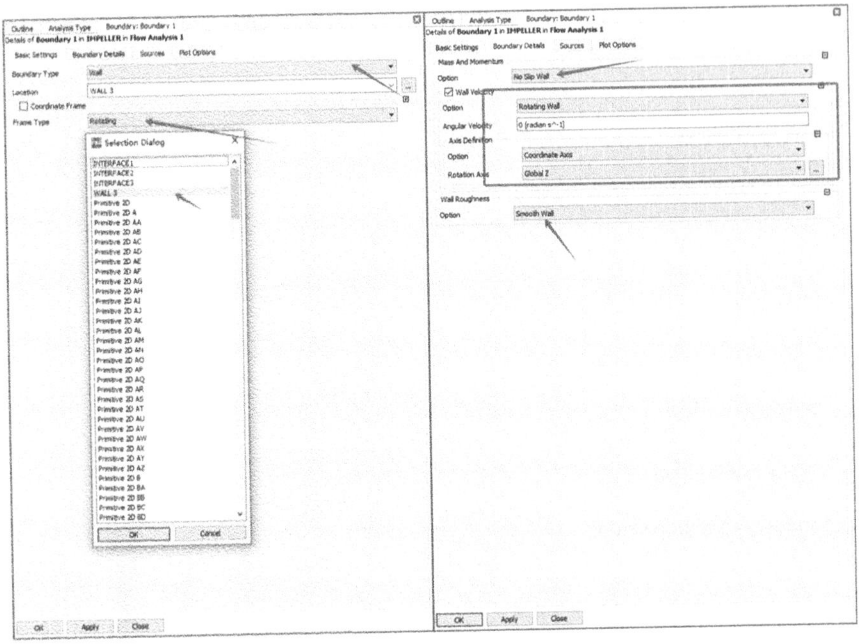

Figure 9.30 Impeller wall boundary condition setup.

the [Location] option, select the corresponding wall from the pop-up dialog box, and set the reference coordinate system [Frame Type] to [Rotating]. Then go to the [Boundary Details] tab and select No Slip Wall as the boundary condition type in the [Mass And Momentum] section. For [Wall Velocity], select [Rotating Wall] and enter the value of 0 [rev/min], and set the rotation axis to [Global Z]. Wall Roughness is selected as No Slip Wall, but can be adjusted according to the actual situation. After completing the settings, click [Apply] to define the wall conditions of the rotating calculation domain. The specific steps are shown in Figure 9.30.

In the definition of the fluid motion domain of the wall needs to be set for the wall of the stationary computational domain, first right-click on the stationary computational domain, click [Insert] → [Boundary], in the pop-up dialog box, enter the name of the wall, click [OK], the left-hand side of the control tree pop-up boundary conditions properties editing tab. In the first tab [Basic Settings], select [Boundary Type] as [Wall]; in the Location dialog box, select the corresponding wall; the reference coordinate system [Frame Type] defaults to the global static coordinate system. Click the second tab [Boundary Details], in [Mass And Momentum], select No Slip Wall as the boundary condition type; do not check the [Wall Velocity] option; Wall Roughness (Wall Roughness) is selected as a nonslip wall (Smooth Wall) (The reader can define the size of the roughness according to the actual situation); click [Apply] to complete the definition of the wall conditions of the stationary computational domain. The rest of the wall of the stationary computational domain, including the inlet pipeline wall, the outlet pipeline wall of the boundary conditions, are defined similarly, the stationary computational domain wall conditions are set as shown in Figure 9.31.

The definition of the interface is crucial in numerical simulation to ensure the smooth data transfer between different computational domains. When dealing with fluid dynamics problems involving both dynamic and static computational domains, the application of the multiple reference system (MRF) method becomes particularly critical, so it is especially important to set up the data interface between the dynamic and static computational domains in a reasonable manner.

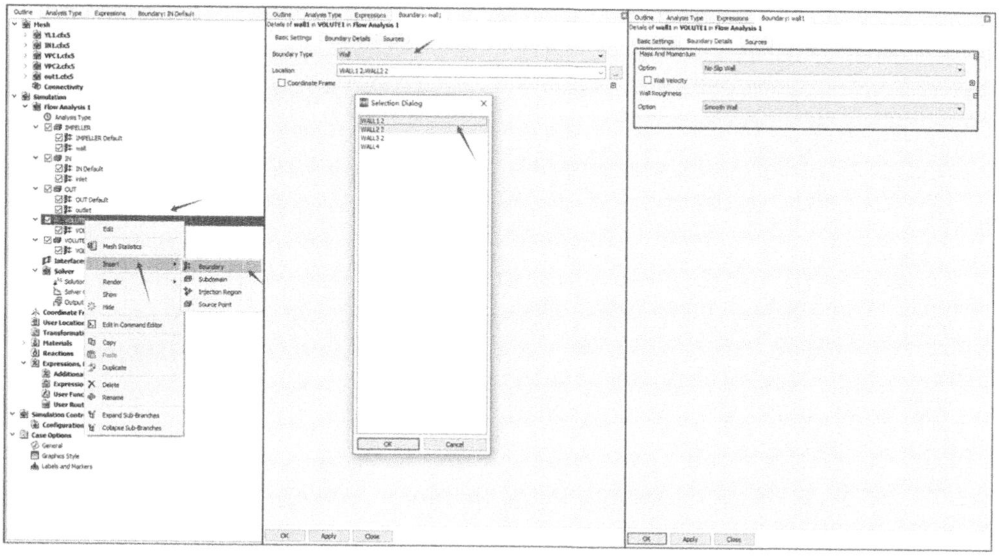

Figure 9.31 Stationary domain wall boundary condition setup.

In the numerical simulation of the vortex pump, the interface of the dynamic-static calculation domain mainly includes the interface of the front and rear passages of the impeller and the worm case, i.e., Interface 1 and Interface 2. Take the definition of the interface of the front passage of the impeller and the worm case as an example, select [Insert] → [Domain Interface], and enter “Interface 1” in the pop-up dialog box and confirm it. In the [Basic Settings] tab, the two sides of the interface need to be configured in detail. For [Interface Side 1], select “VOLUTE1” for [Domain], and select the surface where the front passage of the volute is connected to the impeller in [Region List]. For [Interface Side 2], select the impeller calculation domain “IMPELLER” for [Domain], and select the surface where the impeller is connected to the front passage of the worm case in [Region List]. Select [General Connection] as the [Interface Model] for the connection model. The [Frozen Rotor] is used as the [Frame Change/Mixing Model] to simulate the relative motion between the dynamic and static calculation domains. The area ratio [Pitch Change] is set to [Specified Pitch Angles], and the specific values are input respectively, in which Pitch Angle Side 1 and Pitch Angle Side 2 are set to 360°, indicating the complete matching of the interface. After completing the settings, click [Apply] to complete the impeller and volute front passage dynamic—static calculation domain interface definition. Similarly, the definition of the dynamic-static computational interface between the impeller and the rear passage of the volute also follows the same steps, and the specific details are shown in Figure 9.32.

In the numerical simulation setup, in addition to the dynamic-static computational domain interface, the setup of the static-static computational domain interface is also crucial, which mainly involves the interface between the inlet and outlet tubes and the front and rear volute flow passages. Take the static-static calculation domain interface between the inlet pipe and the rear passage of the volute as an example, the definition process is as follows: First, click [Insert] → [Domain Interface], enter “Interface 3” in the pop-up dialog box, and click [OK] to confirm. Next, in the [Basic Settings] tab, configure both sides of the interface in detail. For [Interface Side 1], click on [Domain], then select the inlet calculation domain “IN” in the pop-up dialog box, and select the surface where the inlet pipe is connected to the rear passage of the volute in the [Region List]. For [Interface Side 2], select the calculation domain “VOLUTE2” in [Domain], and select the surface where the inlet pipe is connected to the rear passage of the volute in

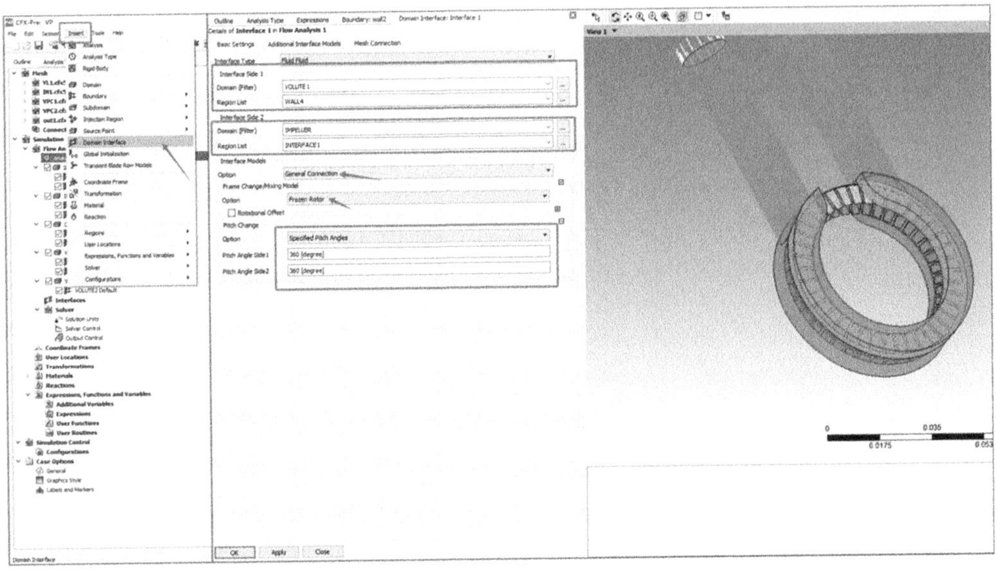

Figure 9.32 Rotating and stationary domain interface setup.

[Region List]. Select [General Connection] as the [Interface Model] for the connection model of the interface. Since this is a static-static computational domain interface, [None] is selected as the [Frame Change/Mixing Model] for the coordinate system transformation. Meanwhile, the area ratio [Pitch Change] is also set to [None]. After completing all the above settings, click [Apply] to complete the definition of the static-static calculation domain interface between the inlet pipe and the rear passage of the volute. Similarly, the definition of other static-static calculation domain interface also follows the same steps, the specific details can be referred to Figure 9.33.

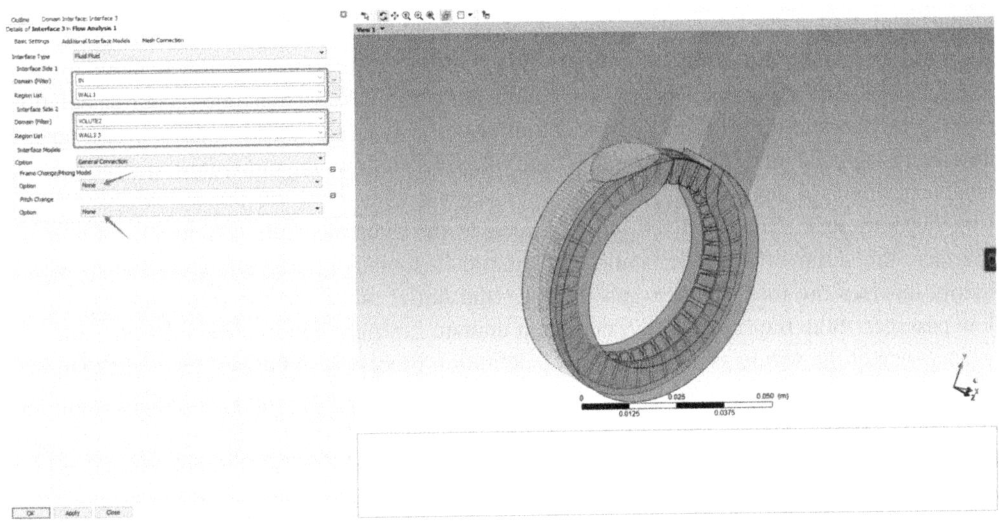

Figure 9.33 Stationary-stationary domain interface setup.

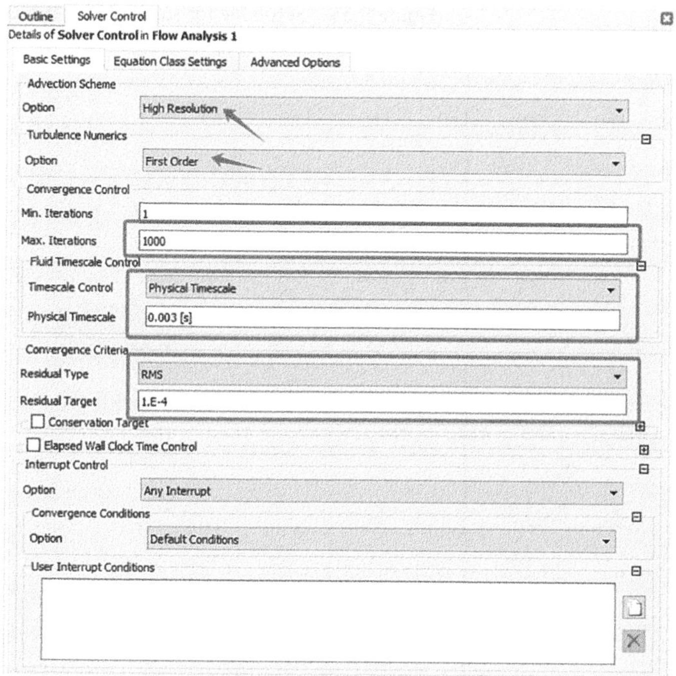

Figure 9.34 Steady-State calculation setup.

9.3.3 Solution Settings

When setting up the solver, there are certain differences between steady and transient flow field calculations. For steady solver settings, the first step is to define the solver parameters. The specific operation is as follows: double-click [Solver Control], which will open the solver parameter editing interface. In the [Basic Settings] section, select [Advection Scheme] as [High Resolution], set the turbulence numerical scheme [Turbulence Numerics] to [First Order]. Next, set the total number of iterations by entering 1000 in the [Max. Iterations] field under [Convergence Control] (this value can be adjusted based on actual needs). For the transient solver parameters, choose [Timescale Control] as [Physical Timescale], and input the value 0.003 seconds (typically set to the inverse of the impeller rotational speed). For the convergence criteria, set [Convergence Criteria] to the average [RMS] value, and input the threshold of 1.E-4 (this value usually satisfies typical engineering requirements). Finally, click [Apply] to complete the definition of the solver parameters. Detailed steps are shown in Figure 9.34.

9.3.4 Calculation Settings

The initial flow field is usually not required for steady calculations. Therefore, for the setup of CFX file of steady calculation, select [Define Run] on the right side of the toolbox and click it to select the path of .def file in the pop-up dialog box. If necessary, turn on parallel computing, select [Intel MPI Local Parallel] in the check box of [Run Mode], and click on the following tab to increase the number of parallels. Select the working directory to save the results in the [Run Environment] below. Finally, click [Start Run] to start the calculation. Specific operation is shown in Figure 9.35.

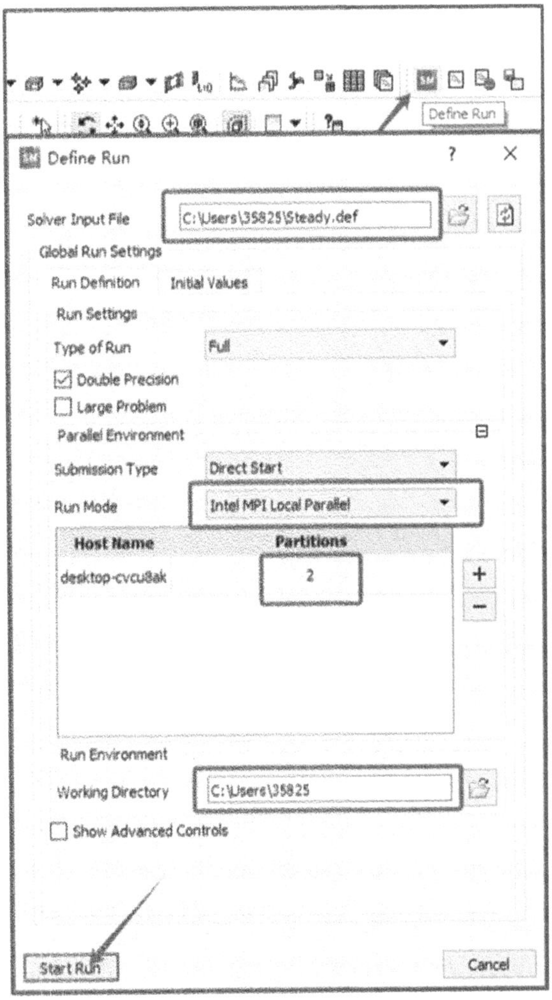

Figure 9.35 Setup file for steady calculation.

9.4 Transient Flow Simulation Setup

The transient simulation setup is almost similar to the steady simulation setup with only a few differences. Therefore, the transient calculations can be implemented with minor modifications based on the steady calculations [2]. The specific modification steps are as follows:

9.4.1 Transient Flow Setup

Find and double click [Analysis Type] on the model tree to enter its attribute editing interface. For transient calculations, change the value of [Analysis Type] to [Transient]. Next, the total time for the transient calculation is defined. Generally, in order to obtain a reliable solution, the transient calculations need to cover at least 6-10 rotation periods of the impeller. Input the required maximum time steps in the [Max. Timesteps] input field under [Maximum Number of Timesteps]. Then, the number of calculation steps per rotation cycle is defined. This can be understood as a calculation to determine how many degrees the impeller rotates. For example, in this example, set the impeller 1°per revolution as a time step of 0.000059524 seconds. Finally, click [Apply] button to complete the setting of transient calculation. Detailed operation steps are shown in Figure 9.36.

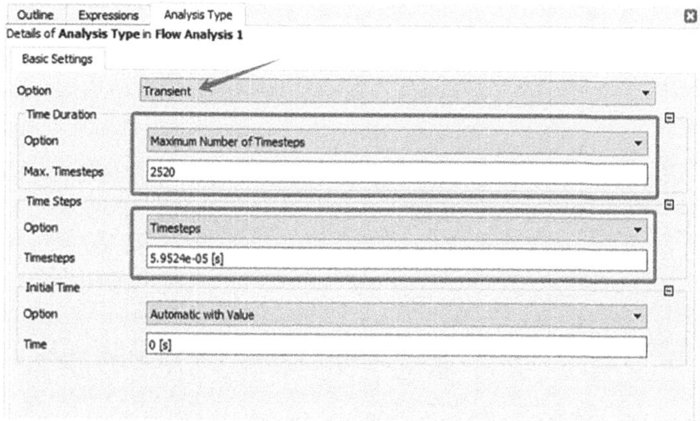

Figure 9.36 Time step setup for transient calculation.

9.4.2 Interface Model Modification

The interface setting of transient calculation is significantly different from that of steady calculation, which is mainly reflected in the connection model configuration of the interface between dynamic and static calculation domains. Double-click the interface between the dynamic and static calculation domains in the vortex pump. For the connection model of the interface, although [Interface Model] is still selected as [General Connection], the coordinate system transformation [Frame Change/Coupling Model] needs to be changed to [Transient Rotor State] to meet the requirements of transient calculation. The remaining settings are consistent with the steady calculation. When the above configuration is completed, click [Apply] to complete the data interface definition of dynamic and static calculation domains in transient calculation. Detailed operation steps are shown in Figure 9.37.

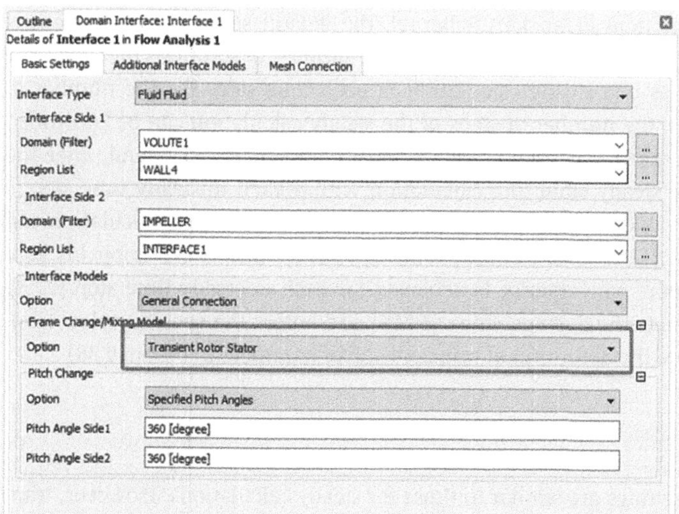

Figure 9.37 Interface setup of transient computation: dynamic and static computational domains.

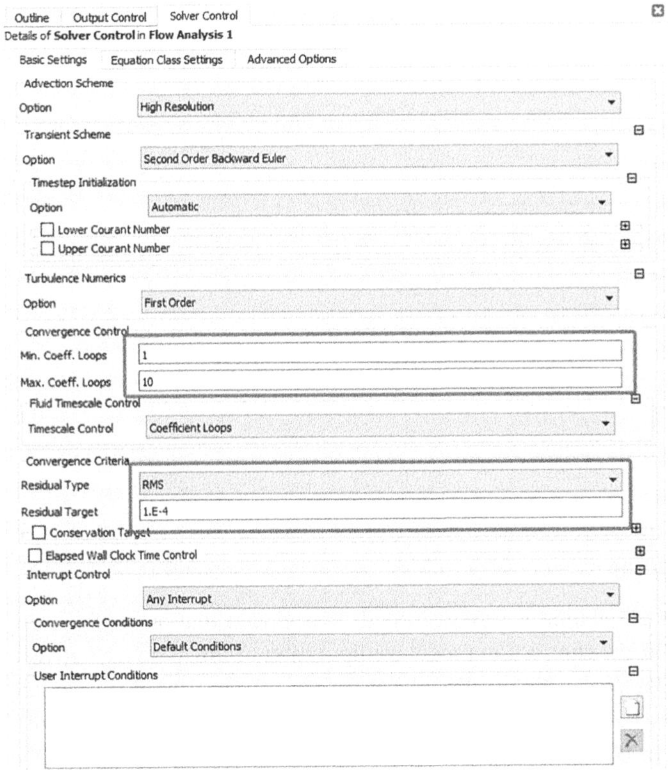

Figure 9.38 Transient solver settings.

9.4.3 Solving Settings

There are some differences between the steady and transient calculations in the solution setup. During transient calculation, the setting of solver parameters is significantly different from that of steady calculation, mainly reflected in the definitions of [Transient Scheme] and [Convergence Control]. For the transient time item [Transient Scheme], the default setting of the system is generally acceptable. The inner circulation calculation sets the number of solutions for each time step, which can be considered as one steady calculation in each time step. For the number of internal circulation calculations, the number of steps of the steady calculation can be controlled by adjusting the [Min. Coeff. Loops] and [Max. Coeff. Loops] parameters. In general, after the transient calculation enters the steady state, the calculation within each unsteady time step is easy to reach the convergence value 1E-4, so the number of internal circulation calculation can be reduced accordingly. Usually [Max. Coeff. Loops] is set to 10 and [Min. Coeff. Loops] is kept at the default value of 1 to ensure convergence is achieved for each unsteady time step [2, 3]. [Convergence Criteria] still selects the average value [RMS] and enters the specific value 1E-4. Finally, click [Apply] to complete the definition of transient solver parameters (Figure 9.38).

9.4.4 Calculation Setup

For transient calculations, the settings are similar to those for steady calculations. However, transient simulations typically require an initial flow field, which is the result of the steady calculation. Therefore, the foundation of transient calculation is a completed steady calculation. As shown in

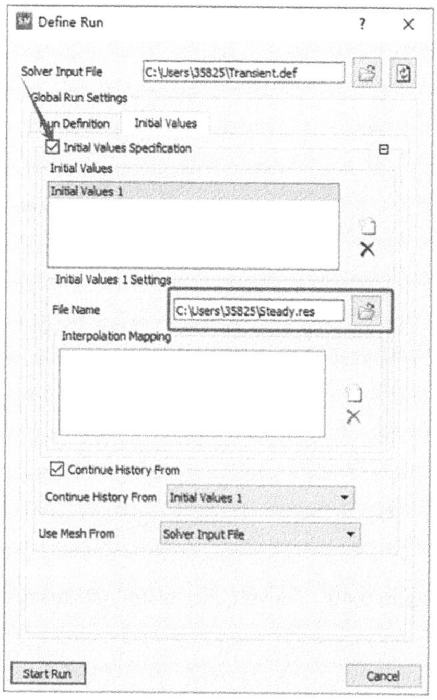

Figure 9.39 Transient calculation setup.

Figure 9.39, to set up the transient calculation, first check [Initial Values Specification], which will display [Initial Values 1 Settings] below. Then, locate the previous steady result file (.res file), and finally, click [Start Run] to begin the calculation.

9.5 Calculation Results and Postprocessing

Before proceeding with postprocessing steps, the primary task is to verify the convergence of the calculation results. Figure 9.40 clearly shows the residuals trend during the steady calculation. As can be observed from the figure, after more than 300 calculation steps, the residual values successfully meet the predefined convergence criteria, leading to the termination of the calculation process. Subsequently, the steady calculation results are used as the initial condition to initiate the transient calculation.

During postprocessing, the calculation results need to be imported into CFX-Post. First, click on the CFX-Post button in CFX-Launcher to open the CFX-Post window. Then, import the model pump calculation results file by selecting File → Load Result, as shown in Figure 9.41.

Generally, contour and vector plots used to visualize internal flow conditions in the pump are constructed based on cutting planes. Therefore, the first task is to learn how to create a cutting plane. After clicking the "Location" button, a menu with geometric elements such as points, lines, planes, bodies, and surfaces will appear. At this point, select "Plane" to construct a plane, essentially setting a cutting plane or section within the computational domain. Once "Plane" is selected, a dialog box will pop up, prompting to enter a name for the cutting plane. To view the flow state at the middle section, choose *XY* Plane in the dialog box, enter a name, and click "OK." In the "Domains" option, specify the region to be cut by the plane, typically using the default "ALL Domains" option. The Method option allows to select the direction of the plane. To create an *XY* Plane, select the "*XY* Plane" method. The *Z*-coordinate value of the cut position can be entered directly or adjusted using

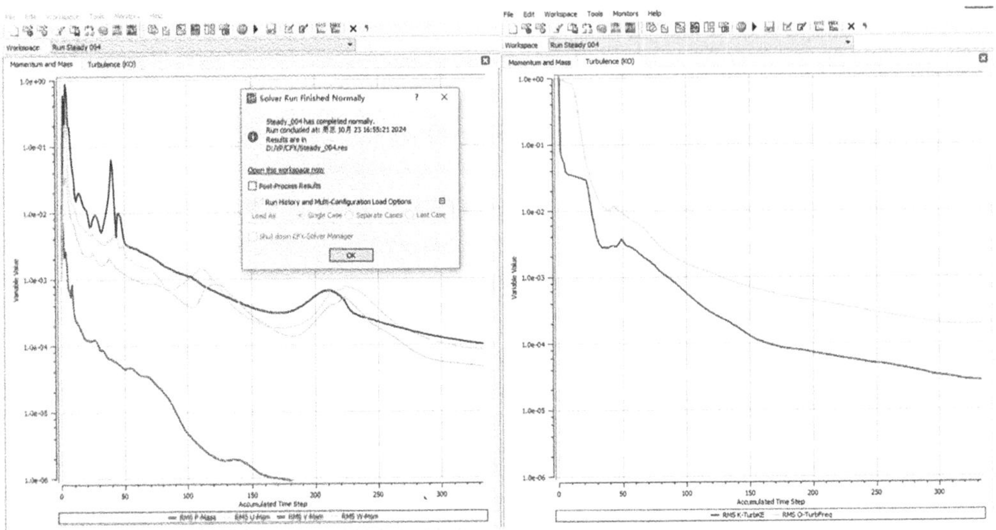

Figure 9.40 Convergence history of steady-state calculation.

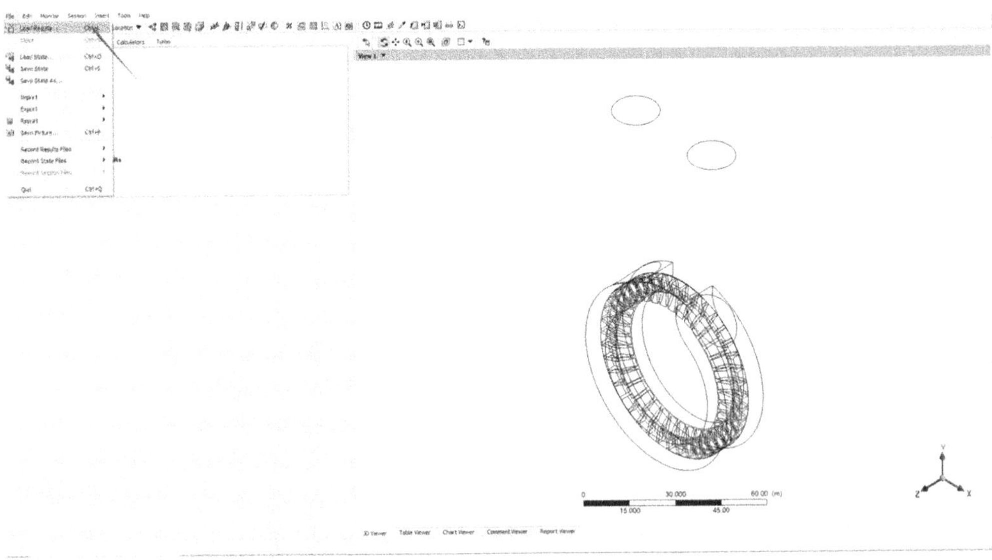

Figure 9.41 Import of calculation results.

the slider, with the default unit being meters (m). After entering the plane parameters according to actual needs, click "Apply" to generate the cutting plane. In addition to setting an *XY* Plane, one can also create a plane based on points and normal vectors. Choose the "Point and Normal" vector method, set the point and normal vector, and click "Apply" to generate the cutting plane. The specific steps are shown in Figure 9.42.

Next, a variable contour plot will be displayed on the created cutting plane, typically used to represent scalar fields. To create a contour plot with Pressure as the variable, select "Insert" from the menu bar and click "Contour" in the drop-down menu. In the pop-up dialog box, enter a name for the new contour and click OK. In the Contour settings panel, the "Domains" option should be set to "ALL Domains" by default or specify the desired computational domain to display the

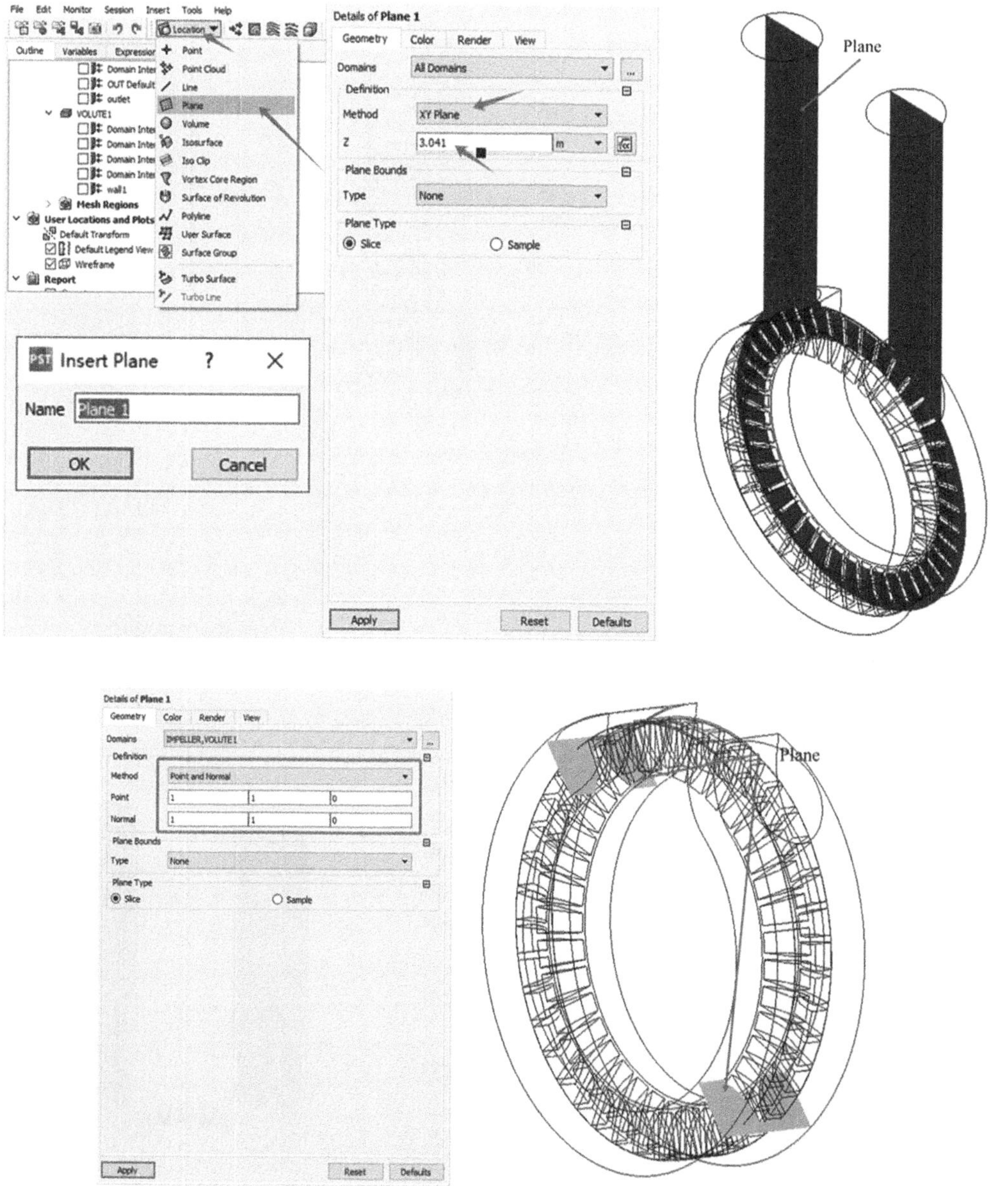

Figure 9.42 Cutting plane settings.

pressure distribution on the cutting plane. Under "Locations," select the previously created plane. In the "Variable" option, choose "Pressure" as the physical quantity to be displayed. The "# of Contours option" at the bottom controls the display precision of the pressure contour plot. By entering different values, one can divide the current Variable (i.e., pressure) into different intervals. After completing the settings, click "Apply" to generate and display the pressure contour plot. The specific result is shown in Figure 9.43.

In addition to contour plots, streamlines can be generated on the plane to observe fluid motion within the pump. To do this, select "Insert" from the menu bar, then click "Streamline" in the drop-down list. In the subsequent dialog box, enter a name for the new streamline and click "OK."

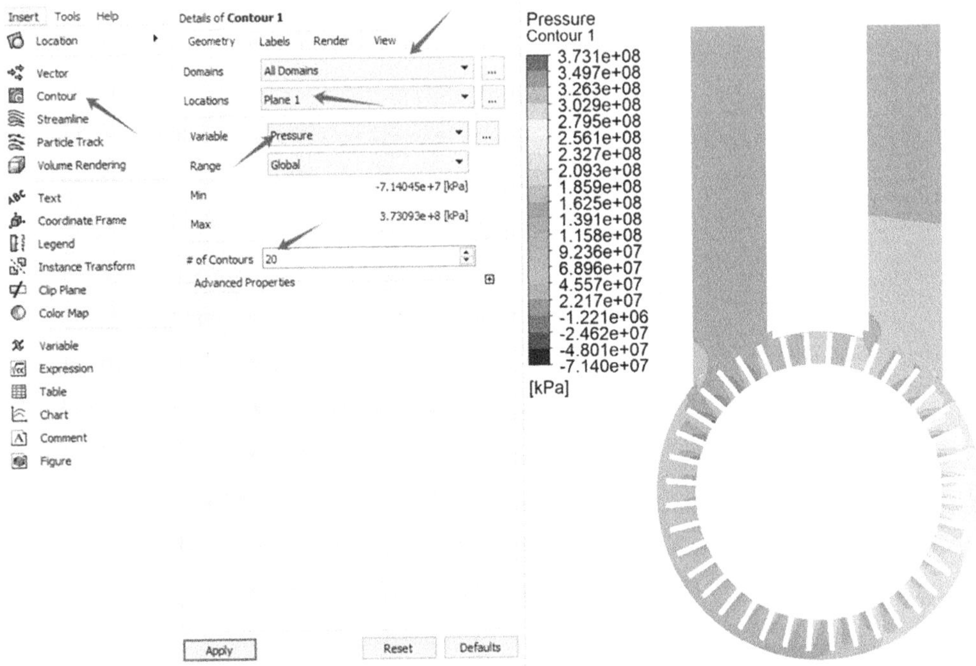

Figure 9.43 Contour settings.

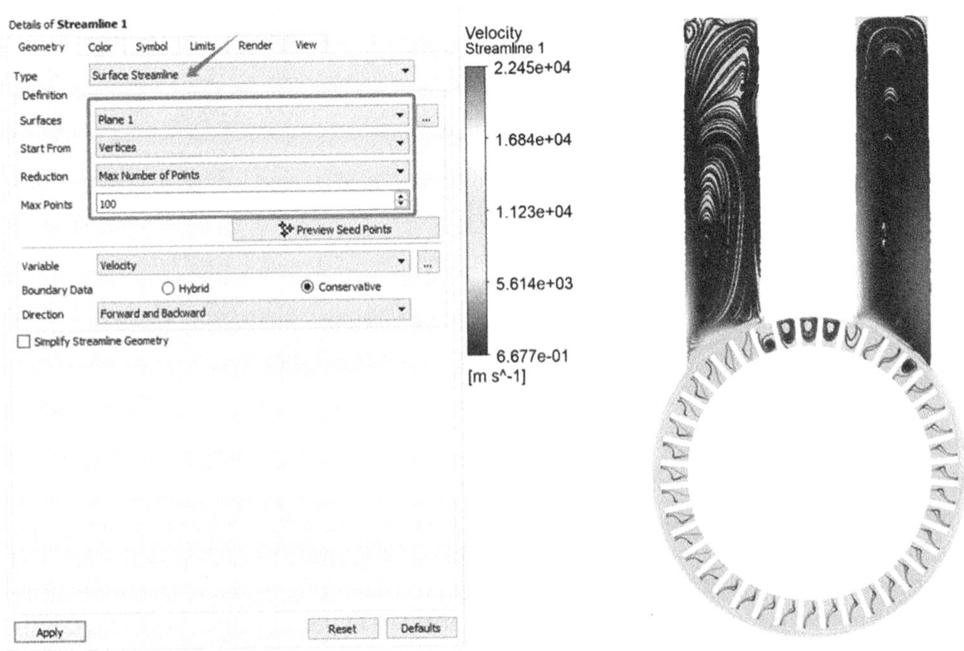

Figure 9.44 Settings for streamline plots.

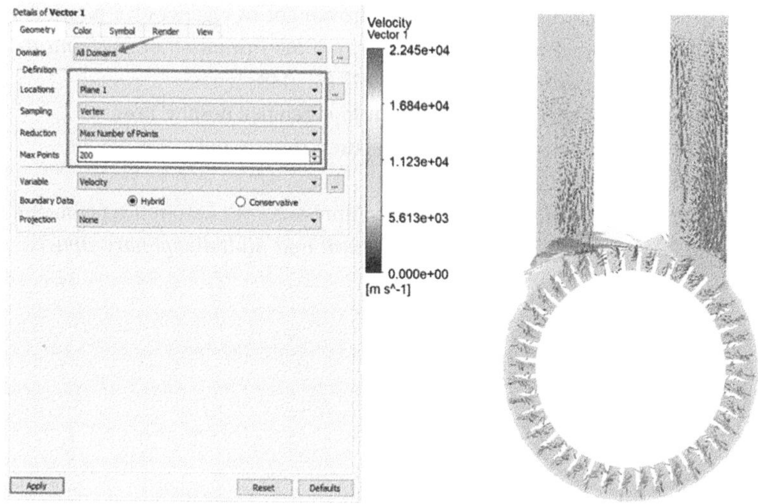

Figure 9.45 Settings for the velocity vectors plots.

In the streamline settings panel, the "Type" option is set to "Surface Streamline" by default, and the Surfaces option should be set to the predefined plane. The Start From option can be configured according to specific requirements. In the "Reduction" section, select "Max Number of Points" and set an appropriate value. The Variable should be set to the default Velocity, with other parameters remaining as default. After completing the configuration, click "Apply" to generate and display the streamline plot, as shown in Figure 9.44.

Next, a velocity vector plot can also be created on the cutting plane [5–8]. This can be done by selecting Insert → Vector from the menu bar or by directly clicking the corresponding icon in the toolbar. In the dialog box, enter a name for the new vector plot and click "OK." In the Vector settings panel, set Domains to the desired fluid domain, typically "ALL Domains." Choose Locations as *XY*, and set Variable to the default Velocity, keeping all other parameters at their default values. Click "Apply" to generate the velocity vector plot, as shown in Figure 9.45.

References

1 Kumar, M., Venkateshwaran, A., Kumar, M.S.S.P., Sreekanth, M., Jebaseelan, D., and Sivakumar, R. (2022). Strength analysis of are generative flow compressor and a pump based on fluid-structure coupling. *Materials Today* 51: 1619–1624.

2 Fan, Z. (2020). Dynamic characterization of vortex structures and their evolution mechanisms in a side channel pump. *Journal of Fluids Engineering* 142 (11): 111502.

3 Kaiyao, Z.H.A.O., Houlin, L.I.U., Zhenming, D.U., Minggao, T.A.N., Sheng, H.U., and Liang, D.O.N.G. (2021). Analysis of flow loss characteristic of vortex pump based on entropy production. *Journal of Drainage and Irrigation Machinery Engineering (JDIME)* 39 (12): 1284–1290.

4 Zhang, F., Chen, K., Zhu, L., Appiah, D., Hu, B., and Yuan, S. (2020). Gas–liquid two-phase flow investigation of side channel pump: anapplication of MUSIG model. *Mathematics* 8 (4): 624.

5 Tahoun, A., Abelsamie, A., and Hafiz, A.A. (2022). Improvement of regenerative pump performance by using blade with combined angles modification. *Engineering Research Journal* 174: M1–M18.

6 Nejad, J., Riasi, A., and Nourbakhsh, A. (2019). Efficiency improvement of regenerative pump using blade profile modification: experimental study. *Proceedings of the Institution of Mechanical Engineers, Part E: Journal of Process Mechanical Engineering* 233 (3): 448–455.

7 Mosshammer, M., Benigni, H., and Jaberg, H. (2019). Maximum efficiency despite lowest specific speed-simulation and optimisation of a side channel pump. *International Journal of Turbomachinery, Propulsion and Power* 4 (2): 1–6.

8 Nejadali, J. (2021). Shape optimization of regenerative flow compressor with aero-foil type blades using response surface methodology coupled with CFD. *Structural and Multidisciplinary Optimization* 64 (4): 2653–2667.

10

Numerical Simulation and Analysis of a Multistage Pump

10.1 Computational Domain Modeling

Based on the three-dimensional impeller, guide vane, inlet, and outlet models of the multistage pump created by Creo, the computational domain of the fluid inside the multistage pump is extracted as shown in Figure 10.1. The fluid domain in the inlet and outlet passages of the multistage pump can also be modeled directly using the **[Extrude]** command for solid modeling, while for the impeller channel extraction, the following steps can be employed to complete the modeling.

Create and name the impeller flow domain.prt file the same way as the impeller model. Import the impeller axial projection into Creo. Click the **[Revolve]** button, and select **[Solid]** in the Revolve dialog box. Enter 360 for the rotation angle. Complete the rotation operation, as shown in Figure 10.2.

As shown in Figure 10.3, click the menu **[Get Data]** → **[Merge/Inheritance]** and select **[Cut]** to open the impeller blade model drawn in the previous stage.

As shown in Figure 10.4, define the relative position of the two solid models by aligning the coordinate system, select the coordinate system of the two models, click [√] to import the impeller blades, determine the position of the blades after completing the import, click [√] on the top of the menu bar to perform the Boolean subtraction operation to complete the impeller water modeling, and get the impeller water model as shown in Figure 10.5.

The fluid domain for the guide vane can be modeled using the **[Extrude]** command with the aid of the diffuser vane sketch, as shown in Figure 10.6.

10.2 Structured Mesh Generation

10.2.1 Impeller Mesh Generation

The structured meshing is usually done with the help of the ICEM software in the following steps:

10.2.1.1 New ICEM File

Import the impeller geometry model into the ICEM software, select **[File]** → **[Import -model]**, select the impeller flow domain file, keep the default settings, click **[OK]**, and the geometry model is successfully imported. Click **[geometry]** → **[surface]** under the model tree to display the impeller surface as shown in Figure 10.7.

10.2.1.2 Boundaries Definition

After successfully importing the model, first define the boundary surface, which is required for subsequent boundary condition setup. Select the model tree **[parts]**, right-click and select

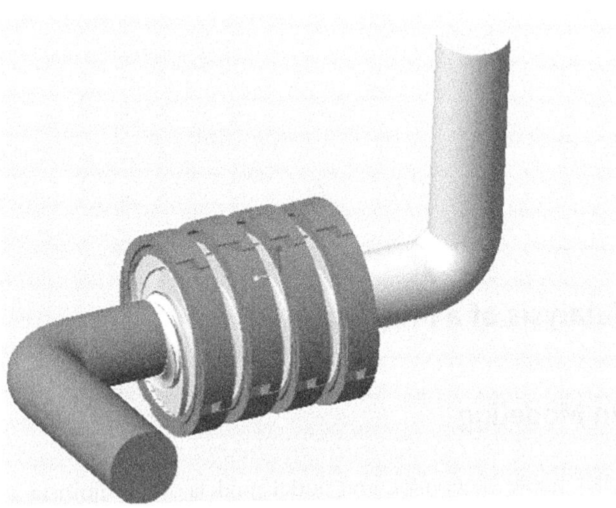

Figure 10.1 Computational domain of the multistage pump.

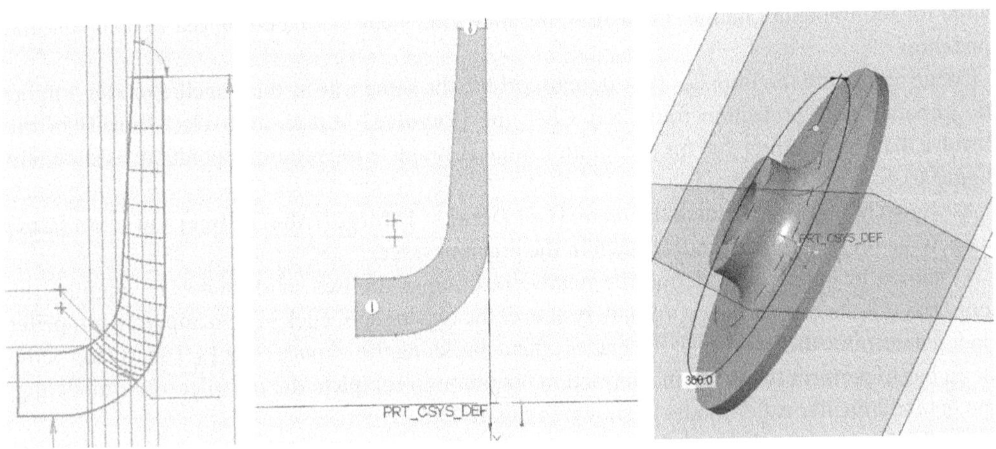

Figure 10.2 Steps in creating the revolving impeller flow domain.

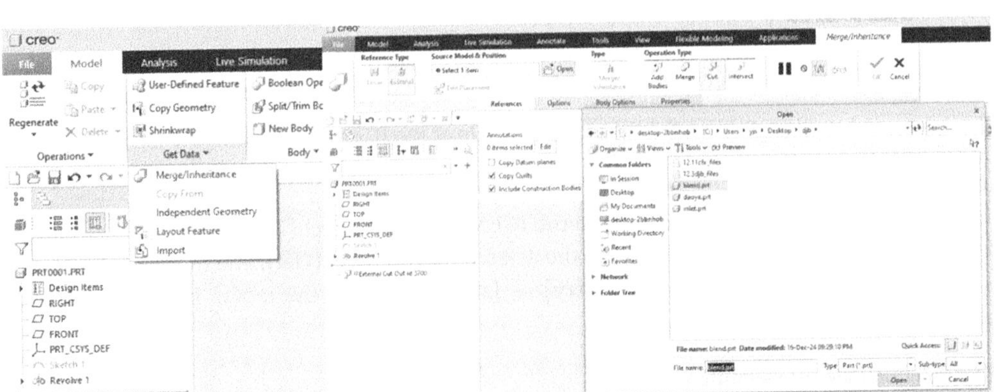

Figure 10.3 Merge/inheritance step.

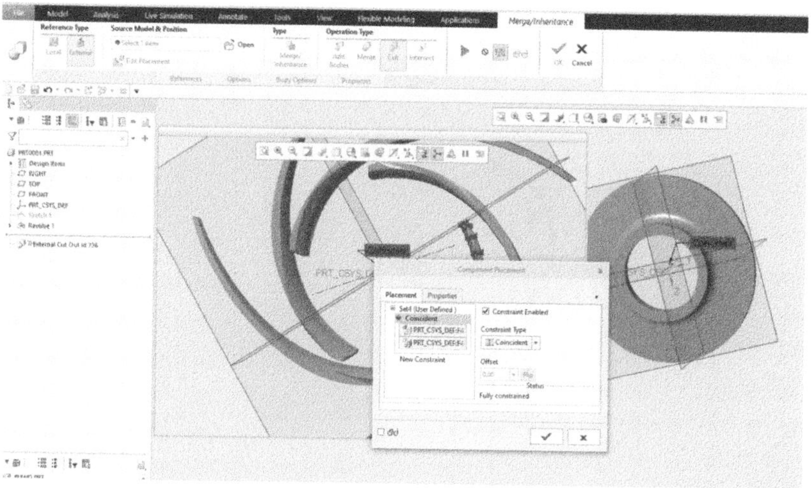

Figure 10.4 External cutting step.

Figure 10.5 Impeller fluid domain.

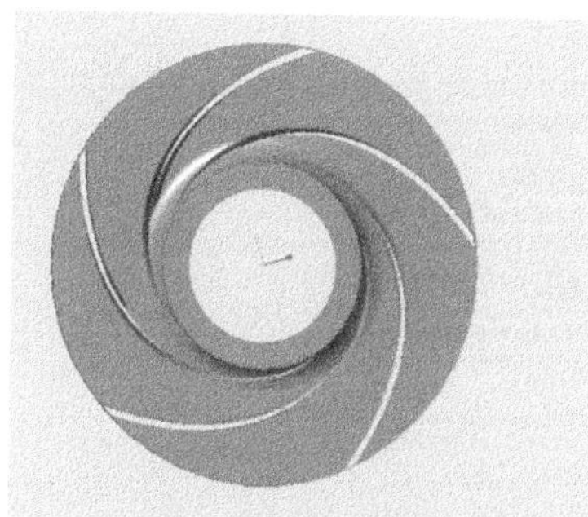

Figure 10.6 Guide vane fluid domain.

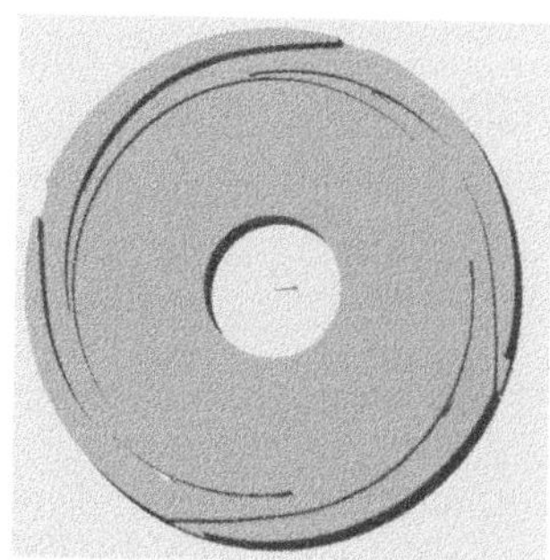

[create part], a dialog box will pop-up as shown in Figure 10.8. Input the name of the part, such as "YP-GZM", and click the entities icon to select the defined boundary surface. After selecting the end, click the center button to confirm and get the part of YP-GZM. Similarly, define other boundary regions, and obtain the result shown in Figure 10.9.

10.2.1.3 Establishment of the Topology

10.2.1.3.1 Checking the Geometry Check the imported geometry before generating the structured mesh. Click **[Geometry]** in the toolbar and select **[Repair Geometry]** → **[OK]** to get the red line edge of the geometry as shown in Figure 10.10.

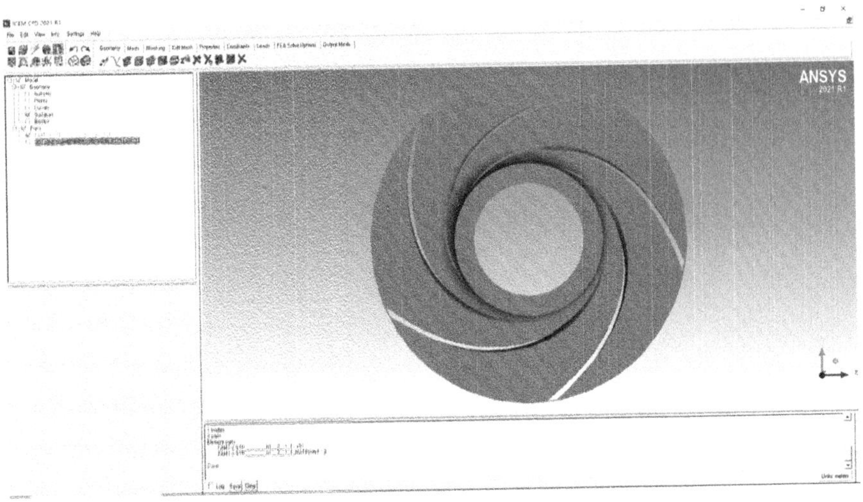

Figure 10.7 Impeller surface model.

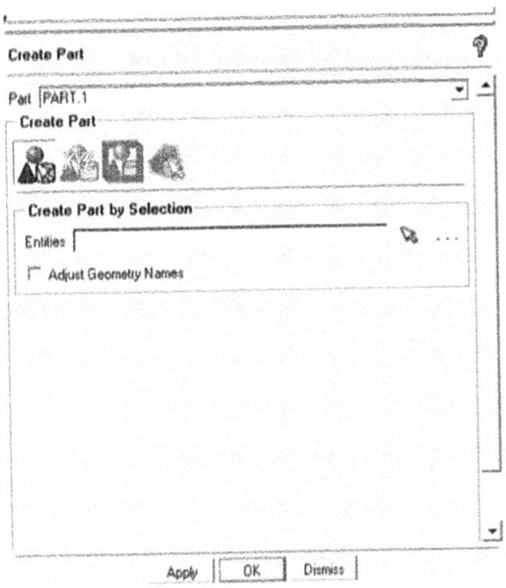

Figure 10.8 Define part dialog box.

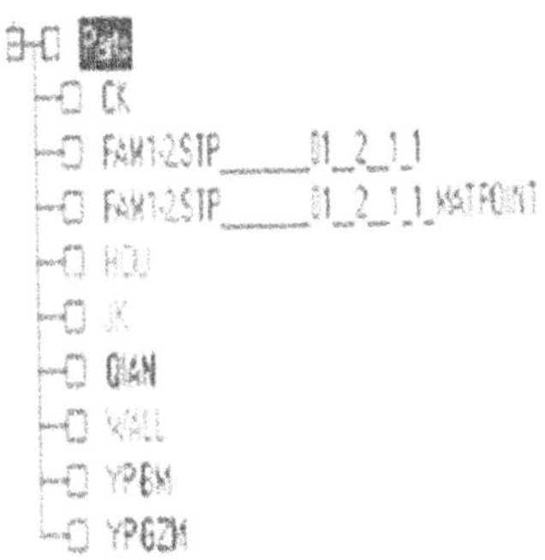

Figure 10.9 Generating the part model tree.

Figure 10.10 Interface after checking impeller geometry.

Check **[Curves]** and **[Points]** in the geometry of the model tree to display the curves and points in the impeller model, and do not display the face or display the face **[transparent]** to facilitate the subsequent division of the block.

10.2.1.3.2 Build Block The impeller meshing method is first create a fundamental block for a single impeller flow passage and generate the mesh for this block. Then, generate a full 360-degree periodic mesh for the entire impeller domain via rotational patterning. Subsequently, check the mesh quality, and finally, export the mesh file. A method for creating this fundamental block is described below:

1) Select **[Blocking]**, click **[Initialize Blocks]** in the dialog box to create blocks, as shown in Figure 10.11. Select any two space curves in the passage to create an initial block as shown in Figure 10.12. This block can be named **[YL]**.

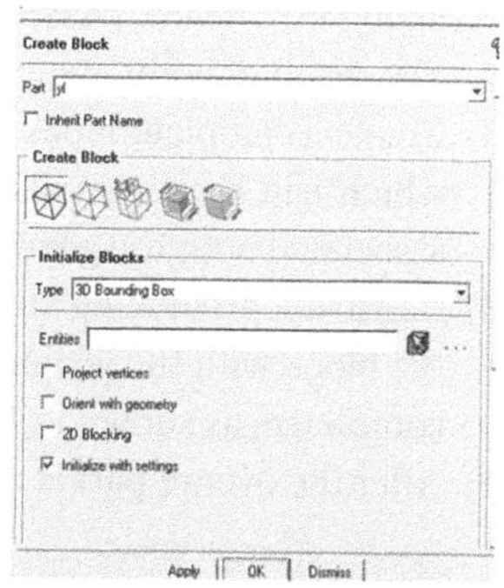

Figure 10.11 Create block dialog box.

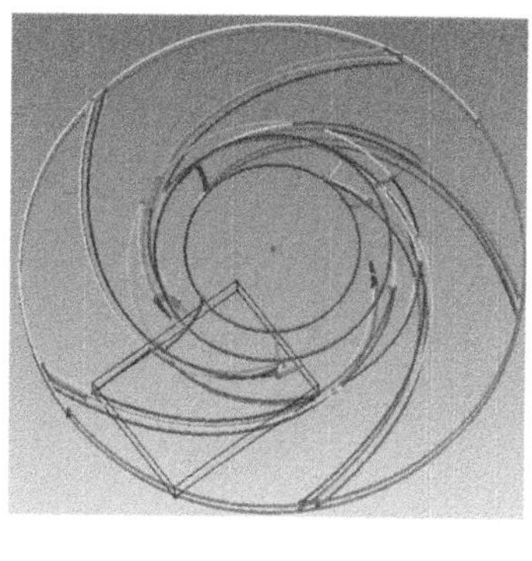

Figure 10.12 New block.

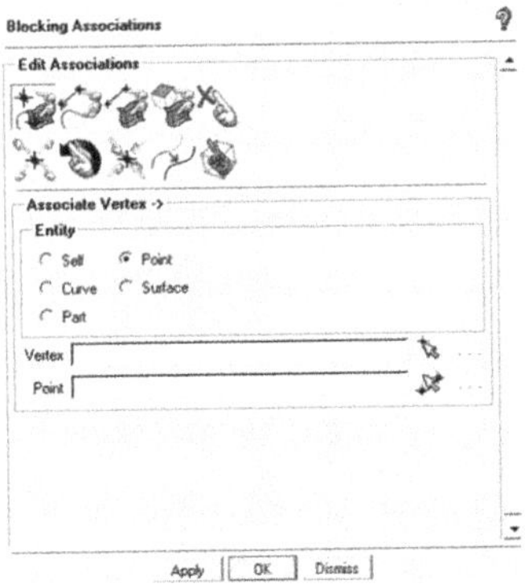

Figure 10.13 Associated dialog box.

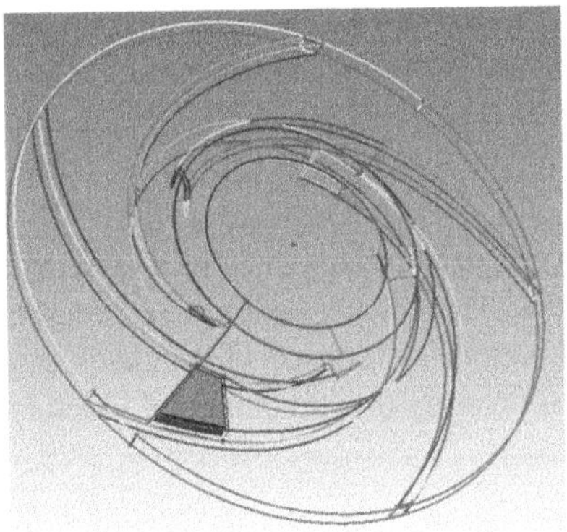

Figure 10.14 Associated geometry.

2) Associate the vertices on the block to the geometry, click **[Associate]** to bring up the association dialog box, select **[Associate Vertex]**. Select **[point]** in **[entity]**, as shown in Figure 10.13. The result of point association is shown in Figure 10.14.
3) Create a complete Block that contains the entire flow in a way that builds on an existing Block, which can be accomplished by means of the **[Extrude Face]** method. The Stretch Block is associated with the geometry in the same way.
4) Construct a blocking structure for the entire passage by extruding the block and associating the block with the geometry. The creation of the N-fraction of an impeller runner block is thus completed, as shown in Figure 10.15.
5) After the overall part is completed.

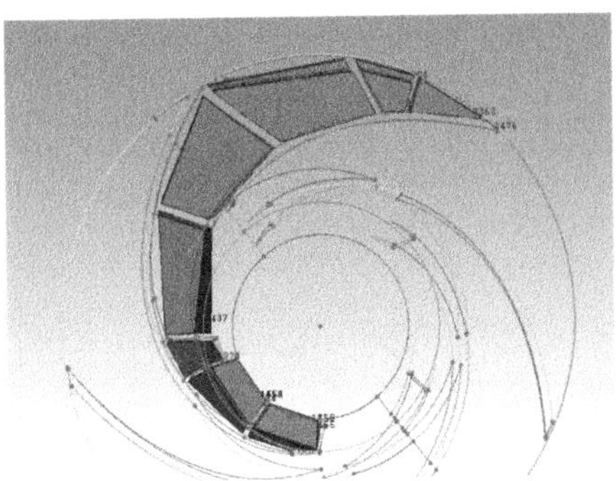

Figure 10.15 1/5 impeller flow passage block.

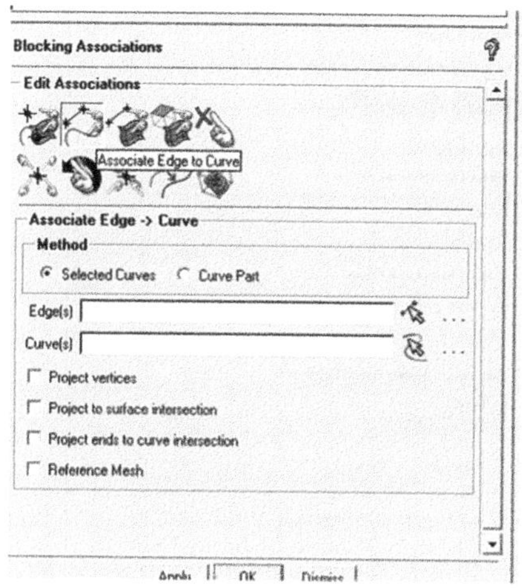

Figure 10.16 Association options.

6) Associate the edges of the Block with the curves of the geometry. Click the [Associate Edge to Curve] function, then select the Block edges to be associated, followed by the corresponding geometry curves as shown in Figure 10.16.

10.2.1.3.3 Handling of Regions Requiring Y-Type and O-Type Cuts Through the above steps, a Y-type topological cut is made within the newly created block. Then, associate the vertices, edges, and faces of the segmented block with the geometry, as shown in Figure 10.17. With these operations, the shape of the geometry is realized to be projected onto the block as a hexahedral mesh.

10.2.1.3.4 Handling of Associated Surfaces Deleting a cycle surface association operates as follows:

1) Select **[Blocking]** → **[Associate]** → **[Disassociate from Geometry]**
   After completing the above operations, perform a grid preview to view the associated issues. The operation is as follows:
2) Select **[Mesh]** → **[Global Mesh Setup]** → **[Global Mesh Size]** to get the dialog box as in Figure 10.18. Set **[Scale factor]** and **[Max element]**.

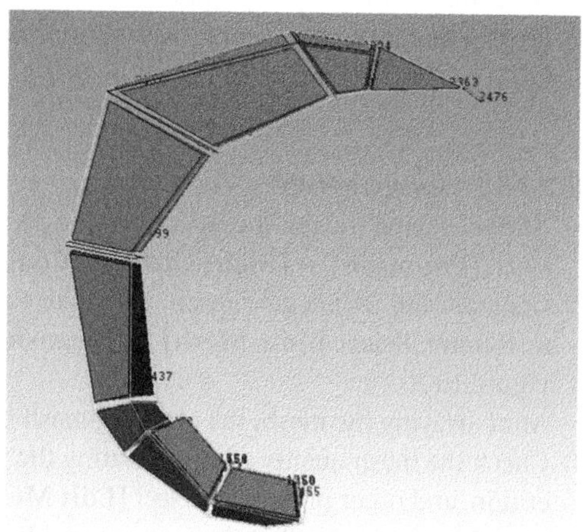

Figure 10.17 Block after association.

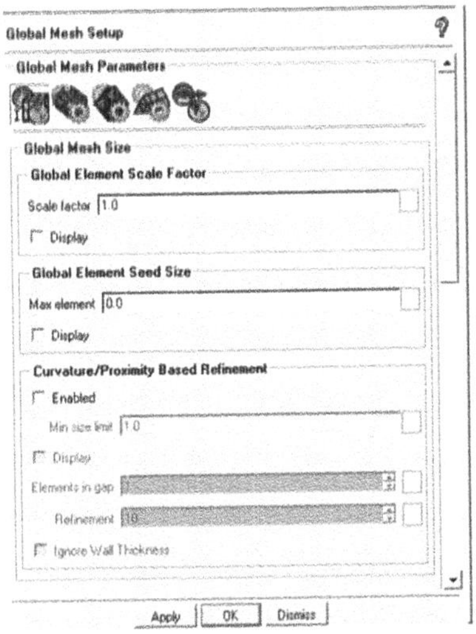

Figure 10.18 Defining mesh dimensions.

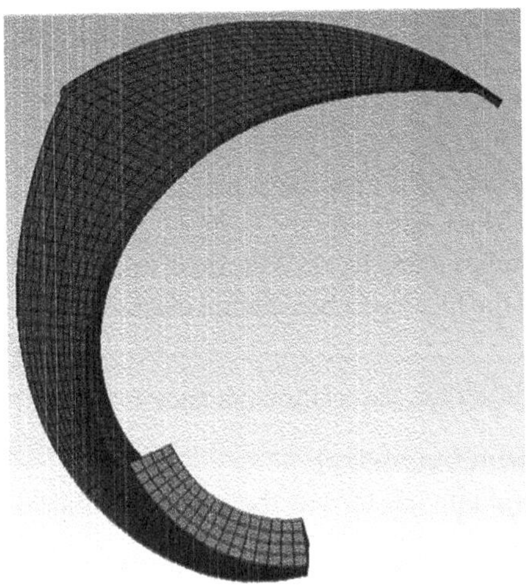

Figure 10.19 Generating preview mesh.

3) After setting the global mesh size, set the preview mesh as follows: Select **[Blocking]** → **[Pre-mesh Params]** → **[Update Sizes]** → **[OK]**.
4) Finally, under the model tree, click **[Blocking]** → **[Pre-mesh]** → **[√]** → **[yes]** to bring up the mesh interface as shown in Figure 10.19.

10.2.1.3.5 Generate Mesh

1) After checking the pre-mesh, convert the previewed mesh to an unstructured mesh as follows: Click **[Pre-mesh]** → **[Right click]** → **[Convert to Unstruct Mesh]** in the model tree.
2) Generate full 360-degree mesh, complete the impeller flow domain mesh. The operation is as follows: Select **[Edit Mesh]** → **[Transform Mesh]**, the pop-up dialog box is shown in Figure 10.20.
   After arraying the mesh, the resulting mesh is shown in Figure 10.21.
3) Check the mesh quality after generating the mesh and modify the negative volume, point repetition, and other problems. Select **[Edit Mesh]** → **[Check Mesh]** → **[Apply]** and output the mesh after checking that there are no problems.

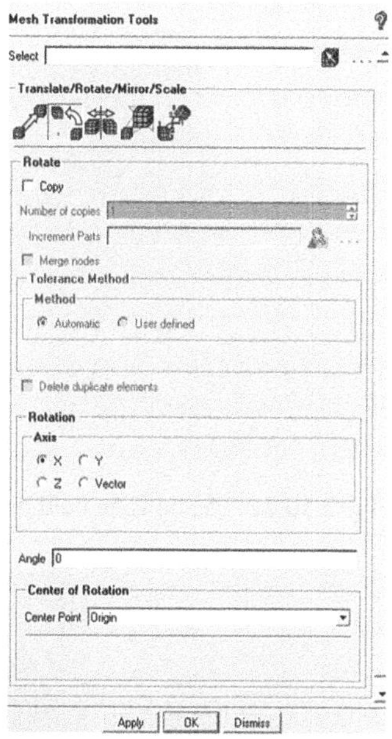

Figure 10.20 Array mesh options.

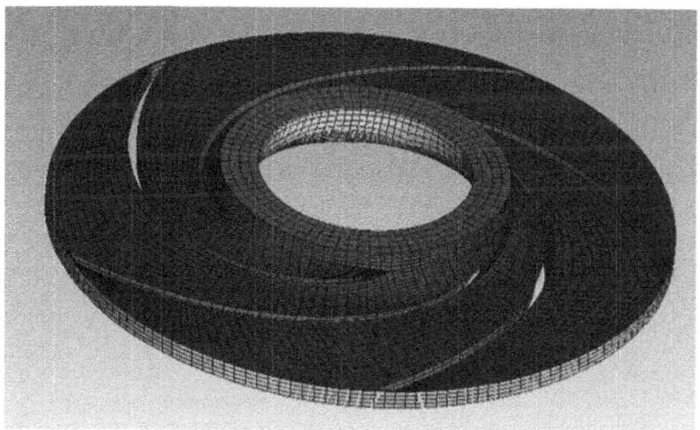

Figure 10.21 Generate impeller mesh.

4) Select **[Output]** and choose a solver in **[Select Solver]** to complete the impeller Structured Grid (Mesh) and output.

10.2.2 Guide Vane Mesh Generation

For multistage pumps, the guide vane is the core hydraulic component, and the guide vane flow domain is imported into ICEM for meshing, as shown in Figure 10.22. The steps for mesh delineation of the guide vane are similar to those of the impeller, and are briefly described below.

1) Definition of the part of guide vane

   The previous part has described the impeller Part definition method, Part definition of the operation process of the guide vane and the impeller is the same. Definition of the guide vane Part is shown in Figure 10.23.
2) Establishment of the block of guide vanes

   The block division method for the guide vane region is consistent with that for the impeller region. When dividing blocks in the guide vane region, avoid creating excessively small angles. Figure 10.24 shows the positive and return guide vane block division, in the triangular block area needs to be Y-type cut to improve the quality of the mesh.

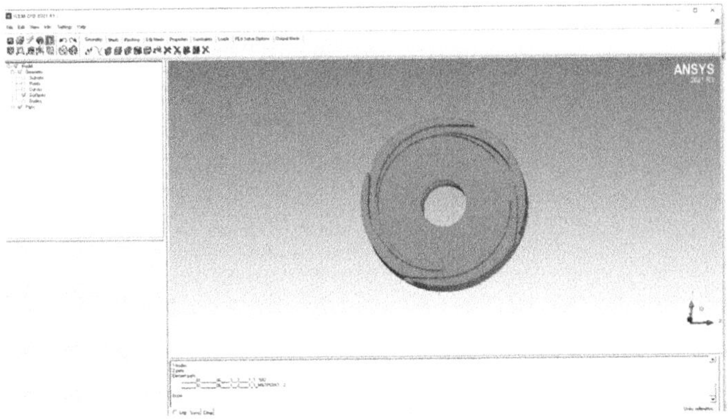

Figure 10.22 Guide vane fluid domain.

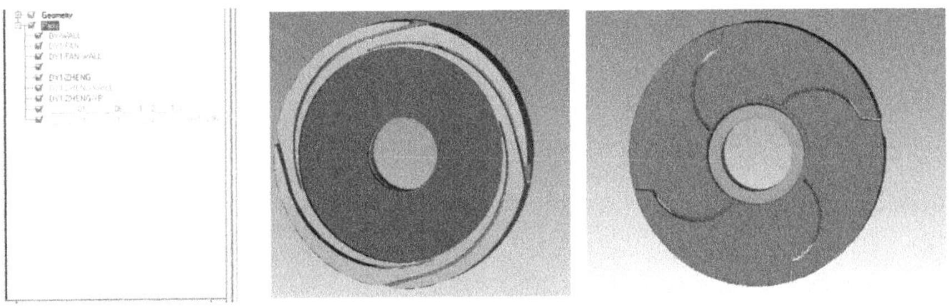

Figure 10.23 Positive and return guide vanes fluid domains.

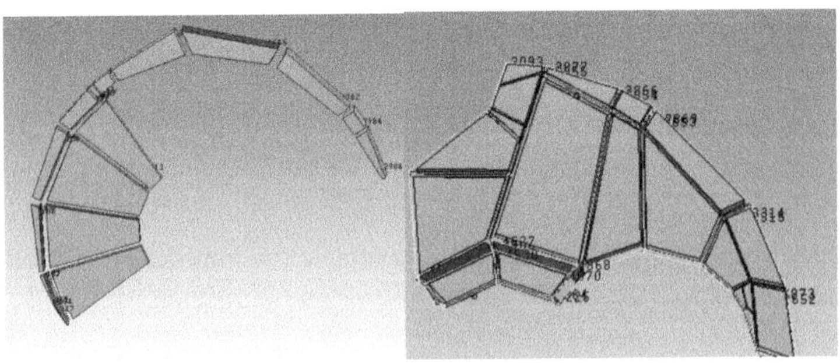

Figure 10.24 Block of positive and return guide vanes.

3) Mesh generation for guide vanes
   The mesh generation steps for the guide vane are the same as those for the impeller, and the forward and reverse guide vane meshes are obtained as shown in Figure 10.25.

Following the steps above completes the guide vane meshing. After meshing, it is necessary to check the grid for issues such as negative volumes, connectivity problems, or insufficient mesh density. After checking the mesh, select the same solver mesh as the impeller output mesh to output the guide vane mesh.

The meshing method and procedure for the impellers and guide vanes of subsequent stages are identical to those of the first stage. After that, the structured mesh is divided and output for the flow domains of the pipeline in the inlet and outlet of the multistage pump, and thus the flow domains mesh of the whole flow channel is obtained.

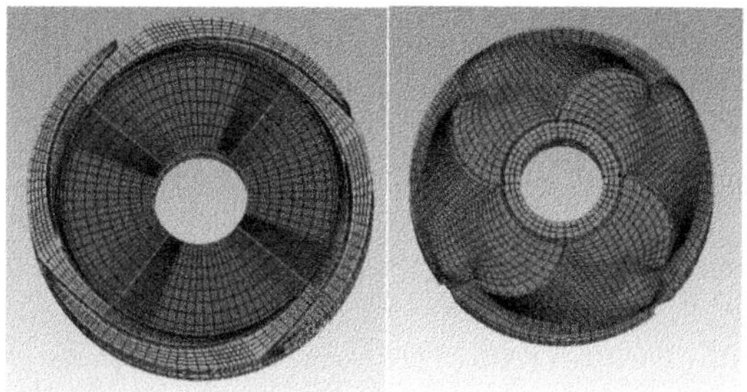

Figure 10.25 Positive and return guide vane grids.

10.3 Steady State Flow Simulation Setup

10.3.1 Importing Mesh Files

After completing the division of the overall mesh and verifying its correctness, the .cfx5 file exported from ICEM can be imported into CFX for further inspection of mesh quality and geometry conformance. The specific operation steps are as follows: First, click **[File]** in the menu bar, then select **[Import]**, then select **[Mesh]**, then an import dialog box will pop up. In the "Files of type" option of the dialog box, select ICEM CFD format (file extension .cfx or .cfx5). Next, select the unit of millimeters (mm) in the "Mesh Units" item on the right side of the dialog box. After that, select the mesh file to be imported from the file list and click the [Open] button, as shown in Figure 10.26.

After completing the mesh, the mesh file will be displayed in CFX as shown in Figure 10.27.

10.3.2 Fluid Domain Setup

After importing the mesh successfully, firstly, select steady calculation, double click **[Analysis Type]** for editing, set **[Analysis Type]** to **[Steady State]**, and click **[Apply]** button to complete the setting of calculation type. The next step is to define the flow domain and boundary conditions for the input mesh as follows:

1) The operation of defining the flow domain is: select **[Domain]**, the pop-up window shown in Figure 10.28, enter the name of the fluid domain, and click **[OK]**, the pop-up dialog box shown in Figure 10.29.

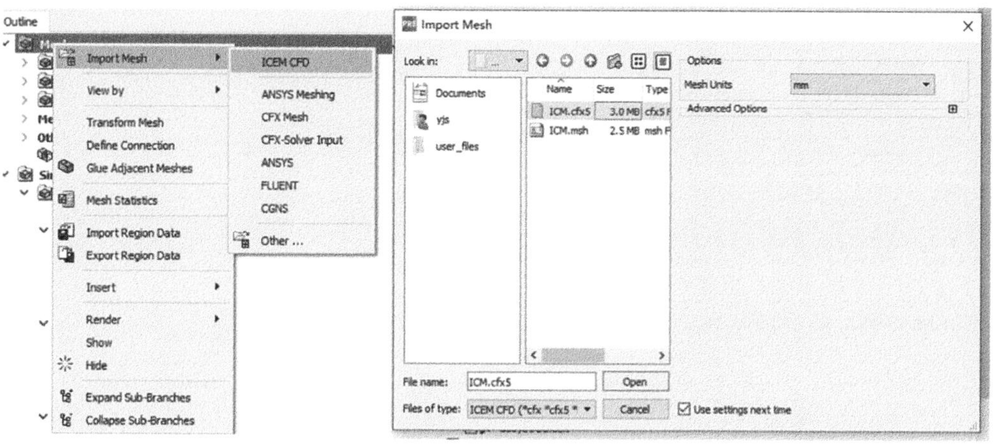

Figure 10.26 Preprocessing selection dialog box.

Figure 10.27 Assembled mesh in CFX.

Figure 10.28 Define domain dialog box.

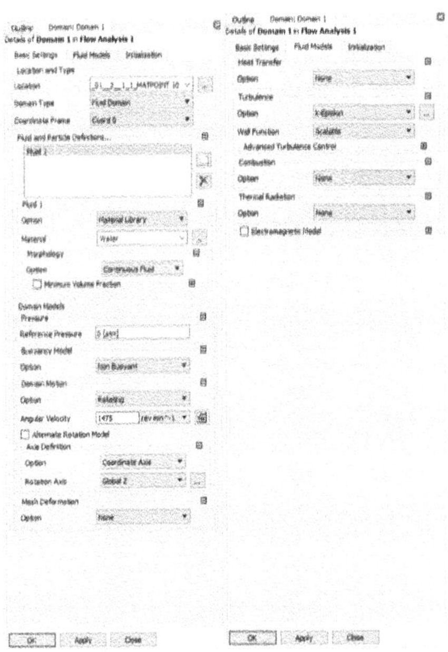

Figure 10.29 Define domain dialog box.

2) In the **[Location]** select the geometry to be defined, in the **[Domain]** after the selection of the fluid domain, in the coordinate system options - select the default, in the fluid medium - select water, in the reference pressure - select an atmospheric pressure; the domain is rotating, select the velocity defined by the right-hand rule, and define the selection axis; in the turbulence model under the menu only need to select the k-Epsilon turbulence model, or other turbulence models want to select, and click **[OK]** after completion.
3) Follow a similar procedure to define attributes for all fluid domains, with the impeller flow domain as a rotating domain and the guide vane flow domain, inlet, and outlet defined as stationary domains.

   After defining the fluid domains, it is also necessary to define the interface, i.e., the interface between the fluid domains, this interface does not exist in practice, due to the need to define the fluid domain of the divided mesh, the entire fluid domain is divided into a number of regions.
4) Define the cross-interface as follows: For the static cross-interface, select **[Domain Interface]** to bring up the window shown in Figure 10.30, define the name of the interface, and click **[OK]**. The dialog box shown in Figure 10.31 will pop up.
5) In the interface type column, select the two flow domains on either side of the interface, if it is static or dynamic fluid domain handover, one needs to choose in the Frame Change/Mixing Model column **[Frozen Rotor]** and in the Pitch Change select **[Specified Pitch Angles]** → **[360 (degree)]**, the impeller flow domain and the inlet passage and the guide vane of the handover are static and dynamic handover. If it is a static fluid domain handover, the option is shown in Figure 10.31; the guide vane outlet and outlet passage handover is a static handover.

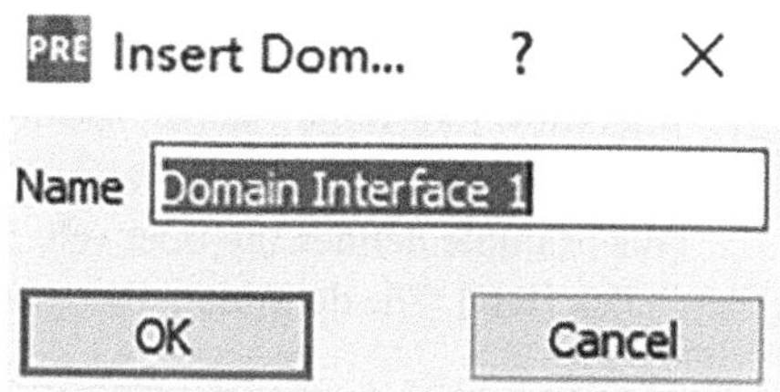

Figure 10.30 Define interfaces dialog box.

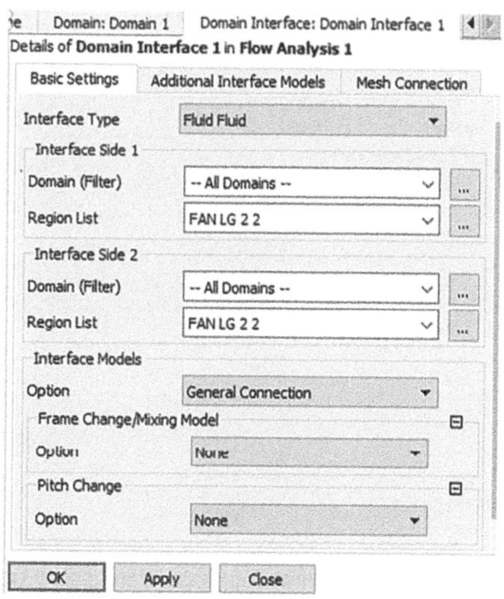

Figure 10.31 Define interfaces dialog box.

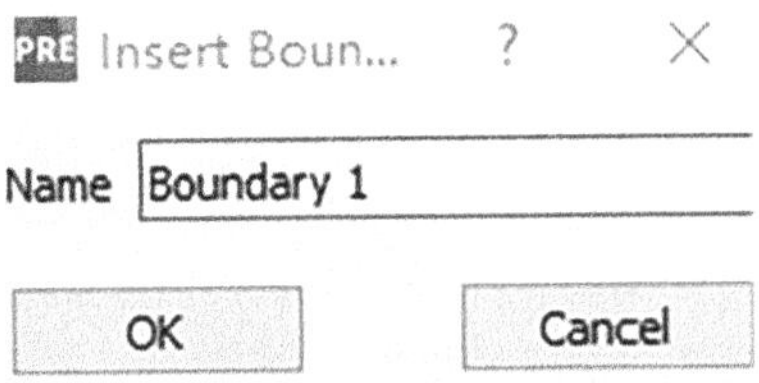

Figure 10.32 Define boundary dialog box.

10.3.3 Boundary Conditions Settings

Following the above method, the definition of all interfaces is thus completed, after which the inlet and outlet of the fluid domain and the wall properties need to be defined. The fluid domain has a total inlet, i.e., water flows into the inlet of the multistage pump, and there is also a total outlet, i.e., water flows out of the outlet of the multistage pump. In addition to the inlet and outlet, there are the fluid domain pump casing, vane wall contact location, which are defined as Wall.

Define the wall and inlet/outlet as follows: Click **[Boundary]**, select the corresponding fluid domain, and then the window shown in Figure 10.32 will pop up, enter the boundary name, and click **[OK]**.

1) Define the boundary type, select Inlet, Outlet, Wall, and the corresponding boundary type is **[Inlet]**, **[Opening]**, **[Wall]**.
2) After selecting the boundary types, select the boundary location. In CFX, the inlet and outlet boundary conditions include mass flow, total pressure, and static pressure, and the commonly used boundary condition types are:
   ① Inlet: total pressure, outlet: flow rate;
   ② Inlet: flow rate, outlet: static pressure;
   ③ Inlet: total pressure, outlet: static pressure.
   This example defines the inlet velocity and outlet pressure, select default for other options and click **[OK]**. The definition dialog box is shown in Figure 10.33.

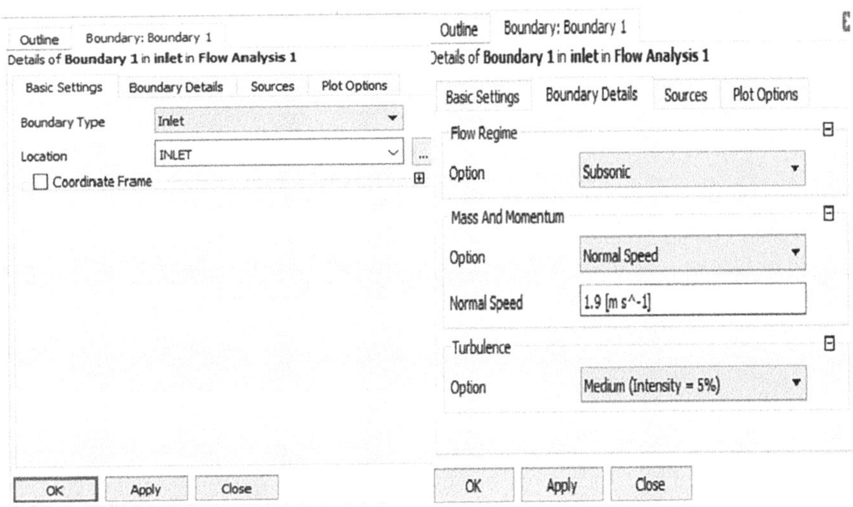

Figure 10.33 Define inlet boundary dialog box.

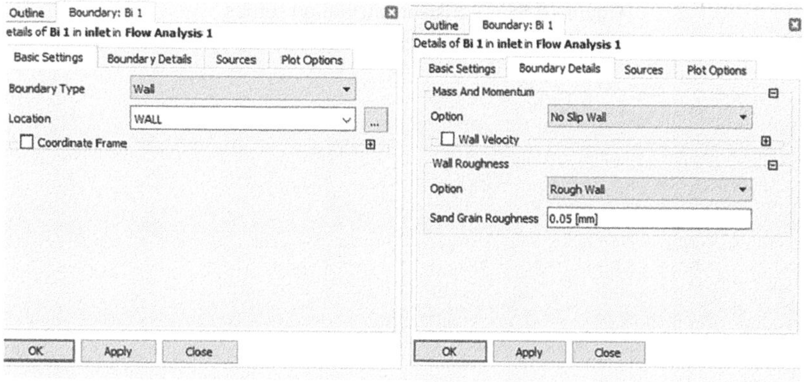

Figure 10.34 Defining wall dialog box.

3) Define the wall boundary condition, as shown in Figure 10.34, select the boundary condition Wall, and specify the corresponding wall surfaces in the Position menu bar. Set the wall to be no-slip, and set the wall roughness to 0.05 mm.

When the surface roughness is defined by the casting process, a surface roughness of 0.05 mm is acceptable for a typical casting process.

10.4 Transient Flow Simulation Setup

Change the **[Analysis Type]** to **[Transient]** in the model tree. Next, define the total time for the transient calculation. Typically, transient calculations need to cover 6–10 rotational cycles of the impeller in order to obtain a reliable solution. In the **[Max. Timesteps]** input box under **[Maximum Number of Timesteps]**, enter the maximum number of timesteps. Then, define the number of calculation steps within each rotation period. This can be interpreted as determining the number of degrees per rotation of the impeller for which the calculation is performed. For example, if the speed of the impeller is 1475r/min, the time it takes to rotate for one week is 60/1475 = 0.04068s, and the total time step to rotate for 5 weeks is 0.20339s; defining the time required to perform one calculation for every 4° of rotation in one cycle is 0.000452s [1] (Figure 10.35).

There are some differences between the interface settings for transient simulations and those for steady-state calculations, primarily reflected in the configuration of the connection model for the rotor-stator interface. Specific operation is, double-click the multistage pump in the dynamic–static calculation domain interface; change the coordinate system transformation **[Frame Change/Mixing Model]** to **[Transient Rotor Stator]**, the rest of the settings are consistent with the steady calculation. After completing the configuration, click the **[Apply]** button to complete the definition of the data interface of the dynamic–static calculation domain in the transient simulation.

In transient simulations, the solver parameters need to be defined for the transient time term **[Transient Scheme]**, as well as for the inner loop calculation **[Convergence Control]**. The Transient Scheme can be set by default. [Convergence Control] is the number of times for each time step, which is equivalent to a steady computation in each time step, and the number of times for the inner loop calculation **[Min. Coeff. Loops]** and **[Max. Coeff. Loops]** are adjustments to the number of steps of the transient simulations, and in general, the transient simulations will easily

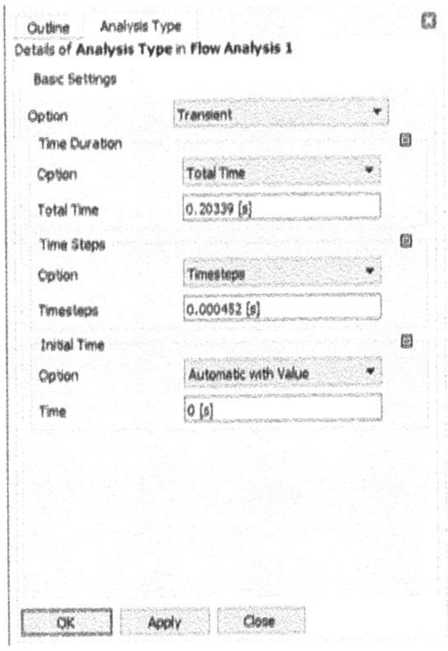

Figure 10.35 Transient calculation setups.

reach convergence in each time step when it enters the steady state. In general, after the transient simulation enters the steady state, it is easy to reach the convergence value of 10^{-4} for each time step, and the number of inner loops can be very small. Generally, **[Max. Coeff. Loops]** is set to 10, and **[Min. Coeff. Loops]** is set to 1 to meet the calculation requirements. The other settings, such as convergence criteria, are the same as those of the steady simulation.

10.5 Solution Setup

After defining the fluid domain, interface, and boundary conditions for the flow domain, the solver control settings need to be configured. In the model tree, select **[Solver]** → **[Solver Control]** to

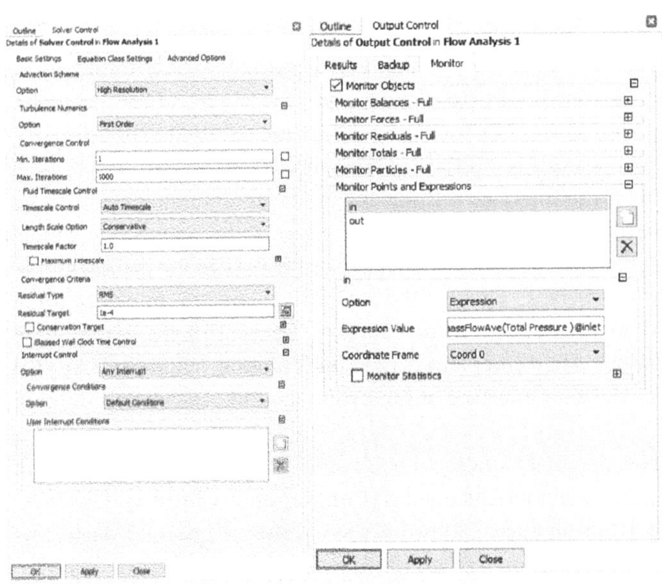

Figure 10.36 Define calculation dialog box.

bring up the dialog box shown in Figures 10.3–10.10. In general, the maximum number of iteration steps needs to be defined should be above 1000 steps, and the convergence residual is recommended to be as shown in Figure 10.4 or even smaller to ensure the accuracy of the calculated results.

After configuring the solver control settings, it is necessary to set up monitor points to track key parameters, typically including the inlet pressure, outlet pressure, and pump head. The operation steps are as follows: Select **[Output Control]** →**[Monitor]**, pop-up dialog box shown in Figure 10.36 right, click **[Add new item]** named in the Option to select Expression, in the expression enter massFlowAve(Total Pressure) @inlet, indicating the total pressure to monitor the average mass flow at the inlet. Similarly, define the outlet monitoring as massFlowAve(Total Pressure) @outlet, and the head is defined as (massFlowAve(Total Pressure)@outlet-massFlowAve(Total Pressure) @inl -et)/1000[kg m^{-3}]/g.

After completing the above settings, start to perfrom the calculation, pop up the dialog box shown in Figure 10.37, define the directory of the calculation file, define whether to calculate in parallel or not, and then select **[Start Run]**.

The same setup is used for transient simulation, but an initial flow field is typically required. Here, the initial flow field is the result of the steady simulation. Conducting a steady-state

Figure 10.37 Steady state calculation dialog box.

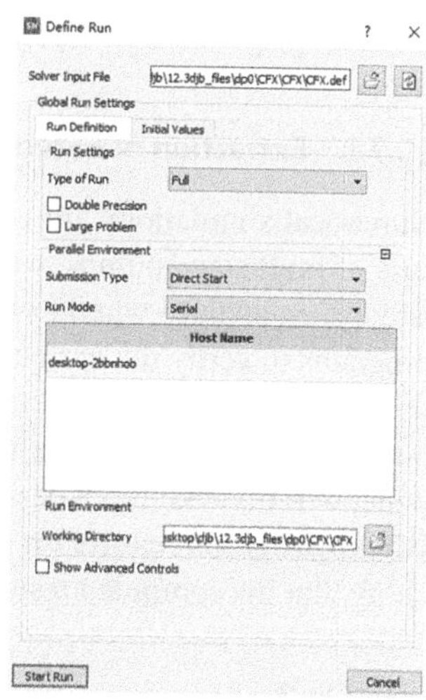

Figure 10.38 Transient calculation dialog box.

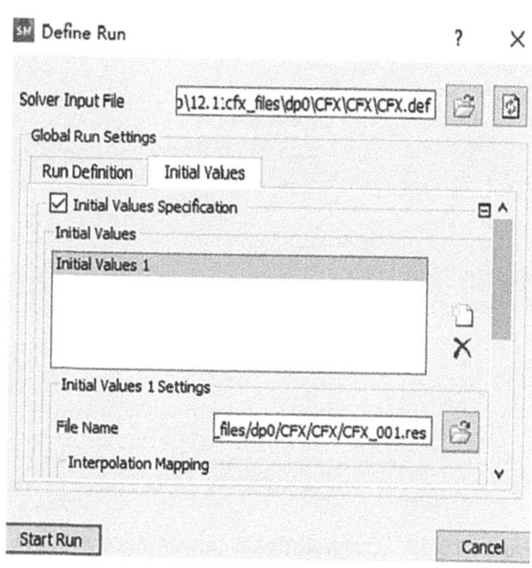

simulation is the basis for the subsequent transient calculation. Perform the specific operation as shown in Figure 10.38, first check the **[Initial Values Specification]**, the bottom of the **[Initial Values 1 Settings]**, to find the results of the previous steady calculation, that is, the .res file, and finally click **[Start Run]** to start the transient simulation.

10.6 Computations

The process described above completes the preprocessing and CFD solver setup for the simulation and initiates the calculation. Once started, the solver monitor window will open. The solver will automatically stop when it reaches the specified convergence criteria or the maximum number of iterations, and will generate the result files simultaneously. The calculation convergence monitor window is shown in Figure 10.39.

10.7 Computed Results and Postprocessing

After completing the structured mesh and solver setup for the multistage pump CFD model, postprocessing operations are performed next.

10.7.1 Extraction of External Characterization Results

Numerical simulations are conducted to obtain results similar to those in the experiments. In case of multistage pumps, we generally extract torque from the pump rotor components, calculate forces, analyze whether the axial force is balanced, and compare the results across different programs to verify the correctness of the results [2–4]. Specific steps are: select **[Calculators]** in the model tree, pop-up dialog box shown in Figure 10.40a, click [Function Calculator], and then pop-up dialog box shown in Figure 10.40b. As can be seen from this figure, the lower part of the window is divided into a parameter configuration section and a results output section, as well as **[Clear previous results on calculator]** and **[Show equivalent expression]**, allowing users to verify that the computed results are as intended.

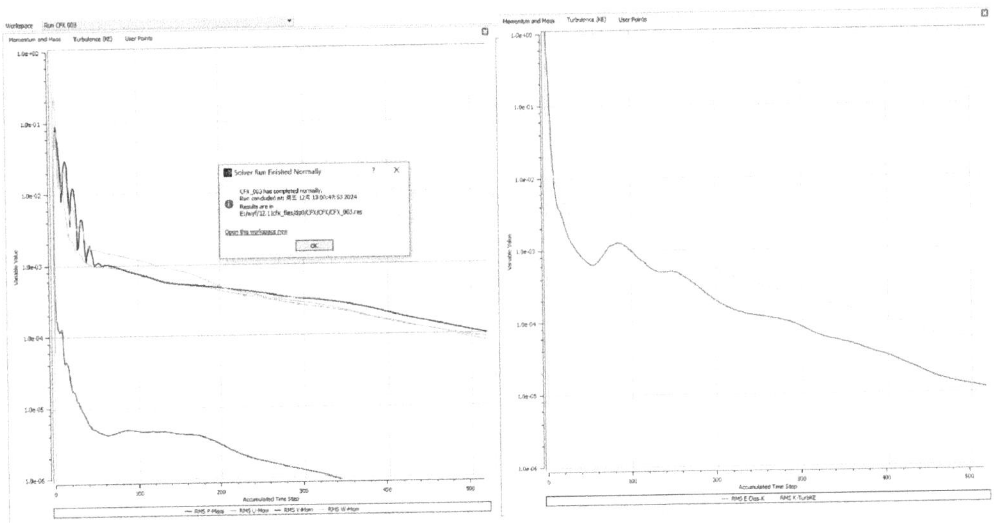

Figure 10.39 Computational convergence window.

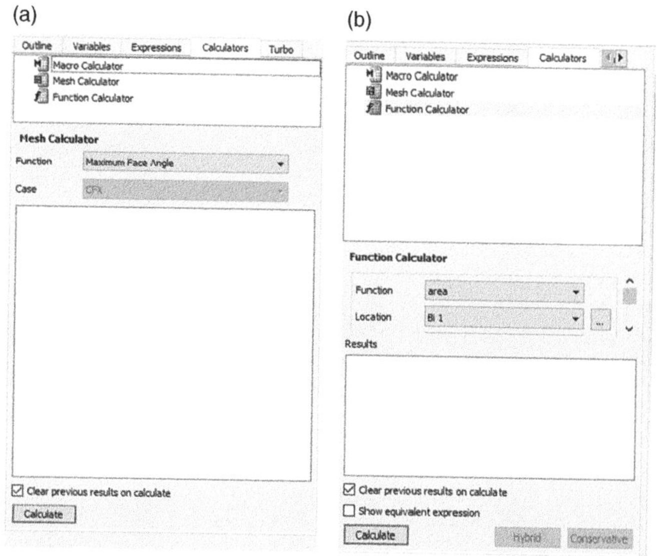

Figure 10.40 Results extraction dialog box: (a) Calculator type selection; (b) function calculator.

In the drop-down menu bar, **[Function]** indicates the calculation function, **[Location]** indicates the geometric location for the computation, and **[Variable]** indicates the physical variable to be calculated. For multistage pumps, which are generally more concerned with the pump head, efficiency, and power, user can extract the impeller import and export pressure and torque in the **[Function]**, to achieve the purpose of calculating the head, efficiency, and power. Specific operations are: in **[Function]**, select the type of extraction results for the massFlowAve, on behalf of the average of the integral on the opposite side, in **[Location]**, select the extraction location for the inlet, in the **[Variable]**, select **[Total Pressure] as the variable**, other defaults, click **[Calculate]**. Figure 10.41 shows the results of the inlet and outlet pressure. Similarly, user can compute the torque of the impeller by the same method, as shown in Figure 10.42. Select torque in **[Function]**, select the impeller's axis of rotation, Z axis of the rotation axis of the rotary axis, other defaults, and click **[Calculate]**.

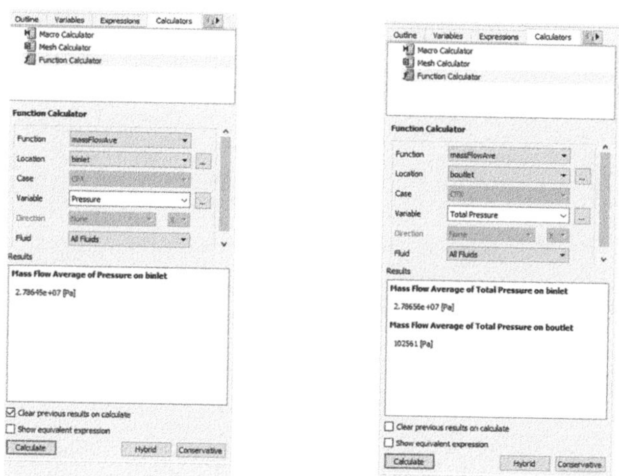

Figure 10.41 Extracting the total inlet and outlet pressures dialog box.

Figure 10.42 Extracting rotor component torque dialog box.

10.7.2 Streamline and Particle Dynamic Tracking

In order to understand the internal flow structure of the multistage pump, the internal flow can be analyzed through streamlines. Specific steps are: Click **[Streamline]**, pop-up dialog box shown in Figure 10.43 to name the streamline, click **[OK]** pop-up dialog box shown in Figure 10.44, **[Type]** to rotate the 3D streamline, from the inlet to the beginning of the whole domain of the streamline, and adjust conditions such as density as needed. click **[Apply]** to get the three-dimensional streamline.

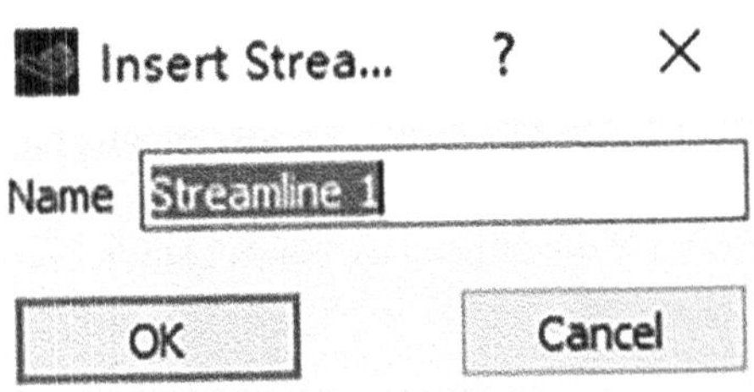

Figure 10.43 Define streamlines.

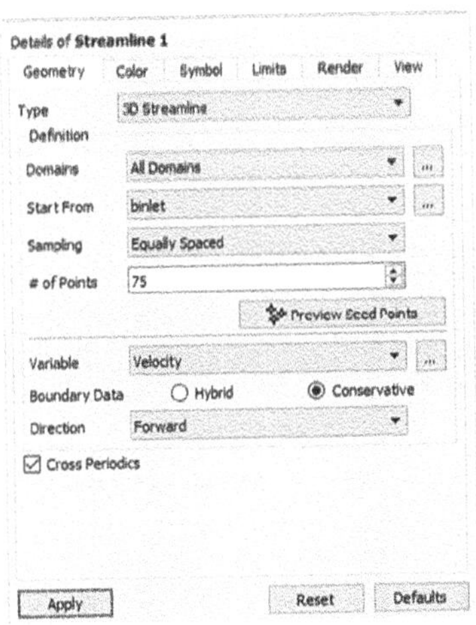

Figure 10.44 Define streamlines dialog box figure.

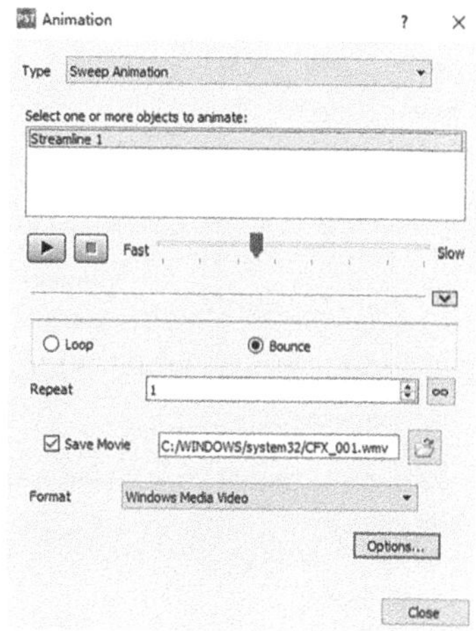

Figure 10.45 Define dynamic tracking particles dialog box.

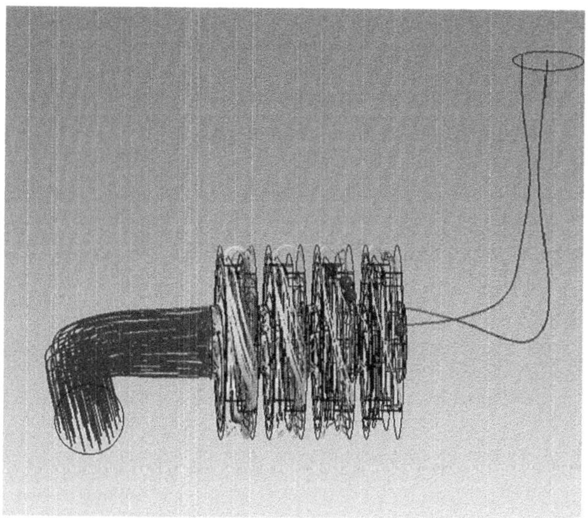

Figure 10.46 Streamline diagram and particle tracking interface.

Figure 10.47 Define contour dialog box.

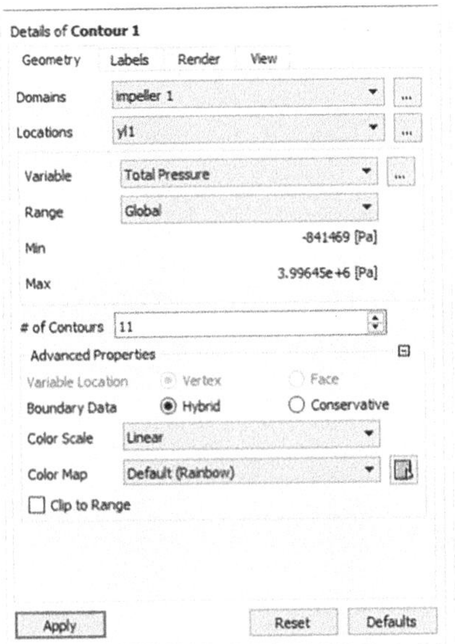

Figure 10.48 Define contour map dialog box.

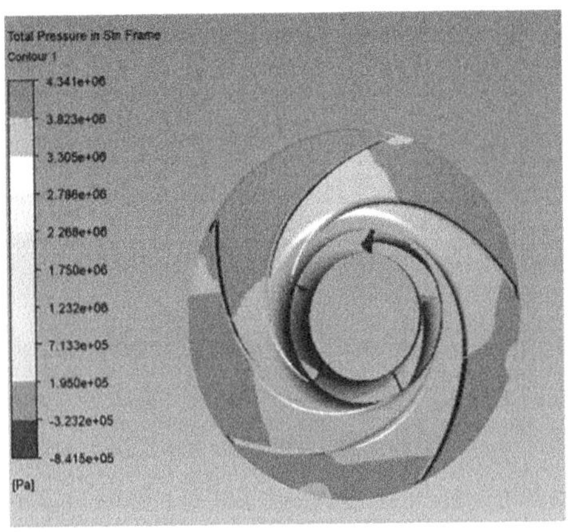

Figure 10.49 Contour map results.

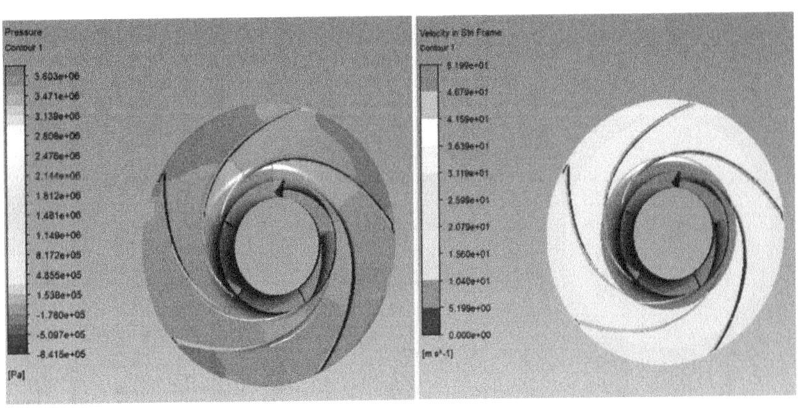

Figure 10.50 Other contour mapping results.

After getting the streamline, a dynamic particle trace animation can be generated, the specific operation is: in the menu bar, rotate **[Animation]**, pop-up dialog box shown in Figure 10.45, select the generated streamline, click **[Play the animation]**, get the view of the particle flow along the streamline shown in Figure 10.46. Simultaneously, the default settings in the [Definition]dialog box can be accepted to generate a .wmvvideo file.

10.7.3 Contour Diagram Generation

In order to further analyze the performance of the pump, investigation can be conducted by examining internal flow fields such as velocity and pressure. Generating the contour diagrams of the flow variables inside the pump allows for further optimization of the pump [5, 6].

The specific operational steps to derive the contour diagrams are: click **[Contour]**, open the pop-up dialog box shown in Figure 10.47 to define the contour diagram, click **[OK]**, open the pop-up dialog box shown in Figure 10.48. In the dialog box, define the contour region as follows: set **[Domain]** to the impeller fluid domain, set **[Location]** to the impeller surface, and select **[Variable]** as total pressure for the contour type. Click **[OK]**, get the contour map results as in Figure 10.49. Similarly, other types of contour plots, such as for static pressure and velocity, can be generated using the same method. The result is shown in Figure 10.50.

References

1 Chen, C., Hu, X., Deng, F., Wu, K., Zhang, Z., and Si, Q. (2024). Study on the transient flow characteristics of multistage centrifugal pumps during the startup process before system operation. *Water* 16 (13): 1876.

2 Shi, L., Yao, S., Zhao, J., Mou Jiegang., Shen Binjie., Zhang Hui., Wu Tianxin., and Wu Denghao. (2024). Research on axial force balancing device for large vertical multistage centrifugal pumps. *Fluid Machinery* 52 (07): 56–62.

3 Qian, C., Luo, X., Yang, C., and Wang, B, and Wang, B. (2021). Multistage pump axial force control and hydraulic performance optimization based on response surface methodology. *Journal of the Brazilian Society of Mechanical Sciences and Engineering* 43 (3): 136.

4 Marjan, G., Dusan, F., and Bran, S. (2002). Hydraulic axial thrust in multistage pumps-origins and solutions. *Journal of Fluids Engineering* 124 (2): 336–341.

5 Qi, M., Li, G., Mao, J., Zhao, D., Luo, Y., and Hou, T. (2024). Analysis of flow loss characteristics of a multistage pump based on entropy production. *Water* 16 (20): 2974.

6 Kock, F. and Herwig, H. (2005). Entropy production calculation for turbulent shear flows and their implementation in CFD codes. *International Journal of Heat and Fluid Flow* 26 (4): 672–680.

11

Conclusions and Future Prospects

11.1 Challenges and Opportunities of Current Pump Technology

In the context of the current accelerated global industrialization, the development trend of the pump industry, as a crucial fluid transportation and pressurization equipment manufacturer, has attracted considerable worldwide attention [1]. Pumps, with their unique operating principles and structural characteristics, are widely used in many major industries of the national economy, such as chemicals, petroleum, construction, agriculture, electricity generation, and papermaking etc. [2–8]. They not only support the daily operations of these industries but also play an indispensable role in promoting industrial upgrading and technological progress [9]. According to statistics from the International Pump Manufacturers Association (IPMA), the global pump market reached $45 billion in 2022 and is expected to grow at a compound annual growth rate of 5% per year until 2027 [1]. Despite significant progress in pump technology over the past few decades, it still faces many challenges in energy efficiency, materials, intelligence, and other aspects. Here, pump intelligence refers to the integration of AI and advanced software to enable automated monitoring, predictive maintenance and optimized performance of pump systems, which will reduce unplanned downtime, cut energy losses, and enable remote and proactive management. At the same time, with improvements in environmental protection requirements and the rise of intelligent manufacturing, the pump industry also faces new opportunities. This chapter discusses the main challenges currently faced by the pump technology, including energy efficiency, material selection, intelligence, and environmental regulations, and analyzes the opportunities brought about by these challenges.

As an important component of general machinery equipment, the development level of pumps is directly related to multiple aspects of the national economy [10]. From simple water circulation to complex technological processes, pumps demonstrate their unique value. Especially driven by large-scale equipment renewal policies, the demand for rotating machinery such as motors, pumps, and compressors continues to grow, further highlighting the fundamental position of mechanical pumps in industrial production [11]. This growth in demand is not only reflected in the increase in quantity but also in the higher requirements for product quality and performance, prompting the mechanical pump industry to continuously plan and develop high-end and intelligent (smart) solutions [12].

Facing the increasingly fierce market competition and ever-changing customer needs, technological innovation has become the key to the sustainable development of the pump industry. After experiencing a development process from simple to complex and from inefficient to efficient, the traditional pump industry is gradually moving toward high-end, intelligent, and green designs of pump systems [13]. By introducing advanced manufacturing technologies and design concepts, the energy efficiency ratio and reliability of products are significantly improved [14]. Combined with modern information technologies such as the Internet of Things (IoT) and big data, the intelligent upgrading of pumps has been promoted, realizing functions such as remote monitoring, fault diagnosis, and predictive maintenance [15]. The enhancement of environmental awareness

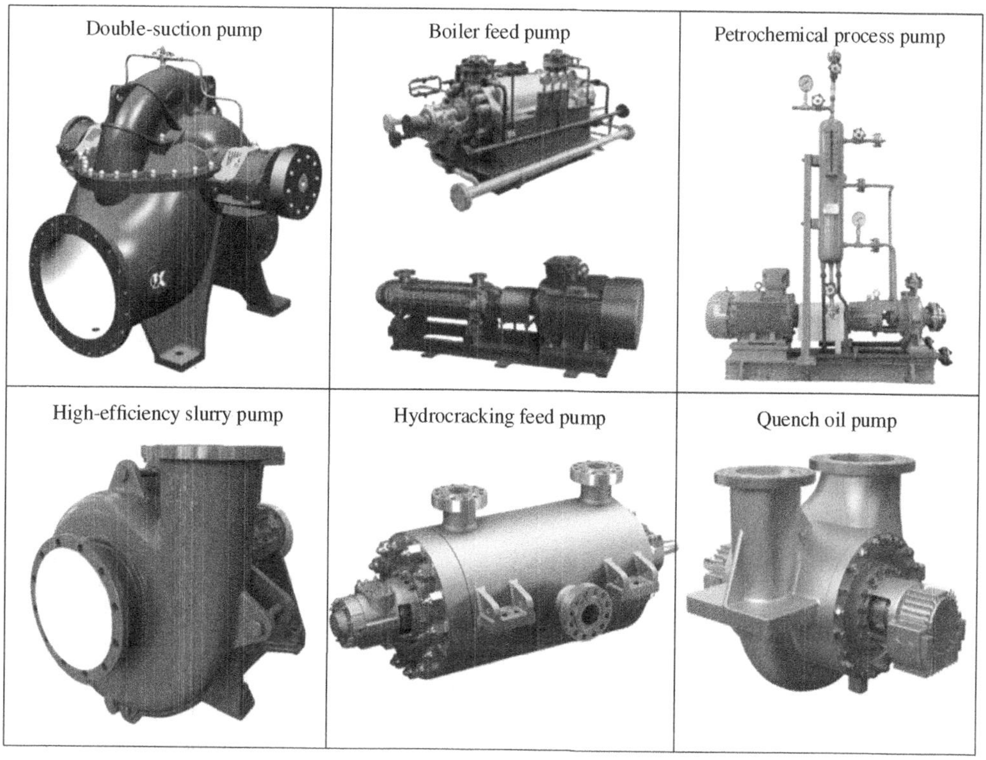

Figure 11.1 Variety of pumps employed in different applications. *Source:* Shanghai Kaiquan pump (Group) Co., Ltd.

also prompts the pump industry to pay more attention to energy conservation, emission reduction, and recycling in material selection, manufacturing processes, and product design.

As an important support force for national economic development, the pump industry has broad and full opportunities for development. Driven by market demand, policy support, and technological innovation, the pump industry is expected to usher in a more prosperous future. Figure 11.1 provides a glimpse of the variety of pumps employed in different applications.

11.1.1 Current Challenges of Pump Technology

11.1.1.1 Technical Challenges

11.1.1.1.1 Energy Efficiency and Energy Saving With the intensification of the global energy requirements, energy conservation and emission reduction have become global issues. Pump systems, as major energy consumers, urgently need to improve their energy efficiency. The energy consumption of pumps accounts for a significant portion of total industrial energy consumption. According to statistics, improving pump energy efficiency has become the top priority of the industry. How to reduce energy consumption while ensuring required flow rate and pressure has become an urgent problem for pump manufacturers to address. Furthermore, the demand for high-efficiency pumps is increasing in the market, making traditional pump designs face enormous challenges.

According to reports from the International Energy Agency (IEA), the energy consumption of pump systems accounts for 20%–25% of global industrial electricity use [16]. The efficiency of traditional centrifugal pumps is usually between 40% and 70%, indicating significant potential for improvement in energy efficiency. Research shows that pump systems using variable frequency drives (VFDs) can improve energy efficiency by 15%–50%. However, a large number of pump systems have not yet achieved energy efficiency optimization, requiring innovation in intelligent control and system integration.

11.1.1.1.2 Application of New Materials and Technologies During operation, pumps often need to handle fluids with corrosive and abrasive properties, which places higher requirements on pump material selection. Although the development and application of new materials can improve the durability and performance of pumps, they also increase production costs. Additionally, the sustainability and environmental friendliness of materials have become important factors to consider in today's manufacturing industry.

The performance of core pump components, such as bearings and seals, directly affects the working efficiency and stability of pumps. However, the application of new materials is not yet widespread, especially in high-temperature, high-pressure, and corrosive environments. Furthermore, the research and application of new technologies, such as new thermal cycle construction, efficient compression, and heat exchange equipment, also lag behind and require further breakthroughs [17].

11.1.1.1.3 Maintenance and Fault Prediction The management of maintenance of pumps is crucial for their long-term reliable operation. Traditional maintenance methods often rely on periodic inspections, making it difficult to achieve real-time monitoring. According to statistics from research institutions, downtime caused by pump failures accounts for 15%–20% of total production downtime. Therefore, the adoption of smart sensors and IoT technologies for real-time monitoring and fault prediction has become the key to improving the operation reliability of pumps [18].

11.1.1.2 Environmental Challenges

With the global increase in environmental protection awareness, governments of various countries have introduced stricter environmental protection regulations, placing higher requirements on the environmental performance of pumps [19, 20]. This not only promotes the research and development and promotion of pump products with energy-saving and environmental protection characteristics but also accelerates the pace of technological innovation throughout the industry.

11.1.1.2.1 Increased Environmental Protection Requirements As environmental awareness continues to grow, industries have increasingly become aware of high environmental protection requirements. As fluid transportation equipment, pumps inevitably produce some degree of noise, vibration, and leakage during their use. Therefore, pump manufacturers need to continuously optimize pump designs to reduce noise, decrease vibration, and improve sealing performance to meet the environmental protection requirements.

11.1.1.2.2 Impact of Climate Change Global climate change has led to frequent natural disasters such as water shortages, droughts, and floods, placing higher demands on installed pump systems. Ensuring the normal operation of pumps under extreme climate conditions has become an important challenge for pump enterprises.

11.1.2 Opportunities for Current Pump Technology

11.1.2.1 Intelligence and Digital Transformation

With the advancement of Industry 4.0, the intelligence and digital transformation of pumps have brought new opportunities to the industry. According to a report by market research company Mordor Intelligence, the smart pump market is expected to reach $8 billion by 2025, with a compound annual growth rate of about 10%. By integrating IoT technology, pumps can achieve real-time data monitoring and intelligent control, significantly improving the energy efficiency and reducing the failure rates [21].

11.1.2.2 Environmental Protection and Sustainable Development

The global emphasis on environmental protection provides new opportunities for the development of pump technology. According to data from the United Nations Environment Programme (UNEP), energy conservation and emission reduction have become important policy goals of governments worldwide. The development of low-energy consumption and low-emission pump technologies not only meets environmental protection requirements but also enhances market competitiveness. For example, the application of circulating water pumps and regenerative water pumps is becoming increasingly popular, opening up new market areas for the pump industry [22].

11.1.2.3 Demand in Emerging Markets

With the development of the global economy, especially the rapid industrialization of the Asia-Pacific region, the demand for pumps continues to grow. According to research by Mordor Intelligence, the pump market in the Asia-Pacific region is expected to grow at a compound annual growth rate of 6.5% over the next five years. Seizing this market opportunity, pump enterprises can expand their market share by developing pump products suitable for local needs [23].

Currently, pump technology faces many challenges in energy efficiency, material selection, and maintenance, but the growth of intelligence, environmental protection, and market demand brings new opportunities to the industry. Pump enterprises should actively respond to these challenges, seize the opportunities brought about by industry changes through technological innovation and adjustments to market strategies, and achieve sustainable development. Researchers and engineers should actively explore new technologies, materials, and methods to address challenges, seize opportunities, and promote the sustainable development of pump technology. In the future, innovations in the pump industry will provide strong support for the progress and development of various industries. Driven by the dual forces of the global economy and environmental protection, pump technology will usher in broader development prospects.

11.2 Prediction of Future Development Trends in Pump Technology

Pumps, as general equipment widely used in many industries, e.g. agriculture, construction, chemical and municipal fields etc. continually need to address and incorporate the new technological innovations to meet the market expectations and demand. With global population growth, economic development, and technological progress, the pump industry is undergoing profound changes. The following sections aim to analyze the current status of the pump market, explore the main development trends of future pump technology, and predict the coming changes in intelligence, energy conservation and environmental protection, industry segmentation, and international cooperation.

Pumps, which convert mechanical energy into kinetic and potential energy of liquids, are widely used in water transportation, wastewater treatment, agricultural irrigation, industrial manufacturing, air conditioning systems, and other fields. In recent years, with the acceleration of global urbanization and industrialization, the demand for pumps has continued to grow. The following sections describe the future trends in pump technology from the perspectives of technical intelligence, energy conservation and environmental protection, industry segmentation, and international cooperation.

11.2.1 Current State of the Pump Market

Upon in-depth analysis of the pump industry supply chain, it is evident that it is complex and highly interconnected, encompassing a number of links from supply of raw material to final product. The upstream industry serves as the cornerstone, aggregating suppliers of various raw materials such as steel, nonferrous metals, and plastics. The quality and supply stability of these basic materials are directly related to cost control and quality assurance in pump manufacturing. Through precision processing and assembly, these raw materials provide a solid material foundation for the midstream industry, namely pump manufacturing enterprises [21].

Midstream pump manufacturing enterprises are the core driving force of the entire industry chain. They not only bear the responsibility of design innovation but also transform raw materials into efficient and reliable pump products. This process involves complex production processes, strict quality control, and continuous technological upgrades to ensure that products can meet the diversified needs of downstream sectors. At the same time, these enterprises are also responsible for market expansion and sales services, continuously optimizing product portfolios and improving the service quality to consolidate their market position and explore new application areas [22].

The downstream industry showcases a wide range of application scenarios for pump products, including but not limited to chemicals, petroleum, construction, agriculture, electricity, and paper-making. These sectors have varying demands for pumps, with some pursuing high efficiency and energy conservation, while others emphasize corrosion resistance and long service life. Therefore, pump manufacturing enterprises need to closely monitor downstream market dynamics and promptly adjust their product strategies to meet customer needs. The healthy development of downstream industries also provides stable market demand and broad development space for the pump industry [24].

The rapid development of the electric vehicle sector has brought new growth opportunities to the pump industry. Especially in the field of electric vehicle electronic water pumps, which achieve coolant circulation through ECU intelligent control, the electronic water pumps are becoming an essential part of the electric vehicle thermal management system. As the electric vehicle market continues to expand, the demand for electronic water pumps will also continue to grow, presenting new market opportunities and challenges for pump manufacturing enterprises. This not only requires enterprises to continuously improve product performance and quality but also to strengthen research and development innovation to adapt to the specific needs of the electric vehicle market.

The various links in the pump industry chain are closely connected and interdependent, jointly constituting a collaborative development ecosystem. In future development, enterprises should further strengthen cooperation and exchanges to jointly promote technological progress and market expansion in the pump industry. At the same time, it is also necessary to closely monitor industry development trends and changes in market demand, and flexibly adjust strategic directions to ensure long-term stable development.

11.2.1.1 Market Size and Growth

According to data from multiple market research institutions, the global pump market has shown a steady growth trend in recent years. It is estimated that the global water pump market size will reach RMB 460 billion by 2024, with China accounting for about 50%, or RMB 236 billion. By 2025, the global water pump market size is expected to exceed USD 100 billion, with an average annual compound growth rate maintained at a high level [25].

In the current context of global economic recovery and transformation and upgrading, pumps, as one of the core pieces of equipment in the manufacturing industry, are experiencing diversified and high-end trends in market demand. Strong demand from the manufacturing industry, increasingly stringent environmental protection policies, and rapid development in emerging fields jointly constitute new drivers of growth in the pump industry.

As a pillar industry of the national economy, the rapid development of the manufacturing industry has led to a continuous increase in demand for pumps. Especially in high-end manufacturing sectors such as automotive, aerospace, and petrochemical industries, high-quality, high-performance pumps have become indispensable in production. With the booming development of the new electric vehicle industry, electronic water pumps, as an essential part of the electric vehicle thermal management system, have seen their market size expand rapidly. According to estimates from professional institutions, the global market size for electronic water pumps used in electric vehicles has grown from USD 1.45 billion in 2024 to a significant level in the subsequent years, with an average annual compound growth rate exceeding 20%. This trend indicates that the application of pumps in the manufacturing industry will continue to deepen and expand [26].

11.2.1.2 Geographical Distribution

The Asia-Pacific region is the main contributor to the global pump market, accounting for more than 50% of the market share, with China being the largest pump manufacturing base and exporter globally. The European and North American markets are also maintaining growth, especially in countries like Germany and the United States, which occupy important positions in the global pump market.

11.2.1.3 Product Types

The pump market offers a wide variety of products, which can be classified into vane pumps (such as centrifugal pumps and axial flow pumps) and positive displacement pumps (such as reciprocating pumps and rotary pumps) based on their working principles. Among these, centrifugal pumps occupy an important position in the global pump market due to their wide range of applications and high efficiency.

11.2.2 Future Trends in Pump Technology

11.2.2.1 Intelligence and Automation

With the continuous development of the IoT, big data, artificial intelligence, and other technologies, the pump industry will accelerate technological innovation and the intelligence process. Smart pumps will feature adaptive control functions, remote monitoring capabilities, and fault prediction, thereby improving operational efficiency and reliability and reducing maintenance costs. For example, through IoT technology, the operating status of pumps can be monitored in real-time, and faults can be predicted and addressed in advance, realizing unmanned or low-manpower-dependent operations in smart factories.

Intelligence has become an important driver of transformation and upgrading in the pump industry. By integrating advanced sensors, controllers, and intelligent algorithms, pumps have gained capabilities such as remote monitoring, real-time data analysis, fault diagnosis, and automatic adjustment. These functions not only significantly improve the operational stability and reliability of pumps, reducing production losses caused by downtime due to faults, but also significantly reduce maintenance costs. Intelligent control technology also makes pump operation more flexible, allowing automatic adjustment of working conditions based on actual needs to achieve optimal matching of energy consumption and efficiency. With the continuous integration of technologies such as the IoT, big data, and artificial intelligence, the intelligence level of pumps will continue to improve, providing more efficient and intelligent fluid transmission solutions for industrial production.

11.2.2.2 Energy Conservation, Environmental Protection, and Green Manufacturing

Environmental protection and sustainable development have become global foci. The pump industry will pay more attention to energy conservation, environmental protection, and green development, promoting the use of energy-efficient and environmentally friendly pump products to reduce resource consumption and environmental pollution. For example, stainless steel pumps, as a new type of energy-saving pump, consume about 10%–20% less electricity than traditional cast iron pumps and will be widely used in water conservancy infrastructure and other fields. In addition, water pump products that are low-noise, low-energy consumption, and high-efficiency will be favored by the market.

Facing increasingly severe environmental pressure, the pump industry is actively researching and applying environmentally friendly materials and technologies to reduce negative impacts on the environment. Using corrosion-resistant, wear-resistant, nontoxic, and harmless materials to manufacture pump bodies and key components effectively extends the service life of pumps and reduces waste generated from replacing parts. Optimizing the operating mechanism of pumps reduces leakage and noise pollution, enhancing the overall environmental performance of pumps. For pumps used in specific industries, such as wastewater treatment pumps, special attention needs to be paid to their environmental performance to ensure that they do not cause secondary pollution to the environment during the treatment process. The application of environmentally friendly materials and technologies not only reflects the pump industry's commitment to social responsibility but also lays a solid foundation for the sustainable development of the industry.

11.2.2.3 Industry Segmentation and Customized Services

In the face of diversified industry backgrounds and complex application scenarios, the pump industry will segment the market more finely and provide tailored solutions. For example, in the new energy, biopharmaceutical, and fine chemical industries, higher requirements are put forward for the corrosion resistance, hygiene standards, and sterile cleanliness of pumps. Pump manufacturers will launch industry-specific products based on the characteristics of different industries and specific operating conditions, such as new energy lithium battery-specific pumps and small, versatile mining pumps.

With the rapid development of emerging industries, e.g. in the areas of new energy and new materials, the application fields of pumps continue to expand. In the new energy sector, pumps are widely used in key links such as battery manufacturing and solar photovoltaic, providing strong support for the rapid development of these industries. In the environmental protection field, pumps have become one of the important pieces of equipment in sewage treatment, waste gas treatment, and other links, playing a positive role in improving environmental quality and protecting the ecological system. With the continuous integration of technologies such as the IoT and big data, the emergence of new products such as smart pumps has also brought new growth points to the industry. The demand from these emerging fields has not only brought new market space to the pump industry but also prompted the industry to accelerate its transformation and upgrading to adapt to changes in market demand.

11.2.2.4 Miniaturization and Integrated Design

With the progress of microelectronics technology and miniaturization manufacturing processes, small and compact pumps will be increasingly used in various fields such as scientific research laboratories, medical equipment, and household appliances. Through integrated design, the interfaces between pumps and other equipment or systems will become closer, enhancing overall efficiency.

11.2.2.5 International Cooperation and Competition

As globalization accelerates, international cooperation as well as competition in the pump industry will become more intense. Pump companies therefore need to actively engage in international cooperation to enhance the product competitiveness and brand influence, while simultaneously facing challenges from competitors worldwide.

The future development of pump technology will exhibit diversified trends such as intelligence, energy conservation and environmental protection, industry segmentation, and international cooperation. The pump industry needs to keep pace with technological advancements and changes in market demand, and should continuously innovate, improve product performance, and service quality to meet the diversified needs of different engineering fields. At the same time, strengthening international cooperation and enhancing global competitiveness will be important to guarantee the long term sustained development of the pump industry.

References

1 Avinash Singh. March 2025 Industrial Pumps Market Size - By Pump Type, By Power Source, By Flow Rate, By Power, By Technology, By End Use Industry, By Distribution Channel, Growth Forecast, 2025 - 2034 (10553: GMI10553). Global Market Insights Inc. Retrieved February 15, 2026, from https://www.gminsights.com/industry-analysis/industrial-pumps-market.

2 Global Industrial Pumps Market Report 2023. Yahoo Finance. (7 September 2023).

3 Yao, Z., Wang, F., Qu, L., Xiao, R., He, C., and Wang, M. (2011). Experimental investigation of time-frequency characteristics of pressure fluctuations in a double-suction centrifugal pump. *Journal of Fluids Engineering* 133 (10): 101303.

4 Barbieri, E.S., Melino, F., and Morini, M. (2012). Influence of the thermal energy storage on the profitability of micro-CHP systems for residential building applications. *Applied Energy* 97: 714–722.

5 Verma, J. and Taiwade, R.V. (2017). Effect of welding processes and conditions on the microstructure, mechanical properties and corrosion resistance of duplex stainless steel weldments—a review. *Journal of Manufacturing Processes* 25: 134–152.

6 Qu, M., Abdelaziz, O., and Yin, H. (2014). New configurations of a heat recovery absorption heat pump integrated with a natural gas boiler for boiler efficiency improvement. *Energy Conversion and Management* 87: 175–184.

7 Manek, E. and Haydary, J. (2019). Investigation of the liquid recycle in the reactor cascade of an industrial scale ebullated bed hydrocracking unit. *Chinese Journal of Chemical Engineering* 27 (2): 298–304.

8 Adegoke, A.A., Stenström, T.A., and Okoh, A.I. (2017). Stenotrophomonas maltophilia as an emerging ubiquitous pathogen: looking beyond contemporary antibiotic therapy. *Frontiers in Microbiology* 8: 2276.

9 Industrial Pumps Market Growth, Size, Share & Trends | 2032. SkyQuest.

10 The Global Industrial Pump Market Expected to Reach $87.4 Billion by 2030. QEEHUA PUMP. (31 May 2024).

11 Pumps Market Size, Share, Growth & Trends Report, 2032. MarketsandMarkets.

12 Global Pump Market Outlook 2024. Oxford Economics. (16 July 2024).

13 Industrial Pumps Market Size, Share & Trends. Grand View Research. (January 2024).

14 Advanced Manufacturing: A New Narrative. (2023). World Economic Forum.

15 Industrial Pumps – Global Market Analysis by Pump Type, Application, and Region 2025–2030. GlobeNewswire. (28 November 2024).

16 Advancing Industrial Fluid Handling: Innovations in Advanced Pumping Technologies. Fluid Handling Pro. (11 December 2023). https://fluidhandlingpro.com/fluid-process-technology/pumps-pumping-systems/advancing-industrial-fluid-handling-innovations-in-advanced-pumping-technologies/.

17 OEM Horizontal Multistage Centrifugal Well Pump In China. May. 25, 2025. OEM Horizontal Multistage Centrifugal Well Pump In China.

18 Lindsey Walker. The Future of Pumps, IoT, and Industry 4.0. December 3, 2024. The Future of Pumps, IoT, and Industry 4.0 | IoT For All.

19 Pumping systems consume 20% of global electricity, producing approximately 1,370 million tons of CO_2 annually. Energy-saving measures can reduce pump system energy consumption by an average of 40%.

20 The European Union's Ecodesign Regulation has raised the energy efficiency requirements for variable speed drives (VSDs) and low-voltage induction motors to meet the IE3 premium efficiency standard.

21 Mordor Intelligence Research & Advisory. (2026, January). Smart Pumps Market Size & Share Analysis - Growth Trends and Forecast (2026 - 2031). Mordor Intelligence. Retrieved February 15, 2026, from https://www.mordorintelligence.com/industry-reports/smart-pumps-market.

22 UNEP (2024). Energy Efficiency | UNEP – UN Environment Programme. Energy | UNEP - UN Environment Programme.

23 Mordor Intelligence (2025). Asia-Pacific Pumps Market Size & Share Analysis - Growth Trends and Forecast (2025 - 2030). Mordor Intelligence. Retrieved February 15, 2026, from https://www.mordorintelligence.com/industry-reports/asia-pacific-pumps-market.

24 Pump Castings Explained: Manufacturing Processes, Materials, and Applications (2024). Investment Casting. https://investment-casting.net/pump-castings-explained/.

25 Business Research Insights (2025). Water Pumps Market Size, Share, Growth, and Industry Analysis, By Type (Centrifugal Pump, Diaphragm Pump, Reciprocating Pump, and Rotary Pump), By Application (Industrial, Household, Agricultural, Municipal and Others), and Regional Forecast to 2035. Last Updated: 26 January 2026 |Base Year: 2025|Historical Data: 2022-2024|No of Pages: 140 Region: Global|Format: PDF|Report ID: BRI118500|SKU ID: 28189546. Source: https://www.businessresearchinsights.com/market-reports/water-pumps-market-118500 Source: https://www.businessresearchinsights.com/market-reports/water-pumps-market-118500.

26 Expert Market Research (2024). Water Pump Market Analysis by Centrifugal Pumps, Positive Displacement Pumps, and Other Pumps for Oil & Gas, Chemicals, Power Generation, Water & Wastewater, and Others from 2024 to 2034. FACT9917MR, Mar-2024. Water Pump Market Size, Industry Share | Growth Statistic 2034.

Index

Printed and bound by CPI Group (UK) Ltd, Croydon, CR0 4YY
17/09/2026
14964237-0003